ECOLOGY AND PALAEOECOLOGY OF MARINE ENVIRONMENTS

WILHELM SCHÄFER

ECOLOGY AND PALAEOECOLOGY OF MARINE ENVIRONMENTS

TRANSLATED BY IRMGARD OERTEL

EDITED BY G. Y. CRAIG

THE UNIVERSITY OF CHICAGO PRESS

The University of Chicago Press, Chicago 60637
Oliver & Boyd, Edinburgh

The present work is a translation of *Aktuo-paläontologie nach Studien in der Nordsee*, © 1962 by Dr. Waldemar Kramer

Published 1972
Printed in the United States of America

International Standard Book Number: 0-226-73581-8
Library of Congress Catalog Card Number: 72-81105

Dedicated to my friend
KONRAD LÜDERS

CONTENTS

PREFACE TO THE GERMAN EDITION

THIS book attempts to describe the ecology of the organisms in one shallow sea and to determine their influence on the marine sediment, whether they appear as preserved hard parts of the organisms themselves, as their moulds, or as traces produced by their life activities. It is meant for the use of palaeontologists and geologists. The book was written in response to frequent suggestions by colleagues who visited me in my research area, the southern North Sea with its tidal flats and islands. Their comments and discussions helped me realise which questions about the organisms of a Recent sea are of the greatest interest for palaeontologists and geologists; I also learnt from such discussions which aspects of marine life are least understood, and which geologically important marine events are most likely to be completely overlooked by geologists who have not themselves worked on problems of Recent marine geology. Since the book is written in the form of marine biology for geologists and palaeontologists it contains many facts and figures which may at first appear strange or beside the point and which, in fact, cannot be applied directly to fossil problems. But how can geologists and palaeontologists reconstruct fossil oceans and their organisms unless they have gained some insight into the animal world of at least one Recent ocean with its almost overwhelming variety of numbers, shapes and ways of life? This book is meant to contribute to such an understanding.

The book is essentially confined to the description of organic processes in the North Sea. There, the variety of organic processes is certainly great, but very far from all-encompassing. The same, however, could be said of any other Recent sea. What is discussed in this book is therefore only a limited selection of what could and eventually should be done, a world-wide survey of the Recent oceans, from the palaeontological point of view.

I thank joint investigators who visited us during work at sea and on the tidal flats, and also my hosts when I was invited to work with them on the mainland; I also thank all my collaborators for help, contributed ideas and advice. I dedicate this book to Mr Konrad Lüders in recognition of the good care he continuously took to facilitate my work at sea and at headquarters. I owe thanks to the Deutsche Forschungsgemeinschaft for decisive support in the past years and for its continued interest in marine research; the same institution provided financial help for the printing. I thank the publishers for their understanding of special problems and for the quality of reproductions.

Wilhelmshaven
5th January 1962

PREFACE TO THE ENGLISH EDITION

ALTHOUGH the German edition of Professor Schäfer's book entitled *Aktuo-Paleontologie* is widely known to English-speaking geologists and biologists, few have been able to master sufficiently the Germanic construction in order to appreciate the wealth of information that the book contains. Mrs Irmgard Oertel, a German biologist, has undertaken the considerable task of translation, so making the text available to a much wider audience, who I am sure will be most grateful to her for the clear way in which she has expressed the concepts and facts.

I have no doubt that the book is the most remarkable synthesis yet attempted on the life habits, death and burial of a complete range of animals living in a shallow sea—the North Sea. The text is illustrated with numerous artistic and meticulously drawn diagrams, most of them done by the author himself. The ideas developed by Professor Schäfer serve to remind English-speaking geologists of how differently parts of science can evolve in different countries where language barriers exist. The important research work carried out by German geologists over the last fifty years on trace fossils and the processes of life, death and burial is only now becoming appreciated abroad: and our understanding of this research is now greatly advanced by the publication of the English edition of Schäfer's encyclopaedic work.

In this English edition, the original text has been somewhat shortened (and the type size increased), the diagrams have either been given a scale or the size of reduction has been indicated in the legend, and a new set of photographs prepared. The opportunity has been taken to make minor revisions to the text and bring the references up to date.

G. Y. CRAIG

PART A

The purpose of marine actuopalaeontology

ACTUOPALAEONTOLOGY is a branch of actualistic geology, and as such it has existed since K. E. E. van Hoff (1771–1837) and Charles Lyell (1797–1875). These two were concerned with physical and chemical processes, but later workers included observations of organisms and life processes. Richter (1929a) proposed the terms actuogeology and actuopalaeontology for these subjects of study. Actuopalaeontology is the subject of this book, at least as far as it can be demonstrated in the North Sea.

Geology, and as one of its branches, palaeontology, is both historical and empirical. The empirical aspect makes geology a science in the sense it has had since the Renaissance, when Leonardo da Vinci and Galileo first used experiments to establish 'laws of nature'. Experimentation, measurement and count, and the establishment of causal relationships, are essentially ahistoric, and are applications of the 'scientific method'. In geology the scientific approach has variously been called the principle of actuality or the 'ontological method' (Walther 1930).

Geology, and thus palaeontology, often applies methods and knowledge from related sciences. The science that makes the greatest contribution to actuopalaeontology is marine biology, which can be said to have become established around 1850 when A. Dohrn founded the Zoological Station in Naples. Since then many other marine stations have been started. The work of these institutions was supplemented by oceanographic expeditions, many of which procured samples of marine life from all depths in all oceans.

The main purposes of marine biology are the description of marine organisms and the study of their mutual relationships in local communities and in the larger systems of ecological food chains and food networks. Another purpose of marine biology is to establish the dependence of organisms on their environment. These aspects of marine biology lead to the use and sometimes to the development of such concepts as plant and animal communities, ecosystems, biotopes, and biocoenoses. All play important and sometimes decisive roles in this book.

The science of actuopalaeontology, often combined with the geology of the Recent oceans, has a short history. During the last few decades North Americans have begun to study marine geology in the oceans surrounding their continent. In Europe the Senckenbergische Naturforschende Gesellschaft at Frankfurt founded a station for actuopalaeontological research in Wilhelmshaven in 1928, and began investigations on the tidal flats of the southern North Sea under the instigation and direction of R. Richter. For more than thirty years the author's own work on the actuopalaeontology of the North Sea was done within the same institution. During this time he could always freely draw on the experience of other members of the station (Bucher 1938, Häntzschel 1956, Richter 1929a,b and 1956).

An actuopalaeontologist is naturally interested in almost every aspect of marine biology, so many of which have their uses in palaeontology. However, he has certain additional interests that are normally not shared by his biological colleagues. For instance, he is interested in how recent organisms leave an imprint in recent sediments which he already sees as potential future rocks. Such imprints can consist of the dead body of an organism itself, or at least of its hard parts which are more easily preserved. But the imprint may also be a trace of the activities, and thus of the behaviour, of a living animal.

An actuopalaeontologist must also be an acute observer of the physical processes of deposition, erosion and reworking of sediments and of the processes of transportation which affect both inorganic sediments and organogenic remnants, shifting and sorting them, and comminuting or at least partially destroying many organogenic remnants in the process.

Often when an actuopalaeontologist works on exactly the same problems as a marine biologist, say when he considers the functional morphology of a marine species, or when he considers a species as part of a local community of organisms or as part of an ecosystem or a food chain, he does so with a special emphasis. He is constantly reminded of the fragmentary nature of the fossil record. Generally only hard parts are preserved, many traces of life activities have been destroyed, the record is frequently incomplete because of erosion and non-deposition, and many physical factors of the environment such as salinity, light or temperature can at best be indirectly ascertained. An actuopalaeontologist therefore pays special attention to the interactions of the biotope and its living inhabitants, so that as palaeontologist he may draw the maximum of information from incomplete evidence. He finally translates, while he works on a recent, marine, animated environment, the biological concepts into their parallel but not identical palaeontological counterparts; he wonders how much of a biocoenosis or ecosystem, possibly intermixed with other biocoenoses, might eventually appear as part of a sedimentary sequence which would be recognisable directly in the sediment or indirectly through the fossils. To what extent can the fossil remnants of original inhabitants be distinguished from those of animals that lived far away but were carried into the same area by currents, or dropped into them from the water above?

Palaeontologists have to consider the historical nature of ancient life, the uniqueness of the biological complex and the physical conditions of any given past period. The actuopalaeontologist describes present facts and interprets present interactions. Before he uses such results, a palaeontologist must consider in each case whether he must modify them, or even reject them altogether as irrelevant. He must do so for each fossil, species, facies and past geologic period.

Finally, a few cautionary remarks are required. Although much is known about present oceanic life this knowledge is far from complete. Comparatively little is known about life in the open oceans, in deep polar waters, or at the bottom of deep ocean trenches. Equally incomplete is our knowledge of the less inaccessible shallower epicontinental seas of most of the tropics. Thus the known domains really represent only a small sample of the world ocean as a whole. Very little is understood about the causal relationships of many of the observed phenomena. The 'laws' of actuopalaeontology have been formulated only in the most fragmentary way. Also, little explored as marine actuopalaeontology may be, its terrestrial counterpart is even less well known. This is probably not very serious if one considers the great predominance of marine sedimentary rocks over terrestrial ones, more especially fossiliferous rock units. This book is limited in its observations to the southern North Sea. The reason is simply that this is the region for whose description this author is best qualified, and the volume is already quite large enough as it is. An attempt is made, though, to present results that are widely applicable so that future results from other regions may easily be fitted into the same framework.

PART B

Organogenic hard structures

THE information that a palaeontologist has of extinct organisms is derived almost exclusively from the interpretation of such hard parts as these organisms may have possessed, frequently only a small portion of the whole body. These hard structures may consist of calcareous, siliceous or chitinous material and may have had diverse functions during the lifetime of their possessors. Some are internal or external skeletons, some play a role in feeding, some serve mainly for protection and some even play a role as secondary sexual attributes. The shapes of soft parts of animals are hardly ever preserved and are of no use for taxonomy, phylogenetic history, or any aspects of the functional morphology, ecology or zoogeography of the extinct organisms. The ultimate fate of the disintegrating organic matter is of more interest to geochemists than palaeontologists.

I. Functional morphology

Nowadays palaeontologists can confidently distinguish organic and inorganic objects. They do so by applying, either consciously or unconsciously, a body of knowledge called functional morphology. That this was not always so is shown by the confusion of early geologists who long considered fossils as inorganic 'ludus naturae', or as playful inventions of inanimate nature.

In contrast to inanimate objects all parts of a living body serve specific functions, either internal or external. Internal functions and their interactions establish the unity of the individual organism and give it 'Gestalt' character. The moment this unity is destroyed the organism dies (I. von Uexküll 1919 and 1921). The reaction of an organism to its environment calls for special functional organs, and in the case of animals for a specific behaviour. The function of such organs of external interaction can be understood only when the behaviour is understood; both 'make sense' only in conjunction.

Therefore functional morphology must start with the investigation of living animals, their internal life processes and their behaviour and interaction with the environment. This investigation starts with observation and may extend by experiments, i.e. by wilfully disturbing either internal or external functions and by recording the resulting malfunction or the takeover of alternative functions for those that have been impeded.

The following main functions are common to all animals: (1) ingestion and metabolism, (2) defence against damaging influences of the environment, abiotic or biotic, and (3) reproduction. Many animals also have the additional function (4) of locomotion. There are also secondary functions such as the nervous system, sense organs and glands. In various animals one or the other of the main functions may be more prominent and play a greater role in its internal organisation, and there may also be changes in such prominence during life.

Each function is performed by organs which serve it more or less exclusively; and frequently the principal function of an organ can be deduced from its external shape and internal structure. In spite of a certain overlap one may thus ascribe a functional system of organs to each main function (Benninghoff 1935/6, 1938 and 1949). These functional systems, or spheres of activity as Th. von Uexküll (1949) called them, compete with each other within a single organism for space and for energy. Each species finds its own, specific compromise in this competition. To achieve an orderly interplay of the systems, there has to be what one may call correlative adjustment between them. If such adjustment is successful, as it must be in a viable organism, the different systems do not unduly counteract each other, they collaborate to produce an optimal effect for the whole organism.

The relative importance of the different main functions varies. However, animals with well-developed locomotion are more strongly influenced by this than by all the other functions. The brachyuran crustaceans are an example (Schäfer 1954a). All are vigorous movers, and their shapes are strongly characterised by their principal mode of locomotion, whether they walk, burrow, climb or swim. Similarly, the mode of locomotion can affect the shape and structure of fishes; *Thunnus* is built for fast swimming, *Hippocampus* for gliding through the water, and *Raja* for propelling itself mostly with its pectoral fins.

Animals that move rarely are modified to suit the requirements of one of the other main functions. In endoparasites, for example, the function of reproduction usually dominates, and sessile, aquatic metazoans usually concentrate on metabolism. How sharply an animal is characterised by any of its functions depends to a considerable degree on its specialisation. Specialisation of function is associated with specialisation of behaviour and frequently with a specialisation of habitat, i.e. with increasing dependence on features of the physical and living environment that are found only in rather special, restricted domains. Where specialisation is advanced it usually also produces animals of very characteristic shape and internal features. Such shapes and features are frequently repeated by animals of very different taxonomical position and widely different histories of phylogeny. The features that such unrelated animals have in common constitute an element that can be classified on its own right. This has been done by Remane (1943, 1952 and 1956) who created the concept of 'Lebensformtyp' (constitutional type). According to this author, the similar features which make up a constitutional type are adaptations to a specific biotope and resemble each other because they are due to similar specialisation.

Specialisation does not necessarily imply perfection. Locomotion may be highly specialised so as to allow peak performance for a certain way of locomotion at the expense of all other modes of locomotion; some animals indeed perform quite well in several different modes. For most environments a good all-round performer is best. However, more information can be obtained from the presence of a highly specialised animal in a living or fossil assemblage. An actuopalaeontologist can investigate the specific interplay between the specialised animal and its well-defined habitat, and a palaeontologist can deduce the nature of an ancient biotope if he finds a fossil assemblage containing highly specialised animals. If the palaeontologist turns biostratigrapher he will specifically look for specialised animals so that he may learn more about the physical environment. Also, the inclusion of several highly specialised animals in a thanatocoenosis (an assemblage of dead animals) may indicate to him that they were only brought together after their death. This would be obvious if the animals were specialised but strikingly different or even incompatible.

Another application of functional morphology is the reconstruction, by the palaeontologist, of a whole animal from a few parts. Occasionally, if extinct animals are very different from presently living ones, this may call for considerable ingenuity. Thus von Holst (1957) made a model of *Rhamphorhynchus* whose parts could actually move so that he could study the relationships between the moving parts and the body as a whole. Usually, however, the model is built only in the investigator's mind. He starts by examining individual parts as they were preserved and determines what functional role they may have played. He then fits the parts together into a larger functional whole, finds by comparison with presently living related or sometimes unrelated forms, what type of surrounding such a whole animal would have needed, and what ecological role it could have played in such environment.

Functional morphology may even be helpful in phylogeny. Certain anatomical features of animals and their organs seem to have little purpose at first sight, at least not for the environment in which such an animal presently lives. However, analysed from the point of view of functional morphology, such features turn out to be quite functional, but for an environment once inhabited by the present animals' ancestors (Dohrn 1875). Thus functional features may be transferred during evolution to a new environment, and if they are not actually detrimental there, they may be preserved by heredity for a long time.

II. Hard parts and their function

General biological rules that hold for the structure and organisation of soft parts of an animal also hold for its hard parts. They are thus subject to the laws of structural morphology, and they can express the degree of specialisation, the predominance of certain functions, the dependence on certain habitats or biotopes, possibly the development of functional convergence, and thus they can help to assign an animal to one of Remane's constitutional types.

However, if an investigation is restricted to hard parts only, it is not possible to observe many anatomical parts which are necessary for a complete analysis of function, and there is no chance to study behaviour. But paiaeontologists have to interpret function from hard parts. This is easier it they know as much as possible about the way hard parts function in Recent animals. This is one of the contributions that actuopalae-ontology has to make.

A few general statements can be made about the functional significance of hard parts. They play an essential role in locomotion, especially of crustaceans, asteroids and ophiuroids, and vertebrates. Other locomotory animals, worms and molluscs among them, achieve locomotion exclusively with soft, tubular modifications of the body wall, hard parts occasionally serving as supports.

The next most important function of hard parts is for defence; examples are to be found among echinoids, crinoids, bivalves, bryozoans, and even certain sessile crustaceans. A defence function normally does not exclude a supporting function, and it is frequently combined with locomotory function, but in the extreme these functions become strongly competitive. Heavy armour slows locomotion, and vigorous locomotion precludes too much stiffening and hardening of the body. Hard parts generally play a minor role in metabolism. They are essential for mechanical comminution of food, in the form of teeth, jaws, and various other types of apparatus for chewing, either in the mouth or farther down the digestive tract. Reproduction, finally, only exceptionally involves the functions of hard parts, except for the penis bone and the pelvis of female vertebrates. However, many hard structures, e.g. horns, serve as secondary sexual attributes (Fischel 1947). Some are involved in the mating mechanism or sexually induced fighting, some have 'psychological' significance and help potential mates to recognise each other (Schäfer 1959). Similar seemingly non-functional external attributes, but without sexual significance, also frequently play a social role. They may serve to establish an animal's social rank (Portmann 1948 and 1952).

Apart from the functional significance of hard parts, actuopalaeontologists must study the probability of preservation of hard parts. This may depend on the material of which they consist, calcium carbonate or phosphate, silica, keratin or chitin, on how massive the parts are, on their shape, and on what usually happens to an animal after death and before burial.

PART C

Death, disintegration and burial

MANY remnants of marine animals are preserved bodily. Therefore it is useful to study the causes of death, the phenomenon of death itself, the disintegration of carcasses, the transportation of whole carcasses or their parts, and their final burial. Only protozoans that propagate by simply dividing can potentially live forever. There is no natural death for them, they die only accidentally. All metazoans eventually die, and their carcasses are left, whole or in parts, in the biotope in which the animal used to live. August Weismann (1892) was the first to remark that death appeared with the division of labour into germ cells and somatic cells. Important as the problem of the death or extinction of whole species may be to palaeontology, this problem must be solved by historical methods and does not concern actuopalaeontology.

An individual animal may die accidentally, of old age, as a victim of parasites or other enemies, from lack of food or as a consequence of external forces. Of the many causes of death one may be predominant for a certain species in a given environment. The observation that certain animal remnants are especially frequent in certain facies, or that certain remnants are concentrated in certain localities becomes geologically meaningful only if it is possible to associate these accumulations of animal remnants with specific causes of death. *Vice versa*, knowledge of possible causes of death allows one to predict the numbers and densities of dead to be expected in certain environments. The knowledge of the causes of death also enables the ecological role of an animal in its biotic and abiotic environment to be interpreted. In these respects the study of the processes of death is biological. The processes of alteration, transport and burial that occur after death are the concern of the actuopalaeontologist. The palaeontologist, finally, must combine the biological and actuopalaeontological results when he tries to interpret the biofacies (see also A. H. Müller 1951).

Usually, the death of a marine animal is not followed immediately by burial. The dead body loses all its defences and is subject to chemical decomposition and mechanical destruction. Eventually all traces of life disappear. However, if the dead body is buried before complete destruction its shape may be preserved. If the burial is rapid more of the animal is generally preserved and more can be found out.

Death, disintegration and burial can only be understood if the living animal's organisation, behaviour and ecology are known. Such descriptions for different species follow with a summary of their biotope from the point of view of actuopalaeontology. Density and mobility of populations, their migrations, the ratio of dead to living animals, and the chances of preservation are mentioned in each case, so far as they are known.

I. Vertebrata

Vertebrates have an internal bony skeleton composed of numerous individual parts. Rigid connections between two bones are called synarthroses, synchondroses or synostoses and consist of dense connective tissue, cartilage, or bone-to-bone connections. Where the two bony parts can move freely the connection is a joint (diarthroses). The joint surfaces are separated by a gap enclosed by a capsule. Ligaments, tendons, and muscles have jointly several functions: they move the parts, transmit movement and strengthen the joint. Rigid connections often hold the individual bones together long after the death of the vertebrate. However, flexible joints disintegrate after death as soon as skin, connective tissue, muscles, and ligaments are decayed and the capsule is destroyed. A long time usually elapses—except in very rare cases—between death and burial. Usually enough time has passed for the skeletal parts to have lost their original connections or for the external agents to have separated them. Hence there are only two ways in which complete skeletons of vertebrates can be preserved: either quick burial in areas of rapid sedimentation or burial in places sheltered from transportation forces. The texture and composition of the sediment, and perhaps fossils found in the vicinity, may indicate the method of burial.

In this discussion of the southern North Sea area we are dealing mainly with fishes, birds, and marine mammals (whales and seals), not with amphibians, and only very rarely with reptiles (turtles). According to Mertens (1926) the hawksbill turtle (*Caretta caretta*) occurs in the North Sea as a stray visitor and was caught at the Dutch, Belgian, and English coasts. The leather-back turtle (*Dermochelys coriacea*) also has been recovered along the Atlantic coast of Europe and in the Channel. Greve (1931) reports that in August 1930 a dead individual of *Dermochelys* with a 150 cm long carapace drifted ashore dead at Friederikensiel (East Frisia).

Except for whales, seals, and other marine forms, mammals rarely have contact with the sea (or are buried there). However, roe-deer swim in

Jade Bay (last observation on 27th May 1960; a one-year-old buck was seen in the main channel at the head of the bay near the oil-pier with the tide coming in). In three instances within the last 25 years deer swam to the island of Mellum (over a distance of approximately 14 km across the tidal flats). Those of the East Frisian islands that lie closer inshore are more frequently visited by swimming deer.

(*a*) CETACEA

Fossil remains of whales and dolphins are not important stratigraphically. However, they are among the few organisms living today which can be compared with the marine reptiles whose way of life, death, disintegration, and final burial are of such fundamental significance for palaeontology and stratigraphy (see also Hofmann 1958). The processes of

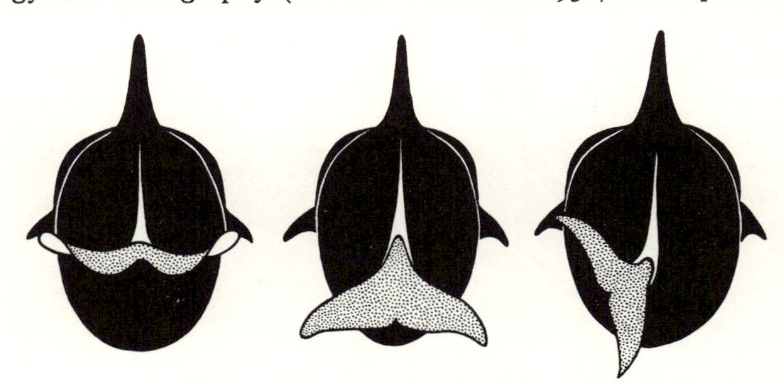

FIG. 1. Three rear views of swimming whale (tail fin dotted). Left and centre: arc-swimming; right: sculling

fossilisation can give us some idea of the circumstances after death which lead to the fossilisation of sea reptiles. The marine turtle *Amblyrhynchus cristatus*, at home around the Galapagos Islands, would be another suitable model, but it has not yet been investigated.

Whales first appeared in the Middle or Late Eocene (Archaeoceti, *Eocetus*, *Zeuglodon*). They are a group of mammals whose members are secondary swimmers: the structure of these tetrapod mammals which were originally built for living on land has been completely changed for active swimming in the ocean. They are steady swimmers and are well adapted to a purely aquatic life. The fossil ichthyosaurs evolved similarly when they changed to an aquatic life. In some respects the adaptation of the Cetacea is more perfect than that of the ichthyosaurs. The hind limbs of the cetaceans are reduced so much that they are externally no longer apparent. Thus the posterior part of the body, especially the tail fin, evolved to form the sole organ for locomotion in water. Otherwise the two types of adaptation in cetaceans and ichthyosaurs are very similar (the same holds for the fossil and the recent Sirenia). Both live exclusively in water, and despite lung breathing, whales usually die soon after leaving the water. The animals are born and nursed in open water. In contrast, the other tail-swimming mammals (*Fiber*, *Castor*, *Lutra*, *Potamogale*) continue to restrict these activities to firm ground.

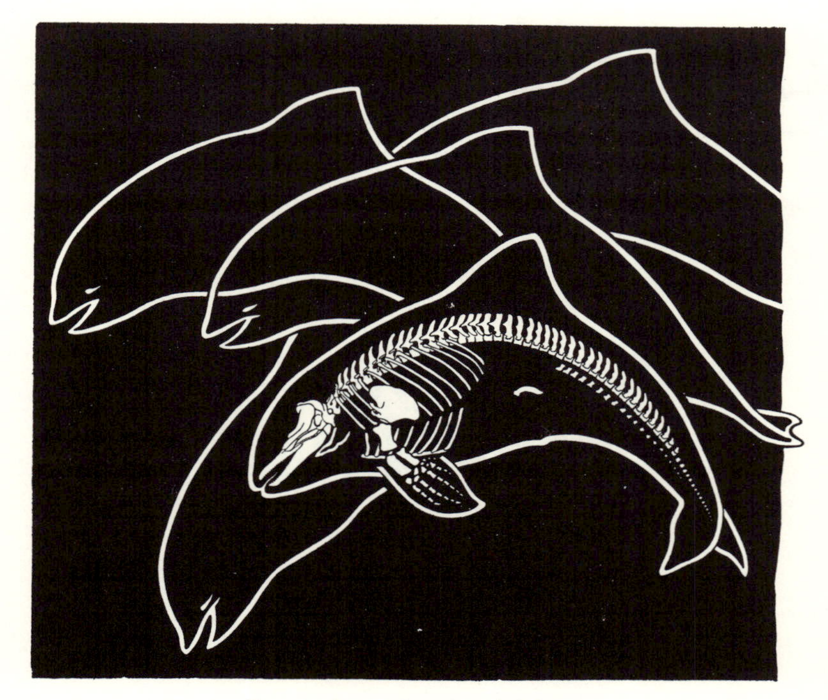

FIG. 2. Arc-swimming dolphins

Much earlier in vertebrate history the muscles of trunk and tail rather than those of the limbs propelled the animal. The upright, symmetrically built tail fin and the dorsal fin (not developed in other marine mammals) are newly acquired. As reported in detail by Roux (1895), the tail fin has no skeleton or muscles but consists of firm, fibrous connective tissue with tough collagen fibres which form three layers. In the two outer layers the fibres run from the stem of the tail fin radially to the outside in the plane of the fin. The fibres of the median layer lie partly in the plane of the fin and run at right angles to those of the outer layer. In other areas the fibres of the median layer are transverse to this plane and connect the two outer layers. The dorsal fin is similarly built.

The function of the tail fin is debatable. Do the tail fins of whales and dolphins move in a dorso-ventral wave, or rotationally in a figure-of-eight? According to Kükenthal (1908, 1909) the asymmetric build of many embryonic tail fins of whales (an upbent, left outer rim of the tail fin and a downbent, right outer rim) makes a screw-like motion seem

FIG. 3. Anterior limbs of whales and marine reptiles. Top: whales, left to right: *Balaena mysticetus*, *Phocoena phocoena*, *Globicephala melaena*; bottom: reptiles, left to right: *Mixosaurus cornelianus*, *Ichthyosaurus communis*

probable. In his view, the swimming angle caused by the asymmetry of the tail fin explains the asymmetry of the skull (see also Abel 1912).

From direct observations and photographs under water we now know that the tail fin can propel the whale in two ways: by sculling and by arc-swimming (see also Böker 1935). In the first instance the tail, with the tail fin held horizontal, beats laterally. During movement to the right and to the left the tail fin is slightly twisted from its normal position resulting in a shearing movement through the water (Fig. 1). In its oblique position either the top or the underside of the tail fin pushes against the water. A tendency to dive downward or to rise up due to this tail action is compensated by the ventral fins.

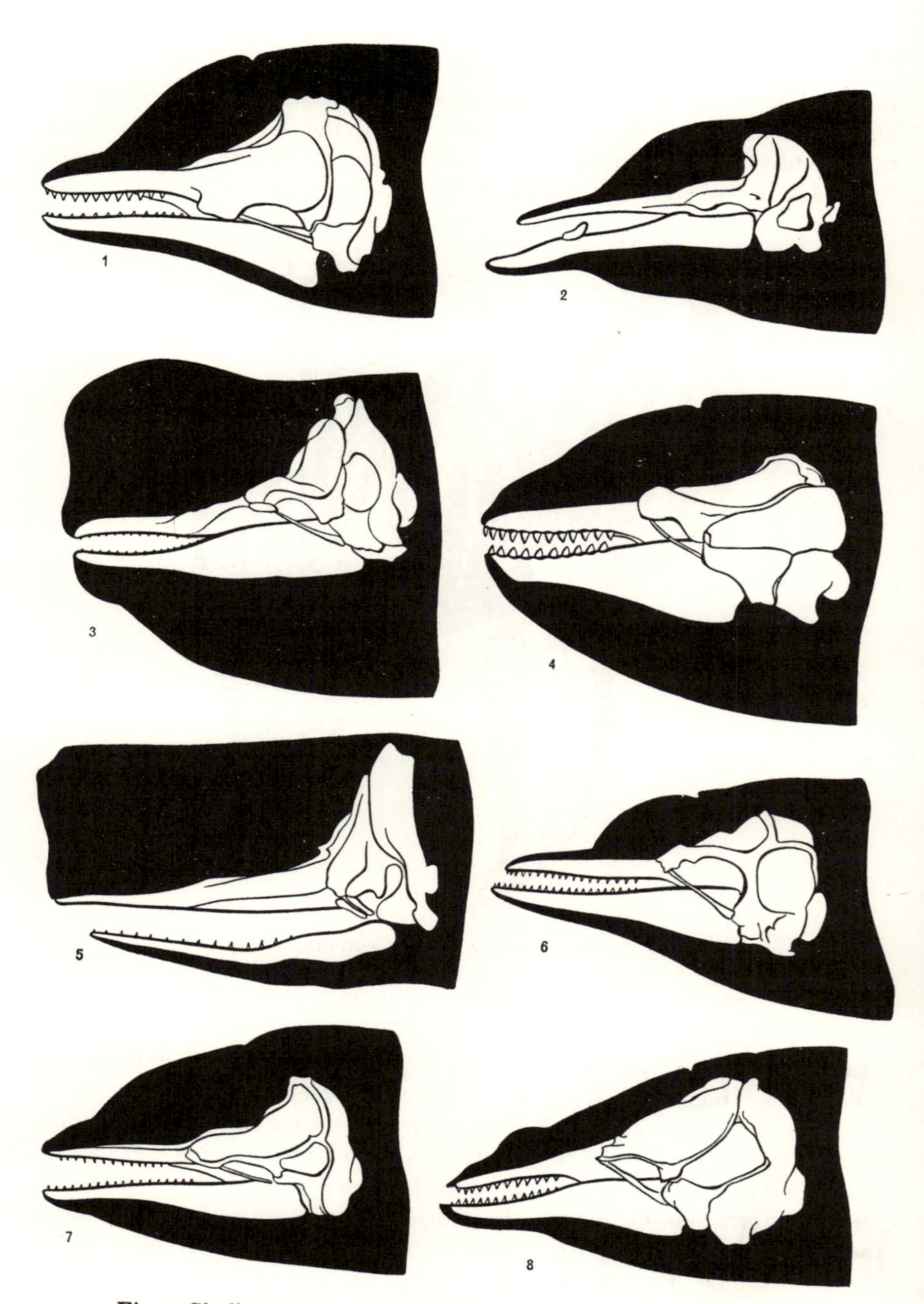

Fig. 4. Skulls and head outlines of toothed whales (after Scharff, 1900). (1) *Tursiops tursio*; (2) *Physeter catodon*; (3) *Mesoplodon bidens*; (4) *Delphinus delphis*; (5) *Globicephala melaena*; (6) *Phocoena phocoena*; (7) *Orcinus orca*; (8) *Lagenorhynchus albirostris*

In arc-swimming the tail and the horizontally held tail fin move up and down in a dorso-ventral wave motion. When swimming in this way just below the surface of the water the body partly breaks the surface with each of its upward arcs (Fig. 2, and Slijper 1958a and b).

The dorsal fin (not present in all species) stabilises the body while the fins which evolved from the forelimbs do the steering. The skeletons of these anterior limbs show clearly that they developed from tetrapod limbs with five fingers (Fig. 3). However, the anatomical adaptations (length of fins, number of fingers, number of phalanges, relative freedom of arm and hand) vary widely. The functional significance of these remarkable differences in the structure of the forelimbs is not yet clear. Several examples of skulls and outlines of heads in Fig. 4 show how problematical reconstructions of marine mammals must be if they are based on their skeletons only (the same probably applies to ichthyosaurs).

Frequencies of species and individuals. The Atlantic Ocean is connected with one of its largest adjoining seas, the North Sea, by the English Channel and by the much wider opening between Scotland and Norway. Consequently the whale populations of the North Sea and of the Atlantic intercommunicate. The natural frequency of whales in the North Sea must have been considerably higher before man had begun to hunt them regularly using modern whaling techniques. This is confirmed by historical reports on the number of species and individuals of whales living in the North Sea, especially in its southern part and in Heligoland Bay. Scharff (1900) lists fourteen species of Cetacea for the Irish Sea: *Eubalaena australis*, *Megaptera boops*, *Balaenoptera musculus*, *B. acutorostrata*, *Physeter catodon*, *Hyperoodon rostratus*, *Mesoplodon bidens*, *Phocoena phocoena*, *Orcinus orca*, *Globicephala melaena*, *Lagenorhynchus albirostris*, *L. acutus*, *Delphinus delphis*, *Tursiops tursio*. The present estimates of density and distribution of whales and dolphins are based on statistics of catches. For centuries Norwegians and Färoe Islanders have hunted small whales in the northern North Sea, especially along the Norwegian coast. Table 1 gives lists of catches (see also Schubert 1954, 1955).

TABLE 1

Catches of small whales near the Norwegian coast (Schubert 1955)

Year	Number	Year	Number	Year	Number
1938	1714	1943	1665	1948	3606
1939	1043	1944	1447	1949	4178
1940	564	1945	1824	1950	2055
1941	2129	1946	1937	1951	2854
1942	2205	1947	2716	1952	3398

According to a list of the Norwegian catches the dwarf whale is mainly hunted (see Table 2).

TABLE 2

Species of small whales taken near the Norwegian coast (Schubert 1955)

Year	1949	1952
Lesser rorqual (*Balaenoptera acutorostrata*)	3520	3366
Bottle-nosed whale (*Hyperoodon rostratus*)	220	17
Killer whale (*Orcinus orca*)	34	13
Caaing whale (*Globicephala melaena*)	3	2

Statistics for the smaller dolphins and also for the large whales in the North Sea are not available because they are not regularly hunted. However, whales have been reported to have drifted ashore, dead or alive, on the Dutch coast. Records exist, starting from A.D. 808 (see Table 3).

TABLE 3

Species of whale stranded on the Dutch coast and recorded in writing between A.D. 808 and 1930 (Van Deinse 1931)

Balaenoptera acutorostrata (lesser rorqual)	*Lagenorhynchus albirostris* (white-beaked dolphin)
B. physalis (fin whale)	*Mesoplodon bidens* (Soerby's whale)
Delphinus delphis (common dolphin)	*Monodon monoceros* (narwhale)
Globicephala melaena (Caaing whale)	*Orcinus orca* (killer whale)
Grampus griseus (Risso's grampus)	*Physeter catodon* (sperm whale)
Kogia breviceps (pygmy sperm whale)	*Tursiops truncatus* (bottle-nosed dolphin)

Obviously such a list cannot be complete. Many of the documents only mention a 'Groote Visch'. However, this list gives us some idea of the whale fauna and its constituent species in the southern part of the North Sea from the Middle Ages to modern times. The dolphin (*Phocoena phocoena*) is numerous in the coastal waters of Holland. The present distribution by year of strandings of this dolphin on the Dutch coast is listed in Table 4.

TABLE 4

Distribution over the year of dolphin (*Phocoena phocoena*) strandings in Holland (Van Deinse 1931)

January	0	July	8
February	3	August	18
March	2	September	4
April	6	October	3
May	8	November	2
June	12	December	4

From 1900 to 1928, forty large dolphins (*Tursiops truncatus*) stranded in Holland, and from 1899 to 1921, twelve finwhales (*Balaenoptera physalis*) suffered the same fate. Similarly, four finwhales are known to have been stranded on the

German coast of the North Sea in the years since 1910. The last of these (22nd April 1958) was about twenty metres long and was stranded in shallow water at the coast of Eckwarderhöhe at the head of the Jade Bay after having been observed by several coastal vessels for five days. In 1942, in the mouth of the river Weser, a finwhale of about the same size was misidentified as an enemy submarine by the crew of an anti-aircraft battery and was shot. In 1952, 1954, 1955, 1956, and in July 1959 several toothed whales, *Tursiops truncatus*, *Delphinus delphis*, and *Lagenorhynchus acutus*, drifted ashore in the same area and at the island of Mellum. At Wangerooge a *Lagenorhynchus albirostris* was stranded twice. On the 10th December 1959, a living dolphin weighing seven hundredweights (species not stated) was thrown on to the sandy beach of the dune of Heligoland during a strong, easterly gale. The number of stranded toothed whales in the southern North Sea area may well be even higher, particularly on the wide tidal flats which are deserted in winter, and where therefore some cases may go undetected.

These compilations are sufficient to demonstrate how many whales (especially toothed whales) used to live and still live permanently in the North Sea or visit the North Sea occasionally. The facts about stranded animals are particularly illuminating because they show how frequently large vertebrates die, disintegrate, and are embedded under natural conditions in the North Sea and along its coasts; large quantities of hard parts are constantly intermingled with the sediments (for world-wide statistics on whales, see Schubert 1954).

Causes of death. Senility, illness, parasites, and hazards at birth must be considered as the natural causes of death for whales and dolphins. Predators attack them frequently and particularly select the weaker individuals of a school. Therefore old age and illness are rarely the direct causes of death. The worst enemy is the killer whale (*Orcinus orca*). These vagrant cosmopolitan animals move in schools through all the oceans including the North Sea and kill many small whales, dolphins, seals and sea birds. The killer whale is up to eight metres long, and is robust and thickset; each of its jaws carries twelve or thirteen conical teeth, approximately 90 mm long (Fig. 4). The killer's favourite morsel is the attacked whale's tongue (Schnakenbeck 1947). Dead, large whales that are made fast alongside processing ships are another favourite prey. The frayed bodies of the killer whale's victims soon shed their more or less broken skeletons over the sea floor. Attracted by the scent of fresh blood, sharks soon reduce the remains left by the killer whale to clean skeletal parts. The Neaumari shark (*Lamna nasus*), the picked dogfish (*Squalus acanthias*) and the tope (*Galeorhinus galeus*) play the role of commensals of the killer whale in the North Sea. The large schools of killer whales (exceptionally up to fifty or even a hundred animals) with their need for huge amounts of food must cause significant losses of the populations of small whales and dolphins. This single species of predators must also be responsible for large amounts of food remains which finally reach the sea floor.

On 20th November 1956, a living *Orcinus*, 6·5 m, was stranded on the northwest beach of Wangerooge island, trapped at low tide in the tidal channels which run parallel to the coast. After many attempts the islanders finally killed the animal. The skeleton is kept in the museum of the Institut für Meereskunde in Bremerhaven.

As these reports indicate, whales and dolphins quite often swim into the shallow coastal waters. Their return to the sea may be cut off by the ebbing tide, and the animals are stranded. During strong gales this can easily happen to whales who swim close to the coast. Their freedom of movement is limited by the shallow water, and the dangers of grounding and of total, if temporary, emersion are heightened by alternating surf and backwash. The heavy body may be thrown on to the sand where it is unsupported by the water. The condition of the frightened animal is then aggravated by panic reactions and poor eyesight outside the water, and the animal remains stranded. Although the atmosphere which as a mammal it needs for breathing envelops it completely on land it is doomed to die by suffocation. The musculature of the thorax is adjusted to a body supported by water, the animal has no limbs to support its thorax, and it is incapable of sustaining normal breathing movements against the weight of the body. However, the slowly suffocating animal continues to snort, grunt, and bellow for hours and days on end. Over and over again it whips its tail and churns the sand with rolling, twisting, and trembling movements of its strong body.

Dr A. Lang (of the Isle of Juist) informs us of an entry in the Staatsarchiv Stade on the 2nd October 1723, which describes the stranding of a whale in the tidal flat of Wurst: 'Bevor er gestorben, hat er ein grausames Geschrey angefangen' (Before he died he gave forth gruesome yells). According to Eggerik: 'Im Vorsommer des Jahres 1549 wurde auf Norderney ein junger Wal, der bei Ebbe in Bedrängnis geraten war, gefangen . . . Davon wurde sovile Tran gesotten, wie zehn Wagen wegfahren konnten' (In the early summer of 1549 a young whale, trapped at low tide, was caught on Norderney island. . . . So much oil was boiled from it that ten carriages were required to drive it away). In 1580 a whale was stranded at the Dorumer Grode. An entry in the Staatsarchiv Oldenburg, 1645, mentions a stranded whale whose head was '16 Holtzfuss lang gewesen' (having a length of 16 units of an old measure approximately equivalent to a foot); likewise in 1589 and 1663. We know from Dr Kreuzberg (of Den Helder) that the sensitivity of the different species of dolphins to exposure outside the water during land transport varies widely. *Phocoena phocoena* must be placed on its back after being caught to avoid suffocation and it must be supplied with oxygen. *Tursiops* is more resistant.

Some of the same reasons that lead to death by stranding in individual whales and dolphins may, for some species, cause also the mass-destruction of whole schools of animals.

The Caaing whales (*Globicephala melaena*), particularly good swimmers who travel in large groups of up to 1000 individuals in the North Atlantic, frequently end up being stranded in entire schools. Probably started off by a growing general panic not necessarily caused by enemies, these whales follow their leaders blindly. Parts of such schools swim on to the beach where they die in masses (Deegener 1918, Schnakenbeck 1947, Rühmer 1954).

Drevermann (1931) reported two large-scale strandings of the black small whale *Pseudorca crassidens*. In October 1927, 127 animals stranded in the Dornoch Firth in NE Scotland. The animals were found along a 48 km long stretch of the coast (see also Hinton 1928). A connection was suggested between this event and the sudden appearance of enormous schools of octopuses that had entered the firth with an unusually strong influx of Atlantic warm water. The sudden abundance of food may have attracted whales in great numbers into the firth. Crowded and

Fig. 5. Mass-death of whales (*Globicephala melaena*), drawn from a photograph

excited by the rich hunt they were stranded and died. Perhaps the crowded whales were in turn hunted by killer whales. Another mass stranding of the same species happened in August 1929 on the beach of Valanae Island at Katys in North Ceylon. There 167 animals were stranded, including adults, adolescents, and also pregnant females, some of whom gave birth while they were dying. The *Münchner Illustrierte* magazine of 10th July 1950, reported mass deaths of *Globicephala* on the Scottish coast. Van Deinse (1931) mentions two strandings of 37 and 61 animals of the same species at the Dutch coast (at St Annaland and Ouddorp). Weigelt (1927) showed a picture of a stranding of 75 black whales at the coast of Massachusetts (Fig. 5).

On 22nd December 1723 (Dr Lang), a sailor observed 26 whales (species not stated) being stranded at normal high tide and in quiet weather on the tidal flat of Wurst. With the outgoing tide 19 of them 'beat themselves to death . . . which was terrible to watch so that the sailor, afraid that they might come too close to his ship, felt he had to weigh anchor and to move further away'. Seven whales refloated in time, all were of approximately the same size: 52 to 60 feet long and 10 to 12 feet thick.

Kellogg & Whitmore (1957) compiled a number of documented strandings of whales. They, too, mention especially observations of mass deaths of *Globicephala* on the coasts of the British Isles, Prince Edward Island, New England, Florida, Australia, Tasmania, New Zealand, and the Indonesian islands. They also mention a mass stranding of *Pseudorca* on the coast of South Africa. These particular toothed whales seem to die frequently in groups as well as individually. The tidal coasts with sand ridges running under water parallel to the beach (as at the German coast of the North Sea) seem to be especially dangerous for whales at ebb tide (Hill 1936, South African coast; Peacock 1936, Tay Estuary, Scotland). Kellogg & Whitmore (1957) noted strandings of large numbers of whales at the coasts of Scotland, Zanzibar, Ceylon, and the Chatham and Galapagos Islands.

Psychically induced mass death in social communities (see Schäfer's discussion of the concept of panic, 1956d, p. 20) is a rare phenomenon. It requires social groups of highly organised individuals. Experiences with cetaceans, kept and trained in large sea water aquaria for long periods of time, leave no doubt that they are highly organised social animals *par excellence*. Even so, generally only extraordinary meteorological, volcanic, or oceanic events, i.e. catastrophic circumstances, lead to mass death. Hence in most palaeontologically significant cases it is externally induced and not self-inflicted as with *Globicephala* and *Pseudorca* (Schäfer 1955c).

Drift, decay, and final burial of the carcass. Carcasses of marine mammals that die from natural causes drift for weeks on the surface of the sea. Whales with high fat content drift at the water surface from the moment of death while those with low fat content (the majority) first sink to the sea floor and surface only after decomposition gases have developed in the body cavity.

Carcasses sinking into anaerobic waters of protected ocean troughs, where conditions are unfavourable to bacterial activity, produce insufficient decomposition gases for surfacing. Therefore such carcasses stay on the sea floor, and their skeletons are buried complete, usually in their original positions. The shallow North Sea with its many currents has no euxinic basins. Photographs from the bottom of the Atlantic Ocean have shown that complete mammalian skeletons occur there at depths of 3000–5000 m. The necessary conditions are as yet unknown.

The surfaced carcasses drift about for weeks. In a large ocean, such as the Atlantic, even after weeks of drifting the body may not be beached, and finally the strong skin ruptures. The skeletal parts have in the meantime separated from each other by decay and are loosely held between connective tissues and muscles, and they fall to the ground one by one as from a drifting sack. The individual skeletal parts are spread over miles of the sea floor.

A complete history of the slow disintegration of a drifting seal carcass is available from recorded observations. For whales and dolphins these stages have not yet been completely observed. We know from observation of a dead dolphin, seen at the mouth of Jade Bay, that the integument tears first where the tensile stresses are particularly strong: above the roof of the skull, at the outer rims of the lower jaw, above the scapulae, and in the tail section. The connection of skull and body stays intact for a long time supported by the strong trachea, but finally the whole skull drops from the loose integument through the much widened opening in the skin at the back of the head. At this stage the lower jaws have long lost their connection with the skull (Fig. 6).

FIG. 6. Drifting dolphin carcass, observed in outer Jade Bay

If a whale dies in a small, more or less closed ocean basin of about the size of the North Sea (an area of 400 by 800 km), the corpse generally reaches a beach before it disintegrates. The surf then washes it on to the sand. Sometimes a whole body is embedded, and the skeleton remains complete. Frequently a carcass is refloated by a higher tide and is set adrift by the backwash of the surf. In the southern part of the North Sea carcasses float with the west–eastward tidal currents along the coast and are carried into Heligoland Bay after being stranded and refloated many times. There, on the wide sandbanks of the Alte Mellum or on the tidal flats of the Knechtsand, they finally come to rest. The sand areas of the island of Mellum which recede more than the other islands have for ages been a kind of resting place for drifting bodies.

Decomposition soon causes such high gas pressure inside the body that the corpse turns into a firmly inflated, torpedo-shaped barrel. The whole body and also the skeleton are stretched straight by this uniform internal pressure. The tail section becomes perfectly straight, the forelimbs spread horizontally from the body, and in the male animals, the internal gas pressure results in an erection of the penis (Figs. 7 and 9). In this state the whale body is easily moved by the surf and is rolled to and fro on the beach by waves. During this period the smooth, strong integument and the easy mobility of the barrel-shaped body protect it from mechanical destruction. After approximately 20 to 30 days, depending on the temperature of the sea water and the air, the gas escapes, and the body collapses into a limp mass. If the body is again moved it is liable to be bent and twisted in a way that could not have happened earlier. The skeletons of bodies buried in the beach sediments when inflated will be found after their fossilisation in a straight and orderly position. If, however, the body was embedded only after all the gas had escaped, then it is bent, the individual parts have changed their relative positions and are telescoped together, the vertebral column may be distorted and no longer in its original disposition.

FIG. 7. Stranded sperm whale with penis erected by decomposition gases. After an engraving, Scheveningen, Holland, 1598

The gas does not escape through the oesophagus or the anus; both openings are blocked by distension of tissues. It escapes through breaks in the skin which occur first where the body is in contact with the ground. There the skin stays moist, even on dry sand. Mummification and hardening do not happen in moist areas. Because the symphysis of the two branches of the lower jaw becomes very easily separated the originally strong connection of the jaw sections is soon lost. The area of the lips remains constantly moist and also tears. The base of the tongue splits open, and the hyoid is exposed. As a consequence it is rarely preserved in fossils. Weigelt (1927) called this the 'law of the lower jaws'. The head soon separates from the rest of the body during renewed shifting because the neck section splits open at an early stage. Head and torso, once severed, react differently to transporting agents and thus follow different paths. When a corpse is on its side (this is most frequent) the tail fin (horizontal in the living animal) instead of being forced into a vertical position, is twisted through 90°. This causes stresses in the skin and musculature of the tail stem. As a consequence the vertebrae and disks of the tail stem soon drop like coins and become widely scattered, while other parts of the skeleton remain complete. Thus the destruction of the vertebral column begins at a place which apparently carries no particularly great load in the dead body. This destruction should not be mistaken as evidence of predation; rather, it is due to an early loosening of the skeletal structure, and its early disintegration is caused by the particular position of the tail fin when the body is lying on the beach. The weight of the slowly sinking body causes the false ribs on the lower side of the body to press against the integument from the inside, and eventually pierce it. Once the connections with the vertebral column and sternum are destroyed, most of the ribs finally tip over sideways. Weigelt calls this the 'rib law' according to which the ribs are supposed to take up positions in a certain order. However, modifications of this rule, as it might more properly be called, are frequently observed.

When the body cavity opens, the gases and liquids that have collected during decay escape from the carcass together with the highly viscous oils derived from the subcutaneous layers. The outflowing oils form a dark brown halo around the body which stays discoloured for a long time. The oil-soaked sands eventually become slab-like concretions which may outlast transportation as individual, round-edged pebbles. The same oils also penetrate the decaying tissues, prevent their rapid liquefaction and protect the body from maggots. While the body is exposed on the beach the weather determines the length of decay. In warm, wet summers the described events happen in rapid succession. The rainwater speeds up decomposition by repeatedly removing the oil from the impregnated tissues and thus allows the intestinal bacteria to spread from the intestines and to take part in decomposition of the carcass. Cool or even wintry temperatures interrupt the process of decay completely; dry heat coupled with strong winds and drifting sands leads to mummification of the soft parts. Because conditions in nature vary and because different parts of the body may be simultaneously exposed to different conditions, mummification of the top of the body may be combined with complete chemical decomposition of the underside. This results in half-mummies which then are filled up completely with sand blown into wide openings below and at the sides. High salt concentration in the corpse, as well as the oils from the subcutaneous layers hardened by sand, help to kill bacteria and to preserve the soft parts. If the body is completely mummified the skeletal connections including the fibrous tissue and the tendons harden, and the skeleton does not collapse completely. Supported by the vertebral column and the ribs, the body remains arched high, the skin turns leathery, or even often hard as stone, as folds and wrinkles over the skeleton. The cavities are rapidly filled with blown sand. If mummification does not occur, decomposition continues; the skeleton, once three-dimensional, is projected on to one plane, as it were. After intensive drying and shrivelling the gums, consisting of fibrous tissue, and the periosteum of the branches of the jaw both separate together from the jaw and pull the teeth with them from their alveoli. The teeth remain in the dried gums in rows. Especially in young animals the alveoli, although consisting of bone, can widen so much that the teeth can be shaken from the gums by rain or wind and may be widely scattered. The individual bones of the skeleton are not equally well preserved when they are exposed to atmospheric influences, blowing sand and transportation. The bullae osseae tympanicae and the stronger parts of the cranial skeleton turn out to be particularly resistant. The bullae, the heaviest parts of the skull, soon break off as a unit when subjected to frequent transportation. Therefore skulls of dolphins which have retained the bullae are rarely found. On the other hand, single bullae are found in the sediments of the coast or in the swash marks (see also Pompeckj 1922). Bullae from mystacocetans that were found in the red deep-sea clays in the equatorial Pacific Ocean were described by Eastman (1903 and 1906; also Dietrich 1957). If sand-blasting is strong it soon works through the outer, compact substance of the bone and the spongiosa is laid bare. Sand and mud fill the porous structure. The weight of the skeleton thus increases considerably, and the bones can no longer be carried by wind (Thenius 1954a and b). The high resistance of the bone structure of whales is demonstrated convincingly by fences and tombstones in Holland made from bones of whale which have lasted for centuries.

Death of whales in prehistoric times. Our description of the stages of decay of carcasses deals with the general case and should therefore apply in similar circumstances in the geological past; here we shall consider processes in the prehistoric past around the North Sea. For many centuries the stages of postmortal disintegration of whales have been recaptured in pictures, such as the remarkable Norwegian rock engravings created about 6000 years ago by the hunters and fishers of the coast (Schäfer 1955b, 1956c). Those early observations confirm our own. Certainly at that time the intention in picturing whales was not actuopalaeontological. Instead, it was a wishful attempt to attract whales and dolphins to the coast by means of pictures endowed with magic powers. Nothing less was needed to influence animals that were beyond their grasp in the open sea but highly welcome as stranded prey. Probably these representations were addressed to a powerful god, and it was natural that the desired event was pictured in the strictest possible realism. Adam von Bremen who lived in the Middle Ages reported (Janssen 1937) that people in Norway attract whales to the shores by magic words. This appears to have been an old and long-maintained custom. In prehistoric engravings images of whales and dolphins are found together with those of fishes, sledges, humans, reindeer, and arms. They cover surfaces of rocks polished by Pleistocene glaciers. These images of whales and dolphins etched on to rock were misinterpreted by some archaeologists as 'fishes' (Kühn 1952); other investigators interpreted them correctly as marine mammals. They were considered, however, to represent catches like those of the other animals. This led to the conclusion that prehistoric man was capable of hunting whales and dolphins at sea. As the result of our own studies we now regard the pictures as images of decayed carcasses on the beach inflated with gas, mummified, crumpled or headless; they are not fresh prey. Obviously the stranded whale was the desired object, but it played this role only as long as the living whale was out of reach as a prey. This situation would have changed at the very moment when healthy whales could be hunted on the open sea. These pictures evoke the old habits of beachcombers

FIG. 8. Decaying dolphins, Norwegian prehistoric rock engravings. Top: Rodoy; bottom: Evenhus

who are first out on the beach after a stormy night when perhaps the desired prey may have been washed up. Even a prey that was already decaying was certainly worth possessing because it provided bones and hide, and oils to burn in huge quantities; if the prey were stranded alive it also provided blubber and meat for many months. But what actually do the old engravings show?

(1) Engravings at Rodoy and at Evenhus, Fig. 8. Decayed and degassed carcasses of dolphins which must have been transported and redeposited several times are crumpled and telescoped, the tail section is kinked, the dorsal fin is drawn distinctly separated from the body, the ventral fins are drooping loosely, the neck section is drawn in, folds and skin destroyed at the back of the head indicate that the head is about to separate from the rest of the body. Hallström (1938) misinterprets the line behind the head as a hint of a gill plate, and he concludes that this is

not a representation of a whale but of a large fish. The mouth is distinctly split.

(2) Engraving in Askollen (Vestfold) and Skogervejen, Figs. 9 and 10. Extensive shrinking and reticulate folds of the dry skin are characteristic

FIG. 9. Mummified dolphin, prehistoric rock engraving, Åskollen, Vestfold, Norway

of mummification. The reticulate design on the whale is therefore not just a decorative addition of the artist intended to fill the space within the body outline of a whale, but is correctly observed.

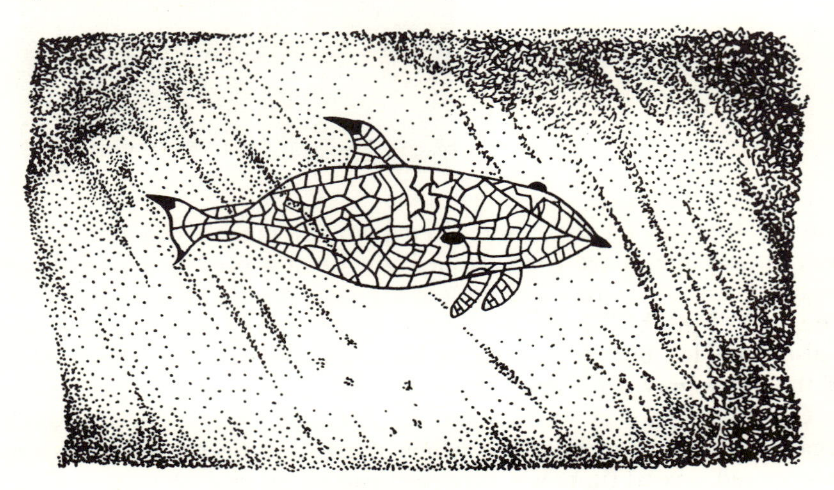

FIG. 10. Mummified dolphin, prehistoric rock engraving, Skogervejen, Norway

The tail fin and the dorsal fin are dried, often with shrivelled tips (densely ruled in the rock engraving, rendered as black in the illustration). The Askollen (Vestfold) drawing represents a male animal whose penis is erected by internal gas pressure. In the Skogervejen drawing the right side of the chest appears to have an opening which must have resulted from the left, false ribs piercing from within. The tension of the shrinking skin must have subsequently widened the opening. Gas pressure in the skull caused frothy liquids to emerge through the nostrils.

The tail in this particular engraving deserves special attention: the tail-stem shows a pad-like swelling close to the tail fin. Hallström wonders whether this swelling may possibly be a correction by a later artist. Our observations of recent whales indicate, however, that such

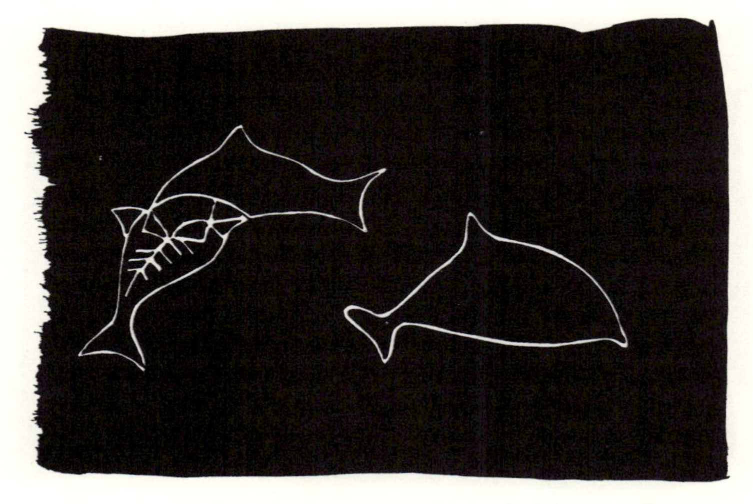

FIG. 11. Prehistoric rock engraving of dolphins, Evenhus, Norway. Left: two carcasses pushed into each other on the beach by the surf; right: jumping dolphin

swelling is due to the twisting of the tail fin through 90° because the body is lying on its side. This leads to early distintegration of the tail.

(3) Rock engraving in Evenhus, Norway, Fig. 11, left. Two dolphins (probably *Phocoena phocoena*) have been pushed on to the beach by the surf and lie parallel to the coast. The integument of the carcass in the foreground is destroyed so that the backbone with the spinal processes and the ribs are exposed. In the last stages of disintegration the head is frequently missing, the tail fin is separated, the skin is like a sack, open at both ends, in which the partly damaged vertebral column can easily be turned and twisted by external forces. Often it slips so far sideways that the spinal processes produce a ridge of the integument in the flank of the animal (Fig. 11 far left).

Not only dead whales and dolphins were pictured by prehistoric man on Norwegian rocks but he also drew living animals. We can thus compare how the artist saw and rendered them dead and alive.

(1) Rock engraving in Evenhus, Norway, Fig. 11, right. Even today dolphins jumping out of the water can frequently be seen from the beaches. Roman and Greek artists have often sketched the silhouette of a jumping dolphin, and so did the prehistoric artists of Norway. The jumping dolphin is well proportioned and has all the characteristics of a strong, moving body.

(2) Rock engraving after Hallström (1938), Fig. 12. Whales and dolphins are most frequently observed when they break surface. Occasionally, however, they stay for longer periods close to the surface,

FIG. 12. Prehistoric rock engraving of surfacing toothed whale (Halström, 1938)

especially when they feel danger and stay close inshore. Then a fisherman in his kayak may have a chance to observe them. Fig. 12 shows such a meeting. Probably this is the Caaing whale, *Globicephala melaena*, with a slightly gaping mouth baring the teeth. Its tail whips the water into froth, the ventral side which is immersed in water is partly omitted in the drawing.

(3) Engraving in Meling-Rogaland, Norway, Fig. 13. This is the report of a memorable event. Four well-manned boats go out to sea and meet a large whalebone whale. Just before reaching the boats it dives downward. The men in the first vessel see the whale in the clear water below, the diving head appears more pointed because the water distorts the wide, round shape. The second boat coming up somewhat behind the first is struck at the bow by the diving whale and damaged, the third vessel is lifted by the surging waves that are whipped high by the whale,

the fourth boat, however, is struck and destroyed by the beating tail fin, and the men and the remains of the boat drift in the water. Although such stray encounters at sea between man and whale obviously happened at that time, nowhere do we find any representations of hunting.

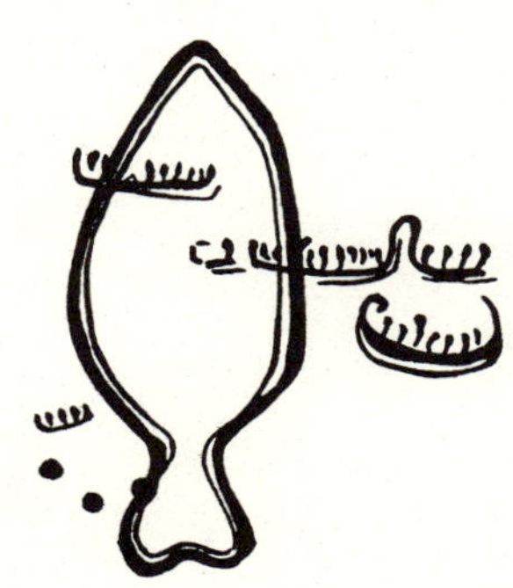

FIG. 13. Prehistoric rock engraving of diving whalebone whale and four manned boats, some damaged; Meling-Rogaland, Norway

There is no reason why whaling scenes should not have been found among the rock engravings if whaling had been undertaken at all. Hunts for large terrestrial mammals have been depicted in a number of rock engravings. Swimming and emerging cetaceans were certainly observed from boats; but whaling was not yet possible.

(*b*) PHOCA VITULINA

Seals (Phocidae) as well as whales are suitable models for interpreting the process of death, disintegration and final burial of carcasses of higher marine vertebrates. While whales are entirely aquatic, seals have not completely lost their tie with the land and the coast. They live like amphibians. In our North Sea area only the common seal (*Phoca vitulina*) is indigenous, although the larger grey seal (*Halichoerus grypus*) and the

FIG. 14. Cross-section through the foot of *Phoca vitulina* at the level of the phalanges, folded and spread. (Actual size × 0·8)

FIG. 15. Diving seal (*Phoca vitulina*)

harp seal (*Phoca groenlandica*) appear occasionally as stray visitors at the southern North Sea coast. As mammals with amphibian habits the seals spend long periods of time resting on firm ground in spite of being prevalently aquatic. They give birth and nurse their young on sandbars inaccessible from the mainland, and there the herds also gather to rest and to warm themselves in the sun.

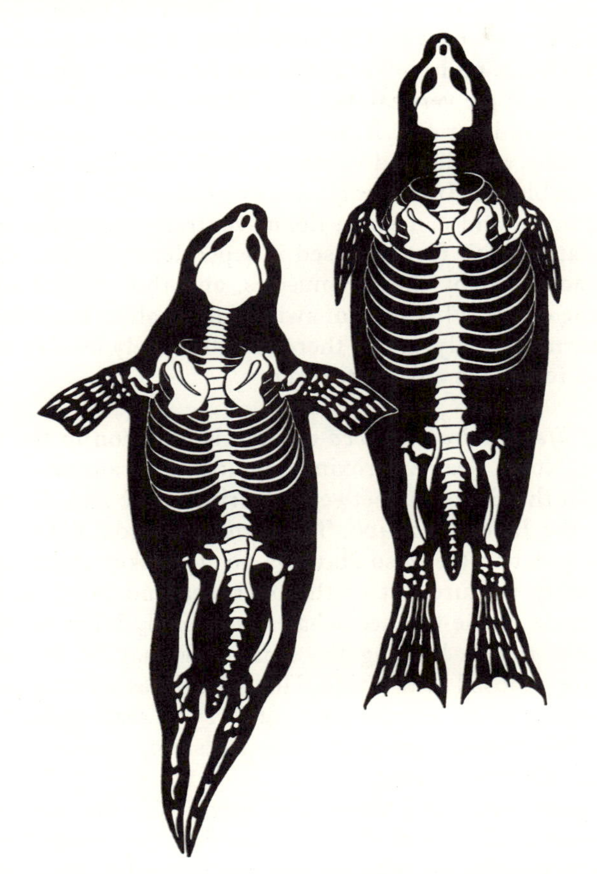

FIG. 16. The two swimming modes of the seal (*Phoca vitulina*). Left: lateral undulating; right: dorso-ventral undulating

Although the forelimbs and the hind limbs are changed considerably the animals are able to move on firm ground (for a discussion of their tracks and bedding traces, see Reineck 1959). Locomotion is commonly achieved by dorso-ventral undulations involving the whole rump without using the forelimbs. The hind limbs are unfit for locomotion outside the water because they cannot turn forward (the same holds for the sea lion and for the walrus). These hind limbs, built like broad, strong fins with elongated foot sections, extend far beyond the end of the body. Their five elongated toes are webbed; the first and fifth toes are of equal length and longer than the intermediate three toes. The outline of the foot is thus markedly concave. On land the toes are usually held closely together (Fig. 14). The soles of the feet can be turned downward (ventrally) as well as inward. Seals can swim with the hind limbs in either position (Figs. 15 and 16). If the soles point inward they can be joined behind the short tail stump to form a single vertical 'tail fin'. This enables the seal to swim like a fish by lateral undulations. If, however, the feet are turned through 90° so that their soles are directed ventrally, both hind limbs together form a symmetric double fin. It lies horizontally at the end of the body like the tail fin of a whale and can move dorso-ventrally using the whole body. As an alternative both feet are used independently as fins that move in succession, activated by the leg muscles, and thus propel and steer the body as in the crawl. This style of swimming enables the animal to move fast and to turn suddenly, and is therefore effectively used in fish hunting. The webbed forelimbs aid in steering.

Modes of life and death. Phoca vitulina is common in the tidal flats of the southern North Sea. Approximately 300 adult animals are estimated to be living in the tidal flats between the river Elbe and the river Weser; the estimate for Jade Bay is 400. The remaining tidal flats have a similar density of population (see also Havinga 1933, Junker 1940a, Mohr 1955). The seals rarely venture out to the open sea and north of the islands, possibly because of the toothed whales (especially killer whales) hunting there. Another reason holding the seals back may be the lack of sand-banks that are exposed at low tide. Heligoland has its own population of seals. According to an estimate of 1960, 1300 seals were living at the coast of the North Sea near Schleswig-Holstein, 172 seals were hunted and killed in this area in 1959.

For the seals with their amphibian way of life the contact with land and the temporary drying of the tidal flats does not bring deathly danger as it certainly does to the whales. On the contrary, seals even seek the land when death other than by predators seems to be approaching (old age, accident at birth, parasites). Then they try to find protection on sand-bars and periodically dry flats. Carcasses of seals are therefore no rare sight near the coast and on the tidal flats; they may either be found drifting inflated with gas in tidal channels or lying on the beach in all stages of decay.

Carcasses of young animals are more frequently encountered than those of adults. Every year during June and July many young animals die which were born in the same year (see also Junker 1940b, Reineck & Reineck 1956 and 1958, M. Reineck 1959). They may have been born during periods of adverse weather or perhaps they were left by their mothers in the first days of life if more than one young was born (Mohr 1955). Animals which are in the nursing stage and frequently still carry the umbilical cord, wander about alone for several days and finally die from hunger. The number of baby seals that die in this way each spring in the area of the East Frisian islands is estimated to be 40 to 50. More young animals, victims of parasitic nematodes (metastrongylids and especially *Otostrongylus circumlitus*), die each year near Jade Bay. These young seals contract the disease which affects the lung and the heart muscle, after the change to a fish diet. The yearly number of carcasses of young seals in the tidal flats is thus rather high, and skeletons are frequently found in the sediment.

Transport and decay of seals' carcasses. Seals, like whales, sink to the sea floor at the moment of death; several days later they begin to float when gas has collected in the body cavity. They may then drift about for several weeks. From our own experiments with seals we know exactly the stages of decay and their duration (Schäfer 1955c). Observations were made over a period of 57 days, counted from the day of death:

After 5 days the carcass emerges at the surface of the sea.

After 16 days. The body is firmly inflated, the eyes protrude, and the lips disintegrate.

After 22 days. The fur of the face disintegrates, and the lower jaws become loose at the symphysis.

After 28 days. The lower jaws lie open and are held loosely in the tissue. The skin at the back where it rises above the water level tears, and the exposed subcutaneous layers contain maggots.

After 36 days. The small intestine emerges through an opening at the flank and floats in the water. The face is completely disintegrated, and the lower jaws have already dropped to the sea floor. The skull hangs loosely in remains of fur, its occipital region and the bullae osseae tympanicae have already sunk. The skeleton of the forelimbs is exposed up to the carpals, and some of the phalanges are missing. The skeleton of the hind limbs is still held together by ligaments and tendons.

After 41 days. The carcass is still afloat but it now resembles a loose sack. Bits of neck fur hold a few remaining fragments of the skull.

After 48 days. The fur is now torn at both ends of the body but it still contains the vertebral column with the rib cage, the shoulder blades, the pelvis, and the thighs. The carcass continues to float.

After 57 days. Only a few pieces of fur are drifting about. The last skeletal parts have separated and, one by one, have dropped from the integument to the sea floor.

Thus carcasses of seals, just as those of whales, cannot be embedded complete in disturbed waters unless they have been stranded and are rapidly embedded in the shore sediments.

The stages of decay of stranded dead seals that are constantly exposed to the atmosphere are also known in detail but only for bodies of adolescent animals. For example:

After 9 days. Decomposition gases have noticeably inflated the body which is lying on its ventral side with outstretched neck and hind limbs.

After 12 days. The mouth and the anus discharge oily liquids.

After 15 days. The lips and the anus region begin to decay, and maggots abound in both areas.

After 20 days. From the anus region decomposition expands and increases. Oily solutions are now also discharged from the abdominal area and form a halo in the sand around the carcass. The mandibles are exposed and separated at the symphysis.

After 24 days. The fur over the occipital region and the shoulder-blade tears. The neck area disintegrates. Heavy rains wash loose hairs off the body and leave them in selvages along the edges of small rainwater runnels or puddles around the carcass. Individual whiskers are scattered around on the ground close to the head. The neck area and the pelvic regions collapse after their connections have loosened (Fig. 17).

After 30 days. The integument of the underside of the body is destroyed and the abdominal cavity is open. The anus and neck regions contain thousands of maggots. The rump has collapsed, and individual spinal processes pierce the skin from the inside.

After 37 days. Maggots have loosened parts of the skeleton at the skull, the neck, and the pelvis, and they drop from their original positions. The carcass begins to dry. The halo of sand that surrounds the body and is cemented by oils, dries and forms a hard crust with desiccation cracks.

After 44 days. The mandibles and parts of the facial skull (nasale in connection with maxillare and praemaxillare with turbinalia) have been separated by showers of rain, and some of the vertebrae of the neck are washed away. Slab-like concretions, often containing hair, have formed from fragments of the oil-soaked, hard sand between the desiccation cracks. The slabs rapidly become rounded at the edges and turn into pebbles which are transported by rainwater and wind.

Fig. 17. Seal carcass in dry sand, 24 days after death. Black, oil-encrusted sand

After 57 days. The teeth drop from the alveoli and are scattered around the carcass by renewed rains and high winds (Fig. 18).

FIG. 18. Skull of adolescent seal in oily sand, 57 days after death. Area 100 by 120 cm. Thin, white lines—fur hair

This record does not do justice to the importance of maggots during the disintegration of the corpse on the beach. They are particularly important when frequent rains regularly wash out the impregnating oils and when the tissues cannot mummify. Not only do the maggots avoid oil-impregnated tissues, but the flies do not lay their eggs on them. The individual maggots originating from one cluster of eggs stay together and eat all organic material from an area about the size of a hand. In most cases the maggots touch each other, and their heads are pointed in the same direction. They usually start cleaning off a bone at its lower end and then eat upwards. The resulting uniform movement of many hundreds of maggots can rotate light parts of the skeleton. Thus a bone may be dislodged by maggots from its original position as a first step in its ultimate transportation. If the lumps of eggs are laid close to each other, the individuals of different clusters intermingle which results in 'eating communities' that consist of many thousands of individuals (Weigelt 1930, 1935a and b, 1940; Walcher 1933; Deegener 1918). In the end only the upper side of the carcass remains because it is not affected by maggots. According to our observations the maggots resist temperatures down to −5°C under natural conditions (compare laboratory observations by Steiner 1941). Shortly before changing into chrysalises, maggots creep over the sand in search of suitable places for transformation, but most of them die before they pupate. A few of the maggots change into pupae in the carcass while it is mummifying. While brachyurans and pagurids take part in disintegrating corpses on tropical beaches, crustaceans are absent from carcasses on the North Sea coasts. All higher crustaceans indigenous to the North Sea feed only in water.

Modes of preservation. The limbs are especially prone to mummification because they dry fastest, whenever they become wet, and decomposition fluids hardly ever reach the limbs. Parts of limbs or even entire limbs with their connections intact can thus frequently be found in beach sediments and along swash marks.

The stomach, the lungs and the aorta descendens resist decomposition for some time. Frequently the stomach still contains some remains of food like bones of fishes and vertebrae, mainly from flatfishes. The lungs disintegrate except for the thin arteries and bronchiae that look like a tough, spongy web of tubes in mummified carcasses. Foerster (1933) reported that similar processes have been observed after death in lungs of newly born humans. The aorta descendens with its strong walls is connected with the vertebral column by resistant tissues. Even when the connections between the vertebrae themselves have been severed, they are held in place by their more resistant connections with the aorta. Occasionally, when a decaying carcass is flooded by water, a chain of lumbar and thoracic vertebrae, held together by the aorta, is lifted out of the carcass, and the assemblage is carried away.

If bones stay exposed to the air and are not covered up by sediments, the periost and the bone marrow as well as all organic substances (osseine of the bone) soon disappear. Only calcium carbonate and calcium phosphate remain; as a result the fine bars of bone in the spongiosa become clearly visible. The bones thus lose up to two-thirds of their weight and are easily transported by the wind. On the wide sand flats of Old Mellum we saw vertebrae of *Phoca* being carried by the wind, racing and jumping over the slightly salt-encrusted surface of the plain, while much smaller single valves of the lamellibranchs *Macoma* and *Mactra* and the round shells of *Littorina* were lying immobile on the sand.

The bones of the limbs disintegrate early into their epiphyses and diaphyses. This happens frequently at a stage when the joints between neighbouring bones of the limb are still connected by the joint capsule, ligaments, tendons, and their sheaths. Two bones connected by a joint are thus frequently separated even before the organic substances are destroyed.

Comparisons of resistance of skeletons and their parts lead to the conclusion that bones of adult animals are considerably better preserved, both in water and in air, than those of adolescent animals. The entire vertebral column and the skull as a whole, which are the two most resistant parts of the skeleton of an adult animal, are not yet hardened during the first year of life, and disintegration in young animals sets in very early in areas where ossification has been incomplete. Hence in carcasses of young animals, the pleurapophyses, neurapophyses (or haemapophyses) become detached from all but the cervical vertebrae. The tympanica connected with the squamosum breaks from the original position in the skull and, consequently, the occipital region loosens so much that its connection with the base of the skull is soon lost. These phenomena of disintegration are similar to many of those already described for the dolphins. The vertebrae and skulls of adult animals may show traces of abrasion on their outer surfaces after prolonged transportation.

(*c*) AVES

The number of species of birds in the fossil record is not as small as is usually assumed; in addition to the fossil remains of bodies there are numerous trace fossils. Lambrecht (1933) listed approximately 700 fossil species. They are more frequently obtained from fresh-water sediments of arid areas or from cave deposits than from marine sediments. In the Late Cretaceous different orders, such as fowl, rails, snipes, sea gulls, storks, geese, and sparrows were already in existence. According to Franz (1924) and Lambrecht (1933) the majority of orders of birds that live today had representatives in the Eocene.

Numbers of species and individuals. In European latitudes birds are most abundant near the sea coasts. The swarms of migratory birds that fly over the southern coast of the North Sea and Heligoland Bay in spring and in autumn, and the birds that remain for longer periods of time without breeding, far exceed in number those that are indigenous. Only seventeen species of birds breed directly on the coasts, eleven of which are species breeding in colonies. Not less than 110 930 breeding individuals of the seventeen species have been counted on one occasion at the German coast of the North Sea. Birds of sixty species that breed elsewhere, however, spend a large part of the year (mainly the late summer, autumn and winter) on the coasts, tidal flats, and off-shore islands of this area and represent the bulk of its bird population.

Reliable counts of individuals of only a few species have been made in this area. Some species are represented only by rare visitors, others, however, by tens of thousands of individuals who stay every year for a considerable time. Ducks are the only birds that have been counted with reasonable accuracy in Germany (Requate 1954).

Those members of migratory species that do not spend the summer in their native northern habitat constitute another group of birds populating the German North Sea coasts. These birds may be sexually immature, they may be older birds whose brood was destroyed, or adult birds that missed the departure of their swarms. There are also some individuals that have lost their migratory instinct and others that are diseased (Remmert 1957). Numbers of individuals that did not migrate but stayed on the island of Trischen are given in Table 5, and estimates for the island of Mellum for June and July in Table 6. Remmert estimates that 250 000

TABLE 5

Frequencies of migratory birds spending the summer near the island of Trischen (Remmert 1957)

AT STORM TIDES:	
Calidris alpina	15 000
Haematopus ostralegus	2000
Melanitta nigra	600
Chlidonias niger	400
Larus argentatus	200
L. marinus	25
AT NORMAL TIDES:	
Calidris canutus	several hundreds
C. alpina	600–800
Haematopus ostralegus	600–800
Numenius arquata	1000
Arenaria interpres	6
Larus marinus	15
L. canus	10
L. ridibundus	several hundreds
Anas penelope	4
Spatula clypeata	2

TABLE 6

Estimated frequencies of migratory birds near the island of Mellum in June and July (Goethe 1939)

	June	July	Year
Calidris alpina	10 000	15 000	1933
C. canutus	100	500	1928
Numenius arquata	200	150–200	1928
Larus argentatus	3000	5000	1933
Haematopus ostralegus	600	200	1934

dunlins stay at the coasts of the German islands of the North Sea. Estimates for migratory birds who are natives of the north but were living in September 1955 in the strip of mainland coast between Emden and Jade Bay are listed in Table 7.

TABLE 7

Estimated frequencies of migratory birds on the coastal strip between Emden and Jade Bay in September 1955 (Bub 1956)

Tadorna tadorna	6450
Ducks	18 550
Calidris alpina	13 700
Recurvirostra avosetta	4057
Numenius arquata	6780
Haematopus ostralegus	4580
Larus canus	2150
L. ridibundus	24 300

Other migratory birds that breed mostly inland in Norway, Sweden, Finland, and in arctic regions interrupt their flight for longer periods in Heligoland Bay. Unfortunately, there are no estimates of the number of birds flying over Heligoland Bay during one entire period of migration. The only estimates that have been carried out on Heligoland are restricted to a few species of migratory birds over a few hours, days and nights; several hundred thousands of birds have been observed passing over the island within a few hours (Drost 1925, 1928a and b). Graphic statements by Weigold (1930) concerning migration over Heligoland should be mentioned here as well as the monthly tables of the 'Vogelwarte' of Heligoland and its field stations at the coast. These tables have been compiled since 1924 in order to assess the stock of birds periodically. The total number of individuals of one species that migrate over the island of Heligoland, let alone over the German sea coasts, during an entire migration period cannot possibly be estimated at present. Such figures, however, would be useful because particularly bad weather has catastrophic effects on these rather vulnerable migratory birds which pass over the sea, yet are natives of interior regions. We do not know how long the phenomenon of migration has existed. It is possible that similar phenomena developed in the geological past. If any mass-migrations of birds had taken place in earlier times, death and burial of large numbers of animals would have been a natural consequence (see also Franz 1924).

Places of life and of death and causes of death. The number of carcasses and their relics must be abundant where enormous numbers of large vertebrates exist and where members of the individual species constantly move in large groups. Carcasses of birds can be found the year round along the coastal swash marks of the islands and of the mainland. The death rate increases catastrophically at times of extreme weather conditions especially in the spring and autumn when many birds are migrating. Gales accompanied by cold rains, sudden cold fronts, cold mists, and frost are the main causes of mass death of migratory birds. Usually such disasters are only noticed if the carcasses reach the beaches and the swash marks, but remain undiscovered if the bodies stay in the open sea. The ornithological literature is rich in reports on mass deaths. Only a few observations are quoted here: Wiman (1914) stated that dead razorbills were seen in huge numbers on the ice of Spitzbergen as well as floating skeletons of the same species and of fulmars. Lambrecht (1933) reported that heavy mists with following frost caused mass deaths among starlings, fieldfares, song-thrushes, mistle-thrushes, redwings, blackbirds, wood larks, chaffinches, skylarks, linnets, wood-pigeons, hooded-crows, jackdaws, lapwings, bramblings, long-eared owls, shore larks, mergansers, guillemots, and diving ducks. In all, 214 dead birds were found along three kilometres of beach. According to Stresemann (1930) 509 dead oystercatchers were counted on the beaches of the Dutch coast in the winter of 1928–9; the figure he gives for a certain section of the English coast are 240 golden plovers, 30 lapwings, 200 redwings and starlings and 100 song-thrushes. Thienemann (1925) and Geyr von Schweppenburg (1930) mention similar events. Kuhk (1956) gives hail as a cause of death of storks. Thienemann (1939) mentions numerous deaths over the Baltic Sea in spring 1918 and again in 1923 (2520 birds were counted along 35 km of beach). Our own team observed mass deaths of coots at the edge of grassy coastal marshes of the upper Jade Bay as well as dead starlings and dead larks that were lying in batches at the beach of Wangerooge.

Individual dead birds are common; we found dead guillemots of Heligoland at the coasts of the East Frisian Islands, the carcass of a gannet at the north beach of the island of Mellum, several hooded-crows, starlings, and other songbirds rather frequently, also ducks. Sea gulls, terns, oyster-catchers, and redshanks are regularly encountered along the swash marks. In these cases death is probably seldom due to the weather but rather to normal old age, disease, or accident. Concentrations along the beaches of animals which die over the open sea in groups, or individually, and are then washed ashore after long periods of drifting, may be considered as secondary assemblages.

The breeding colonies of sea gulls, sea-swallows, and other coast-breeding birds represent periodical concentrations of the living animals themselves. A concentration of carcasses of adult and young birds from such a colony may be called a primary assemblage. We are mainly concerned with the breeding colonies of the herring gull (*Larus argentatus*), common gull (*L. canus*), common tern (*Sterna hirundo*), arctic tern (*S. macrura*), sandwich tern (*S. sandvicensis*), and little tern (*S. albifrons*). The colonies of all these species are at the very edge of the sea so that the dune deposits on which they rest may eventually cover them. These deposits can still be considered as belonging to a marine facies. However, sediments from epilittoral and supralittoral areas are the most likely marine sediments to be subsequently destroyed. If the sea regresses the new deposits are exposed to subaerial erosion; if, on the other hand, the sea transgresses, most of the deposits are removed before they are covered and protected by new marine sediments. This may well be the reason why fossil remains of birds are so rarely encountered—as was already mentioned above—in marine and particularly marine-littoral sediments. Once more it can be seen that favourable conditions for burial and fossilisation in themselves are no guarantee for the final preservation of a fossil.

Nevertheless it is useful to form an idea of life and death in such colonies and to estimate the numbers of individuals found in them. The herring gull forms the largest colonies of the southern North Sea coasts. The colony of the island of Mellum has (or had) up to 6000 breeding pairs, that of the island of Memmert up to 8000 breeding pairs (see also Leege 1935), that of the island of Borkum up to 20 000 breeding pairs in 1875, and that of Langeoog island 10 000 breeding pairs; these are (or were) probably the largest colonies with the highest numbers of individual members in that entire area. Many smaller colonies with a thousand or less could be added to the list. Today 30 000 pairs are estimated to be living at the German North Sea coast. As one brood of the herring gull generally consists of three eggs, up to 25 000 individuals of about chicken size live in a colony with 5000 pairs on barely a quarter square kilometre at the time when the young are growing up; the number of males that do not breed but stay close to the other animals are not included. Obviously, the number of deaths among young and old animals cannot be small. Large numbers of carcasses that are gradually decaying or being fossilised are constantly found in and around the colony. As the years go by, huge amounts of skeletal remains collect in the area of such a colony.

Goethe (1956a) has given a detailed account of the causes of death in herring gulls. He mentions that foxes and white-tailed eagles must formerly have played important roles as enemies of these birds. According to the same author (1956b), herring gulls which spend their nights on the tidal flats close to the coast are occasionally killed by foxes even today (in November 1956, fifty to sixty herring gulls were killed by foxes on the Heppenser Groden near Wilhelmshaven). The marsh harrier attacks young birds that are almost at the fledgling stage; according to Goethe the victims are in most cases plucked at the back and cut open from there. Gyr falcons occasionally attack herring gulls. Young animals may be killed and eaten by members of their own species; in certain summers 90 per cent of the adolescent animals of some sections of the colonies are estimated to have been killed in this way. The victims are not eaten if the killing constitutes an act of defence of territory by the older birds against larger, adolescent ones. This kind of cannibalism occurs only in very densely populated colonies where territorial rights are from necessity frequently violated by members of the same species.

According to Goethe, herring gulls frequently catch epitheliosis, a virus disease which is non-fatal in mild cases. Its symptoms are glossy swellings at the beak, at the rims of the eyes, and at the joints of the feet. These tumours can grow to the size of a small potato. Fungal diseases (especially *Aspergillus niger*) are thought to be fairly common, particularly in wet summers. In 1954 hundreds of young animals fell victim to this disease in a colony of the island of Juist. Paratyphoid A and B have also been confirmed. In addition there is a large number of parasites. Requate (1951) observed the fly *Lucilia sericata*, an ectoparasite, that lays the eggs in the beak of a young bird. The animal dies soon after the maggots have developed. According to Niethammer (1942) 23 species of trematodes, 16 species of tapeworms, and 10 species of threadworms occur as endoparasites. Their intermediate hosts are mainly shellfish, crayfish, and fishes. The parasites live in the digestive tract in huge numbers, and birds with a poor constitution often die as a result. All these diseases occur more frequently and more intensely at the time when the animals are closely crowded together in colonies. This explains the high number of carcasses in and near colonies. Drost & Schilling (1940) state (cited from Goethe) that up to 70 per cent of the herring gulls die during the first year of their life. According to these authors the average life span of the herring gull is six years; Goethe, on the other hand, thinks that the average life is longer than that. He points out that nothing is known about the fate of 90 per cent of the ringed birds and that statistical calculations are therefore based on too small a number of individuals to be significant.

Little is known about the causes of death among different terns during their life in the breeding colonies. These colonies never reach the size of those of the herring gull. The sandwich tern forms colonies of 1500 to 2000 pairs; but the little tern builds colonies of barely 200 pairs. The young animals are more dependent on the weather conditions than the young herring gulls. The tern parents catch the food by rapid, vertical diving from flight. Bad weather over the ocean can thus prevent them altogether from catching small fish, and the young birds in the colony are weakened by hunger.

As already mentioned, the colonies of the sea gulls and of the terns are built at the edge of the sea in the epilittoral and supralittoral regions of the islands. These areas of low dunes, grassy marches, and beaches are constantly altered by wind-blown sand, and over a long period of time by epeirogenic movements.

Whenever a seaside locality changes, so do the plant and animal populations. Changes in the topography of the islands are thus accompanied by variations in the biotope for breeding birds. As we have discussed in 1941b dune islands can be considered as young, mature, or ageing, and the regular sequence of their geological development is accompanied by parallel biological changes. A number of successive stages can be distinguished in the development of an island between the first appearance of a small mound of broken shells and the high dune island where erosion has already set in. All these successive stages in their regular sequence are characterised by well-defined biotopes. The newly forming dune island forms an open, predominantly subaerial biotope of tidal flats. As the island grows the biotope becomes one of oligohaline green plants. This in turn is followed by herbaceous, green, fresh-water plants, and the final biotope is that of woody, green plants. Under favourable conditions such a period of development can take fifty years or even less (compare the development of the island of Trischen, Wohlenberg 1950).

A certain animal community corresponds to each of the biotopes mentioned, and it is obvious that the breeding locations which were once chosen by a certain species of birds cannot be kept indefinitely. The same place is usually the base for a breeding colony for decades, but the species vary. They change in regular succession according to the biotope. The little terns build the first colonies, forming the lowest level of a succession. The animals breed in loose, sparsely populated colonies. They favour sandy areas or surfaces covered by shell fragments that are bare of any vegetation and often subject to intensive sand drift. The common terns and the sandwich terns breed in rather crowded colonies which lie higher above the mean high tide line. A loose or even dense cover of plants already exists where they breed. As soon as the vegetation becomes too luxuriant and too high the terns depart, and the herring gulls move in and build a colony in their thousands. This process of change-over among different species in the same place is known in detail of the island of Mellum (Schäfer 1954b), and can be confirmed for other islands. Such changes are reflected in the carcasses and skeletal remains that are embedded in the sands according to the sequence of the former colonies. A rapidly changing landscape like that of the sandy and muddy tidal coasts is the prerequisite for rapid changes in biotope, followed by equally rapid alterations in the animal population. The colonies of rock-breeding birds which are not affected by biotope changes within such short periods of time are therefore more stable and last for many decades, probably even for centuries. On one cliff on the island of Heligoland the guillemot (*Uria aalge albionis*) has continuously bred for over a hundred years; according to Gätke (1900) the breeding places have been known since

1837. Such places are only given up if the cliff is undercut and falls, as happened several times in the past to other colonies of the guillemot on the island of Heligoland and has been vividly pictured by the painter Gätke.

There is little prospect of the carcasses being finally preserved if they are embedded in the island dunes as discussed above. However, purely marine species of birds which rarely approach the coast except for breeding, probably die mostly on the open sea (see also Alexander 1959). Their skeletons are more likely to be preserved in sediments far from the coast.

The following species occur in the open North Sea, all of them in small numbers: the storm petrel (*Hydrobates pelagicus*), the fulmar (*Fulmarus glacialis*), the little auk (*Plautus alle*), the puffin (*Fratercula arctica grabae*), the kittiwake (*Rissa tridactyla*), the skua (*Stercorarius skua*), the pomarine skua (*S. pomarinus*), the long-tailed skua (*S. longicaudus*), the arctic skua (*S. parasiticus*), the great shearwater (*Puffinus gravis*), the Leach's petrel (*Oceanodroma leucorrhoa*). All these species must be considered as marine; only a few are indigenous to the North Sea. Some of them are occasional visitors during gales, others are regular visitors, mainly in winter. The diet of these birds demonstrates in particular their strong ties with the open sea. Several of these species feed on macroplankton (*Oceanodroma*, *Puffinus*, *Rissa*, *Tubinares*); others live mainly commensally by snatching the food caught by other birds (*Stercorarius*); again others catch fish near the surface by scooping them up in full flight (*Fratercula*, *Fulmarus*).

In addition to these purely oceanic species there are the inhabitants of the continental shelf; these species occur more frequently in the North Sea and its adjoining seas rather than in the Atlantic. Even so they spend the main part of their life on the sea far from land. They include the gannet (*Sula bassana*), the razorbill (*Alca torda*), the guillemot (*Uria aalge*), the common scoter (*Melanitta nigra*), and the eider (*Somateria mollissima*). *Sula* dives from the air like most members of the genus *Sterna*, *Alca* dives by swimming under water, and *Melanitta* and *Somateria* are diving ducks in the shallower coastal waters.

Transport and decay of birds' carcasses. When a bird dies on the sea or when a carcass is driven out to the open sea by waves and by the currents of the outgoing tide, the body does not sink immediately to the sea floor as do carcasses of fishes, reptiles, and mammals but drifts on the surface for a considerable time. The air in the quills, between the down feathers, and in the tubular bones of the limbs keeps the body floating. According to our observations the carcasses of the herring gull disintegrate in the following way:

The carcass drifts initially on the actual surface and is soon immersed somewhat deeper, once the feathers are thoroughly wet.

After 4 days. Young maggots abound in those parts of the carcass that jut out of the water. Although the tissues are soaked with salt water the maggots thrive and grow rapidly.

After 13 days. All skeletal parts above the water surface are now bare of musculature and fibrous tissue. The large flight-feathers which spring from the periosteum of the wing bones have fallen out of their sockets and are lying loosely on the carcass or drift individually away from it.

After 27 days. The carcass is still afloat, the head hangs into the water and is well preserved, even skin and feathers. The legs, however, have dropped away; the sternum, too, has separated from the skeleton and has sunk to the sea floor.

After 38 days. The body sinks to the sea floor.

After 65 days. The carcass which is still held together by muscles and ligaments lies on the sea floor. No part of the body floats up again.

Unlike the carcasses of the mammals (and as we shall see later, unlike the bodies of fishes) the bodies of birds can be embedded complete on the sea floor even under aerobic conditions. In the shallow sea, however, the skeletal parts which have still remained in their original places, are scattered about by bottom currents. Many carcasses of birds which have died on the open sea reach the beaches before sinking because the bodies drift for weeks. One observes them occasionally on the North Sea from the low deck of a small vessel.

In addition to the carcasses that drift to the beaches, many birds die on land, and subaerial fossilisation must occur frequently. Once the carcasses of birds lie on dry sand, mummification sets in. It is accompanied by characteristic, postmortal contractions of certain muscles. These contractions persist even if temporary submersion or further rain wets the body. The skin shrinks especially at the chest and on the skull when the body begins to mummify. The feathers ruffle up. When the carcass dries further the muscles of the back and of the neck contract. These contractions cause the head and the neck to bend backward and the tail with its feathers to bend up (Fig. 19).

Occasionally it has been contended (e.g. Deecke 1915, p. 125) that the actions of an animal immediately before its death are reflected in the position of the fossil body; it is maintained then that a certain body position caused by all the complex actions of muscles, groups of muscles and nerve impulses could suddenly be arrested at the moment of death. We mentioned in earlier publications that positions after death are not true images of attitudes and positions of living animals at the moment of death; rather, postmortal contractions of the musculature or displacement by external forces determine the final configuration of the fossil. Both factors most influence the final position in those animals whose bodies are well articulated and which have long limbs, tails, and necks (terrestrial mammals, birds, ophiuroids, long-legged arthropods). Their influence is less pronounced on the compact bodies of fishes, whales, or ichthyosaurs.

The back, neck, and head of dead birds typically bend backward whenever the resting place of the carcass is somewhat dry, and the sun shines for several hours a day. The strong backward bend of neck and head forces the shrinking trachea of the dead animal from the original ventral to a dorsal position. There it remains and solidifies by dehydration. Bone-like rings of the trachea are known in fossils. The postmortal

FIG. 19. Mummified common tern (*Sterna hirundo*). Head and tail bent backward by postmortal muscle contractions (Schäfer, 1955c)

contraction of the strong musculature of the chest pushes the upper arm away from the body. As a result the wings of mummified bodies lie always turned away from the body, and the distal section of the wing is usually pulled toward the humerus or twisted. The body can be completely covered and preserved in such a position by gently blown sand. In most cases, however, the covering sand—particularly when it gets wet—compresses the body. This causes fractures of the joints and new relative displacements of the parts, which have nothing to do with any possible life attitude. Predation and decay of the mummifying carcass of the bird always set in on the underside of the body which stays moist for long periods, and where insect larvae and amphipods feed, protected from the sun. The upper side dries out rapidly and becomes resistant to decomposition and predation. Blowing sands invade and fill the abdominal and the chest cavity from below. In carcasses of vertebrates which are embedded on the sea floor, on the other hand, the underside is usually better preserved because it is protected by sediments while the upper side may be exposed to destruction for a long time.

No postmortal contractions of the muscles occur if the carcass stays wet before it gets embedded in the beach sediments, if it is repeatedly flooded and shifted by water, and if it is finally deposited in wet sand. The joints remain mobile, and the final positions of the limbs, neck, and head are exclusively determined by the current flow. As the disintegrating body is repeatedly lifted and dropped by the water, the hind limbs finally separate from the trunk, the pelvis from the lumbar vetrebrae, and are transported away. Wings, coracoid, clavicle, and sternum continue to hold together as a unit for a long time. The individual parts are frequently twisted longitudinally relative to each other. The head, with the cervical vertebrae, separates from the body, and the tough trachea is left as the only connection between head and body. The skull breaks first between nasal and frontal bones so that the upper part of the beak is lost, and the lower part separates later from the joint. The strong flight feathers and tail feathers remain connected with the skeleton on the mummy for a long time because they are firmly anchored in the periosteum. The down which is rooted only in the skin becomes detached when the skin is destroyed.

Two counts of skeletons and parts of skeletons of birds on the sand flats of the Old Mellum, along its swash marks, and also on its vegetated area, are given in Table 8.

R. Heldt (1960) compiled a table containing the members and species of bird carcasses along the swash marks of the coast of Eiderstedt. He counted a total of 5926 birds in the years 1953, 1954, 1959, and 1960—236 thrushes, 2055 ducks, 843 broadbilled sandpipers, and 1818 sea gulls were among the carcasses.

Many of the carcasses found on the beaches and in the dunes contain gastroliths. They help to grind food in the stomach. They are discarded from time to time through the anus or mouth and are replaced. According to Lambrecht gastroliths have been observed in fossils; they are also occasionally encountered in the shrunken stomachs of carcasses of birds recently embedded. The following species of sea birds carry individual gastroliths in their stomachs or an assemblage of small stones, frequently mixed with fragments of mollusc shells (partly personal communication, F. Goethe): the ducks *Melanitta nigra*, *Anas acuta*; the geese *Branta leucopsis*, *Anser fabalis*, *A. brachyrhynchus*; the divers *Podiceps auritus* and *P. ruficollis*, *Colymbus immer* and *C. arcticus*; the sea gulls *Larus ridibundus*, *L. marinus*, *L. canus*; the black-tailed godwit *Limosa limosa*; the curlew *Numenius arquata*; the plovers *Charadrius hiaticula*, *C. alexandrinus*, and *Haematopus ostralegus*. Probably the number of sea birds which carry gastroliths is considerably larger.

Table 8

Counts of bird remains on the island of Mellum

(I) 18th May 1959	(II) 24th November 1959
(A) FRESH CARCASSES OR COMPLETE SKELETONS	
Larus argentatus, adult, 4	*Larus argentatus*, adult, 9
L. argentatus, adolescent, 9	*L. argentatus*, adolescent, 14
Sterna hirundo, 2	*L. ridibundus*, 1
Haematopus ostralegus, 1	*Fulica atra*, 2
Colymbus stellatus, 1	*Numenius arquata*, 1
Melanitta nigra, 2	*Tringa totanus*, 1
Anas crecca, 1	*Calidris alpina*, 4
Tadorna tadorna, 1	*Haematopus ostralegus*, 1
Sturnus vulgaris, 4	*Vanellus vanellus*, 1
	Melanitta nigra, 4
	Anas crecca, 1
	A. platyrhynchus, 2
	Ardea cinerea, 1
(B) SKELETAL PARTS	
Larus argentatus:	*Larus argentatus:*
pelvis, 7 (1 with pygostyl, 1 with left femur)	skull, 2
skull, 6 (2 with cervical vertebrae)	sternum, 7
coracoid, 1	pelvis, 3
sternum, 6 (2 with clavicula)	anterior limb with coracoid, 2
leg, 1	trunk skeleton with legs, 1
hand wing, 1	*Oidemia nigra:*
Sterna hirundo:	pelvis, 1
skull, 1	
pelvis, 1	
sternum, 2	

(*d*) PISCES

Strangely enough, little work has so far been done on the processes leading to fossilisation of fishes.

Nevertheless, we know that only in a few cases can complete fishes be fossilised (with the skeleton in original position and with all the scales). Fish carcasses are more vulnerable to decay than other vertebrates. This has one advantage: once we are familiar with the restricted conditions that permit total preservation, our conclusions about mode of death, living conditions, and fossil preservation of extinct fish species will be more precise. We will thus be able to predict where complete skeletons can be expected, or where skeletons with displacements are likely to be found, or where only unimpressive remains of skeletons can be expected as traces of a former fish population. One approach to the problem is to examine fossil fishes and the sediments in which they are embedded; a second way is to investigate living fishes and to study the burial of their carcasses in sediments. Both studies are complementary but we confine ourselves here to the latter.

The southern North Sea is fairly uniform as an environment for life and for burial. Climate and sedimentation vary little, and so the choice of marine conditions which determine fossilisation is limited. Only a few possible forms of fish remains need to be investigated under these circumstances. Experimental fossilisation of fish has provided a wider range of factors that do not normally exist in the North Sea.

Numbers of species and of individuals, modes of death. The North Sea is particularly rich in fish; the number of species, however, is exceeded by that of most subtropical and tropical shallow oceans. Indeed, the number of fish species that live regularly in the North Sea is remarkably small. Only 55 species of teleosteans and 13 of elasmobranchs inhabit the southern North Sea regularly in large numbers of individuals; another 25 teleostean and 8 elasmobranch species live temporarily in the area. Rare stray visitors in small numbers represent some 27 teleostean and 11 elasmobranch species.

Only the numbers of fishes that are marketed provide an estimate of the number of fishes that live in the sea. The species that are dealt with by the commercial fishing industry are few in number; the frequency of inedible fishes is unknown. The statistics of catches reveal that the few species which are regularly caught in the North Sea occur in enormous numbers of individuals.

Figures of catches in 1951 from the north and west European oceans are listed in Table 9. Rühmer (1954) provided the following additional figures: *Clupea*

TABLE 9

Commercial catches of fish off northern and western Europe and in Atlantic (Rühmer 1954)

Area	Millions metric tons	Per cent
North Sea	1·8	30
Kattegat and Baltic Sea	0·4	6
Iceland and Färoe Islands	0·8	15
British waters	0·3	5
Norwegian waters and Barents Sea	1·6	28
Greenland and Newfoundland	0·3	5
Biscay to Morocco	0·6	11

harengus (herring), total catch in northern Europe yearly 1 200 000 to 1 500 000 metric tons. *Sprattus sprattus* (sprat), total catch by the north European fishing industry in 1950 31 000 tons. *Merlangius merlangus* (whiting), catch by German fishermen in 1952 598 tons. *Pollachius pollachius* (pollack) in the North Sea and in the Kattegat 46 tons. *Anarhichas lupus* (sea wolf) in the North Sea and in the

Kattegat in 1952 36 tons. *Cyclopterus lumpus* (lump sucker) in the North Sea in 1952 760 tons. *Scophthalmus maximus* (turbot), total catch in Europe in 1950 approximately 8000 tons. *Hippoglossus hippoglossus* (halibut) by the north European fishing industry in 1950 16 240 tons. *Pleuronectes platessa* (plaice) 3·3 tons in 1952. *Limanda limanda* (common dab) in the North Sea in 1950 5832 tons. *Solea solea* (sole) in the North Sea in 1952 1200 tons. *Lophius piscatorius* (angler) in the North Sea and in the Kattegat in 1952 31·3 tons.

Generally only catastrophic events are reckoned to cause death and fossilisation of fishes (Eberle 1929, Gunter 1941, 1947, Kaiser 1930, Ladd 1959, Kristensen 1956, Nümann 1957, Jüngst 1937, and many others).

The catastrophic death of great numbers of fishes is rare in the North Sea. There are no volcanic eruptions (to name, with L. R. David 1957, the most important kind of disaster causing death among fishes) nor poisonous gases; floods and their after-effects are lacking; oxygen is always present on the sea floor, and nowhere in the North Sea is there an overabundance of H_2S or iron sulphide, or of organic material. There is also no sudden sedimentation of finest muds and, finally, there are no areas without life.

Only in one case can fishes die in masses in the North Sea. Such events are caused by sudden massive reproduction of certain phytoplankton which produce toxic waste products; the sea water subsequently changes chemically and thus causes the fish population to die with abnormal symptoms as soon as the plankton die. The simultaneous decomposition of so much plankton leads to a considerable decrease in oxygen and to increases in phosphates and in nitrites in the lower layers of the water body (Nümann 1957a). These organisms are free-swimming algae (mostly dinoflagellates, but also diatoms); their almost explosive development in nutrient-rich upwelling water has frequently been described (Brongersma-Sanders 1944, 1948, 1949, Falke 1950, Richter 1950). They may also develop in any nutrient-rich water without upwelling. When they occur in large quantities the water turns a striking reddish-brown and is then generally called 'red tide'.

The last time that such a red tide has been noticed in the North Sea was in 1958 (English Channel, Plymouth); for the Dutch coast see also Fonds and Eisma (1967). Nümann (see above) describes in detail two such events (in the Bay of Luanda at the coast of West Africa in 1951 and in the Bay of Jzmir-Smyrna in the Aegean Sea in 1955). Nümann assumes that in the case of the Bay of Luanda the fishes which died between a depth of 10 m and the sea floor had suffocated. In the upper water layers above 10 m depth, however, the fishes died from poisonous waste products of the plankton (phytoplankton, consisting of 69 to 92 per cent of *Exuviella baltica*, 10 per cent of *Prorocentrum micans*, 13 per cent of *Peridinium* sp.). In the red tide of the Aegean Sea another Peridinea,

Gymnodium sp., caused death in fishes supposedly by suffocation. According to Nümann phytoplankton bloom appears to set in when trace elements, enzymes and other catalysts have been carried by fresh water into the sea. At the same time, nutrients must be present in the quantities needed for such a massive development of plankton. However, we do not believe that phytoplankton in great quantities can only occur in the vicinity of the coast. If the waste products of the plankton alone can kill the fishes, the lack of oxygen is an added cause of death for the nekton as soon as the oxidisation of the decaying plankton uses up most of the remaining oxygen. Nümann also says that even in regions with abundant oxygen the gills of the fishes become irritated by the countless organisms and finally close.

The remaining causes of death in the North Sea are the normal ones, predation (with increasing size of the fish these cases decrease), old age, and parasites. Even where large numbers of individuals are assembled, as in herring schools, the number of dying animals is always so small that the decomposition of the decaying carcasses does not influence the chemical processes at the sea floor. Situations may arise in the North Sea which can lead to the death of fish by cold shock (according to a personal communication by the captain of a cutter). Such shocks may be caused by bodies of very cold water that occasionally advance in late summer from the Norwegian trough into the warm, shallow waters of the central North Sea (for discussion of shocks from cold, see Nümann 1957b).

Disintegration of the carcass and ways of preservation. A fossil can be deciphered only as long as some hints of a shape are left. The presence of organic material may well demonstrate that former life existed but the species that produced the organic compounds cannot really be identified. The soft parts are soon destroyed by chemical (and bacterial) decomposition. Mechanical destruction of the body follows. Even when organic decomposition is slow, characteristic shapes are seldom sufficiently well preserved to be identified (see also Hecht 1933).

Thus we depend on the fossil record of those parts which are only slowly attacked by chemical decomposition, that is, in particular the skeleton. A complete skeleton can be preserved in marine sediments only under special circumstances. Vertebrates are composed of many individual parts and the connections of these parts in fishes are destroyed especially rapidly by chemical decomposition. Many different factors determine whether a skeleton will be preserved in its complete original structure in marine sediments or whether only individual parts of the skeleton will reach the sediment. The factors which influence the preservation of the skeletons are anatomy, oceanographic-chemical conditions, and geological circumstances. Experiments can elucidate these relationships and can help to solve many of the questions that can be asked about fossilisation. In what period of time do carcasses of different species disintegrate? Does this length of time vary significantly among different species? On what, exactly, do these possible differences depend? What influence have the properties of the surrounding water on the processes that are going on in the carcass of the fish? Which hard parts of the body of the fish are most resistant?

I have observed the kinds of disintegration and different lengths of time taken by carcasses of eight species of fishes. The experimental conditions have differed, and the measurements are not exact.

(1) *Myoxocephalus scorpius* (sea scorpion). The carcass sinks to the sea floor with its back directed downward (Fig. 20). Water temperature + 18°C.

FIG. 20. Stages of decay of *Myoxocephalus scorpius* carcass. (1) inflated stomach; (2) floating at the surface, bubbles of decomposition gas under the skin; (3) gas partially escapes through tears in abdominal cavity and throat area; (4) carcass sinks to the bottom. This whole process takes four days (Actual size × 0·33)

1 day after death. The lenses of the eyes turn opaque, the anus protrudes, the mouth closes to a narrow slit. Twelve hours later a bright bubble of gas is seen under the skin on the right side of the abdomen which soon starts bulging.

2 days after death. The gas bubble breaks through the abdominal skin and rises to the surface. Simultaneously, new gas bubbles form in other places under the abdominal skin. The epidermis drops in pieces from the sides of the body. The carcass slightly changes position, and due to this movement the tail fin produces scraping traces in the mud.

After 4 days. The carcass floats to the surface with its ventral side up. The body then touches the surface with its ventral side while the carcass is bent backward. Many individual gas bubbles shine under the abdominal skin and pull the integument upward. A few hours later the abdominal skin tears. The entire back of the body droops downward; the body is held at the water surface only by its head and chest section which is filled with gas. The stomach is still unbroken and is shifted anteriorly. At this stage, the unpaired fins drop their rays which sink to the bottom. An hour later the carcass sinks to the sea floor again and collapses.

The carcass has kept intact nearly all the parts of its skeleton during its surfacing and renewed sinking; the skeleton, however, does not lie naturally any more because the body has been severely telescoped. In the last stage, the remaining

FIG. 21. Skeletal parts of *Myoxocephalus scorpius* (same as in Fig. 20) after three months in agitated water. (Actual size)

organic material decomposes on the sea floor. After approximately three months the skeleton lies on the bottom and has been stripped of organic remains.

In the course of the experiments the partial pressure of oxygen in the water constantly decreases and that of H_2S increases. If oxygen is added artificially, the course of events is the same for the carcasses of *Myoxocephalus* but the stages follow each other more rapidly. In such cases body parts that have disintegrated on the sea floor may surface again. They may lift sections of the skeleton with them, and, as soon as they sink down, the individual skeletal parts may be scattered. Small sections of the skeleton stay connected (vertebral column, root of the tail) and are thus embedded (Fig. 21). Many parts lie out of order, others are missing (frequently the neurocranium with its otolith is missing because the skull resurfaces more frequently than other parts and drifts away).

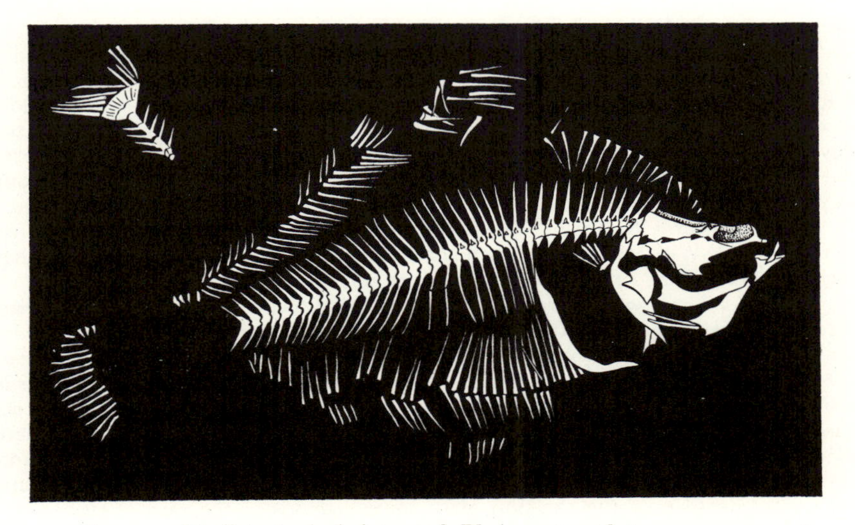

FIG. 22. Undisturbed skeleton of *Pleuronectes platessa*, carcass not refloated during decomposition. (Actual size ×0·4)

(2) *Agonus cataphractus* (pogge). The carcass sinks to the bottom with its ventral side down (water temperature +18°C). After 4 days the body surfaces and floats with the back upward.

For 28 days the neurocranium and the dermal bones of the back float at the surface while the remaining parts of the skeleton (the dermal bones of the ventral side, rays of pectoral fins, pelvic fins, anal fins, vertebrae and viscera) sink to the bottom in a disorderly heap. The skeletal parts are scattered as they do not drop vertically but veer off more or less, according to their shape; the deeper the water, the more widely they are scattered.

(3) *Pleuronectes platessa* (plaice). The carcass sinks to the bottom (water temperature +18°C).

The body remains at the bottom until it is entirely free of organic material which takes three months. The skeleton stays thus almost in its natural position, except that minor bottom currents may disturb some fin rays (Fig. 22).

(4) *Callionymus lyra* (gemmons dragonet). Its carcass sinks to the ground

ventral side downward and decays there; the body is not moved at all. If there are no currents at the bottom the skeleton remains lying in the original position. The carcass disintegrates within seventeen days if the water temperature is 16°C.

(5) *Syngnathus acus* (sea needle) sinks to the bottom with its ventral side downward. After two days in well-aerated sea water at a temperature of 15°C the pectoral section of this snake-shaped fish begins to arch upward due to the development of gases in that part of the body. The head and the posterior part of the body with the tail lie on the bottom. During the following three days the entire anterior part of the carcass rises, and only the tail touches the bottom. The epidermis swells up at this stage and sinks to the bottom in pieces. The skin armour, however, holds the body together in its original form. After three days the anterior part of the fish sinks back to the bottom. The plates of the exoskeleton remain in their original positions in the structure for another eighteen days, thereafter the scales drop one by one to the sea floor.

(6) *Ammodytes lancea* (lesser sand-eel). The carcass of this pencil-thin fish sinks to the bottom in oxygenated sea water at a temperature of 16°C; the back is directed downward. After five days the body cavity opens and the intestines float upward. The carcass remains at the bottom in the same position for another ten days, then the musculature of the trunk and of the tail disintegrate. Thirty-three days after death the skeleton is free of any fibrous tissue and musculature. As long as the skeleton is untouched it remains undisturbed where it was first deposited.

(7) *Trigla corax* (tub fish). The carcass with its back directed downward sinks to the bottom. With water at 16°C, the body remains at the bottom, and after thirty days it is free of any soft parts. The skeleton stays in its natural position.

(8) *Galeorhinus galeus* (tope). The carcass sinks to the bottom with its ventral side down (temperature 18°C). The neck region begins to rise on the next day, and on the third day, the carcass floats with its head and tail drooping. On the fifth day the body sinks back to the ground and disintegrates rapidly. Pieces of skin hover in the water for a while and then sink back to the ground on the leeward side of the carcass where they form disorderly shreds of skin. After these shreds are completely decomposed, only disorderly heaps of placoid scales are left. The abdominal cavity breaks open, and the liver escapes in individual, floating crumbs. Twenty days after death only the skeleton, with some remains of the rays of paired and unpaired fins, the jaw-bones and mouth teeth, is left, and heaps of placoid scales lie scattered in the sediment (Fig. 23). The vertebrae are very light and are rolled away by gentle currents.

These examples are sufficient to demonstrate that the carcasses of different species of fish do not decay in the same way despite almost equal external conditions. Instead we can discern two groups: (1) Those fishes whose carcasses in given circumstances are refloated by gas that develops in the abdominal cavity and sink to the bottom only after the wall of the body cavity is destroyed and the gas has escaped. (2) Those whose carcasses remain at the bottom during the whole period of decay because their body cavity and the quantity of developing gas are too small to float the carcass. The ratio of the size of the body cavity to the mass of the body thus determines the mode of decay in the carcasses of fishes in similar environments.

This rule, as derived for the few described species, however, is valid

FIG. 23. Skeleton remnants of the shark *Galeorhinus galeus*. Fine white dots: dermal teeth; top: jaw teeth. (Actual size × 0·4)

only for certain temperatures and salinities of the water. If the water temperature is so low as to slow down the decay considerably, the fishes of species which do float up at moderate temperatures (+15°C) may not come up to the surface. If, on the other hand, the water temperature is quite high, the development of gas in the body cavity sets in almost explosively. Even carcasses of those species will then rise—which would not do so at moderate temperatures when the gas development is slow. Similar considerations apply to the salt content. We performed our experiments in sea water of approximately 2·9 per cent salt (water of the upper Jade Bay); if we increase the salt content to 3·5 per cent salt (the average content of the great world oceans) decay is delayed relative to the described experiments. Hecht (1933) states that, probably, most of the bacteria which are usually active during decay cannot live in such high salt concentrations and that this may be partly the reason for slower decay. Whenever the decomposition is slowed down, gas develops only slowly; carcasses of the species of fishes which would certainly rise to the surface in water of a somewhat lower salt content, will either not float at all, or only for a shorter time wherever the salt content is higher. Lack of oxygen in the water has a similar effect; many bacteria cannot live under anaerobic conditions, and the decomposition slows down. The place where the organic substance of the fish carcass decays, whether at the bottom or near the surface, determines the subsequent fate of the carcass and the way in which it is embedded.

Carcasses that float never reach the sea floor intact; their skeletons disintegrate, and some parts are always lost. The fins and their rays are normally detached while the carcass drifts in open water, and certain sections of the vertebral column become disarticulated, dermal bones and scales loosen, parts of the visceral skeleton are shifted, and the gills disintegrate. When fish carcasses rise from the sea floor from the effect of the decomposition gases, they do not necessarily float to the surface. As the rising carcass passes through zones of lower water pressure than at the bottom, the gas that is enclosed in the body cavity often bursts the body wall before the fish has reached the surface. The gas escapes, and the torn body sinks back to the sea floor. Whenever the rising carcass of the fish happens to get into the upper zones of agitated water, the body is rapidly completely destroyed. Carcasses of fishes, once seized by an ocean current, travel considerable horizontal distances before they reach the sea floor again. Such skeletons are generally imperfect.

The sea floor from which carcasses of fishes rise may be in the region of still water (below the agitated zone and at a depth of more than 100 m). After gas develops in the body cavity during decay the carcasses leave the sea floor for a short time and return there in a fragmentary state. Destroyed fossil fishes whose skeletons are incomplete do therefore not necessarily indicate the presence of bottom currents at the place of deposit. And it is plausible that completely preserved skeletons occur beside destroyed ones in one and the same facies, even on the same bedding plane, and where the sediment is fine-grained, thinly bedded, and free of ripple marks. The fossils which are preserved complete are derived from carcasses that have not floated; the fossils which are broken up, however, as well as individual scattered parts are derived from refloated carcasses. Unfortunately, neither the fossil skeleton nor the outline of the body of fossil fishes gives any indication which species of fishes do or do not float under given conditions of salinity and temperature. Without experiments we cannot even discern the two groups among recent fishes. What is more, within the same species some individuals may rise and others may remain at the bottom, depending on whether their stomachs or intestines are full and therefore develop more gas, or whether they are empty.

If a fish carcass has remained at the bottom during the period of decay, it is possible that a complete, undisturbed skeleton may result. However, buoyancy is not the only factor; there are scavengers, wave action, and bottom currents, all of which may affect the decaying carcass and disintegrate it completely. The more quickly a carcass is finally buried, the better are the chances for preservation. As soon as it is covered by sediment, only compaction of the sediment and of the carcass itself can lead to minor displacements of the skeletal parts. The conditions necessary for complete preservation of fish skeletons improve with increasing depth of the sea; a deep sea floor is less influenced by wave action and currents than a shallow one. But scavengers remain a danger even at depth, or possibly more so because a carcass at depth is as a rule longer exposed, due to the general decrease in the rate of sedimentation with depth and with the distance from land.

Only completely closed, very deep basins are quite devoid of currents and are stagnant. They turn into oxygen-free zones by decomposition of planktonic and other organisms that drift downward. In this process hydrogen sulphide develops and poisons the water and the sediment on the sea floor. The necessary conditions for skeletons to be preserved in their natural order are thus guaranteed; moreover, not only the skeleton but also the organic soft parts can be preserved under these conditions. The corresponding sediment is a mud rich in fine, dispersed, organic matter, a sapropel. The boundary between the oxidising and the underlying reducing region lies normally in the sediment; however, in this extreme case it lies in the water, high above the sea floor. (In experiments such a boundary frequently forms a sharply defined plane between transparent water and dark blue, turbid water.) This not only excludes normal benthonic life, such as bivalves, worms, snails, crustaceans, and echinoderms, but a zone above the benthonic one becomes temporarily or permanently azoic except for anaerobic bacteria. In the absence of oxygen the organic material of a carcass breaks down in a different way. Gas does not develop in the body cavity, and the carcass therefore remains where it was deposited. That explains why most fossil fish with well-preserved skeletons are found with remnants of their soft parts, and that the enveloping sediments bear all the characteristics of a former sapropel.

Brongersma-Sanders (1948 and 1949, see also Kollmer 1961) drew attention to certain periodical cases of mass-mortality of fish, other vertebrates, and pelagic invertebrates in unprotected shelf seas adjacent to deep ocean basins (African and South American coasts, coast of California and of the Arabian Sea). Upwelling, nutrient-rich water in these areas permits explosive phytoplankton blooms which may turn into red tides and cause widespread deaths among nekton and plankton. The effect is all the more devastating since the bloom itself attracts large swarms of fishes and other marine animals. Large amounts of organic matter accumulate on the sea floor as a consequence and may well cause depletion of oxygen on the sea floor and formation of sapropel muds. Nutrient-rich, ascending water allowing explosive phytoplankton bloom is not restricted to coastal waters but occurs also in the open ocean. Consequently, one may expect organic-rich muds to form in areas of periodic upwelling in the open ocean, independently from the existence of enclosed, euxinic basins. Areas that extend over many hundreds of nautical miles may receive more organic material than can be oxidised even by the generally abundant oxygen supply of the deep sea floor. The sediment becomes a rotting gyttja-mud whose surface lies only occasionally within a region of oxidation and which is well suited as a repository of undisturbed fish skeletons together with soft parts. The low rate of sedimentation of inorganic material in these depths combined with a high rate of introduction of organic material, leads, as pointed out by Richter (1950), to the preservation in the sediment of the very organic material which also causes the oxygen depletion.

Clearly, these conditions do not exist in the North Sea. The southern part of the North Sea is hardly anywhere deeper than 50 m, and wave action caused by winds above force 8 reaches the sea floor everywhere; the tidal currents have the same effect, and the entire sea floor is well aired. The sediments are constantly transported, and benthos and nekton are rich in species as well as individuals. Each of these conditions would be sufficient to prevent the preservation of complete fishes or of their organic material in the southern North Sea. Each fish carcass that reaches the bottom is mechanically disintegrated within a short period of time by transportation over the sea floor, or by refloating, or by scavengers; and abundant oxygen in the water and in the sediment on the sea floor leads to rapid decomposition of organic material. Hard parts with enough mechanical strength to withstand repeated displacements, and resistant to chemical decomposition until they are finally embedded, are all that can be preserved. They therefore deserve special attention.

Hard, preferentially preserved skeletal parts. After the death and disintegration of a fish, individual parts of the skeleton are usually transported and finally embedded individually. Repeated transport of skeletal fragments together with the sediment of the sea floor soon breaks up most of them or destroys them by abrasion. Only certain parts of the skeleton are especially hard and suited by their shape to resist destruction. They are the parts that are most frequently found in marine sediments and ancient sediments, and are increasingly studied by micropalaeontologists. Otoliths, mouth teeth, gill-raker teeth, dermal bones, scales, vertebrae, and fin spines are the more common hard relics.

(1) *Otoliths* (also called statoliths) lie in each of the two labyrinths of the living fish. There are three to each labyrinth, situated next to three corresponding patches of receptor tissue. The otolith which lies in the sacculus is called sagitta, is generally the largest one, and is therefore most frequently found in marine sediments. It also has a most distinct shape and structure. The second otolith, called lapillus, lies in the utriculus, and the third otolith, the asteriscus, lies in the lagena. Both are of less importance than the sagitta because they are smaller and less easily identifiable.

Otoliths are the last parts of a dead fish to be freed and to be transported on their own, because the labyrinth is enclosed by the bony neocranium. This part of the skull remains long intact, and, as we know from experiments, it is frequently lifted from the rest of the skeleton by decomposition gases and drifts away from where the carcass has first sunk.

When even the neocranium is broken up, the otoliths lie freely on the sea floor and are no longer transported over great distances because of their compact shape. It is therefore rare for certain areas to be subsequently enriched by preferential transport of otoliths; if concentrations exist one must conclude that they are due to a concentration of fish carcasses, or at least of skulls.

Otoliths are important to fishery biologists because the growth of a fish is recorded in these hard parts by growth zones. Vertebrae, opercular bones, and scales also show growth-related zones. Age-determination of fishes by otoliths has been done especially on *Pleuronectes platessa, Platichthys flesus* and *Limanda limanda*; in the case of *Solea solea,* scales are better. One yearly ring consists of two zones one of which is formed in spring and summer and is light and opaque, and contains more organic material than the other which is dark and transparent and produced in autumn and winter.

The shape and the sculpture of the sagitta are by no means known for all

Recent fishes. Figs. 24 and 25 show the sagittas of fourteen species of the North Sea. Few of these forms have been found in the sediments of the North Sea, yet they all must exist in great numbers (see Table 9). Otoliths appear to be more frequent where other skeletal parts, scales, ossifications of the skin, and teeth are also found. However, in recent as well as in fossil marine sediments the occurrence of otoliths is not tied to any particular sediment. Because they originate from vagrant animals that pass over wide areas, otoliths may be found in sands as well as in muds, in shell layers and in layers of plant fragments.

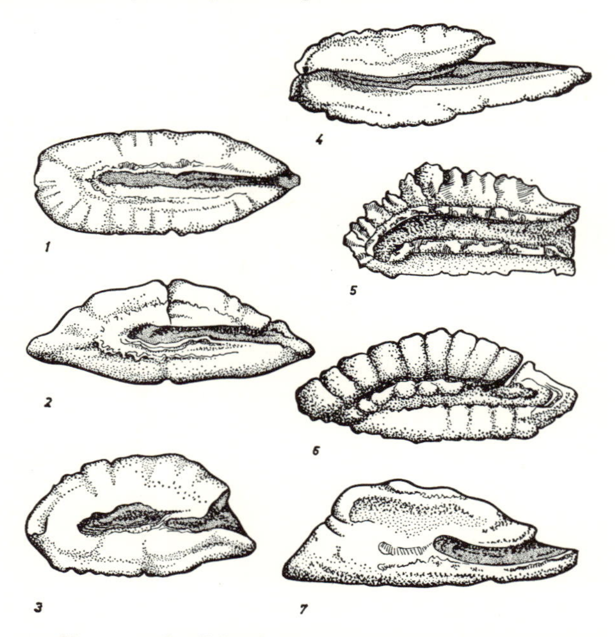

FIG. 24. Otoliths (sagittas). (1) *Ammodytes lancea*, 3·2 mm; (2) *Agonus cataphractus*, 4 mm; (3) *Zoarces viviparus*, 1·5 mm; (4) *Clupea harengus*, 6·5 mm; (5) *Scomber scombrus*, 4·1 mm; (6) *Myoxocephalus scorpius*, 6·5 mm; (7) *Callionymus lyra*, 3·1 mm

The range of variation of the sagittas of one species can be wide. Fig. 26 shows the two extreme forms from a sample of thirty sagittas of Norway haddock (*Sebastes marinus*). Otoliths always vary in many points (serration of the dorsal and ventral margins, relative width of the body, sculpturing of the planes, size, shape, or existence of a central nucleus etc.; for a discussion of these details in herrings see Bohl 1960). As the sagittas also change their shape and their sculpture in the course of ontogenesis, the variety of forms within one species becomes unfortunately so great that sagittas of closely related species are difficult to distinguish. Tables of generic characteristics of the recent fauna have not been published yet, but Chaine and Duvengier (1934) have described in detail the otoliths of twenty-seven European species of teleosts. Otoliths in recent and fossil sediments occasionally show breakage or abrasions. As their shape is generally simple it is possible to reconstruct fossil specimens.

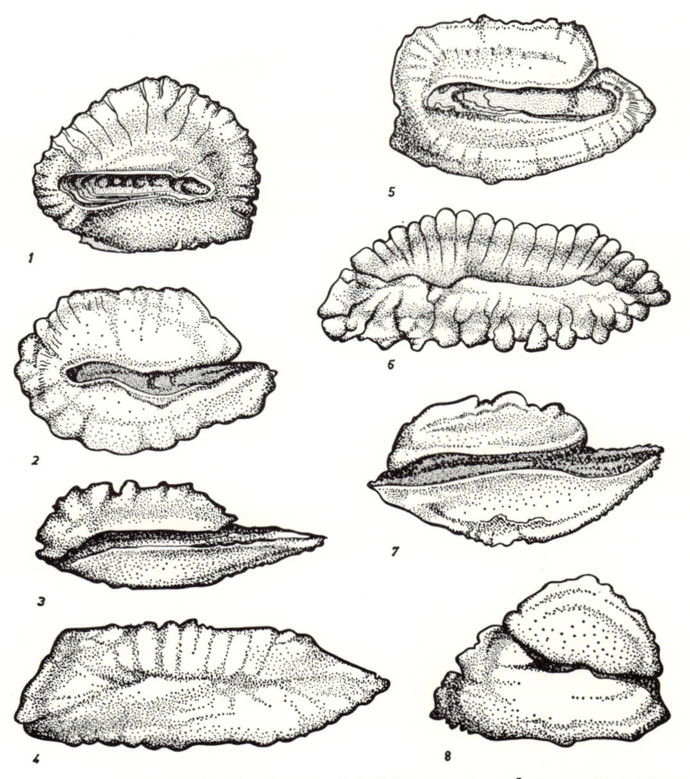

FIG. 25. Otoliths (sagittas). (1) *Pleuronectes platessa*, 7·5 mm; (2) *Scophthalmus maximus*, 9 mm; (3) *Trachurus trachurus*, 11·1 mm; (4) *Pollachius virens*, 18 mm; (5) *Hippoglossus hippoglossus*, 7·8 mm; (6) *Melanogrammus aeglefinus*, 15 mm; (7) *Osmerus eperlaunus*, 6·4 mm; (8) *Trigla corax*, 5·1 mm

FIG. 26. Different shapes of otoliths of *Sebastes marinus*, 13 mm

(2) *The jaw teeth* of the fishes, like their otoliths, are especially hard and resistant. While each fish produces only one pair of sagittas, which are freed after its death, it releases several full sets of its numerous mouth teeth as they are replaced during the course of its life.

We know that almost all Recent fishes replace their teeth if they have teeth at all; however, mouth teeth of sharks and rays are the only ones that are found frequently in marine sediments (see also Breder 1942). Both possess an almost unlimited ability for replacement of the teeth at the rim of the jaws, independently of changes in diet and in seasons (Marquard 1946, Ifft & Zinn 1946, 1948, James 1923). The teeth are

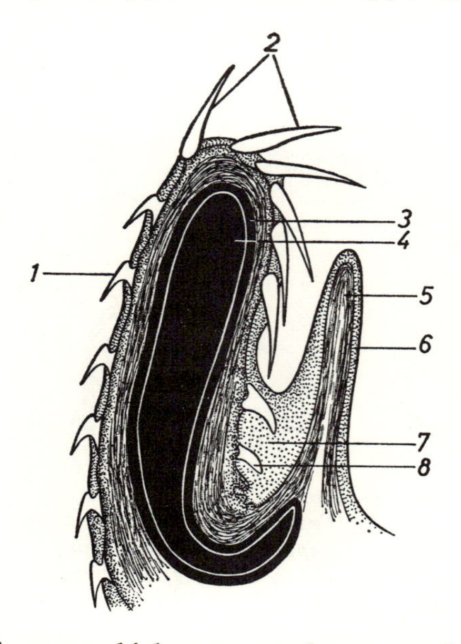

FIG. 27. Cross-section through the lower jaw of shark (after Rauther, 1940). (1) dermal teeth; (2) mouth teeth; (3) calcifying zone of jaw cartilage; (4) jaw cartilage; (5) palisade of connective tissue; (6) mucous membrane of mouth; (7) epithelium; (8) first growth stage of tooth. (Actual size × 0·6)

set in rows which correspond to successive generations on the rims of the cartilaginous jaws, and they are discarded periodically or aperiodically at the labial side, being replaced from the lingual side by the corresponding number. However, they are not replaced rapidly enough to remain unaffected by usage. Frequently the durodentine breaks at the tips, and at the upper sides of the teeth, into the pulp cavity; moreover, several cases are known of cavity-forming organisms attacking the living teeth (Bernhäuser 1956, with bibliography). In the development of a tooth a small tooth shard is first formed embryonically on the tooth ridge; later, a base is built under the shard of trabecular dentine. The lingual part of the tooth ridge remains active throughout life in the production of replacement teeth. During growth, and as long as the tooth is used, it moves either continuously or in distinct advances in the labial direction, together with its supporting connective tissue (Fig. 27). As the tooth approaches the labial skin, the connective tissue enveloping the root is gradually resorbed and thus flattens. As soon as the tooth touches the placoid teeth of the labial skin, its root starts to loosen, the enveloping connective tissue regresses and both the shard and the base of the tooth are bared. Then individual resorption sinuses form under the base of the tooth and loosen the connection between the teeth, the supporting connective tissue and the mandibular cartilage, and thus prepare the separation.

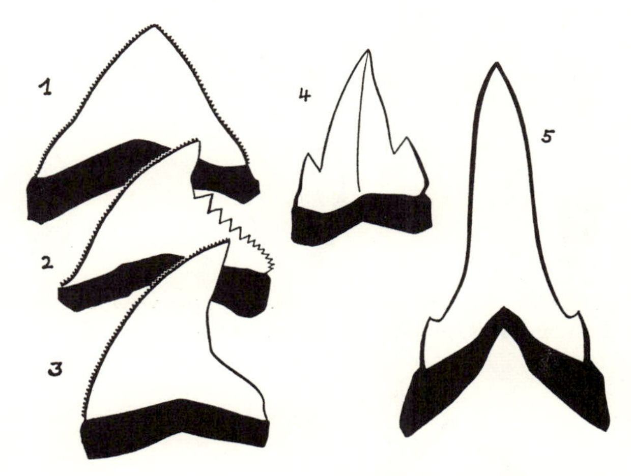

FIG. 28. Incisors. (1) *Carcharhinus*; (2) *Galeocerdo*; (3) *Prionace*; (4) *Triaenodon*; (5) *Lamna*. (Actual size × 2)

Either several rows of teeth of several generations, or only one row, function at any one time on the edge of the cartilaginous jaw. Most of the dentitions of sharks and rays are heterodont; as a consequence, not every single tooth which is found in the sediment is characteristic of the species to which it belongs.

Teeth of sharks and rays can best be classified according to their function and morphology: (a) incisors, (b) grasping teeth, and (c) grinding teeth. Each type includes several forms of teeth (see also Jakobi 1939).

(*a*) The incisor is set with its narrow base parallel to the jaw-line. The triangular, exposed part of the tooth has approximately the same width as the root below (Fig. 28). The upper part of the tooth can be strictly symmetrical or asymmetrical; the tip of the tooth may thus lie in the middle above the base or it may be shifted sideways. Teeth whose tips are shifted to the right side are on the right section of the jaw, those with their tips shifted to the left are on the left part of the jaw. The exposed edges of the teeth may be sharp like a knife or they may be serrated. Such a serrated edge may be formed on either one or both sides. The

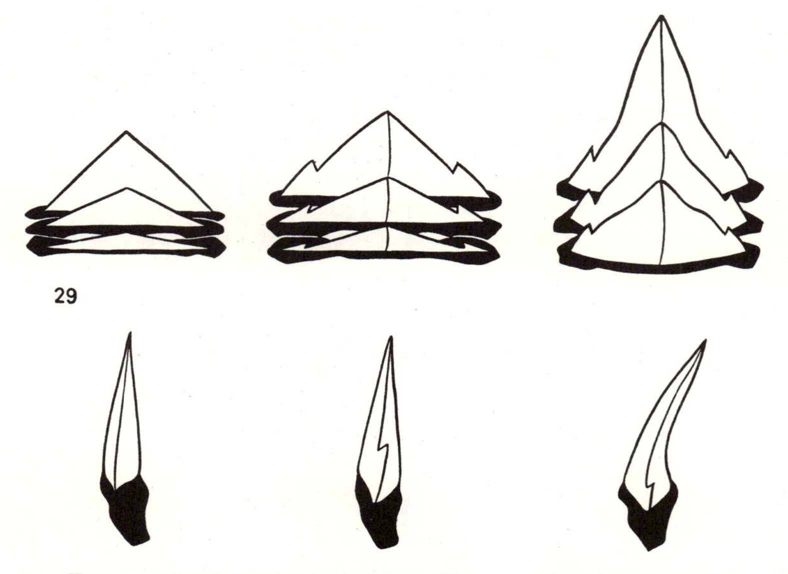

FIG. 29. Mouth teeth of selachians. Top: top view; bottom: side view. Left: incisors, grading toward grasping teeth, right

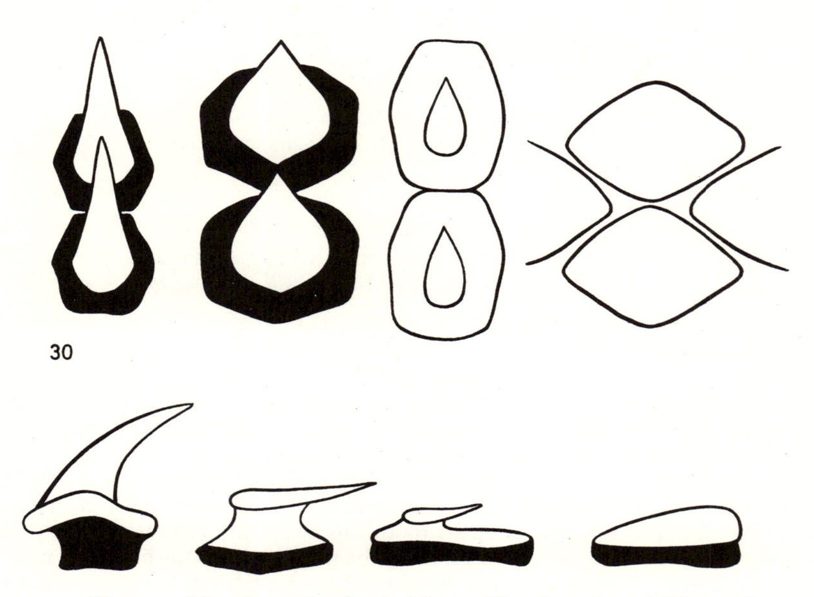

FIG. 30. Mouth teeth of selachians. (Continuation of Fig. 29.) Top: top view; bottom: side view. Left: grasping teeth, grading toward grinding teeth, right. (Actual size × 2)

flat teeth are either straight or bent slightly backward. The incisors may be reinforced by augmentation of the back of the tooth, producing a triangular shape in horizontal cross-section. Other incisors are narrower and more pointed; usually their main tip is flanked by a small tip on each side, and all three tips lie in one plane. These teeth are set vertically and are not bent backward (Fig. 29 and 30). Numerous transitions occur between the narrow and wide types. In a dentition of the incisor type only one generation of teeth functions at one time. It works in close interdependence with the corresponding generation of teeth of the opposite jaw. Therefore it is not possible for successive generations of teeth to move gradually into a working position; one generation of teeth must first disappear completely before it is replaced, and the next generation moves into working position by rotating almost 180° (Fig. 31 and 32). The teeth, therefore, are not functional during this process, and the fish does not eat. The length of the fast depends on the

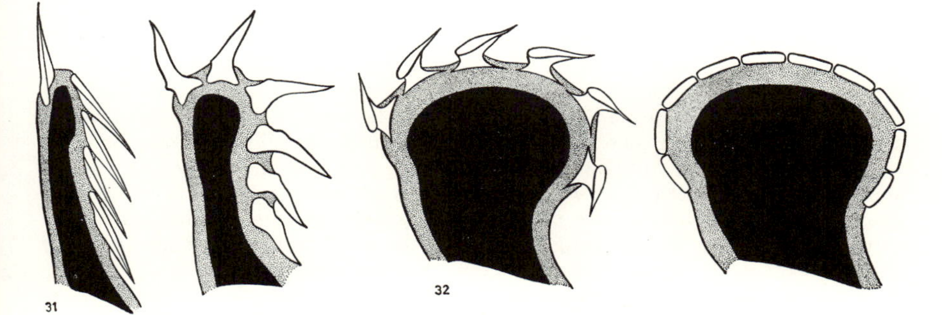

FIG. 31 (*left*). Mouth teeth of selachians. Left: dentition of incisors, one tooth in function; right: dentition of grasping teeth, several teeth in function

FIG. 32 (*right*). Mouth teeth of selachians. Left: dentition of grasping teeth; right: dentition of grinding teeth. Several teeth in function in both. (Actual size × 2)

time needed for the teeth of the new generation to turn into working position. The heterodonty of incisor-type dentitions is generally small. The teeth at the symphysis are the largest ones, and their size decreases toward the angles of the jaws.

Only in a few species do incisors occur in both upper and lower jaws. Usually, the teeth of the upper jaw are incisors, those of the lower jaw are grasping teeth.

Animals with pure incisor dentitions only attack large prey and cut individual (mouth-size) pieces from the musculature of their prey unless they swallow the prey as a whole. This dentition occurs only in certain sharks, never in rays.

(*b*) Grasping teeth are always narrower than the incisors; they are always pointed, often dagger-shaped, and always bent backward. A section through the tooth is oval or round, and the back of the tooth is generally reinforced. Grasping teeth can be asymmetric similar to incisors, with the main point flanked by a smaller subsidiary point either to the right or left. While forces on the incisors act mainly from above toward the base that is elongated parallel to the jaws, the grasping teeth are exposed to horizontal forces producing a tensile stress in them. The root of the tooth is therefore reinforced, and the tip of the tooth is turned inward, away from the main direction of pulling. It is possible to set up a morphologically ordered sequence for grasping teeth, beginning with a high, upright incisor on a narrow base, and ending with a small hook-like tooth which is directed backward and rests on a wide root.

Where the dentition consists of grasping teeth, the teeth of the upper and lower jaws have little if any interdependence. Hence their shape is interrelated only by the fact that they are all directed backward. The teeth of several generations can function on the jaw simultaneously without disturbing each other's action. The newly formed and finished tooth moves slowly into a functioning position with its neighbours; each individual tooth is discarded after transgressing the most active zone of the jaw, its function being taken over by the numerous neighbouring teeth. The flatter the arch which each tooth describes during its labial movement, the earlier it gets into working position, the longer it functions, and the greater is the number of generations of teeth which work simultaneously. The curvature of the arch depends on the shape of the cartilage of the jaws (Figs. 31 and 32). The exchange of teeth goes on continually, and there is no period when the dentition is useless.

Animals with dentitions of the purely grasping type, with small, hook-shaped teeth on wide bases, usually attack smaller prey. The fish approaches the prey broadside, grasps the prey sideways (either from the sea floor or in the open water) and kills it; the attacker then flips the prey to a position parallel to its own body by rapid jerks of its head and swallows it. If the grasping teeth are dagger-like, larger animals can be attacked. Fish with dagger-like teeth bite into the prey, but rather than cutting out a piece of the prey as animals with incisors do, they tear out a section by plucking and twisting motions of the head and whole body (Landolt 1947, called the dagger-like grasping teeth 'catching' teeth).

Both sharks and rays have dentitions of grasping teeth; the rays never have large, dagger-shaped teeth but only small, hooked teeth. Grasping teeth can be associated with breaking or grinding teeth in the same dentition.

(*c*) Grinding teeth serve to break up hard-shelled prey such as bivalves, snails, and crustaceans. The wide, pillow-shaped tooth with rounded edges has a solid root of equal width. On grinding teeth, as on incisors, forces are exerted from above. The grinding teeth function in conjunction with the teeth of the opposite jaw. The cartilage of the jaw which carries the grinding teeth provides a wide basis for several rows of functioning teeth (Fig. 32). The individual teeth of one row, and the rows themselves, form a continuous pavement. Very large grinding teeth are set with their longest extension parallel to the jaw and thus adapted to their migration over the jaw. Where the grinding teeth are large, only one tooth functions for a certain distance along the jaw. Its loss interrupts feeding for a certain period. Only when the next tooth has shifted into working position can food be ground again. If the grinding teeth are small, so that many teeth take part in the breaking up of food, replacement of teeth can go on continually without interruption of feeding.

Durophage dentition can consist of grinding teeth only, or of grinding teeth and grasping teeth. If there are both types of teeth, the grasping teeth are set in the area of the symphysis. The symphysis region of the jaws is then usually projected forward, and the grasping teeth are thus anterior to the grinding teeth. Wherever the grinding teeth are small, the individual teeth have almost identical shapes. Heterodonty is pronounced in fish with large grinding teeth, and in this case, teeth from certain positions can be distinguished unambiguously.

Molariform teeth are large, flat plates which extend between the rims of both the upper and the lower jaws (*Myliobatis aquila*). They are particularly well adapted to deal with hard food (Treuenfels 1896). Each large plate is a single tooth, and not a fusion of separate teeth.

Incisors, grasping teeth, and grinding teeth are all developed from the same elements; an ordered morphological sequence begins with the small-rooted incisor and ends with the wide-rooted grinding tooth. Krukow and von Wahlert (personal communication) consider the grasping tooth to represent the phylogenetically most ancient tooth form, from which the incisor and the grinding tooth evolved in opposite directions. The dermal teeth of sharks and rays are also morphologically grasping teeth.

In summary, certain species of sharks and rays periodically discard a large number of teeth (incisors and large grinding teeth). At the time of the tooth-change the animals are always in the same area because the period of change of teeth is coupled with that of reproduction, and the mating takes place every year in the same region of the sea. However, small grasping teeth and small grinding teeth which are continually and individually discarded anywhere during the entire year, are widespread and are found individually in marine sediments.

The sharks and rays of the North Sea drop teeth of almost all types discussed.

Incisors. All North Sea fishes with incisors have asymmetrical teeth with several points. *Galeorhinus galeus* (tope) is equipped with such teeth and is the most common shark of the southern North Sea. Numerous young, approximately 20 cm long, sharks of this species invade the coastal waters in May and leave them in October or November when they have grown to a length of 70 cm. During the summer, fishing vessels in Jade Bay accidentally catch five to six young sharks each day with the shrimp net. The adult animals (180 to 200 cm long) are quite frequent in the North Sea. The large and the small-spotted dogfish (*Scyliorhinus catulus* and *S. caniculus*) have incisors with several points on a wide base; *caniculus* occurs frequently in large numbers in the southern North Sea, *catulus* is quite rare. *Squalus acanthias* (picked dogfish) occurs in the shallow waters of the southern North Sea as well as in the deeper northern part whenever it is on longer journeys. Other sharks with these teeth, *Somniosus microcephalus* (Greenland shark) and *Hexanchus griseus* (brown shark), are fairly common in the North Sea, *Etmopterus spinax* (lantern shark) and *Sphyrna zygaena* (balance fish) less so. *Lamna nasus* (Beaumari shark) which usually follows the schools of herring is equipped with triple-pointed, slim incisors; the younger animals of this species usually live near the coast. Wide incisors with triangular cross-section (*Prionace glauca*, blue shark) do not occur in the area.

Grasping teeth. *Squatina squatina* (angel fish) has pointed and somewhat bent grasping teeth; almost all rays have short grasping teeth on a wide base, including the male *Raja clavata* (thorn ray), *R. radiata* (starry ray), *R. montagui* (spotted ray), *R. fyllae* (sandy ray), *R. fullonica* (fuller's ray), with quite long grasping teeth, *R. batis* (skate), *Dasyatis pastinaca* (sting ray). The species *R. montagui* and *R. radiata* are frequent in the southern North Sea, *R. batis* mainly in the south-west. *R. clavata* is the most frequent species in the southern North Sea, and also occurs in the shallow waters of the bays.

Grinding teeth. Among the sharks of the southern North Sea, *Mustelus mustelus* (smooth hound) has a dentition of grinding teeth, and among rays the females of *Raja clavata* and *Dasyatis pastinaca* have this type of teeth.

The jaw teeth represented in Figs. 33 and 34 are the most important types of teeth of those that are continuously dropped into the sediments of the southern North Sea. The forms of the species *Galeorhinus galeus*, *Squalus acanthias* and *Raja clavata* are by far the most frequent.

The teleosts, just as the selachians, change their teeth. Ridges on which new teeth develop constantly show teeth in several different stages of growth. Like placoid scales from which they evolved and whose basic structural plan they preserve, these teeth are secreted by tissue cells (odontoblasts) of mesodermal papillae. The dentition of mouth and pharynx of the teleosts is more varied than that of the sharks and rays. Teeth can be formed on the intermaxillaria, the maxillaria, the dentalia, the vomer, the palatina, the ptergoidea, the os entoglossum, and on the dorsal sections of the gill arches where they are usually disposed in large arrays.

The variability in regard to location and appearance of the teeth is increased by numerous teeth set in the mucous membranes. As these tissues do not provide a firm base, teeth are not well rooted and must be replaced frequently. The shape of each of these types of teeth varies greatly (see also Rauther 1940). A sequence of teeth ordered by their function could again be set out; however, this sequence would be less meaningful than with the selachians because each field of teeth has its own task, specific to its place in the mouth. The pronounced heterodonty of the teleosts makes it difficult to interpret the correlations between their tooth morphology and feeding habits. Successful interpretations have been, until now, restricted to fish with highly specialised feeding methods.

Lühmann (1951a, 1951b, and 1954) has analysed such special cases. He investigated the dentitions and the periodical change of teeth in the catfishes *Anarhichas lupus* and *A. minor*, and of *Lycichthys fortidens*, *L. latifrons*, *L. denticulatus*, *L. paucidens*, and *L. parvodens* (see also W. J. Schmidt 1954). Lühmann set up a sequence of dentitions in these species according to increasing specialisation for hard foods, with *Anarhichas lupus* at the end with the best-developed grinding dentition (see also Lüders 1963).

The following bones of catfish carry teeth: the dentalia, the premaxillaria, the palatina, the vomer, and the ectopterygoid bones. In *Anarhichas* two functional groups of teeth can be clearly distinguished, the grasping teeth and the grinding

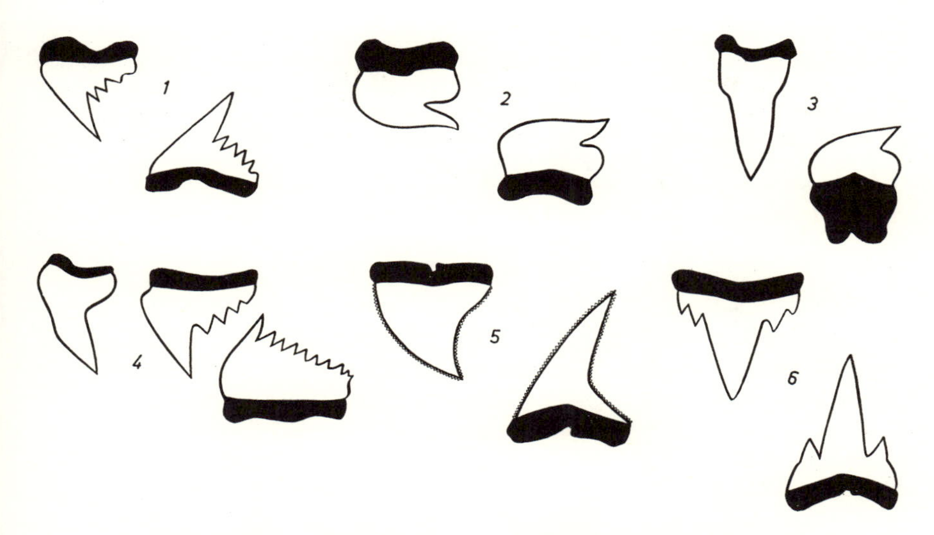

FIG. 33. Mouth teeth of sharks in the southern North Sea. (1) *Galeorhinus galeus*; (2) *Squalus acanthias*; (3) *Laemargus borealis*; (4) *Hexanchus griseus*; (5) *Prionace glauca*; (6) *Scyliorhinus caniculus*. (Actual size × 2)

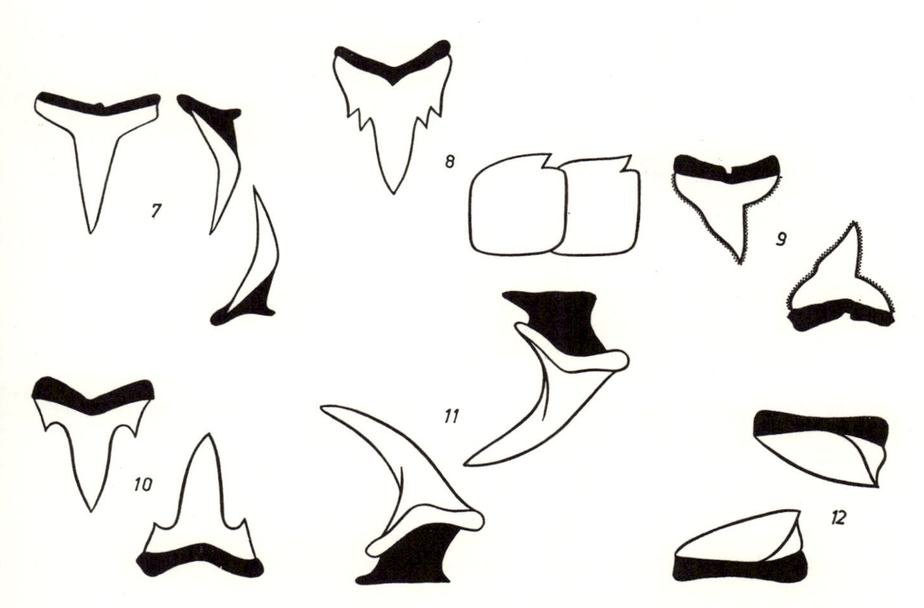

FIG. 34. Mouth teeth of sharks and rays in the southern North Sea. (7) *Squatina squatina*; (8) *Etmopterus spinax*; (9) *Sphyrna zygaena*; (10) *Lamna nasus*; (11) *Raja clavata* ♂; (12) *Mustelus mustelus*. (Actual size × 2)

teeth. The grasping teeth grab prey that usually lies at the bottom (bivalves like *Modiolus*, *Mytilus*, *Cyprina*; brachyurans like *Carcinides maenas*, *Cancer pagurus*, *Hyas araneus*). They have a wide base, are conical and pointed, and they are bent inward with their axes diverging. They are set frontally on the dentale and the premaxillare and protrude in the living animal beyond the lips so that the mouth cannot be completely closed (Figs. 35 and 36).

The grinding teeth are set in two rows; in the lower part of the mouth they are on the dentalia, and in the upper part on the palatina and the vomer. After the shell has been grasped, it is turned until it rests crosswise between the open jaws. Then the dentalia with the grinding molariform teeth press down on the prey

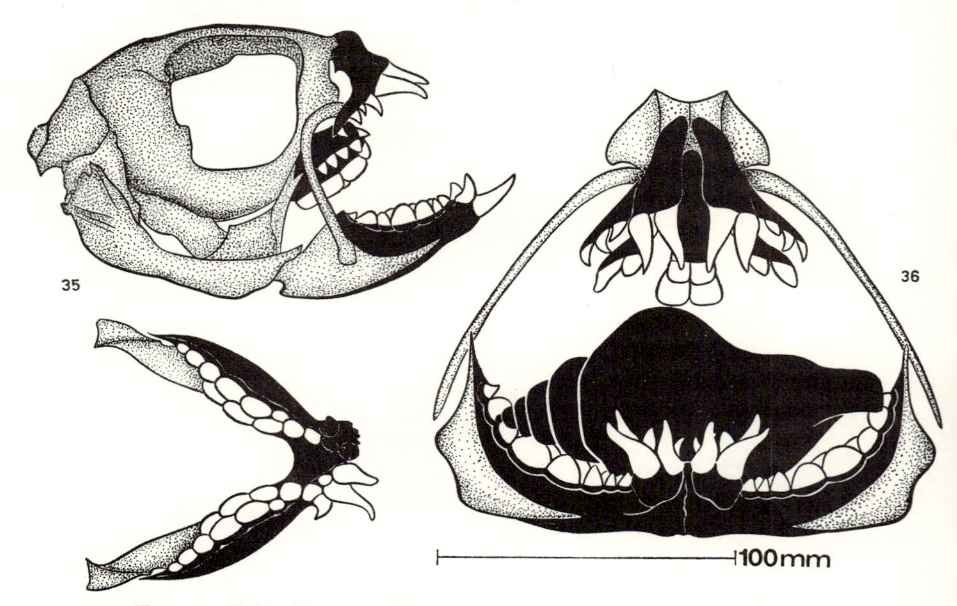

FIG. 35 (*left*). Top: skull; bottom: lower jaw of catfish (*Anarhichas lupus*). Black: tooth-carrying bones

FIG. 36 (*right*). Front view of catfish mouth. Black: tooth-carrying bones and prey, *Neptunea*, before being broken by grinding teeth

against the double row of teeth on the protruding vomer. The prey is finally broken with the additional help of the teeth on the palatina. The animal swallows the broken fragments of the calcareous parts of the prey together with its soft body (for a discussion of the resulting excrement see D II (*b*)).

Changing of teeth is necessary where so much pressure is exerted on them. Since the dentition functions as a unit, the entire dentition is discarded, and complete replacement follows almost simultaneously. The toothless period is short because the new teeth that are formed in the base at the gaps between the old teeth have grown to considerable size when the used ones are discarded. The roots of the teeth regenerate during the change. The dentition is renewed once a year.

According to Lühmann's investigations, the change of teeth in catfish begins at the reproduction period when feeding almost stops anyway. The time of spawning differs in the various regions of the sea. Near Iceland catfish spawn (and change teeth) from December to February, in the North Sea from November to January. The conditions for preservation of the strong and large teeth are good in the mud areas of the deeper North Sea.

In teleosts as in selachians the change of teeth is periodical wherever the dentition functions as a whole; this is particularly so in fishes with specific adaptations and in those which, like the catfish, eat hard substances, i.e. *Chrysophrys aurata*, the families Sparidae (according to Berg 1958, since Early Eocene), the Scaridae (since Eocene), the Labridae (teeth since Palaeocene) and the Tetraodontiformes (since Early Eocene).

The ability to crush prey before swallowing does not necessarily depend on the existence of grinding teeth. *Myoxocephalus scorpius* breaks the armoured *Crangon* by placing the prey crosswise in the mouth—just as the catfish deals with the hard parts of snails—and pressing on the vomer. The vomer and the dentalia of this fish, however, have only numerous small, conical teeth.

(3) *Dermal teeth* should be considered as dermal bones, equivalent to jaw and pharyngeal teeth, to certain ossifications in the area of the head, and to certain supporting structures in the skin of the fins. Frequently these dermal teeth are particularly hard, resistant, and easily preserved. Because they are so numerous in the skin of a single fish, they are quite frequently found in recent and fossil sediments, and are recognisable by their characteristic shape. They will therefore be discussed in some detail.

The hard structures in the skin of the selachians are usually called dermal teeth, as opposed to jaw teeth, and there exist certain differences in their supporting tissue, formation, and structure. The names 'placoid organ' or 'placoid scale' relate to the basal plate, a part belonging to all dermal teeth; the plate is rooted in the corium, and from its centre, a more or less bent spine or tooth extends through the epidermis. Jakobshagen (1920) made a distinction between conical and mushroom-shaped placoid organs.

The placoid scales lie side by side in the skin; they usually form diagonal rows, and they never overlap. The shape and size of dermal teeth vary from one region of the body to another. They originate in the skin as small accumulations of cells in the loose corium, closely under the epidermis. As Schmidt & Keil (1958) have shown, none of the fishes have any enamel. Until recently, the hard layer covering the surface of teeth of selachians, holostians, and many teleosts has erroneously been called enamel; it is, however, durodentine which changes in steps or gradually from the outside to the normal dentine (normodentine) inside. It is formed by displacement of the collagen coupled with mineralisation. Originally, a tooth-papilla secretes a hollow, hemispherical structure whose outer layer is durodentine and whose inner layer is normodentine.

Subsequently, the mesodermal section of the new tooth grows into the firm corium layer and there produces its basal plate (Fig. 37).

During the entire growth period of the animal, and later as replacements for discarded dermal teeth, new tooth buds appear constantly between the existing dermal teeth. As a consequence placoid organs of many different sizes can be seen in the skin of sharks. Despite the constant replacement and addition of teeth, the regular diagonal rows remain; according to Klaatsch (1890) they correspond to the two directions of orientation of connective tissue in the lamellae of the corium.

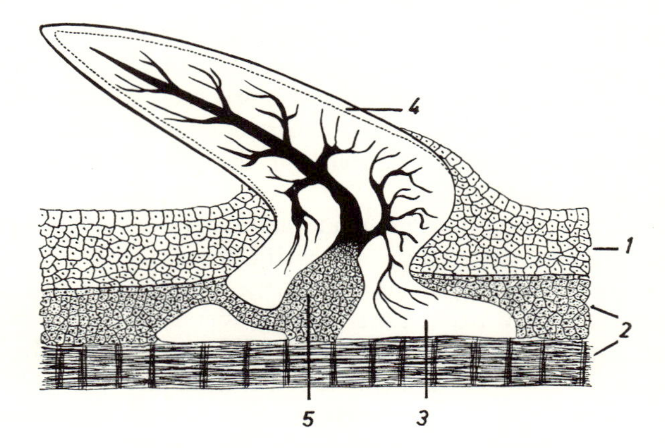

FIG. 37. Selachian placoid scale (after Rauther, 1940). (1) epidermis; (2) corium; (3) basal plate of tooth; (4) duro-dentine; (5) odontoblasts. (Schematic drawing)

The shape and size of dermal teeth is not only influenced by their age. The dentition of the rays differs notably from one part of the body to another. It can be assumed that certain forms of teeth have a specific ecological significance. Only a few cases, however, have been investigated. Frequently, the differences in the shapes of placoid organs depend on their position on the body. Thus the large 'thorns' in the skin of the top and underside of the thornback ray's body have an oval base with an inflated pillow-like appearance; in the region of the tail, however, the basal plates are long and slender, obviously in order to save space, and they have a rather wide perforation for the pulpa.

Fig. 38 gives the most important forms of the placoid teeth on the upper side of *Raja clavata* as well as the position and the distribution of the large 'thorns'. The variety of shapes, however, is greater than appears in the drawing. Both groups of narrow-based teeth, somewhat distal of the centre of the pectoral fins, are particularly significant ecologically. They exist only in males and are used to attract the female's attention before mating (for a discussion of mating behaviour of rays, see Schäfer 1953a). During the disintegration of the carcass the dermal teeth are dropped late because the corium is very resistant. Once the teeth are free they separate rapidly according to their shape and size and end up in different sediments.

The large, placoid organs, so-called 'thorns', lose their projecting parts quite easily, and these parts in turn split lengthwise, so that usually the only part of these teeth

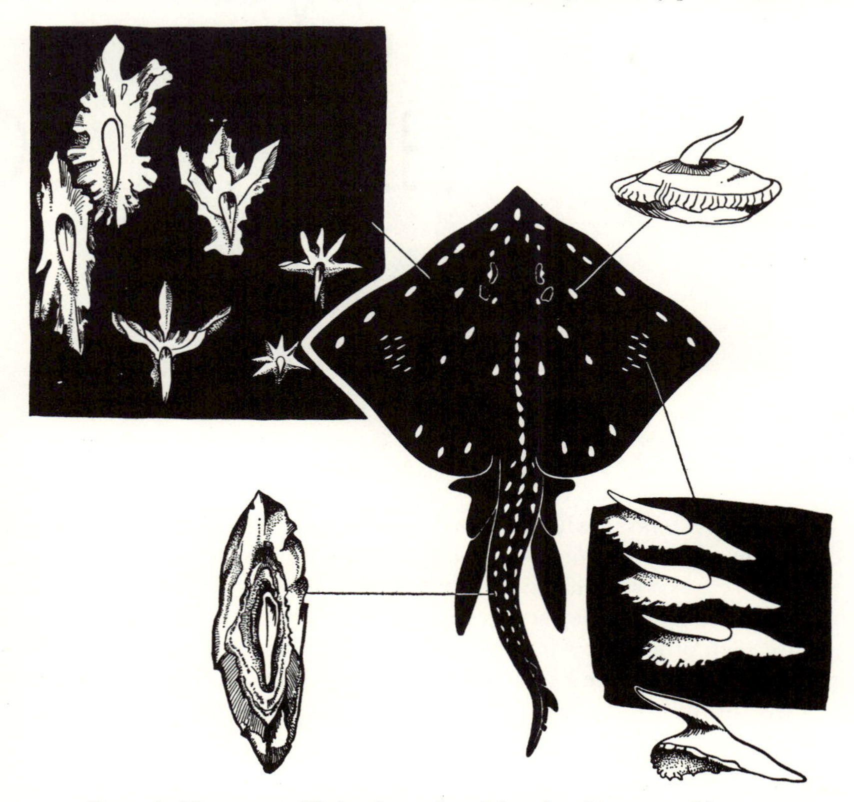

FIG. 38. Thorn-ray (*Raja clavata*) and its placoid organs (drawn to a larger scale)

found in the sediments are the rounded, pillow-shaped, more or less abraded placoid scales. These pieces are occasionally found in shell breccias. One dead ray sheds approximately 130 large 'thorns' into the marine sediment.

The picked dogfish (*Squalus acanthias*) carries two forms of very small placoid organs (Fig. 39) which occur either intermingled (for example on the main sections

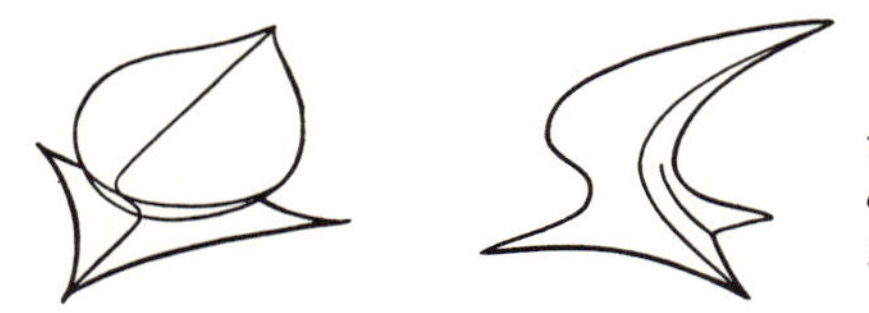

FIG. 39. Placoid organs of the picked dogfish (*Squalus acanthias*). (Actual size × 25–50)

of the anterior dorsal fin) or separately. The underside of the rostrum carries broad, heart-shaped teeth, the sides of the body carry narrow, keeled teeth. Fig. 40 gives the placcid organs of the angel fish (*Squatina squatina*).

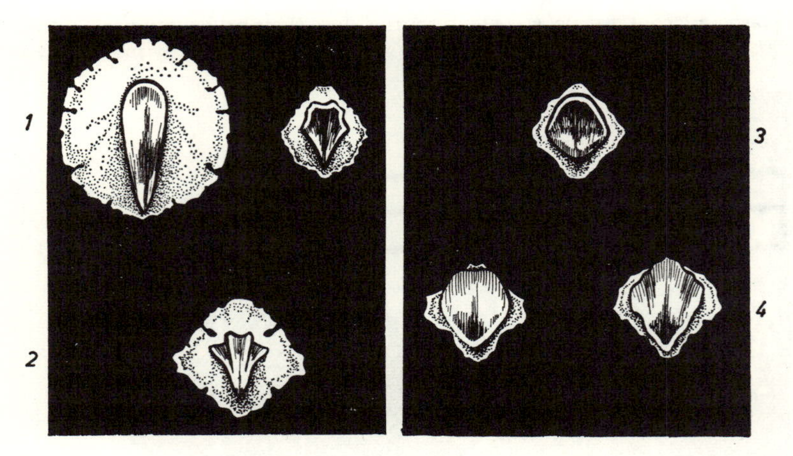

Fig. 40. Placoid organs of the angelfish (*Squatina squatina*). (1) from outer margin of pectoral fin; (2) from dorsal skin; (3) from ventral skin; (4) from side of tail. (Actual size × 3)

(4) *The gill rakers* which are set on the pharyngeal sides of the gill arches as more or less elongate fibrous filaments of the mucous membrane are in many cases supported by hard structures (for a discussion of the function of the gill slits, see Jäger 1876). These gill filaments are set opposite each other but alternate and thus form a dense strainer like two combs with their toothed sides pushed into each other. The water that is taken up through the mouth can leave the pharynx through the slits of this strainer.

In the teleosts the gill rakers usually carry dermal bones on which a group of teeth is set. The selachians, scyllides, spinacides, and *Chimaera* possess dermal bones but they consist only of supporting cartilage and dermal teeth. The long gill rakers of *Rhincodon* and *Cetorhinus*, the only two large pelagic sharks, form a dense and long-toothed strainer (Schnakenbeck 1955).

Teleost gill rakers and their ossifications show a great variety. Generally they are set in double rows in the inside of the four anterior gill arches, one row pointing forward and one backward; the fifth arch has only one row, or none at all. The filaments of one gill arch can be built symmetrically, or the filaments of one row, usually the forward-pointing ones, are larger than those of the other row (Zander 1903, 1906). In several cases only the first gill arch carries gill rakers. The gill rakers may protect the gills from food remains and only permit strained water to enter the gills. They may also retain as food the small animals that have been ingested with the water.

Presumably, both functions operate. Predatory fishes usually carry short, toothed gill rakers, or they are entirely missing (*Lophius piscatorius*, *Buglossidium luteum*). Fishes feeding on plankton and small animals usually carry bristle-shaped gill rakers (Clupeidae, Scombridae, *Mugil*). Rauther (1927) thought that the formation of the gill filaments depends on the size of the fishes; small fishes which ingest small animals carry fewer and more widely spaced gill rakers than large fishes ingesting the same type of food. As larger animals have large areas between the branchial arches due to their body size, larger and more gill rakers are needed.

The strength of the dermal bones in gill rakers varies widely. In many cases these ossifications look like fine, conical hats that are rooted in the connective tissue and whose teeth, alone, extend beyond the skin.

In other cases there are firm, coarse ossifications which certainly can be preserved. Due to their conical shape they are easily transported along the sea floor; dermal bones of the gill rakers are therefore not found near the skeleton to which they belong but are scattered. Probably, they are never accumulated by transportation. One *Gadus morrhua* is the source of approximately fifteen different forms of ossifications of gill rakers in the sediment.

(5) *Fin spines* of the picked dogfish (*Squalus acanthias*) deserve special consideration. The fin spines of fossil selachians are called ichthyoduroliths.

The thorn-shaped fin spines extend up to three centimetres beyond the skin and are set directly in front of the dorsal fin. They are slightly bent in the posterior direction and do not reach the height of the dorsal fin. A cross-section of a spine

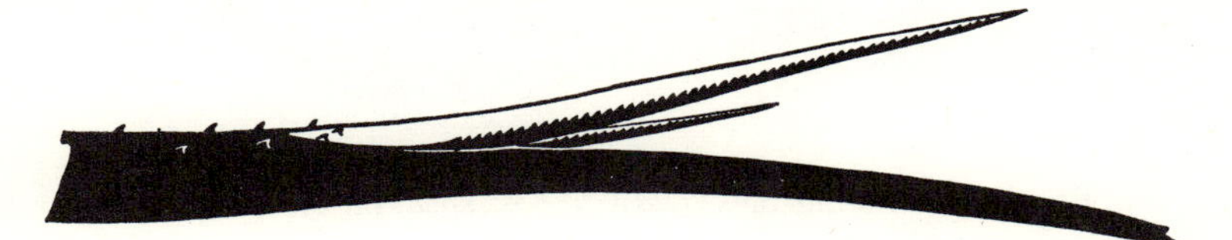

FIG. 41. Tail of the sting-ray (*Dasyatis pastinaca*). The smaller of two serrated spines, one behind the other, is the older one. (Actual size × 0·5)

is triangular; the side facing the dorsal fin is concave. The broad, organic end of the spine lies in the tissue which shrinks as it dries. The structure and development of the spine indicate its affinity to the teeth (Markert 1896, Ritter 1900). The dorsal fin spine of *Squalus* carries a durodentine cover on its upper half (according to Keil 1952 and Poole 1956, cited from Schmidt 1958) like the dermal teeth of the sharks and the 'thorns' in the skin of the rays. The first stage of development of a spine is a depression in the epithelium in front of the first cartilaginous ray of the internal skeleton of the fin. A lost fin spine is not replaced, so that each dogfish generates only two spines.

The Trygonidae, however, carry fin spines for defence. These rays are known to have existed since the Cretaceous and are today represented in the North Sea by the sting ray *Dasyatis pastinaca*. They carry a long, spiny structure instead of a dorsal fin on the dorsal side of the whip-like tail. It is toothed on both sides and supported by trabecular dentine (Ritter 1900, called it explicitly 'defence spine'). Probably these spines are constantly replaced as a developing new spine is usually found anterior to the base of the spine. Hence the sting ray generates more than one spine in the course of its life.

Fig. 41 represents the tail section of a 76 cm long *Dasyatis* from Heligoland Bay which carries two defensive spines, one directly in front of the other. The larger (anterior) one is 16 cm long; the smaller 6·5 cm long. The point of the smaller spine is somewhat bent sideways, that of the larger spine seems to show scars. The spine must have been injured at a time when the entire structure was still in its pouch and could regenerate. Few individual defensive spines have been found in marine sediments. *Dasyatis* frequently digs into the surface of sandy sediments as our observations of captured animals have shown (Fig. 42). It is not known who are the enemies of *Dasyatis* that could be repelled by the spine.

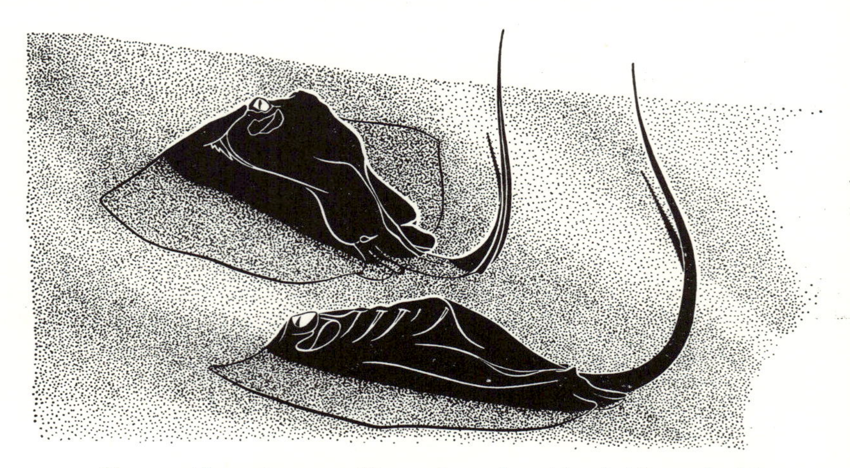

FIG. 42. Two sting-rays (*Dasyatis pastinaca*) buried in sand with erected, spined tails. (Actual size × 0·2)

(6) The sturgeon (*Acipenser sturio*) is the only chondrostean fish with *dermal bones* in the North Sea. It is a large fish, and its dermal bones are so strong that they are preserved as well as any hard parts of fishes.

The dermal skeleton of the sturgeon's trunk consists of five longitudinal rows of large plates that are roughly rhombic in outline. The rows lie in the dorso-medial, lateral, and ventro-lateral positions (Fig. 43). Between the rows lie strips of soft skin which carry small plates. At the anterior rim of the unpaired fins are the so-called fulcra; these are scale-like ossifications, set in a pattern like the tiles of a roof, which are fused in pairs forming a sharp-angled body riding above the notochord. The ganoid plates grow in all three dimensions resulting in good-sized bony plates that are 10 cm long and 1 cm thick in old animals and are easily preserved. At the final stage of growth they almost take up the entire thickness of the firm corium. The bony structure of these plates does not contain ganoine, has branched-out cells, and has Havers' canals in large plates. The cusps on the plates consist of an acellular, homogeneous, bony material. In youth these cusps are particularly pointed and project outward through and beyond the epidermis; they are worn off with age and finally, the plates themselves show traces of abrasion. The plates that have originally been formed with a pronounced, angular relief flatten with age and body growth and move further apart. The number of plates varies widely among individuals. Ten to thirteen dorso-medial plates have been counted, twenty-two to forty lateral plates, and ten to fourteen ventro-lateral plates.

The canal of the lateral-line system runs lengthwise at the level of the lateral plates which have small perforations through which pass the canal itself and its branches.

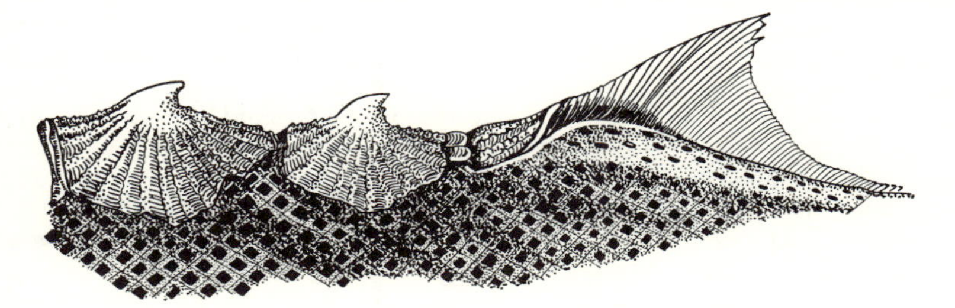

FIG. 43. Fulcres, two ganoid plates and small scales on skin anterior to the dorsal fin of sturgeon (*Acipenser sturio*). (Actual size)

The number of sturgeons in the North Sea and adjoining waters has decreased considerably in the last decades (for a discussion of the causes, see Mohr 1952 who recorded that average yearly catches in this area in the eighteen-nineties were about 7000 (see Table 10)).

TABLE 10

Catches of sturgeon in the North Sea (Mohr 1952)

Year	Elbe and Eider estuaries and tidal sea	Open North Sea (large fishing vessels)
1890	2800	427
1891	3650	615
1892	4272	463
1893	2873	930
1894	3385	802
1895	2343	1088
1896	2322	564
1897	1403	419
1898	1257	338
1899	1350	256
1900	1416	20
1914	71	—
1915	35	—
1916	22	—
1917	17	—
1918	34	—

However, considerable numbers of one- or two-year-old sturgeons are seen every year in the mouths of the rivers Elbe and Weser.

Sturgeons are anadromic, migrating fish which spend a large part of their life in the sea but go up rivers to spawn. From April to June they collect in large swarms in the coastal waters. There are various dangers for these large and heavy animals (up to 6 m long) in the shallow coastal and tidal waters, specially in rough and heavy seas. In such weather some fish are injured by being thrown on stones or sandbars, or are stranded; others are cut off in narrows and thrown on to tidal flats. Mohr (1952) cited Friedrich who reported that on 2nd September 1880, after a stormy night the beach at Oi on the island of Rügen was in places covered with dead sturgeons (see also Weigelt 1927 and 1928 for a discussion of the causes of death and a description of the disintegration of dead ganoid fishes in Smithers Lake, Texas). Because of the sturgeons' prolonged and dangerous sojourn in coastal waters, their hard parts have formerly been found occasionally in the sediments of the regions just beyond the large river mouths (estuaries of the rivers Elbe and Weser and the adjoining tidal flats). In shell deposits bony plates (up to 10 cm long) of sturgeons are still found today, but they are considerably worn and polished. Because of their hardness they remain preserved for decades even after repeated reworking and despite intensive sand-polishing at the bottom of the shallow sea. The burial of complete carcasses of sturgeons, however, is rather improbable in the constantly shifting sediments of the shallow sea. On the other hand, bony plates of sturgeons with perfectly preserved sculpturing of the surface are frequently dug out in large numbers from prehistoric settlements on the coastal plains, proving that these fish were abundant and within easy reach of early man on the shallow, tidal flats (Schütte 1939).

Numerous teleosts in the North Sea have aberrant, large, resistant dermal bones. Normally these fish have much smaller, softer scales. Both the smooth cycloid scales and the ctenoid scales with finely toothed posterior margins are disk-shaped and quite soft and perishable; they are rarely preserved.

Certain hard structures in teleosts are dermal bones in the proper sense. They are securely set in the firm corium and frequently take up its whole thickness. Individual dermal bones, therefore, cannot be lost during life as scales certainly are. In many fishes the dermal bones have both a supporting and a carrying function; in that case they are usually joined to each other or to other parts of the skeleton. They are either parts of a complete dermal armour, or loosely scattered over the corium, or concentrated in certain parts of the body where they have specific functions. In all cases dermal bones become free at a late stage after the death of the animal. Certain dermal bones are so resistant that, besides otoliths, they are often the only parts preserved.

The pogge (*Agonus cataphractus*, Figs. 44 and 45), which belongs to the scleropareid family, carries a completely closed and rather heavy armour of individual dermal bones. The body can still bend, and trunk and tail can undulate, but otherwise its mobility is restricted. The pogge swims by beating its pectoral fins. This adaptation is common to many other species with strong dermal bones. The Syngnathidae (*Syngnathus*, *Entelurus aequoreus*, *Nerophis ophidion*, *Hippocampus guttulatus*) move by undulating the dorsal fin combined with beating of the pectoral fins; the lump fish (*Cyclopterus lumpus*) moves by undulating and beating movements of the unpaired dorsal fin whose position is far toward the posterior end; the sun fish (*Orthagoriscus mola*) moves by sculling with the dorsal and caudal fins. Dermal ossifications or a thick and firm skin full of scattered dermal bones prevent the usual type of fish-locomotion—undulating movements of the trunk and/or tail.

Agonus cataphractus carries eight rows of bony plates along the trunk and six rows along the tail. The individual plates overlap slightly so that a distal process of the first plate lies under the proximal part of the following plate, while a central

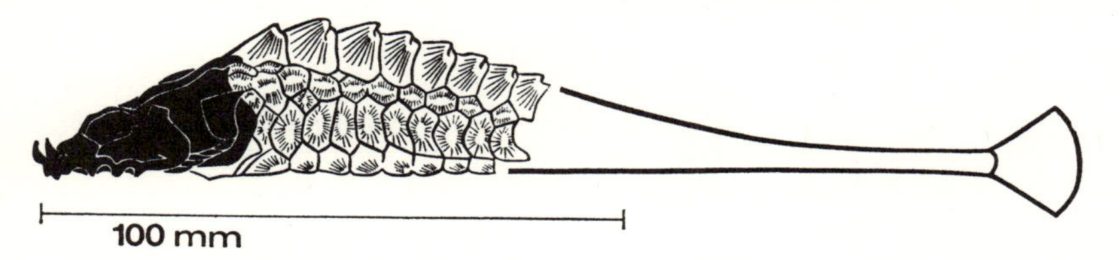

FIG. 44. Pogge (*Agonus cataphractus*), fins removed to show dermal bones.

ridge (not in all plates) of the first plate extends with a tooth-shaped swelling over the following plate. Such tooth-carrying plates occur exclusively in the lateral area of the trunk. Generally, the individual plates are loosely attached to each other; when the trunk bends they move relatively to each other by overlapping as far as the enveloping corium and the covering epidermis permit. Therefore even a complete armour does not make the body wholly rigid.

The bony plates cannot lie flat on the sea floor because of their shape; they are easily carried by currents and frequently transported far from the place of death or disintegration. We have found them several times in the swash marks at the edge of the vegetation-covered part of the island of Mellum.

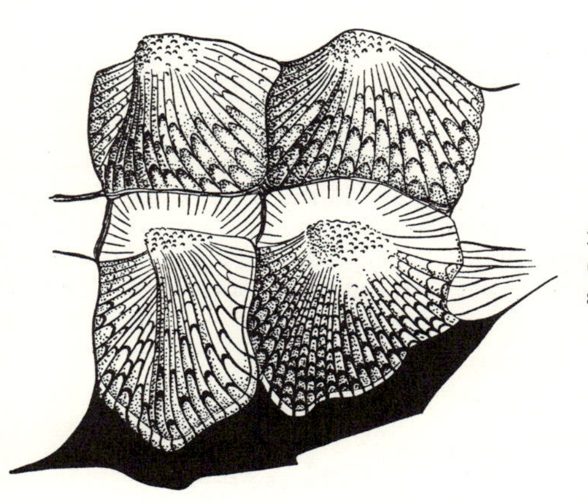

FIG. 45. Four dermal bones from dorsal area of the pogge (*Agonus cataphractus*). (Actual size × 4)

Armours of individual dermal bones surround the bodies of the pipefishes (Syngnathidae) completely. The bony plates form seven rows along the trunk; four rows, also running lengthwise, cover the tail. The anterior margin of each plate somewhat overlaps the body of the preceding plate, and the plates of a longitudinal row overlap the distal margins of the adjoining next higher row. Their connection to the body is strengthened by sinewy processes which are attached to the distal, downward-pointing ends of the plates and reach deep into the intermuscular septs. Parts of the armour can be moved by means of these processes. Due to their deep roots the plates become free from the carcass at a late stage of disintegration. One circle of such bony plates surrounding the body (Figs. 46 and 47) corresponds to each vertebra and the upper plates on both sides of the vertebral column rest on the lateral spinal processes of the vertebra. The two bony plates in the dorsal position (nuchalia) rest on the dorsal spinal processes.

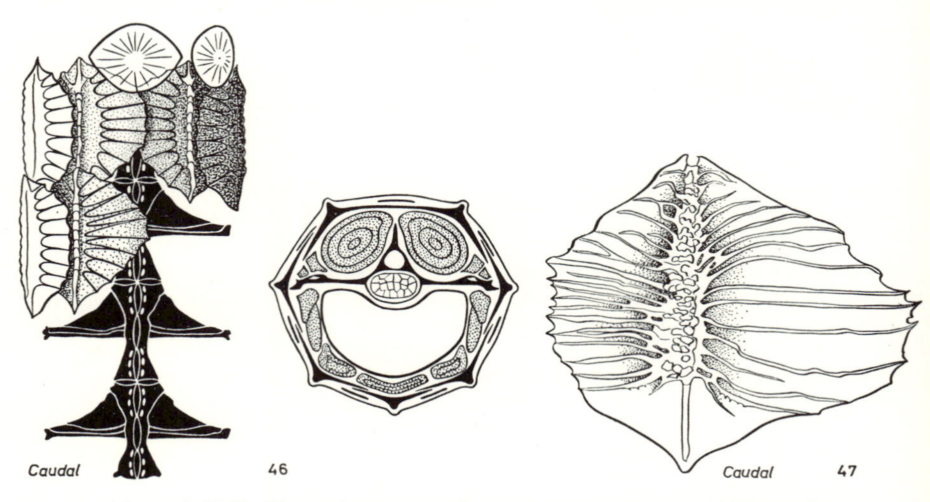

FIG. 46 (*left*). Dermal armour of trunk of the pipe fish (*Syngnathus acus*). Left: top view, black—vertebral column; right: cross-section, black —vertebra and dermal bones. (Actual size × 2·7)

FIG. 47 (*right*). Dermal bone from the trunk of a pipe fish (*Syngnathus acus*). (Actual size × 6)

Since each bony plate is bent and the resulting bony ridge is reinforced, the cross-section of the fish body is angular. It is heptagonal in the trunk and quadrilateral in the tail, according to the number of the existing longitudinal rows of bent, bony plates. The individual bony plates whose shape resembles that of the *Agonus* plates, are transported just as far, once they are shed from the carcass. They are occasionally found in sediments.

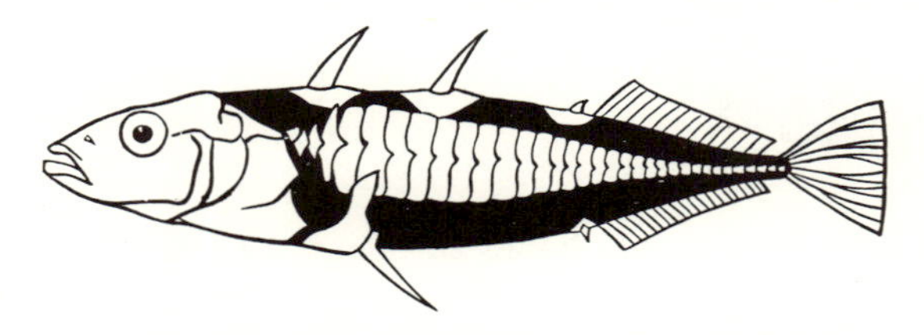

FIG. 48. Stickleback (*Gasterosteus aculeatus*, forma *trachura*). (Actual size × 1·4)

The bony plates on the sides of the body of the small stickleback (*Gasterosteus aculeatus*) are very small and thin (Fig. 48). They overlap in the scale-like fashion. In the sediment they perish rapidly and would not need to be mentioned here if they were not particularly interesting because of the variability of their forms. The number, thickness, and shape of these plates differ according to the habitat of the fishes. According to their lateral armour four varieties of sticklebacks can be distinguished: forma *hologymna* (without lateral armour); forma *gymnura* (with two groups of plates behind the pectoral fin); forma *semiarmata* (with two groups of plates, one behind the pectoral fin and another group on the tail stem); forma *trachura* (with continuous lateral armour). The *hologymna* stage is the first stage of development of all sticklebacks, regardless of the form to which they belong as adults; from that they gradually develop toward their final type. The individuals of the forma *trachura* go through all three preceding stages. The inland waters of Central Europe contain almost only animals of the forma *gymnura*, forma *semiarmata* lives nearer the brackish water, and forma *trachura* in the coastal waters.

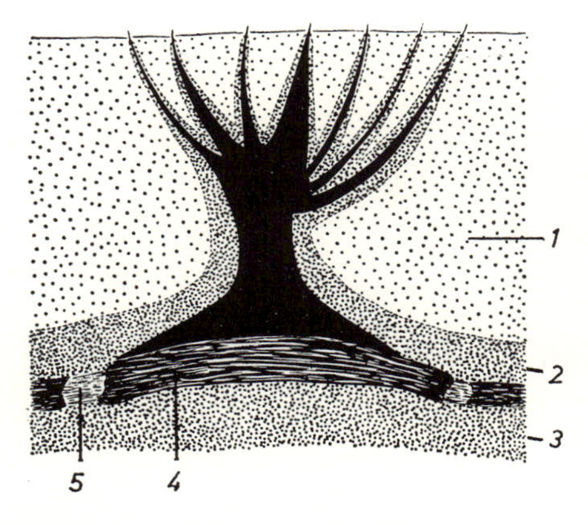

FIG. 49. Spine from the skin of the sun fish (*Orthagoriscus mola*). (1) epidermis; (2) outer corium; (3) inner corium; (4) fibrous plate of scale; (5) connective fibres between neighbouring basal plates of scales. (Actual size × 75)

The dermal bones of the fifteen-spined stickleback (*Spinachia spinachia*) are more resistant. They form four rows along the body, one medio-dorsal which carries the fifteen dorsal spines, two lateral rows which protect the lateral-line system, and one medio-ventral row which is restricted to the tail section. Two pairs of very hard spine-carriers lie in the ventral region.

The cosmopolitan sun fish (*Orthagoriscus mola*), occasionally up to 250 cm in size, carries peculiar dermal bones in its thick, elastic skin. They are calcified plates that lie in the outer layer of the firm corium, they touch each other with their serrated edges, or they are connected by fibres. These plates carry in their centre a spine which is split many times toward the upper end, and the individual points extend just through and beyond the epidermis (Fig. 49). They give a slight roughness to the skin which is several centimetres thick. The young, growing fishes, on the other hand, carry single spines on strong, basal plates that lie scattered in the skin. During adolescence (up to a length of 50 cm) the mouth is toothed in contrast to the adult form. The fish is not common in the North Sea; on the 2nd January 1954, a dead, 98 cm long individual drifted on to the beach of the island of Borkum.

The lump fish (*Cyclopterus lumpus*) has a strong, wrinkly integument that is rich in glands, and numerous, wart-like, complex ossifications of different sizes extend through the integument (Fig. 50). The largest among them reach a length of 2 cm and a height of 1·5 cm in adult animals and are arranged in longitudinal rows; one row is in the dorso-medial position, two in lateral, and two in ventral positions. The cross-section of the body looks therefore pentagonal. Between these rows of large, wart-like, prominent ossifications are many smaller, more or less loosely scattered dermal bones (Hase 1911, talks of four classes of sizes). They are not only on the trunk and the head of the fish but the fins also carry dermal bones of smaller size. Both the large, striking dermal bones and the smaller ones are hollow, cone-shaped, bony structures; they appear to be composed of many bony ridges which run radially toward the principal point and carry small teeth, each with a distinct subsidiary point. Each small tooth has its own cavity which is joined to the large common cavity of the dermal ossification. On the

FIG. 50 (*left*). Front and side views of the lump fish (*Cyclopterus lumpus*). (Actual size × 0·2). Bottom left: part of wrinkled skin with large and small dermal bones.

FIG. 51 (*right.*) Dermal bone of the lump fish (*Cyclopterus lumpus*). (Actual size × 4)

lower parts of the large, bony cone the small teeth are distributed sparsely, but many individual teeth are densely crowded toward the principal point, itself made up by such a tooth on the tip of the cone and piercing the epidermis. The whole structure has developed from simple dermal spines that have coalesced in very early youth; as Hase expresses it—many beginnings of independent origin add up to form one morphological unit, the bony hump (Fig. 51). This bony hump shows no resemblance whatsoever to a scale nor is there any indication of a basal plate; Hase classifies these humps as being close to the placoid organs of selachians. Rauther (1927), however, pointed out that ossifications of *Cyclopterus* originate in the loose corium under the lamellar layer similar to a platelet in its earliest stages of development and unlike placoid organs. The dermal bones are firmly attached to the carcass. If it mummifies, the numerous ossifications in the skin turn into a resistant, spiny armour after the epidermis has dried up. In water the bony cones of the carcass are shed only after the thick dermis is completely decomposed itself, the last part of the body to disintegrate.

Members of the order Scombriformes generally carry small cycloid scales in their skin. However, the lateral-line system of the scad (*Trachurus trachurus*) is embedded into a continuous row of bony plates. They are set in roof-tile pattern, just barely overlapping, and their size decreases from head to tail; where the body is curved, they are slightly shifted in their positions to each other (Fig. 52). A pointed spine which is directed backward provides some connection with the next ossification; in the caudal section of the lateral line, these spines are thicker and protrude more through the skin.

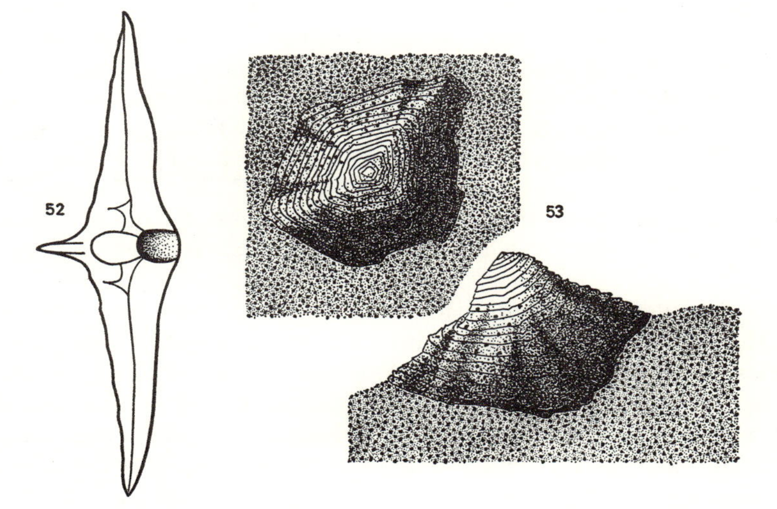

FIG. 52 (*left*). Dermal bone of the scad (*Trachurus trachurus*) covering the lateral-line system. (Actual size × 3)

FIG. 53 (*right*). Top and side views of dermal bones of the turbot (*Scophthalmus maximus*). (Actual size × 10)

The dermal bones of the turbot (*Scophthalmus maximus*) lie scattered over the upper, functional side of the body (Fig. 53). They are peg-shaped, hollow, bony structures that reach a height of up to 5 mm in an adult animal; their base can have a diameter of 8 mm, and its outline has usually four points. The body surface carries approximately fifty such dermal bones. In the living animal they are enveloped by the pigmented epidermis; nevertheless, the concentric bony ridges that encircle the point can be distinguished as more or less transparent zones through the epidermis. In contrast to skeletons, scales, and otoliths of many heterosomata, they are not annual growth rings since many ridges are added within one year. The lateral-line system of the turbot—like that of the scad (*Trachurus trachurus*)—lies under a continuous row of bony plates which are set in roof-tile pattern. On the dorsal side of the free posterior margin, they each carry a strong spine which lies under the epidermis. It is invisible in the living animal but can be felt by touch.

(7) *Vertebrae* of fishes are found in the swash marks of sandy beaches as well as in sediments of the sea floor; they occur particularly in concentrations of shell fragments. Those cases in which the notochord is still the principal supporting organ throughout life (as in the Cyclostomata, Holocephali, Dipnoi, and Chondrostei) will not be discussed here. Only a small number of these fishes live in the North Sea, and the elements of their supporting system are too weak to be preserved over a prolonged period of time, once they are separated from the original structure.

Individual vertebrae of selachians—despite calcification of the cartilage which makes them highly resistant to pressure—are not hard and resistant enough to be preserved in the agitated and often coarse sediments and shell breccias of the shallow sea. Only in fine-grained and undisturbed sediments can the calcareous parts of the vertebrae be preserved; stripped of the cartilaginous parts, however, the calcareous zones have no specific shape. They may be double peg-shaped (cyclospondyle), or ring-shaped (tectospondyle), or they may consist of radially disposed longitudinal ridges (asterospondyle). If these relics occur individually in the sediment, they are usually overlooked, or they cannot be classified if they are observed.

The completely calcified vertebrae of the teleost fishes have characteristic shapes and are strong enough to withstand repeated transport and redeposition in the marine sediments. Single teleost vertebrae are therefore occasionally found. Their shape is usually hourglass-like, and they have a rather spongy texture. Those teleostean vertebrae that are not completely calcified and possess extensive cartilaginous intercalations are soon destroyed in the shallow sea. Short vertebrae (such as in flatfishes) are more resistant than long ones (as in pipe fishes). The upper and lower arches of the vertebrae are also little resistant and become separated from the vertebra at an early stage. The arches of *Anarhichas lupus*, *Gadus virens*, *Trachinus draco* are the only exceptions known to us.

(8) *Spiny fin rays* like those of teleostean species are particularly hard and resistant. However, only the most recent, highly specialised species possess these spines. Hence the number of known fossil spiny fin rays of teleosts is small.

The skeleton of the fin is of ectodermal origin, and each bony fin ray, in a median position in the flat fin, is formed by two paired, lateral components joined along the mid-line. A canal runs along the inside of many spines (Kner 1860–63). The two halves are joined by connective tissue or fibres, or the joining substance between the two parts is calcified. If the connection is not calcified, most of the hollow fin rays and some of the solid ones split lengthwise in two after death, and the parts are too weak to withstand repeated transport and deposition. Spines with calcified connection do not split, and there is a chance that they may be preserved. Most of the spiny fin rays have smooth sides, some have serrated edges. The fin rays can be termed spines where they form an unsegmented, whole structure, are especially resistant, and do not only support and stiffen the fin but have other functions. Hard, spiny rays are preferentially located in the anterior part of the dorsal fin, in the anal, and the ventral fins.

There are few bony fishes with spiny fin rays in the North Sea. It seems that, with a few exceptions, these spines do not serve for defence against enemy species, but for contacts between individuals of the same species, specifically for recognition and choice of mates and for intra-species defence of territories (Schäfer 1959). However, the establishment of territories is dependent on a suitable biotope. An animal must be able to explore a possible territory with its senses, to choose such a territory, to remember its physical landmarks, and to establish precise boundaries through encounters with the holders of neighbouring territories. It also seems that elaborate habits of mate-recognition and of care for the young are often connected with these territorial habits. Appropriate biotopes for the establishment of territories are steep underwater cliffs, coral reefs, areas with dense vegetation or with a dense and varied sessile animal population. Such specialisation for a very specific biotope often also involves a high degree of adaptation such as shape of the body, shape of fins, and mode of locomotion. Such highly special biotopes do not exist in the wide, open expanses of the North Sea. It is mainly populated by fishes congregating in swarms and ranging over wide regions, and by seasonally migrating species that are not concerned with territories and have only the simplest relationships with other members of their species. It seems natural that such fishes do not have organs like fin spines (for a discussion of patterns of behaviour of herrings, see Schäfer 1955d).

The following bony fishes carry fin spines and are frequent in the North Sea: *Gasterosteus aculeatus* has three dorsal spines which are set separately and can be erected, two ventral spines, and one spine in front of the anal fin. Since the forma *trachura* of *Gasterosteus* occurs in millions in the tidal waters during the winter (tons of them used to be caught in Jade Bay for oil), numerous spines up to 7 mm long are deposited in tidal channels. The fifteen-spined stickleback (*Spinachia spinachia*) carries numerous but small spines on the back. In the pink brame (*Ctenolabrus rupestris*) the two first fin rays of the dorsal and anal fins are spines. This species is uncommon in the North Sea; the animals live near coastal cliffs. The same holds for the sea partridge (*Symphodus melops*). In the scad (*Trachurus trachurus*) two spinous rays lie in front of the anal fin. In the dory (*Zeus faber*, Fig. 232) the first ten long rays of the dorsal fin and four rays of the anal fin are transformed into spines. In addition, all dorsal and anal fin rays have two lateral spines at their bases (Fig. 54). In the weever (*Trachinus draco*) and the little weever (*Trachinus vipera*), the first (six and four) dorsal fin rays as well as all the rays of the ventral fin are spines. The angler (*Lophius piscatorius*) carries six hard and large spines on its back (Fig. 232).

FIG. 54. Spines of the dorsal fin and the hard, dorsal, subsidiary spines of the dory (*Zeus faber*). (Actual size × 0·5)

Rare visitors to the western part of the North Sea, mainly near the coastal cliffs of the British Isles, are: the trumpet fish (*Macrorhamphosus scolopax*) which carries a single, serrated, very large, spiny fin ray (Fig. 55); the file fish (*Balistes capriscus*); *Helicolenus dactyloptera*; the boar fish (*Papros aper*), and the wreckfish (*Polyprion americanum*).

The ecological significance of the spinous fin rays of the last-mentioned fishes is not well known. As most of these fishes, however, live in sharply defined biotopes and do not travel in schools, it may be assumed that in these fishes, too, the spines of the fins play a role in the contact between members of the same species. *Lophius piscatorius* whose spines are used in defence against animals of other species (as observations in aquaria have shown) is an exception as are *Trachinus*

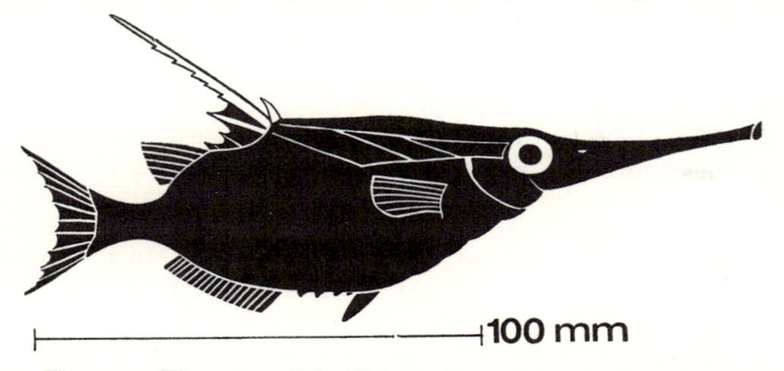

FIG. 55. Trumpet fish (*Macrorhamphosus scolopax*) with serrated fin spine

draco and *T. vipera* whose anal fin spines serve to dig into the sediment (see D I (*b*) (iv) and Fig. 233). Strong fin rays in the dorsal fin which simply erect the fin, but are not really spines, are common. Frequently these fin-supporting rays have a further function in the contact between individuals (for example *Callionymus lyra*). There is a simple transition from supporting rays into spines.

It has already been mentioned that spinous fin rays are a late evolutionary development of the teleosts. This may be because spines often have complex intraspecific functions. Such functions arise only in specialized animals with a highly developed central nervous system, indicative of an advanced stage of evolution.

Hard remnants in the sediments of Heligoland Bay. Pratje (1931) in his exhaustive investigation of the sediments of Heligoland Bay noticed remnants of only one fish skull and a few vertebrae and otoliths of fishes in 211 sediment samples. From these observations he concluded that fishes or parts of their skeletons are hardly ever preserved. After travelling frequently on fishing vessels, and after visits to large fish markets which give an impression of an overabundance of fish in the North Sea, the author could not believe that so little of these large, strong animals with hard skeletons should be preserved in the sediments. He therefore searched over several years systematically for traces of fish with the following results.

Fish remains can be found the year round in the debris along the swash marks of the mainland coast and of the island coasts toward the tidal flats; vertebrae are the most common. Their sizes always correspond to the size of other animal remnants in the same swash marks. Thus in swash accumulations containing mostly *Hydrobia* shells one also finds 5 to 8 mm long crustacean claws, individual limbs of crustaceans, shells of young *Cardium* and *Macoma*, and only such fish remnants that have the same size range. The finds are therefore mainly vertebrae of young flatfishes, of the pogge, sea scorpion, eel pout, ling, pipe fish, and young tub fish. Usually eight to ten fish vertebrae lie on an area of 20 by 20 cm of swash accumulation, occasionally with a few small skull bones. In swash accumulations consisting of objects with elongate, unwieldy shapes, skull bones occur more frequently alongside vertebrae complete with neural and haemal arches, separate ribs or rays, scales, and dermal bones; these fish remnants occur with crustacean limb fragments of comparable size, everything matted with small bryozoa branches, pieces of wood and birds' feathers. In the swash accumulations of *Echinocardium* spines, fish ribs and rays of similar fineness and size are occasionally found as an admixture. They probably come from young herring which live in millions near the islands and in the tidal channels. Large fish parts are not found in the swash accumulations of the tidal flats. In those of the beaches toward the open sea, however, large vertebrae are found together with coarse shell fragments (for example in the swash accumulations of the western beaches of the East Frisian Islands and of the island of Sylt, all exposed to a powerful surf). On the same beaches, swash accumulations of small fish remnants can also be found, but they occur in distinct accumulations. Size sorting is strong in all these accumulations.

Animal remnants washed up by the sea are fully exposed in swash accumulations, and even small pieces are not easily overlooked. Observations are more difficult in deep water. A meaningful survey of the fish content of marine shell deposits of the funnel-shaped tidal channels near the islands can only be obtained by thorough examination of dredged material. A commercial shell dredger can deal with 70 000 cubic metres of shell deposits per year. It can be estimated that such a quantity of shell deposits contains three to four cubic metres of fish vertebrae. Over the past few years we have examined and re-examined such shell deposits from the outer Jade Bay and from the sea near the island of Spiekeroog. Eighty per cent of them consisted of valves of *Cardium edule*, the remainder are mostly valves of *Macoma*, *Ostrea*, and *Mya*. The sizes of vertebrae and valves are similar. The vertebrae seldom carry processes and belong mostly to codfish, mackerel, turbot, and other large fishes. Skull bones are rare; there are no scales, but bony plates of the sturgeon and 'thorns' of the thorn ray are occasionally found. Vertebrae of dolphins occur. Obviously, the fish remnants found in these deposits come not only from fishes that lived in the immediate area of the deposits, but also from several square kilometres of tidal sea floor, carried by incoming tides, transported from west to east along the coast, and ending up in the backwaters of the tidal currents in the outer reaches of the bays—Jade Bay and the bay at the mouth of the river Weser. Only the hardest and most favourably shaped fish parts can withstand constant transportation in this environment.

Until now we have found only few hard remains of fishes in the pure or almost pure sands which cover the sea floor off the East Frisian Islands, in the areas near the Elbe and Weser rivers and off the North Frisian Islands. However, only small grab samples were at our disposal (same as Pratje). Nevertheless it seems that these constantly transported sands are indeed poor in remnants. Any large objects that may temporarily fall into sand (valves of bivalves, gastropod shells, echinoid plates) are sooner or later sorted out by continuous transport and redeposition at deep levels and may either reach shell deposits after repeated transport, or may eventually be destroyed by sand abrasion.

It is different in areas with mud and silt. These sediments are more stable. Once muds are deposited and solidified by compaction they can only be eroded by storms or strong currents. They are eroded in form of platy or polyhedral mud pebbles. Certainly the quantity of organogenic hard remnants is generally smaller in muds than in sands because fewer animals with shells favour a muddy habitat. However, anything that is once embedded in mud will probably remain there. Preservation in mud is thus not rare; moreover, the less resistant hard remnants of fishes are not exposed to sand abrasion or repeated transport. Under these circumstances the hard parts are not sorted according to their weight or size. Generally, the hard remnants of an individual section of a body remain together, or, occasionally, skeletal parts are found in isolation. Otoliths, teeth, spines, scales, and dermal bones are found preferentially in mud deposits. Individual pieces are usually scattered at random; one find does definitely not indicate that more remnants can be expected in its vicinity. This can be explained by the tendency of many fish carcasses to refloat and to shed individual parts one by one on the sea floor; moreover, benthonic animals (mostly scavenging crustaceans) may feed on carcasses that remain on the bottom before they are covered, and scatter the remains over a large area. Scattered over even larger regions are the hard fish remnants contained in faeces of fish-eating fishes which travel over wide distances. In the absence of erosion and redeposition in muddy areas, there is no sorting or concentration of separate fish fragments.

These statements refer to one experience with mud sediments that extend westward approximately from the position of lightship P 10. The muds lie at a depth of 28 to 40 m and represent the southeastern tongue of a wide mud region in the central North Sea (see the map of the bottom sediments of the southern North Sea by Jarke 1956). Sea birds that feed regularly on fish accumulate otoliths in their regurgitated pellets and faeces. On the north mole of Heligoland we found thousands of otoliths on the resting places of gulls in crevices of the masonry and in the surrounding sediments (Schäfer 1966).

II. Echinodermata

Not all classes of Recent echinoderms are represented in the southern North Sea. The crinoids, so common in the early geological periods, are missing. The holothuroids (sea cucumbers), the asteroids (starfish), the ophiuroids (brittle stars), and the echinoids (sea urchins), however, are so abundant and diverse that a comparison of the conditions necessary for fossilisation is possible. Two distinguishing features help in the detection and classification of these animals in sediments: (1) The shape of the whole animal; in echinoderms the generally fivefold radial symmetry can hardly be overlooked, even if the shape is poorly preserved or disturbed; (2) the individual skeletal elements of the body whose shape and internal plates are readily identified—although not all echinoderms provide both distinguishing features together. Thus the sack-shaped holothuroids lack the typical body shape. Their skeletal elements, however, are generally well preserved, can be found easily due to the great numbers in which they occur in each individual, and are easily identified because their structure is simple, yet shows great variability in shape. The asteroids have a typical body outline and numerous skeletal elements of simple structure. The same holds for the ophiuroids. In the echinoids the skeletal parts are very resistant and remain so long in the original structure that both body shape and skeletal elements carry distinguishing features in the fossil.

When a holothuroid dies, only individual skeletal elements can be preserved, without definite spatial relationship to each other. Therefore it does not matter how long after death and where such an animal is finally embedded. Asteroid and ophiuroid fossils are more complete the sooner after death they have been buried. The former existence of a starfish or a brittle star can be deduced from a single skeletal element, but the species can rarely be identified. In the echinoids also, the fossil is the more complete with early burial. The skeleton provides a mould, often made of strong plates resistant to compaction. Internal moulds of echinoids are frequent and are often well preserved. Their individual skeletal elements are suitable for specific identification because of their shape and ornament.

(*a*) ASTEROIDEA

Life and death. Asterias rubens is the most frequent starfish of the southern North Sea. It is common in the tidal sea and in depths to 200 m. Kowalski (1955) described its occurrence in brackish water (according to Remane 1934 and 1952, *Asterias* still occurs in the central Baltic Sea where the salinity is down to 8 per mil). Preyer (1886) investigated and described the arm movements and locomotion of starfish, with many good illustrations. Locomotion is possible on firm as well as on yielding ground. In sand the animal moves by scraping movements of the locomotory podia of the ambulacra; on firm ground suckers on the podia are used while all five arms perform undulating and pushing movements. The animal is also able to dig into the sediment. Large amounts of food can attract numerous individuals of the mobile, red starfish within a short time, as demonstrated by commercial fishing of starfish in the southern North Sea. On a single mussel bed (*Cardium*, *Mactra corallina cinerea*, *Macoma baltica*, and *Mytilus edulis*) one to two metric tons of starfish a day can be fished with a ground net for weeks on end. The quantity of catches decreases rapidly as soon as the bivalves are eaten or covered by sand. It can be concluded that the starfish move rapidly to new feeding places and that their locomotion must be quite efficient on all kinds of ground. *Asterias* seems to live for seven or eight years, but many animals die earlier in the shallow sea by being covered with sand. During gales, 60 cm of sediment can be deposited within half an hour, and *Asterias* is unable to penetrate a layer of such thickness. The animals suffocate rapidly, and more so if crowds of them are covered together. In that case the animals do not lie horizontally and with their arms extended on the bedding plane, but are found in oblique positions and with coiled arms in the sediments because the starfish try to reach the surface of the newly deposited layer. Animals already covered with sediment may also crawl along the bedding plane. Because of the resistance of the sediment they then usually move by advancing one or two arms while the others trail behind, or all arms may be turned backward; they slowly proceed in this way until they are forced to stop and die. In fossils such a position may be mistaken as the effect of the veering action of bottom currents on the arms. The deception may be the more complete if several animals have crawled in the same direction (Fig. 56), suggesting a current that has influenced all the bodies simultaneously.

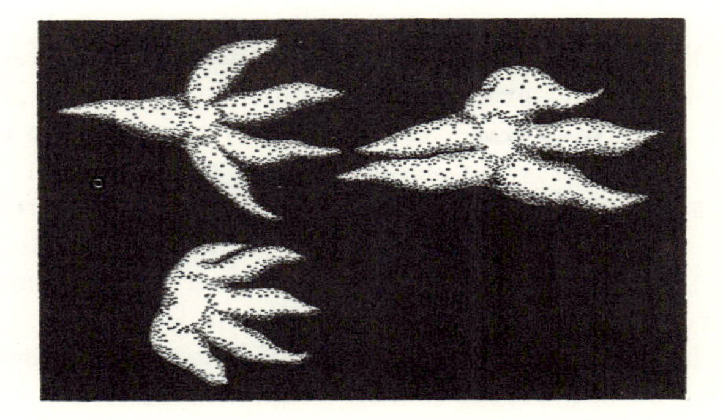

FIG. 56. *Asterias rubens* in creeping position on a bedding plane below fresh sediment. False appearance of carcasses being drifted by current. (Actual size × 0·15)

The sun star (*Solaster papposus*) seems to be the most important enemy of young starfish. Benthonic fishes also often eat numerous starfish of all sizes. Lack of oxygen, decreasing salinity, or rising water temperatures may cause *Asterias* to commit autotomy and cast off one arm, the largest animals being the first to do so. More arms follow if conditions remain bad, and finally the starfish dies. Once this process has started in a stagnant place, entire populations may die. The more numerous the population and the more arms are cast off the faster the animals perish. Young and small animals react less frequently by autotomy to poor living conditions than large ones. An enormous quantity of calcium carbonate is contributed to the sediments of the North Sea by skeletal parts of *Asterias rubens*. No other echinoderms are nearly so common or have a comparable influence on sediment composition.

Asterias mülleri, a close relative of *rubens*, occurs less frequently. The southern boundary of its distribution is near Heligoland, and it occurs to a depth of 800 m. The species does not spread rapidly because its larvae are not pelagic. The eggs are carried by the mother around the oral opening until they develop into fully formed starfish. Areas in which the species has died out due to rapid sedimentation are only slowly resettled. This starfish seems to be restricted to primary hard ground like solid rock or rock-strewn bottoms. In such an environment the skeletal parts can accumulate in crevices and gaps between rocks and be preserved there as loose deposits; completely buried animals, however, cannot be expected on such a sea floor. The sun star (*Solaster papposus*), like *Asterias mülleri*, prefers hard ground; its large and hard skeletal elements are not rare in the sands of the island of Heligoland.

Astropecten irregularis digs for food in the surface layers of the sediment. Even so it cannot reach the surface if only slightly covered by additional sediment, because all its movements are horizontal. The short arms can neither bend nor push (for details about locomotion of *Astropecten*, see D I (*b*) (i)). *Astropecten* lives therefore mainly in areas which have been sediment-free for long periods of time; yet it happens frequently that sediments suddenly cover many animals during a gale so that they die where they have lived. The animals always lie with their rigid arms fully extended horizontally under their cover of sand. They, of all starfish, are most frequently embedded complete.

Little is known about the ecology of *Luidia sarsi*, *Solaster endeca*, and *Henricia sanguinulenta*, except that they exist in the North Sea.

Hard parts. Well-preserved fossil asteroids are easily identified. This is not so if the skeleton is disintegrated, as happens commonly in the shallow sea, and only individual pieces are left. According to Mortensen (1938) individual ossicles, aboral shields, spines etc. cannot be precisely identified, although the highly specific terminal ossicles may be an exception. Nevertheless, the skeletal elements will be described briefly so that the processes of disintegration of the carcass can be understood. Unfortunately, neither the structure of individual skeletal parts nor that of the whole asteroid enables us to interpret the ecology of the living animal.

The starfish skeleton is composed of the dorsal (aboral) skeleton, the ventral (oral) skeleton, and the lateral skeleton (Fig. 57). The aboral skeleton consists of small, flat, shield-like calcareous plates. Some of their margins overlap in such a way as to form a fenestrated, net-like, irregular structure (*Asterias*, *Solaster*); in *Astropecten* calcareous, vertical, barrel-shaped bodies form an orderly mosaic (Fig. 58). The net structure carries spines of different sizes on its outside or little calcareous columns which bear on their free ends circles of spinelets (paxillae). The pedicellariae are also on the aboral side. The oral skeleton (Fig. 59)

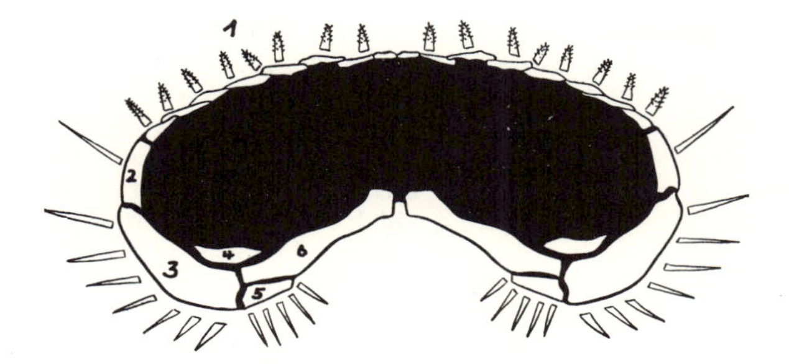

Fig. 57. Cross-section through arm skeleton of an asteroid, showing dorsal, ventral, and lateral skeleton. (1) dorsal skeleton with paxillae; (2) supramarginalia; (3) adambulacrala; (4) supra-ambulacrala; (5) exambulacrala; (6) ambulacrala. (Actual size × 2)

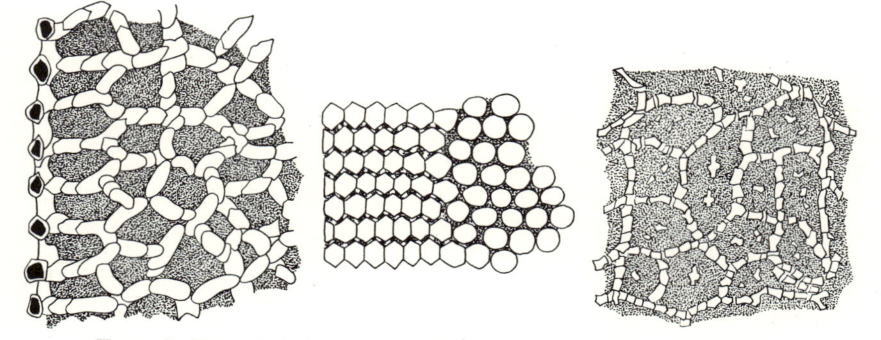

Fig. 58. Dorsal skeletons of asteroids. Left: *Asterias rubens*; centre: *Astropecten irregularis*; right: *Solaster papposus*. (Actual size × 5)

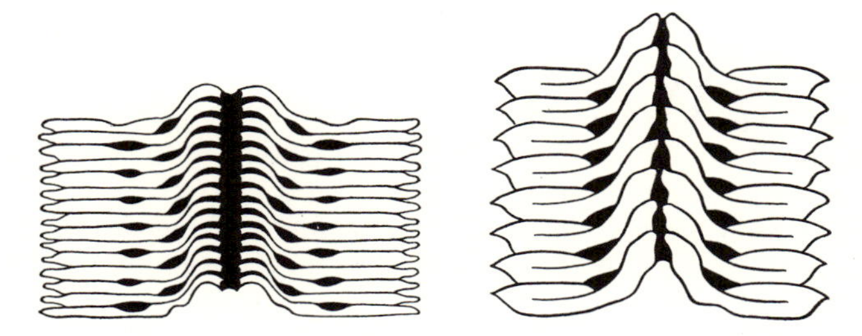

Fig. 59. Ventral skeletons of asteroids. Left: *Asterias rubens*; right: *Solaster papposus*. (Actual size × 0·7)

consists mainly of the rows of ambulacral plates which are closely set but leave room for the ambulacral podia to pass through. The rows of ambulacral plates of the opposite sides of one arm join at an angle, forming the ambulacral groove. The ambulacral ossicles have no spines. The ambulacral rows end in the tip of the arm in an unpaired ocular plate. Lateral to the rows of ambulacral ossicles that form the groove is a row of adambulacral ossicles which carry several rows of spinelets, the adambulacral spinelets. Their number, position, and size are useful for identification. The lateral skeleton which borders the ambulacrals and connects them with the aboral skeleton is not always distinct; in *Asterias* it is missing. Generally a dorsal row (supramarginals) and a ventral row (inframarginals) consist of spine-bearing marginal plates. These rows lie either above each other or overlap. Sometimes supra-ambulacral plates give additional support from the inside between the ambulacrals and the marginals.

The calcareous skeletons of the dorsal, ventral, and lateral sides including the spines, the parts of the peristomial ring and the ocular plates at the arm tips contain approximately 4000 individual pieces. The largest among them are 6 mm long. Their characteristic shapes can easily be spotted and are common in the fine-grained sandy and muddy sediments of the North Sea. The total weight of all dried skeletal elements of an average *Asterias rubens* is approximately 8 g, of *Solaster papposus* approximately 14 to 16 g, and of *Astropecten irregularis* approximately 5·3 g.

Disintegration of the carcass. When an *Asterias rubens* dies on the sea floor, the red colour fades away rapidly while decomposition gases arch the body cover. At a water temperature of 18°C this stage is reached after five days, and bottom currents can now move the entire carcass. The body never rises from the sea floor or refloats like fish carcasses. Due to inflation the skin is stretched, all arms are rigid and extended from the disk and cannot bend in the current. Only during one short period can the position of the starfish be influenced by bottom currents and its arms be turned; that is in the few hours immediately preceding and following death. Only then are the muscles limp enough and the body cavity not yet inflated, so that minor currents can move the arms. Contrary to the opinion of Koenigswald (1930) starfish in more or less oblique or overturned positions have not been moved by bottom currents. Dying and injured starfish (this holds particularly for *Solaster papposus*) usually behave in this way; they regularly coil up, twist their arms and even overturn the central disk. Dead starfish do not get into 'vortex positions' by 'local whirlpools' as Koenigswald assumed.

On the sixth day after death the integument of the dorsal side of one or several arms and the fenestrated structure of the dorsal skeleton begin to lift at the tips of the arms. Within the next eleven days the dorsal skin of all the arms has come loose, and small or large pieces of it drift in the current or slip sideways from the body. Strong currents may transport the dorsal skin, and the adhering skeletal parts. The various skeletal pieces of the dorsal skin lie usually entangled and disorderly because the skin that holds them folds up and coils while it is sinking. During these stages the body with the ventral and marginal skeleton lies undisturbed. After some twelve days decomposition has removed the digestive organs, the water vascular system and the musculature from the ventral skeleton, leaving these skeleton parts undisturbed. The weight of the starfish has made them sink somewhat into the surface of the sediment. If, on the other hand, *Asterias* is killed by burial, the reticulate structure of the aboral skeleton cannot lift, and parts of the dorsal, ventral, and lateral skeleton become wedged together into a tangled heap.

Cohering skeletons of starfish are extremely rare in the shallow sea because sediments with their fossils seldom remain undisturbed; usually displacement and redeposition occur repeatedly. As a consequence the composite skeleton of the starfish is broken, and its individual pieces are transported separately.

Astropecten irregularis, whose skeleton is composed of large and relatively heavy plates, lies on the ground during decomposition. The integument is scarcely inflated and it generally remains tight on the body; twisting and turning of the arms or the disk never occur in these animals.

(*b*) OPHIUROIDEA

The number of species of brittle stars in the North Sea is small. Some species are widely distributed and settle on mud as well as on sand or on hard ground, others are restricted; but wherever a species exists it occurs abundantly. There are also wide regions of the sea floor that are free from any brittle stars.

Ophiura affinis, *O. albida*, *O. texturata*, *Amphiura chiajei* and *A. filiformis* usually live on mud deposits. *Acrocnida brachiata* and also *Ophiura texturata*, *O. albida* and *O. affinis* can be found on sandy ground. *Ophiothrix fragilis*, *Ophiopholis aculeata*, *Amphipholis squamata*, *Ophiura sarsi* live on stony ground and rock, as do *Ophiura texturata* and *O. albida*.

Skeleton. Judging from the large number of individuals, from the resistance of the individual skeletal parts, and from the solid appearance of the armour it could be expected that the ophiuroids would be common fossils. This is not so because the parts of the skeleton are joined only by skin or connective tissue and thus become easily detached and scattered after death. Therefore brittle stars can only become well-preserved fossils where there are no bottom currents, or where sediments soon provide a cover, and where these sediments remain undisturbed. As the last condition is rarely realised, well-preserved brittle stars are rare, although parts of the skeleton can be found as microfossils in sands, mud deposits and shell breccias.

The skeleton of the arms consists of an internal row of vertebral ossicles, their size decreasing toward the tips of the arms, and an outer skeleton of four shield-like plates on each arm, the aboral shield, the oral shield, and the lateral shields. Each internal vertebral ossicle corresponds to two ambulacral plates of the starfish. In the embryo it is paired and disintegrates frequently after death into these original parts. Adjacent vertebral ossicles are connected by joints. The lateral plates carry spines on the distal margin, the so-called arm spines. Apertures between the oral and the lateral shields are frequently encircled by spines and provide for the passage of the podia or tentacles. The aboral side of the circular or lobate body is covered with flat plates; primary plates in the centre are surrounded by radial plates, a pair of them at the base of each arm. Both types of plates are useful for identification. The bases of the five arms lie on the ventral, oral side of the body. Each of the five interradial areas is occupied by a large, shield-like plate, the oral or buccal shield, which is flanked on either side by an elongated, so-called adoral shield. From each buccal shield a jaw protrudes into the mouth aperture which is usually edged by little spines called the oral papillae. The 'tooth' or several tooth-papillae lie in the centre. Two elongated plates in each interradial area surround the genital slits which open into the genital pouches.

According to Mortensen (1938) isolated ossicles, shields, and spines of brittle stars are usually not classifiable because the vertebral ossicles of one individual vary considerably in shape and size from the base to the tip of the arm. Moreover, the ossicles of different genera and families are very similar.

Death and burial. Many fishes feed regularly on brittle stars including: *Raja clavata*, *Acipenser sturio*, *Conger conger*, *Gadus callarias*, *Melanogrammus aeglefinus*, *Molva molva*, *Mullus surmuletus*, *Anarhichas lupus*, *A. minor*, *Scophthalmus maximus*, *Hippoglossus hippoglossoides*, *Pleuronectes platessa*, *Limanda limanda*. Faeces of such fishes consist occasionally of nothing but skeletal parts of brittle stars. The faeces soon disintegrate, and the skeleton parts are released.

Areas with frequent sedimentation are usually free from living brittle stars; in the southern North Sea such areas include most of the coastal regions to depths of approximately 10 m. Larvae which settle in these areas usually die and are soon covered by sediment. On the other hand, where sedimentation is slight and in quiet water, brittle stars do occur, even close to the low-water mark (mainly *Ophiura texturata*). Extensive sedimentation during bad weather usually results in disappearance of brittle stars. If only 5 cm of sediment suddenly cover a brittle star, it is immobilized and dies. Species with long, flexible arms (for example *Amphiura filiformis*) are generally found covered with tightly twisted and coiled arms because this is their position at death. Species with short and stiff arms (for example *Ophiura texturata*) keep them extended at death. The arm positions of fossil ophiuroids can lead to misinterpretation about normal movements or modes of locomotion, an error committed by Klinghardt (1930). Death struggle and the premortal muscle contractions often produce most unlifelike arm positions. If the carcasses lie at the surface bottom currents can move the arms after death. As with starfish, the arms of brittle stars are slack enough to be moved and oriented for a short period after death. Fifteen hours after death the arms begin to disintegrate and fall apart. The carcasses of ophiuroids are never refloated by decomposition gases.

During sedimentation, brittle stars react by moving upward. They succeed if the sedimentation is not too rapid. The time when the water calms down after a gale is particularly dangerous because then all the suspended sediment sinks back to the sea floor. Photographs from a depth of 39 m (southwest of the island of Heligoland) show a rich population of brittle stars on ripple-marks with a wavelength of 20 cm. This proves that the animals can withstand strong and sustained bottom currents.

Species living on mud or on fine-grained sand have evidently the best chance of being preserved. Species that live on hard ground with moderate or no sedimentation are only exceptionally preserved complete. Occasionally, coarse sands which fill the gaps between boulders and shell fragments are rich in skeletal parts of these animals (near the island of Heligoland, *Ophiothrix fragilis*). As the upper layers in the entire shallow sea region are constantly moved and redeposited, well-preserved ophiuroids are rare even in areas covered with mud and fine sands. Under favourable conditions, numerous individuals are usually preserved due to their gregarious habits. Individual skeletal parts are transported easily and over great distances. They occur widely scattered, but are consistently found in muds or fine-grained sands. The hard and compact vertebral ossicles of the arms are found most frequently.

(*c*) ECHINOIDEA

Anatomy and life habits. Only a few species of sea urchins live in the southern North Sea. Mortensen & Lieberkind (1928) listed ten species for the entire North Sea of which approximately six occur regularly in the southern part. The echinoids are subdivided into two sub-classes according to their general organization, the regular and the irregular.

The regular sea urchins of the southern North Sea live either on the surface of the sediment or on hard ground (Hoffmann 1914; Tornquist 1903 and 1911), never in the sediment. The stout, long-stalked ambulacral podia are the organs of locomotion, and are spread over the whole body. Spines, pedicellaria, and the robust armour skeleton protect the animal as it sits or moves slowly over the sea floor. The spines and the pedicellaria can perform orientated movements in response to adverse stimuli (Uexküll 1907). The spines occasionally are used for locomotion by a pushing and levering action (*Psammechinus miliaris*). A strong Aristotle's lantern consists of five segments in the centre of the underside. Species inhabiting rocks in the clear, well-illuminated, coastal waters use these teeth for grazing on algae adhering to the rock surface. Species that live on sand or mud in dark water rich in suspended matter eat detritus, carcasses, and small organisms. The bodies of the rock-living species are often

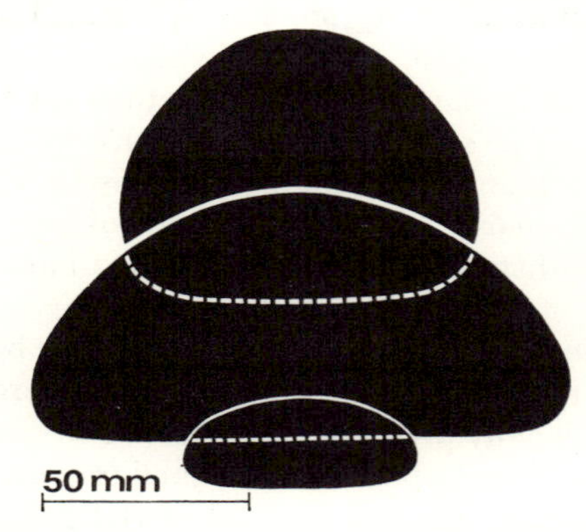

FIG. 60. Three echinoids in profile, showing the relative sizes of the planes of support. Top: *Echinus esculentus*; centre: *Echinus acutus*; bottom: *Psammechinus miliaris*

almost spherical, their spines are strong, straight, and long (*Echinus esculentus*, *E. acutus*). Animals of species that live on sand or mud have usually more flattened bodies. This allows a larger part of the body, bearing more podia and spines, to make contact with the substrate; this prevents sinking into the sediment and also helps locomotion (*Strongylocentrotus dröbachiensis* and *Psammechinus miliaris*, Fig. 60).

Protective spines differ from the others in size but not in shape (Fig. 61). The slender, straight spine rests on a solid base and tapers uniformly toward the tip which can have a distinct cortical cap. Only the base of the spine is surrounded by skin, the calcareous skeleton of the remaining part is bare. The terms 'primary spines' for the larger spines, 'secondary' or 'tertiary' spines for the smaller ones can only be applied as long as the several spines can be compared with each other. As the palaeontologist deals mostly with individual, detached specimens whose original position on the body is unknown this terminology is of no use to him. Moreover, it has no relation to the shape or function of the spine. The regular sea urchins of the southern North Sea carry only pointed defensive spines.

Regular sea urchins with club-shaped, rectangular, spatulate, mushroom-shaped, or beaker-shaped spines exist in other seas; little is known of functions of all these spines.

The irregular sea urchins of the southern North Sea live in the sediment, more or less deeply buried. Protected in this way the animals do not need mechanical or chemical defence organs, and the spines are locomotory. Consequently they are thin and curved, and some have broadened, shovel-like ends (Fig. 62); all can perform circular motions (see D I (*a*) (vii) and D I (*b*) (ii)). The pedicellaria do not bite as they do in the

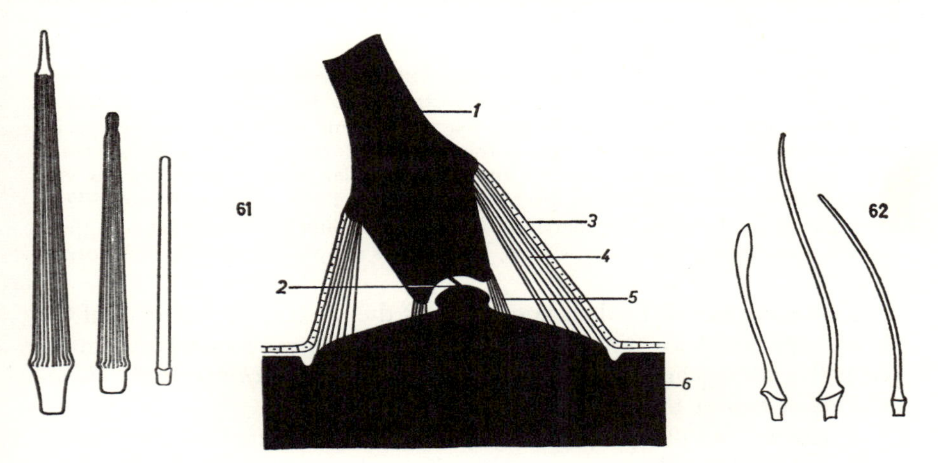

FIG. 61 (*left*). Spines of *Echinus esculentus* (after Hoffmann, 1914). Left: three types of spines; right: mobile attachment of a spine. (1) shaft of spine; (2) ball and socket joint; (3) epidermis; (4) muscle for spine movement; (5) locking muscle. (Schematic drawing)

FIG. 62 (*right*). Spines of *Echinocardium cordatum*. Left: ventral spine; centre: lateral spine; right: spine from the aboral tuft. (Actual size × 4)

regulars but clean instead. The ambulacral podia do not propel the body, those of the oral region ingest small food particles. The shell is thin and weak. The necessity for directional movement in the sediment determines the shape and structure of the irregular sea urchin. Unlike the regular urchins, the irregulars have a functional anterior and posterior end, a left and a right body side. The mouth has generally moved from the central position anteriorly, the anus to a more posterior position. An irregular sea urchin needs no hard dentition; it has a rigid, shovel-like scoop at the posterior margin of the mouth aperture. Irregular urchins have three types of spine. Short-stemmed, spoon-shaped spines on the underside of the body can remove the sediment from under the urchin. Long, bent, round spines occur on all sides, and shovel away the sand brought up from under the animal. A tuft of bristle-shaped spines is located on the upper part of the aboral side; it plays an important role in the construction of a respiration canal through the sediment for the buried sea urchin.

Much is known about the density of populations of the various species (Remane 1940). *Echinocardium cordatum* is the most common. In the fine-grained sand of the East Frisian Islands individuals occur from a depth of 6 to 8 m or more; their average density is from 15 to 20 animals per square metre, and frequently 60–80 animals are found. North of the line Heligoland–Doggerbank *Echinocardium flavescens*, *Spatangus purpureus*, *Brissopsis lyrifera* take the place of *Echinocardium cordatum*, but never reach the same density. *Spatangus purpureus* and *Echinocardium flavescens* settle preferentially in coarse sand and fine shell deposits. *Brissopsis lyrifera* lives in muddy ground.

The regular sea urchins are less frequent and less dense in the southern North Sea. *Echinus esculentus* is almost restricted to the rocks of the island of Heligoland and the surrounding stony sea floor. Some individuals of the species also seem to live on the reef bottom at the island of Borkum. *Echinus elegans* which also favours stony ground is rare in our particular area. *Echinus acutus* and *Strongylocentrotus dröbachiensis* prefer a sandy sea floor; *Psammechinus miliaris* lives mainly in the deep, permanently filled tidal channels of the tidal sea; there it prefers the hard ground (*Sabellaria*, settlements of sponges, and shell deposits).

Death and burial. Little is known about the life span and growth of the various species. According to our observations *Echinus esculentus* seems to reach the adult stage at three or four years, *Psammechinus* and *Echinocardium* at two or three years.

There are two dangers for irregular sea urchins living in the sediment; one is to be covered by sediment, the other to be washed out. If a burrowing *Echinocardium* is suddenly covered by an additional 30 cm of fine sand, the animal is unable to dig itself out. Thus with each gale great numbers of individuals are covered in the shallow areas of the southern North Sea. As the sediments are constantly redeposited in these areas, the tests that have been embedded in their living positions are, sooner or later, washed out again. The fragile skeleton which is meanwhile free of musculature and epithelium disintegrates during transportation. If, on the other hand, the animals are washed out by currents and heavy seas from the sediment that they inhabit, the tests are still held together by their musculature and skin, and usually reach the swash accumulations of the beaches complete and often uninjured. Occasionally large numbers of them are drifted ashore. This shows that these animals live in dense populations and are subject to devastating mass mortality several times each year.

Häntzschel (1936a) reported swash accumulations of *Echinocardium* on the island of Norderney where a 5 km long and 8 m wide area was covered with tests, heaped in places to a height of 30 cm. Several days or weeks later there was no trace of such a swash accumulation, the fragile and thin tests were reduced by the surf to sand-size fragments, and the spines carried away. Irregular sea urchins in the shallow sea are therefore preserved as complete fossils only when the animals are not displaced after burial; otherwise they are broken up and reduced to small fragments. Of these, only the spines are resistant, although they are slender. They are easy to recognise and to identify. As soon as decomposition sets in, the spines become free from the supporting musculature and the sheathing skin, and are transported separately. An individual *Echinocardium* carries approximately 12 300 spines on its test of which 1490 are spoon-shaped. *Spatangus purpureus* bears 31 300 spines, 500 of which are large dorsal spines. Millions of spines are thus daily released into the sediment; hardly a sample of sediment can be taken in the southern North Sea that is free of such spines. They accumulate frequently in marine deposits with transported, fine-grained peat fragments because the settling speed and the transportation characteristics of the peat particles and of the spines are similar. A dense, matted sediment is thus formed. It is usually also particularly rich in foraminifers. Schwarz (1930) has described a sediment consisting of nothing but sea urchin spines with a clay matrix in the Südwatt of Wangerooge. In spring 1959 we found on the beach of the island of Mellum a swash accumulation of peat fragments more than 1 km long with so many *Echinocardium* spines that they gave a conspicuous, silvery shine to the surface. Such accumulations of spines can be found in the remotest corners of the bays (for example in Jade Bay, 35 km from the nearest settlements of living *Echinocardium*). The spines of *Echinocyamus pusillus*, *Spatangus purpureus* and *Brissopsis lyrifera* are similarly well preserved; but as animals of these species do not occur in nearly such large numbers as *Echinocardium*, and as their spines are larger and thicker and therefore less easily transported, the spines are not as common and widely distributed in the sediments.

The regular sea urchins that live on hard ground and rocks can be crushed or injured by pebbles agitated by the surf and currents. Fragments of urchins (for example *Echinus*) which were crushed alive can be recognised by the irregularity of the fracture lines which do not follow the junction between the ambulacral and interambulacral areas. Only in the living animal can the test break in this fashion. During life the spherical test is covered and held together by the epidermis and the musculature of the spines and reacts toward outside forces as a single, brittle shell. Similar irregular borders occur on fragments that have been cracked open by animals with specifically adapted dentitions (e.g. *Anarhichas*). If, on the other hand, a sea urchin's skeleton lies for weeks undisturbed on the sea floor, the test usually begins to disintegrate along the junction between the ambulacral and the interambulacral areas.

A regular sea urchin which is dying can be recognised by the fact that in certain regions at the top of its body the muscles of the spines have ceased to keep the spines erect; instead, they droop and react to stimuli with only a slight re-erection. After death all spines droop downward. In *Echinus* the musculature of the spines is decomposed seven days after death, some spines drop from the body, and the top of the body loses its spines first. Twelve days after death the entire apical system with anal plate and the surrounding plates of the periproct, the genital and terminal plates, and the madreporite plate has become detached, and most of the parts have dropped into the cavity of the test. The oral field (peristome) disintegrates into its component plates and exposes the masticatory apparatus which disintegrates after five more days into its individual parts (pyramid halves, epiphyses, rotules, teeth, and compass parts). All these parts lie inside the test and can only drop out and be transported separately after the test itself is displaced or disintegrated. Usually the spines of the regular sea urchins are too heavy for individual or collective transport over long distances; such spines are always found near the original habitat. Regular sea urchins have fewer spines than the irregular ones. *Echinus esculentus* bears only 5790 spines, *Psammechinus* 2100 spines. Sea urchins that live on hard ground between pebbles and on rocks are rarely preserved. Under favourable conditions individual plates of the test and spines may be preserved in shell deposits.

The chances of preservation are better for regular sea urchins that live on sand. It is part of their adaptation to life in a littoral sediment that such sea urchins (for example *Psammechinus miliaris*) can cope with a sediment cover up to 5 cm thick without changing their normal, horizontal position. If the sediment cover gets thicker a sea urchin uses its spines and ambulacral podia to turn on its narrow side. In this position it can burrow through layers 5 to 20 cm thick (Schäfer 1956a). Layers of new sediment 30 to 50 cm thick are not uncommon in the habitat of these sea urchins, and in such a case the sea urchins die and are preserved complete with all their spines. Fossils of complete sea urchins with spines are nevertheless rare because in areas where rapid sedimentation occurs, deposits are also frequently eroded. Because sea urchins begin to lose their spines quickly after death they must be buried rapidly if they are to be preserved complete with spines. Transported, dead sea urchins have no spine coating.

Mortensen (1938) maintained that individual spines and plates of echinoids can often be identified. However, a palaeontologist concerned with an extinct species might be inclined to allocate to different species, isolated spines of a species with varied spine morphology. Mortensen thought that fossil pedicellaria were unreliable, although they are important in the classification of Recent species and genera as well as of families and orders.

(*d*) HOLOTHURIOIDEA

The rather striking wheel and anchor-shaped, calcareous ossicles (spicules) of the sea cucumbers are amongst the most frequently found echinoderm remains. Their frequent occurrence is due to several factors. They are abundant in the leathery skin of one individual, and so is the number of individuals in a favourable habitat; moreover, these spicules lie often in the interstices between sand grains and are thus protected, and they can move with the sand without being destroyed. If these hard parts enabled species to be identified they would be more useful stratigraphically. In Mortensen's (1938) opinion this is not so.

The North Sea is neither particularly rich in species nor, in general, in individuals. In its southern part only ten species are known (*Cucumaria elongata*, *Echinocucumis hispida*, *Thyone fusus*, *Thyonidium pellucidum*, *Psolus phantapus*, *Leptosynapta inhaerens*, *L. bergensis*, *L. minuta*, *Mesothuria intestinalis*, and *Stichopus tremulus*). Most of them live in mud or in fine sand. *Leptosynapta* has also been found in coarse sand. *Thyone fusus* lives in shell accumulations.

When a sea cucumber dies decomposition gases soon inflate the body, giving it the same shape that the living animal has when its paired, branched-out, respiratory trees are actively filled with water. The gas does not refloat the body, but it is more easily rolled around by bottom currents. When the carcass is completely decomposed (which takes several weeks in the species with hard, leathery skin) the ossicles are finally dropped and are then transported separately. Frequently, however, the slack, leathery skin, as the last organic remnant, still retains the ossicles and is embedded with them. If the sediment remains undisturbed a concentration of ossicles may result which may be interpreted as an accumulation when it is observed in a fossil sediment. Faeces of fishes can also contain accumulated ossicles. According to Liebekind (1928) the species *Thyonidium pellucidum* and *Psolus phantapus* are a favourite fish food. Once freed, ossicles are easily transported over considerable distances. Places where ossicles are found have not necessarily been living areas of sea cucumbers.

III. Arthropoda

Several groups of arthropods have a detailed fossil record. Good preservation, at least of the larger forms, is possible because the chitin skeleton is very hard and resistant, and in many forms it is also calcified.

In spite of these favourable conditions for preservation and identification the fossil representatives of several classes of this phylum, mostly inhabitants of the land or of fresh water, are poorly known, e.g. the classes

Onychophora, Arachnida, and Antennata. The Trilobita, Merostomata, Pantopoda, and Crustacea are purely or mainly marine groups; the trilobites and the eurypterids have no recent representatives. Of the antennates, only the insects must be mentioned because their flying forms occur often in the coastal sediments.

Not all kinds of crustaceans can be preserved in marine sediments; small forms with skin-like, thin exoskeletons and parasitic forms are rarely fossilised. This restricts the number of crustaceans that deserve actuopalaeontological consideration.

(*a*) PANTOPODA

In the southern North Sea the pantopods are represented by few species which, however, occur in large numbers of individuals.

Way of life. The pantopods of that area eat coelenterates exclusively. The Poxichilidae (*Anoplodactylus petiolatus*) graze with their small maxillas on the polyps of hydrozoan colonies; the slender Nymphonidae (*Nymphon longitarse*, *N. grossipes*, *N. strömii*) feed in the same way. The sturdy Pycnogonidae (*Pycnogonum litorale*, *P. crassirostre*) drill with their large proboscis into the basal part of the outer wall of large actinians (*Tealia*, *Metridium*, *Actinia*) and suck their body fluids. A sieving mechanism inside the proboscis prevents tissue fragments from entering the oesophagus. The concentrated nutritive solution that is almost free from hard material is absorbed in a short intestine with eight digestive pouches. They are too large for the short and narrow trunk and extend into the eight (rarely ten), radially placed, locomotory limbs. A short, stump-shaped abdomen bears the rectum whose peristaltic movements form spherical faecal pellets.

These animals do not have to move rapidly because their prey are almost all sessile (however, swimming medusae with attached pantopods are occasionally observed). Their limbs, like those of all arthropods, are segmented and end in sharp claws (or double claws) that are well suited for climbing and holding. The animals climb slowly and often stop for several minutes. *Nymphon grossipes* var. *mixtus* and several other pantopods that do not occur in the North Sea can swim. They do this by slow beats of their locomotory limbs through a vertical plane, moving each anterior pair of legs slightly more energetically than the one posterior to it (Loman 1907, Prell 1911). We have seen swimming *Nymphon grossipes* in Jade Bay.

The following constitutional types of pantopod are known today:

(1) *Pycnogonum* type. These animals have a stout body with short and strong locomotory limbs with strong claws that are suited to hook into the

musculature of actinians. The conical proboscis can suck and bears rasping jaws. There are no accessory limbs for feeding (maxillas or palps). Reefs and hard ground in the tidal sea are the favoured habitat of these pantopods; they occur in the southern North Sea, frequently on *Sabellaria* reefs (Fig. 63).

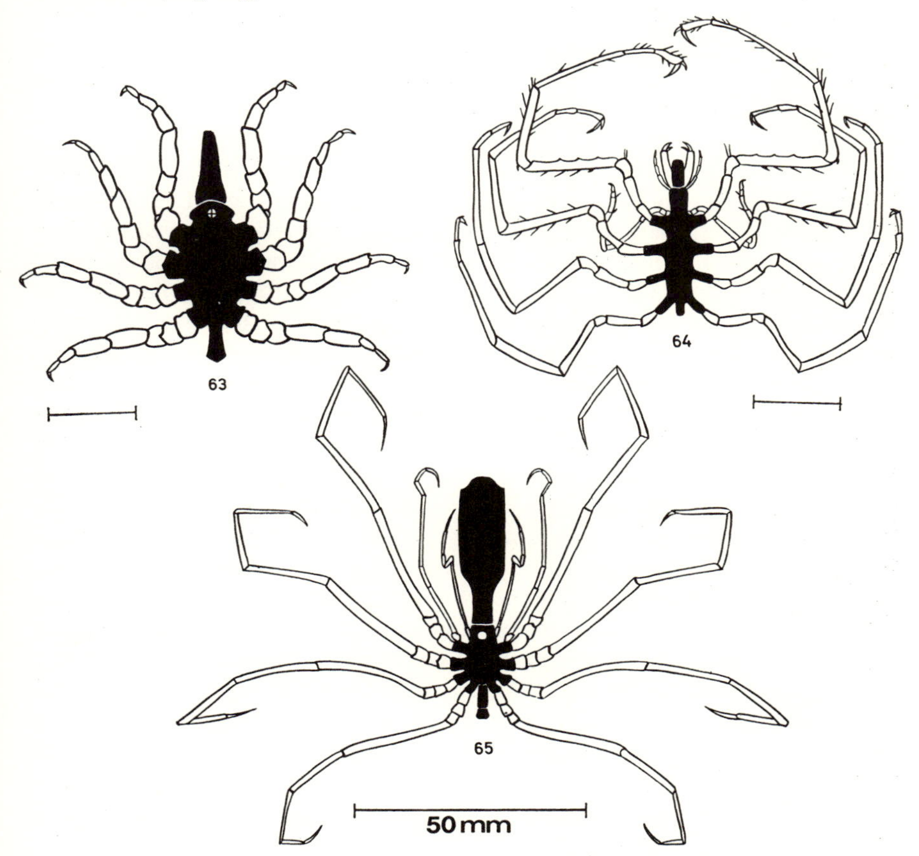

FIG. 63. *Pycnogonum litorale.* FIG. 64. *Nymphon grossipes.*
FIG. 65. *Colossendeis proboscidea*

(2) *Nymphon* type. The slender body has long and thin limbs with weak claws adapted to hold on to swaying, branched colonies of hydroids. The proboscis cannot pierce firm, muscular walls. The animals possess auxiliary limbs (maxillas and palps) for feeding. They live on reefs in the shallow sea. This type of pantopod is capable of swimming (beating with locomotory limbs while drifting in water currents) and occurs in the North Sea (Fig. 64).

(3) *Colossendeis* type. The limbs of these animals are long and thin with weak or no claws. They are adapted for locomotion on mud, especially by the way in which the distal segments of the legs are folded back. The thick proboscis is at least as long as the thorax and is not suited to drill into muscles of coelenterates. The animals have auxiliary limbs (maxillas and palps). Their food, probably sediment, is taken up by the proboscis, and their preferential habitat is therefore muddy ground in deep water, sheltered from currents. This type does not occur in the southern North Sea (Fig. 65).

Death and places of burial. The pantopods that live in the shallow southern North Sea rarely touch the sediment surface itself. They could not resist the impact of drifting sediment particles, nor could they counteract the burying effect of even light sedimentation or hold on to the ground when the currents increase. These forms generally live therefore on hydroid polyps and *Tubularia* stalks or on firm substrates which carry the larger actinians (for example *Sabellaria* reefs, sponges etc.). Preservation in such places is not possible, even for the quite resistant exoskeleton of the *Pycnogonum* type. When pantopods (Pycnogomorpha and Nymphomorpha) which live in the North Sea are seen far away from hard substrates, they are lost, drifting animals outside their own biotope. In such an event *Pycnogonum* spreads its eight limbs radially in an attempt to keep from sinking into the sediment. *Nymphon*-type pantopods in a similar situation are able to progress slowly on fine muds while the long limbs keep the body high above the ground; nevertheless, they also try to reach the dense settlements of hydroids, the sponges or bryozoan reefs where they can hold on and are protected from bottom currents and sedimentation. The chances for preservation in fine mud are good for animals with hard armour; but few of them are found as fossils because most tend to live on hard substrates (Helfer 1933). Where they are found in the sediment other drifted animals are usually associated (dead hydroid polyps, *Tubularia*, fragments of sponges and of other sessile organisms; see Helfer 1933).

The third type, the one that does not occur in the southern North Sea, should preserve well. The animals are large, have a hard exoskeleton and live in fine-grained sediments. A photograph of the sea floor (Owen 1958) south of Cape Cod (Massachusetts), shows a 60 cm long pantopod of this type which crawls between brittlestars on a mud surface full of traces.

(*b*) OSTRACODA

Ostracoda and Cirripedia are two crustacean groups which provide preservable hard structures. All remaining classes of Entomostraca need extraordinary conditions to be preserved since their exoskeletons are fragile. The cirripedes are restricted because of their sessile way of life; the ostracods, on the other hand, are widespread, and almost every type of sea floor is inhabited by certain species of ostracods. Generally, the conditions are conducive to these small crustaceans. The same was true of earlier sea floors, and ostracods are therefore particularly important in micropalaeontology.

Ostracods have a solid, heavily calcified carapace consisting of two valves that enclose the simply shaped, unsegmented body. Only the valves are strong enough to be well preserved (for anatomical details see G. W. Müller 1894 and 1912, Triebel 1941, Hartmann 1955, Mathes 1956, Pokorny 1958).

Modes of life and habitat. There are six habitats in the North Sea where ostracods live: the interstitial spaces in sand, sandy, soft muddy, and hard sea floors, plants, and the water. The regional distribution of ostracods and their way of life are not well known, and only a little of their ecology is understood.

(1) Only very small species of ostracods live interstitially between sand grains; where the sand is rich in detritus and small shells they occasionally occur in large numbers. According to Remane (1940) the most important genera of this type are *Psammocythere*, *Microcythere*, *Microxestolebris* and *Parapolycope*. The animals seem to move about in the interstitial cavities by climbing and bracing themselves against the sand grains. Hartmann (1955) assumed that the polycopids use not only their second antennae but also the furca as locomotory organs. The valves of many of these animals are flat ventrally.

(2) Few species live on sand surfaces (Klie 1929); they are *Polycope orbicularis*, *Hemicythere emarginata*, *Leptocythere lacertosa*, *L. baltica*, *Cytherideis foveolata*, *Cytherura sella*, *C. atra*, *C. bidens*, *C. acuticostata*, *C. clathrata*, and *Cytherois arenicola*. Even these species are shallow burrowers and can sustain themselves in the oxygen-rich water between the sand grains. Locomotion on the surface of the sediment consists mostly of short jumps or intermittent swimming. When resting the animals close their valves and lie on their side. Those that possess strong outer branches of the second antennae can use them for lateral support. These antennae are also the main locomotory organs. According to Klie (1938) *Microcythere affinis* seems to prefer coarse sand. Several forms occur in sand as well as in mud; they are *Cyprideis litoralis*, *Leptocythere pellucida*, *L. ilyophila*, *Cytheridea papillosa*, *Sclerochilus contortus*, *Cytherura nigrescens*.

(3) Remane (1940) listed the genera *Kyphocythere*, *Cytheropteron*, *Bythocythere*, and *Macrocythere* as occurring in the soft muds of the North

Sea. Some have long limbs that can be spread sideways; others are provided with thorny appendages and spines on their valves as *Cythereis dunelmensis* and *C. jonesii.* According to Moore (1931b) and Krogh & Spärck (1936) some bury themselves several centimetres deep in the mud (e.g. *Cythereis*, *Cytheridea*, and *Paracypris*; see a compilation by Klie 1929). The second antennae and the furca seem to be used for digging.

(4) A few ostracods (e.g. the genus *Cythere*) live on hard ground, either rock or the so-called secondary hard grounds of Remane's (1940), such as submarine shell deposits, groynes, stones, and wooden stakes.

(5) Representatives of the genus *Cytheridea* mostly live in the vegetation of the coastal water; they stay firmly attached to the thalli of algae even in strong currents. These forms do not swim but crawl slowly on the plants. Remane (1940) assumed that the genera *Paradoxostoma* and *Cythereis* which live on plants feed by sucking plant juices. Klie (1929) gave the following list of species: *Erythrocypris mytiloides*, *Cythere lutea*, *C. viridis*, *Leptocythere castanea*, *Cytheromorpha fuscata*, *Hemicythere villosa*, *H. angulata*, *Cytherura gibba*, *C. undata*, *C. cellulosa*, *Loxoconcha impressa*, *Xestoleberis aurantia*, *X. depressa*, *Sclerochilus contortus*, *Cytherois fischeri*, *C. pusilla*, *Paradoxostoma variabile*, *P. pulchellum*, *P. abbreviatum* and *P. normani*.

(6) Only few species are able to achieve a pelagic mode of life and to swim freely for any length of time. To keep afloat they constantly move their first and second antennae; a drop of oil in the body of some species helps to make them more buoyant. As soon as they cease moving the animals sink. Free-swimming animals are therefore mostly encountered in areas with strong currents (*Conchoecia elegans* according to Ostenfeld 1931). *Philomedes globosus* rises into surface waters for mating; the females live only in the early summer as plankton. Some Cypridinae and *Sarsiella capsula* are more frequently seen swimming.

Attempts have often been made to find a relationship between valve shapes and life habits of the ostracods; the shape of the valve would then enable us to interpret function and ecology in the fossil ostracods (see Remane 1940, Elofson 1941, Pokorny 1958). Evidently, if reliable relations between function or environment factors such as salinity and morphology could be demonstrated, the information obtainable from the shape of fossil ostracod valves would indeed be significant. However, these attempts and comparisons only show that the valves of sand dwellers are on the average the shortest, and those of animals living in soft ground are the longest, while those of inhabitants of algae are intermediate; the sand inhabitants have the narrowest valves, inhabitants of soft ground have broader valves, and the animals living on algae have the broadest valves. Even from these simple rules there are too many exceptions to apply to fossils.

Clear relationships between structure and function of the body, desirable as they may seem to the palaeontologist, cannot exist in ostracods. Life on the sea floor or at the bottom of fresh water requires that the body of the animal must not be too narrowly specialised; the animals must be able to cope with varying sedimentation with rapidly varying grain sizes. Animals that can swim, dig, or walk over the sediment surface as circumstances demand are best adapted to such a varying environment. They are not tied to a specific habitat by a single, preferred way of locomotion.

The body structure of the ostracods is excellently suited to fulfil these requirements. The hard, bivalved carapace has a dorsal hinge and can be closed around the primitive arthropod body. The animal propels itself by rhythmic pushing or rowing movements of its outstretched limbs. All known Recent and fossil ostracods have or probably had this type of locomotion. The variations in the longitudinal and transverse diameters of the valves do not decisively influence the locomotion of the animals. Where specific locomotory adaptations are found—such as oil drops in some pelagic forms—the shape of the valve is unchanged. Probably, the other, often striking features of the valve such as the sculpturing of the dorsal and hinge margins, the distribution of pore canals, the bristles, the sculpturing of the valve surfaces with spines, thorns etc. have no bearing on locomotion but seem to be connected, as far as presently known, with nutrition, protection, and reproduction. In ostracods as in lamellibranchs with their similar symmetry, proportions of valves, sculpturing and shape of the hinge do not enable us to determine life habits or habitat.

Death and burial. Sedimentation is a problem for ostracods as for all benthonic animals. Many forms can dig through layers of up to 8 cm of fresh sediment. The oxygen-rich water of the sea floor which fills the interstices of the sand permits the animals to breathe while they dig upward slowly and intermittently. Mud is less permeable than sand; therefore the animals can only dig through thin layers, and rapid mud sedimentation is fatal. During an average natural lifetime of one year (Klie 1929), a marine ostracod living at a depth of approximately 15 m is in danger of being killed by covering sediments nine to eleven times whenever the wind rises to force nine or above: but many species can escape by swimming. The distribution and density of populations of benthonic ostracods probably depend largely on how well they can cope with moving sediment.

Death due to natural causes is probably much more frequent than accidental death, because the natural life span is short, and generations follow each other rapidly. Krógh & Spärck (1936) counted in Danish waters 600 to 1600 dead ostracods per square metre at a depth of 17 m.

Moore (1931b) reported 125 ostracods per 100 cubic centimetres of mud sediment. Thus the number of preservable carapaces from adult animals that are constantly released into the sea is considerable. Occasionally fish faeces contain many ostracods; adolescent, benthonic fish are the main consumers of ostracods. Scott (1902) found ostracods in the stomachs of the following fish—*Lumpenus lampetraeformis*, *Melanogrammus aeglefinus*, *G. merlangus*, *Callionymus maculatus*, *Onos mustela*, and *Drepanopsetta platessoides*.

It can be assumed that dead ostracods are rarely concentrated and only within small areas, because the hard parts are similar in weight to the surrounding sediment particles. If fragments of lamellibranch shells are frequently stirred up by shaking, ostracod valves slip into the gaps between the shell fragments and collect at or near the base of the shell deposits. In mud deposits the ostracod valves can be filled with solidifying mud (of the type that may also give rise to internal moulds) and thus become heavier than the surrounding fine mud flakes; this difference in weight aids accumulation by sorting.

Gocht & Goerlich (1957) reported that they used dilute hydrochloric acid on calcite-filled, fossil ostracods and could recover chitinous remnants of the skeleton other than the carapaces, mainly limbs and segments of limbs. If this method could be applied generally, the resulting knowledge of chitinous skeletons could lead to a reclassification of fossil ostracods.

(*c*) CIRRIPEDIA

Several suborders of the Cirripedia are common in the North Sea. The Lepadomorpha (gooseneck barnacles) and the Balanomorpha (barnacles) as bearers of calcareous valves are especially important. Both are sessile forms, and their structure and way of life are therefore different from that of other crustaceans.

The first cirripedes are found in the Early Silurian. The Lepadomorpha are first known from the Late Jurassic, the Balanomorpha appear in Cretaceous times but become frequent only in Late Tertiary.

(i) *Balanidae*

Of the Balanomorpha the family Balanidae is especially important. After several pelagic stages the larva of *Balanus* (barnacle) attaches itself to a suitable, solid object for the rest of its life (Kühl 1950b and 1952b).

Structure and life habits. The structure of the crustacean body is entirely adapted to a sessile life.

(1) Since the animals do not move, parts of the exoskeleton protect the animal from predators, mechanical forces, and chemical effects. A skin-like mantle developed from a double fold of the carapace surrounds the head, the six sections of the thorax, and the stump-like, unsegmented abdomen. From local calcification in the cuticula of the mantle the calcareous shell is secreted. While most crustaceans moult, barnacles do not. The outer layer of the mantle which carries the calcareous shell does not cast off its skin, but the body inside grows in the same way as that of other crustaceans, forms new skins and sheds them. These skins (exuviae) are discarded through the upper aperture of the shell, the orifice. The calcareous armour grows by accretion on the individual plates of the armour. They can therefore never coalesce completely but remain

FIG. 66. Armour plates of *Balanus*. Basal plate, six ring plates, four cover plates. (Actual size × 4)

mobile with respect to each other although ensuring a tightly closed and strong shell at all times (for a discussion of the calcareous shells of barnacles, their growth and reconstruction with changing conditions, see Gutmann 1960, 1961 and 1962).

The larva in the cypris stage attaches itself by secretion; ducts lead from cement glands to the somewhat broadened, second-last segments of the anterior antennae of the head. Six calcifying zones in the cuticula of the mantle join to form a ring (Fig. 66), shaped like a truncated cone. In many species it sits on an approximately circular, calcareous basal plate. In its centre the cypris larva originally fixed its antennae. In some species the basal plate is missing. The orifice, the upper circular aperture of the shell ring, can be closed by two mobile pairs of plates (the scuta and terga).

(2) In adaptation to sessile life, the food (plankton organisms of the surrounding water) is swept toward the mouth, grasped and sieved by

the legs, and swallowed (Nilsson & Cantell 1921–2). The thoracic limbs that once had locomotory functions have now turned into feeding organs. Their numerous segments are now longer, the claws are split. By forming a hemispherical basket outside the orifice these legs act as a sieving mechanism. For feeding they are stretched out through the orifice, and with rhythmic, wide sweeping movements they can create a water current that carries the plankton toward them.

(3) The lack of mobility has another consequence; the animals cannot approach each other for copulation. The cirripeds, as crustaceans, have retained internal fertilisation while other sessile animals (coelenterates, annelids etc.) shed sperm and normally also eggs into the water for subsequent fertilisation. The barnacles (they are hermaphrodites) transmit sperm by means of a penis to another member of the species. The

Fig. 67. Small clusters of *Balanus balanus* settling on shell

penis is normally rolled up ventrally at the vestigial abdomen, but can be extended into the orifice of a neighbouring animal. The individuals of the settlement must sit close together to allow neighbouring animals to fertilise each other. It is evidently important for the future of the animal and for the existence of the whole species where the cypris larva settles. It must be such that the future sexually mature animal will be within reach of other members of the species. The behaviour of cypris larvae in choosing their places for settlement has long been known and was considered to be for the immediate benefit of the larvae themselves and ease of attachment (hard and clean substrate, moisture even at low tides, abundance of food, certain position relative to the surface of the sea etc.). The advantages of certain locations were thought to lead to crowded settlements. A single wooden post, pebbles in a tidal flat, valves and shells and stranded branches are certainly objects that attract the larvae (Caspers 1949b) because they are the only places for settlement in an unfavourable environment. However, a close inspection of the settlements reveals that the individual barnacles always sit crowded together wherever they occur, even if the substrate is not completely covered by them, and if the same conditions prevail widely. This behaviour is particularly striking in species whose members never live in large settlements (Fig. 67). Only a small settlement will inhabit the valve of a lamellibranch, or the exoskeleton of a crustacean, for instance. If there are only two animals, they sit together. One rarely sees single animals that sit alone as mature animals because they have chosen 'wrongly' when they were larvae.

Since the barnacles require densely crowded settlements for their existence, the often described modified growth forms of the shell can be seen in a different light. The high barnacle shells (Fig. 68) that occur frequently in *Balanus balanoides* and in *B. crenatus*, less so in *B. improvisus*, is not just a chance adaptation forced on the individual by a crowded environment as Ulrich (1927) and many others assume. Instead, high

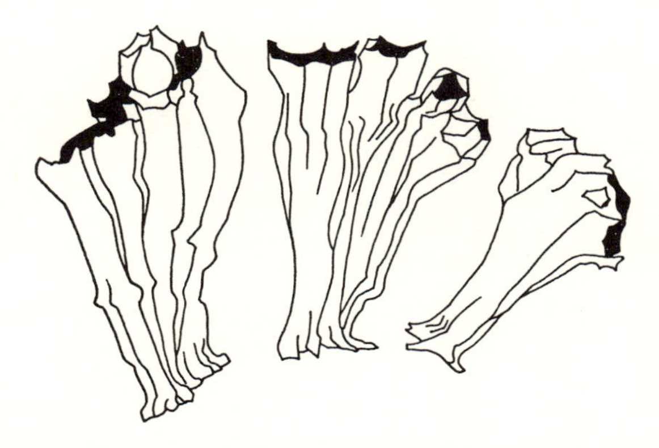

FIG. 68. *Balanus crenatus*. Individuals from closely crowded colony with tall growth.

shells are a meaningful potential adaptation which does not always come into play but is useful whenever very dense settlement prevents the growth of the normal, broadly conical shape. As long as the single individual is exposed to external mechanical forces from all directions, this conical shape is important and advantageous. Its importance decreases as the individuals live closer together and support and protect each other by growing upward. The armour of such animals is also much thinner than that of those with a conical shape. For a discussion of the *Balanus* settlement as a community, see Schäfer (1956d). The larvae thus settle for present and future needs. Vertical growth guarantees survival of many individuals in crowded conditions.

The ability to grow in a certain orientation is another means of coping with unfavourable conditions and is necessary in areas of sedimentation.

Oriented growth starts from the originally conical shell; it is controlled by the individual, not by the colony. Those animals settling on objects partly embedded in sediment frequently grow upward, for example in barnacle settlements on the valves of living *Cardium edule* that are partly buried in sediment, on the shells of living *Mytilus edulis*, or of snails that live in the sediment, or on rocks protruding from the sediment with walls exposed to deposition. However, barnacles settled on other vertical rock walls, not exposed to sediment, as on the side of a mole, do not show upward oriented growth; the axis of the conical shell forms a normal right angle with the substrate. Only sediment approaching from the side causes the orifices to turn upwards (Fig. 69).

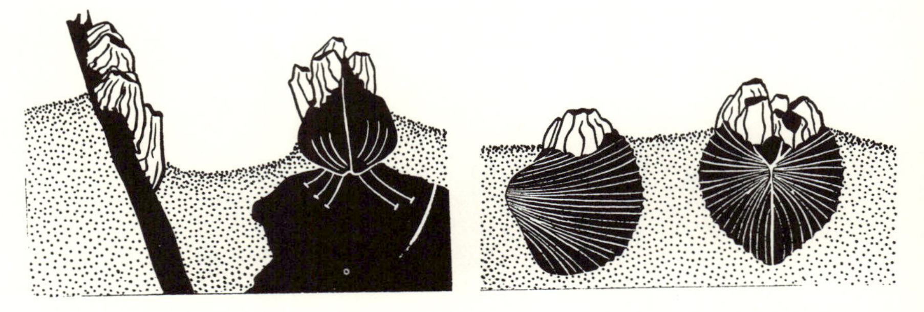

FIG. 69. Oriented growth of balanids on stones, planks and lamellibranchs.

By giving up locomotion altogether the crustacean has lost the ability to escape from extensive, rapid sedimentation that occurs frequently in the shallow sea where habitats for barnacles are limited. Two biotopes remain: surfaces on mobile animals which avoid sedimentation for their own sake, or surfaces on hard objects in areas free from sedimentation. Such areas lie either under deep water where sedimentation occurs rarely, or in regions where strong currents prevent sedimentation (*Verruca strömia* on rocks), or in surf areas wherever there are firm substrates. In the coastal area of the southern North Sea such firm substrates exist only since man has provided them in the form of groynes and sea walls. As a result, several species have become common in the last two hundred years. Therefore the composition of the plankton during May and June when the larvae of the barnacles swarm, differs from that of former times when conditions at the coast were natural.

Four species of Balanidae are common in the southern North Sea and its coasts. *Balanus balanoides* lives in the tidal zone, from the line of spray to below the low-tide mark. This barnacle settles mostly on groynes made of stone, wood, or concrete, rarely on organic calcareous material, almost never on shells of its own species (Schäfer 1952). In crowded settlements this species may grow to a height of 3 cm. The larvae begin to settle at the end of May. The shells grow conically until they touch each other. At this point the height of the shells increases almost equally in all animals of the colony. Occasionally, extensive colonies show regularly distributed bulges and swellings; the animals at the centre are the tallest while the surrounding animals remain slightly smaller. The animals between the bulges are the smallest. The diameter of such bulges is approximately 10 cm. The bulges are probably formed in order to increase the surface area available for the apertures (Trusheim 1932, Schäfer 1948).

Balanus improvisus lives mostly below the water line at low tide (only up to 60 cm above this line). This species lives on all substrates that may be available, on valves of lamellibranchs, shells of snails, crustacean armours, wood, rocks, algae, leather, metal, and plastics (Kühl 1950b). It rarely settles on calcium carbonate secreted by its own species. It usually forms small colonies of five to eight animals.

Balanus crenatus avoids regions with low concentrations of salt and lives therefore only on the sea coasts. It settles above and below the water line at low tide, frequently on the shells of its own species. In spite of that it is capable of tall growth but colonies of individuals sitting above and below each other show a knobbly, disorderly growth.

Balanus balanus is the largest species of the southern North Sea with cones up to 2 cm high. It lives only below the water line at low tide and in the open sea at depths of more than 10 m. It grows mainly on living snails (*Neptunea*, *Buccinum*) and bivalves (*Cyprina*, *Modiolus*) or on coarse shell deposits (*Ostrea*) which cover the sea floor. These barnacles form small colonies of three to forty individuals. The animals sit frequently, and even preferentially, on top of each other, but they do not overgrow the orifices of the animals below them.

Elminius modestus is a barnacle that has invaded the southern North Sea since 1945 (Kühl 1954, with bibliography). This species from Australia and New Zealand was first noticed by Bishop (1947) in British waters. The advance of this species through the North Sea is now well known. In 1955 the first specimens appeared in the area of Jade Bay. Within the following three years they multiplied so much that they are now the dominant species. Their ring-structure consists of only four plates, unlike all other species of that region. The base is not calcified. This barnacle accepts all available substrates in the tidal area such as wood, stone, metal, and calcareous shells.

Verruca strömia lives near the island of Heligoland on submarine screes and on large shell fragments. The shell is asymmetrical.

. . .

Causes of death. Natural death is rare among barnacles living in the tidal sea because most individuals are destroyed long before they reach their potentially maximum age.

Ice drifting near beaches and on tidal flats, as it does almost every winter, rubs most barnacles from their substrates (pictures of drifting ice in the area of Jade Bay in Reineck 1956). Only in mortar joints of walls and in sheltered gaps between rocks are the animals protected from the scraping ice and continue to live for a second and third year.

Balanus balanus which lives at depth does not suffer from drift ice; possibly a larger proportion of these animals live for several years and die naturally.

Balanus balanoides is likely to be attacked from the end of its first year of life with increasing intensity by algae that bore into the calcareous shell (Schäfer 1938b). The colour of the colony turns to greenish-grey. The healthy, new growth zone of the shell is still white and smooth and contrasts strikingly with the older, algae-affected and corroded shell material. Kölliker (1860) was the first to report drilling algae in *Balanus balanoides*; Cruvel (1905) and Cotten (1912) later confirmed the observation. Parke & Moore (1935) noted as drilling organisms the green algae *Gomontia polyrhiza* and *Ostreobium queketti* and the blue-green algae *Hyella coepitosa*, *Plectonema terebrans*, *Mastigocephalus testarus*, and *Microchoete grisea*. Although even advanced corrosion does not pierce the thick plates or destroy them completely, slight damage to the margins of the plates, especially of scutum and tergum, prevents a tight seal of the shell at low tide. The animals thus suffer from loss of water and dry out during that period, particularly in sunshine. While they are under water, however, animals with damaged shells are vulnerable to attacks from starfish and periwinkles (*Littorina*) which live in barnacle settlements and feed almost exclusively on these animals (Schäfer 1950a). Therefore *Balanus balanoides* rarely reaches an age of three or four years. *Balanus improvisus*, living in the same biotope, is generally free from boring algae; *Balanus crenatus*, which lives in water of higher salinity, and *Balanus balanus*, a deep-sea dweller, are also unaffected by algae.

If the substrate of a barnacle colony is a living animal, the settlement—advantageous as such a habitation may be—is exposed to the same dangers as the host. Bivalves and snails that live on the surface of the sediment as well as large crustaceans (*Homarus*, *Cancer*, *Carcinides*) are the most frequent hosts of barnacles. Healthy crustaceans seem to be able to free themselves from larvae of barnacles that are trying to settle on them. If crustaceans (*Carcinides*) are affected by the parasitic cirriped *Sacculina carcini*, the carapaces have usually a particularly dense population of settlers such as *Balanus improvisus* or *B. crenatus*. If the host stays alive the barnacles can live there for several years, because *Carcinides*, once it is affected by *Sacculina*, is unable to moult. When settling on brachyurans, the barnacles prefer to attach themselves to the frontal parts and margins of the carapace because only these parts of the body remain outside the sediment when the host is burrowing. *Balanus improvisus* seems to be particularly adapted to the conditions in the open tidal sea and its floor where there is often danger of being temporarily covered by sediment. Thus *Balanus improvisus* can be found sitting on top of *Cardium* settlements whose siphonal end hardly protrudes from the sediment (Fig. 69). Upward oriented growth of *Balanus improvisus* occurs also on drifting or anchored wooden branches; barnacles that sit on the sides or below have an asymmetric structure. *Balanus balanus* can grow in a similar way, and also occurs on shell accumulations, living snails (*Buccinum* and *Neptunea*) and bivalves (*Modiolus*). When the host dies the settled barnacles tend to die, too, because there is no protection from the dangers of sedimentation.

Where barnacles live on inanimate substrates, colonies are frequently covered by sediment, especially if they are newly established. If the sediment layer is thin, the barnacles succeed for a time to keep a channel to the surface of the sediment open by beating with their limbs. Heavy sedimentation is fatal.

Disintegration of the armour and formation of shell deposits. If the barnacles are not buried alive, the shell begins to disintegrate soon after death. The various species of barnacles react differently during the period of disintegration (the destruction of calcareous shells by boring algae is discussed by Kolosvary 1942).

The plates of *Balanus balanoides* are firmly attached to the substrate although in the absence of a basal plate they touch it only at the narrow margins of the ring plates. Strong sideways pressure can either detach the entire ring of plates from the substrate or separate the individual plates of the ring. No trace of the former settlement remains on the substrate. The animals in a complete and generally large settlement often die from the same causes, and almost simultaneously, because these settlements are, as a rule, composed of animals of the same age; they thus provide great quantities of shells. 150 cm thick shell deposits have accumulated at the groynes of the main dam in Jade Bay where almost every year the majority of barnacles is killed by boring algae or by ice. These shell deposits consist almost exclusively of *Balanus balanoides* plates with only a few admixed, broken valves of *Mytilus* from the same area. The coarser shell deposits contain more *Mytilus* fragments, the finer deposits almost none. This is due to the greater resistance of the *Balanus* plates. The plates show traces of intensive abrasion on the convex side; the protected, inner sides, however, keep their fine sculpture for a long time.

After the death of the animal the individual plates of the ring of

Balanus improvisus separate. The circular, radially ribbed, basal plates remain fastened to the substrate and remain for some time as evidence of a former colony. The basal plates may be buried with the substrate. If they sit on crustacean exoskeletons they reinforce these delicate structures. The calcareous plates of *Balanus improvisus* are widely scattered, and no deposits consisting only of plates of this species are found because the animals settle so frequently on mobile hosts. Parts of the shells can be found in all sediments of the tidal flats and in its deep channels.

Balanus crenatus, like *B. balanoides*, is inclined to be tall. Frequently, young *Mytilus* settle on colonies of *B. crenatus* and cover large areas with their byssal threads. After disintegration of the barnacles and their separation from the substrate, the thin, elongated ring plates of *B. crenatus* are held together by the byssal threads and form matted bundles of calcareous splinters; these bundles may be transported and buried. Individual plates are rapidly destroyed because they are so thin. The resulting minute shell fragments become mixed in the coastal sand and lose their identity.

Balanus balanus shells lose the mobile, muscle-held terga and scuta after the animals have died; but the plates of the ring and the basal plates generally remain fastened to the substrate. Only a sharp impact can loosen the ring plates from the calcareous base. A plate often breaks before it separates from the neighbouring plates. *Balanus balanus* therefore only releases the scuta and terga. Parts of the armour that are transported are subjected to abrasion together with the accompanying sediment (mostly shells of *Buccinum* and *Neptunea*, and valves of *Modiolus* and *Cyprina*) and are finally embedded with it.

(ii) *Lepadidae*

The Lepadidae (gooseneck barnacles) are sessile crustaceans like the balanides, and their body structure is adapted to this way of life.

Structure and life habits. The free-swimming larvae attach themselves to a substrate using the secretion of a cement gland at the first antennae of the head. The elongated body of the adult animal lies in a bilaterally symmetrical mantle which is derived from the cuticula; the individual calcareous plates are secreted within the mantle. This almost enclosed body section (capitulum) sits on a strong, flexible stalk (pedunculus, developed from the upper region of the head) that is enveloped in skin and can be moved passively. The broad base of the stalk is fastened to the substrate (Fig. 70). For feeding, as in the Balanidae, the elongated thoracic limbs protrude far through the slit of the mantle and have biramous claws. When the limbs are spread out they form a sieving basket which is swept through the water to catch plankton; the food is then passed to the mouth parts inside the mantle.

In contrast to the Balanidae, the Lepadidae live in the open sea. They do not survive in water of even slightly subnormal salinity. Their mode of life is sessile-pelagic; they settle preferentially on floating or actively swimming animals. Driftwood is a favoured raft. Today, colonies of gooseneck barnacles are found mainly on flotsam, so abundant along the main shipping routes (see also Ankel 1962).

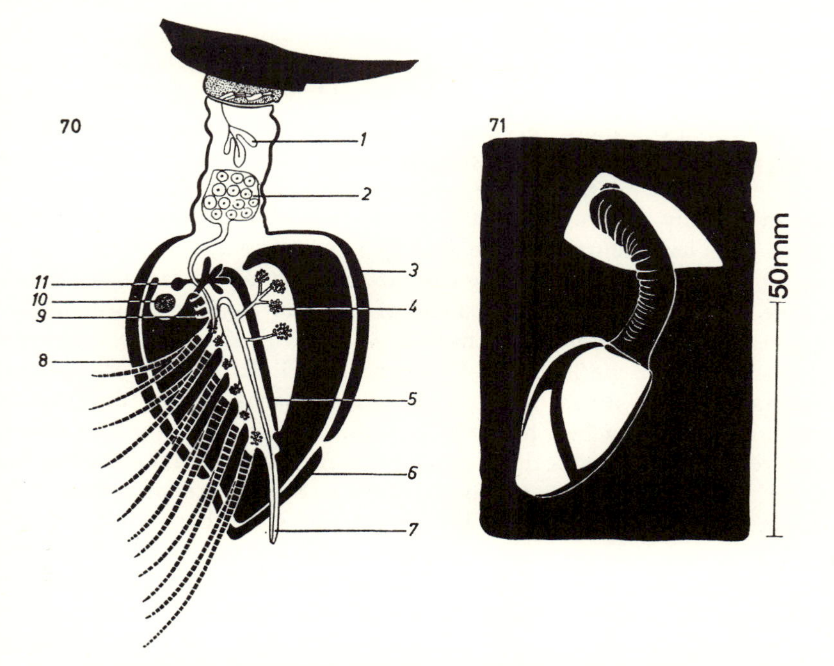

FIG. 70 (*left*). Schematic cross-section of a *Lepas* (after Remane, 1940). (1) cement gland; (2) ovary; (3) carina plate; (4) testis; (5) intestine; (6) tergum plate; (7) penis; (8) scutum plate; (9) jaw; (10) closing muscle; (11) brain.

FIG. 71 (*right*). *Lepas anatifera* attached to calcareous plate of its own species. After the carrier-animal's death the plate with attached animal drops to the sea floor

Species of the southern North Sea. Lepas anatifera is the most frequent species in this region; *L. hilii*, *L. anserifera*, and *L. pectinata* are rare. They differ mainly in body size, length of stalk, and in the shape and consistency of the shell sections. All settle on drifting material. *Lepas fascicularis* which also occurs in the southern North Sea is the only species which can swim by using a bubble-raft (Ankel 1950). First, it attaches itself to floating seaweed using the secretion of a cement gland; then it begins to form at the stalk a gas-inflated raft that increases gradually in size and eventually can carry the barnacle when the seaweed has died.

Death, disintegration and burial. The sessile-pelagic way of life encourages a wide distribution of the species (*Lepas anatifera* and *L. fascicularis* are cosmopolitan) but diminishes the density of populations. Therefore *Lepas* shells are never found as accumulations in shell beds on the sea floor. The individual skeletal parts are not firmly attached to each other. After death they therefore drop individually to the sea floor. They are of great stratigraphic significance because the animals die in all parts of the sea, and their hard remnants are widely distributed. The animals can be buried complete only under special circumstances: if for instance living animals reach the sea floor and are quickly buried, or if they are stranded. Wood floats for only a limited period of time, and branches or logs eventually sink to the bottom together with any epizoans settled on them. This has occurred in the geological past and also happens today in the North Sea. Stranded wood with attached lepadids can also be found on the North Sea coasts. At the end of October 1958, great quantities of seaweed and driftwood, all densely covered with living *Lepas anatifera*, were stranded at the island of Wangerooge (H.-R. Henneberg, personal communication). Burial of such gooseneck barnacles on driftwood in swash marks of beaches, however, is unlikely, and burial on the sea floor is therefore more significant. The animals are not always fastened directly to the driftwood but may also be attached to the calcareous skeletal plates of other gooseneck barnacles. The directly attached barnacles frequently die first, because they are older and also limited in their food intake. After a carrier barnacle has died and a calcareous plate with a living barnacle on it drops off, it takes the living animal with it to the sea floor, the stalk still fastened to the plate (Fig. 71). It is impossible for these pelagic animals to cope with sedimentation. *Lepas fascicularis* floats with its bubble-raft only so long as the animal continues to secrete gas; as soon as the secretion diminishes or ceases the bubble collapses, and the animal sinks to the bottom.

(*d*) MALACOSTRACA

The higher crustaceans (Malacostraca) are known to have existed since the Carboniferous. Not all the species are easily fossilised. Several orders are represented by animals so small and with such delicate exoskeletons that they are preserved only rarely. Of the Peracarida, the Isopoda are significant from our point of view, and of the Eucarida, the Decapoda are most important. Other malacostracans can be preserved (for example the

Stomatopoda) but they do not occur in the North Sea. The Decapoda, as an example, represent a well-defined group, anatomically as well as phylogenetically; however, various subdivisions differ distinctly in their structure, life habits, and in the way they are preserved. Morphological features that are common to all of them are three pairs of maxillar limbs, developed and modified from the most anterior thoracic limbs; the next five pairs of thoracic limbs, the pereiopods, are efficiently used as locomotory limbs. The carapace covers the head and the entire thorax above and on the sides. Little more can be said about the group in general because all other features vary considerably. The distribution of this class is world-wide due to the capability to develop numerous adaptations to different environments.

The suborders Natantia and Reptantia represent two ways of life, mainly swimming, and mainly walking and crawling. The swimming Natantia are all long-tailed and quite uniform in their type. However, the individual groups of the Reptantia (Palinura, Astacura, Anomura, Brachyura, listed in order of increasing complexity of morphology and function) vary widely and lack common morphological and ecological features. It is actuopalaeontologically important that all Reptantia have a resistant, often calcified exoskeleton that can be fossilised, and that their size exceeds that of most other crustaceans. Their chances of preservation are therefore good. No Palinura live in the southern North Sea. The northern border of their habitat is just south of the English Channel.

(i) *Natantia*

The Natantia have an elongated cephalothorax and a well-developed abdomen. Locomotion on a substrate is achieved by ten walking limbs which also serve occasionally for digging. The animals swim by beating with the abdominal limbs (pleopods) while the abdomen is held in a normal straight position. But in escape, the abdomen with its broad, fan-like tail is suddenly folded (Fig. 132). The Natantia frequently form large swarms and they migrate seasonally. Their food consists of small animals, dead organic material, and detritus.

Structure and way of life. The suborder Natantia occurs in two constitutional types with different modes of life.

First type. As long as the animals stay on a substrate they rarely walk; normally they stand high above the ground on their pereiopods, moving only their long antennae and eye stalks or possibly their legs to explore the substrate for food, or for grooming: they never burrow. They are fast swimmers, especially in their backward movements by suddenly recurving the abdomen. The cephalothorax is laterally flattened, the rostrum is often toothed, and the eyes sit on long stalks. The epimeral plates cover the abdomen well down on the sides; frequently the abdomen is slightly bent at the fourth segment. The locomotory limbs are long and thin and set close together, separated only by a narrow sternum. If they

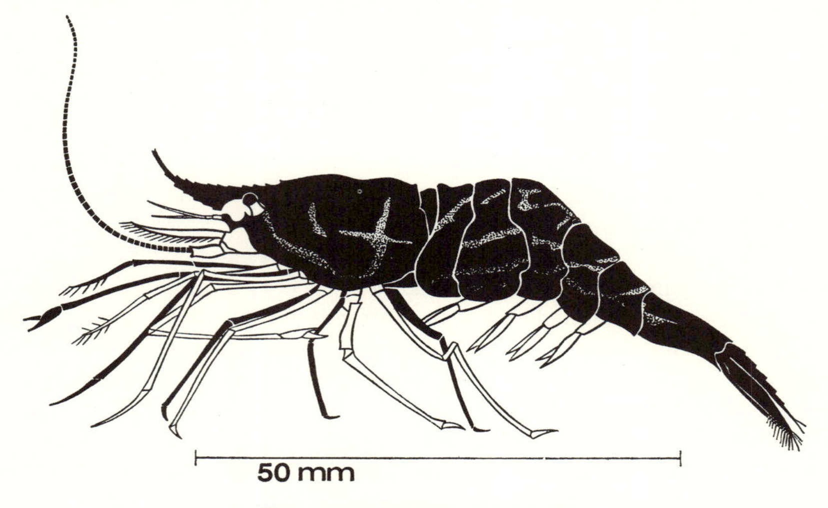

FIG. 72. *Pandalus annulicornis*

have limbs with claws, they hold them folded at the joint between merus and carpus and tucked under the cephalothorax. The claw-bearing limbs are highly mobile and can reach all parts of the body (Figs. 72 and 73). Occasionally (e.g. *Pandalus*) the left and right pereiopods differ in length; this asymmetry indicates that the animals rarely walk and usually rest. The animals are only faintly pigmented and often translucent. If there is any pigmentation it occurs in stripes whose optical effect breaks the

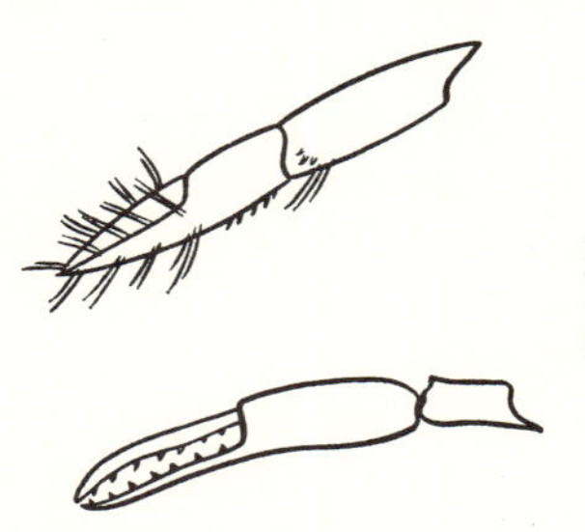

FIG. 73. Claws of *Leander serratus*. Top: first pereiopod; bottom: second pereiopod. (Actual size × 5·5)

continuity of their outline. Examples from the North Sea are the genera *Pandalus*, *Pasiphaea*, *Pandalina*, *Leander*, and *Palaeomonetes*.

Second type. The animals live mainly buried and only swim if they are going to move far. They run rapidly over the substrate. Their cephalothorax and abdomen are flattened dorso-ventrally, the carapace has a short rostrum or none at all. The eyes sit side by side on the upper side of the head and have short stalks. The mouth-limbs are adapted to life in the sediment, the pairs of pereiopods and of pleopods are separated by a broad sternum and thus set wide apart. The claws of the first pereiopods are modified to serve for digging and grasping (Fig. 74). The body is heavily pigmented, matching the substrate in colour and pattern. Examples from the North Sea are the genera *Crangon*, *Sclerocrangon*, and *Pontophilus*.

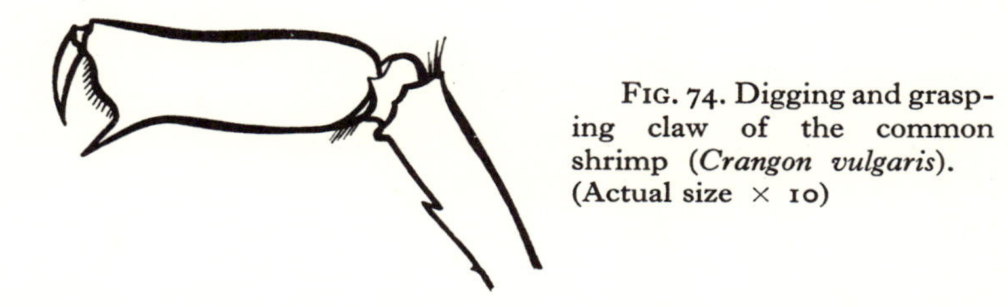

FIG. 74. Digging and grasping claw of the common shrimp (*Crangon vulgaris*). (Actual size × 10)

Death, disintegration and burial. The Natantia with their swimming mode of life have a thin, elastic exoskeleton of chitin. Few Natantia are therefore preserved as fossils compared with the Reptantia; yet, in living populations, the Natantia are usually far more numerous than the Reptantia.

Nothing is known about the causes of death. Most individuals are probably eaten because they are the most frequent and most important food for many species of fishes. Repeatedly, we found fragments of the skeleton of the common shrimp (*Crangon vulgaris*) almost exclusively in faeces of the thorn ray (*Raja clavata*), of the tub fish (*Trigla corax*), of the lump fish (*Cyclopterus lumpus*), and of the striped sea snail (*Liparis liparis*). Meyer-Waarden (personal communication) reports the same fact of the cod (*Gadus morrhua*) and of the whiting (*Merlangius merlangus*). Occasionally pellets disgorged by the herring gull are full of fragments of these Natantia. According to figures from the Bundesforschungsanstalt für Fischerei, Hamburg, the life span of the common shrimp is approximately three to four years. After one year the females are approximately 4 cm long, after two years approximately 6 cm. There are relatively few two-year-old animals. By far the majority of the population and of the catches consist of 1–1½-year-old animals, at the most, two-year-olds. Males are approximately 4 cm long after their first year; they remain considerably smaller than the females. The water temperature has little influence on the adult populations, but it partly determines the arrival and departure of the shrimp in spring and autumn and the duration of their stay on the tidal flats. However, the number and size of the animals of one year is certainly determined by water currents and temperatures during the larva stage. The life span of the other species is not known, but it is similar to that of *Crangon vulgaris.* Complete bodies of Natantia are rarely preserved because the living animals do not become covered by sediment; a shrimp that is buried in sediment leaves its resting place as soon as sedimentation starts. As a rule only exuviae are buried complete. Most of the known finds of fossils seem to be exuviae, because they frequently show a pronounced bend between the cephalothorax and the abdomen, and the carapace is displaced upwards; such positions never occur in carcasses (see also Abel 1935, p. 285, and Fig. 251 of this book).

(ii) *Astacura*

The Palinura, which do not occur in the North Sea, and the Astacura are long-tailed Reptantia whose general organisation still resembles that of the more primitive Natantia.

Structure and mode of life. For animals with a sturdy and heavy armour, walking is the most important means of locomotion. As in the Natantia, the Palinura and the Astacura walk only on the four posteriormost pairs of thoracic limbs; the pair anterior to the locomotory ones (first pereiopods) serves for feeding and protection and is provided with large claws (Fig. 125). In the Astacura, the following two pairs of walking limbs carry small claws, mostly used for grooming. Swimming is restricted to escape movements, produced by suddenly curving the thick abdomen with its broad tail fan; this is the same, highly perfected mechanism as in the Natantia. The number of the abdominal appendages (pleopods) with their branched feet is not always complete. They are shortened and not used for swimming. In the females they carry the eggs and oxygenate them by rhythmic movements. There are only two species in the southern North Sea, *Homarus gammarus*, the common lobster, and *Nephrops norvegicus*, the Norwegian lobster.

The common lobster lives generally on rocks. In the southern North Sea it is found mainly on the submarine sandstone cliffs of the island of Heligoland in gaps between rocks and among boulders. Several tens of thousands of them are caught each year in this limited area (Rühmer 1954). Since stone groynes and other structures to safeguard the coasts have been built (extensively on the Minsener Oldeooge in the Jade Bay area) the lobster has also settled there. The size of its population in these places is not known. The animals also occur in the wide, sandy areas of the open sea floor and are occasionally caught in drag-nets. On the east coast of Britain the lobster is quite frequent. The animals grow rapidly. After twelve to fourteen days of planktonic life they are approximately 16 mm long and descend to the sea floor. After their first year they reach a length of approximately 4 cm, and in consecutive years they reach lengths of up to 8, 12, 16, and 23 cm. Animals 50 to 60 cm long are rare.

The Norwegian lobster is smaller and more slender than the common lobster, and it dwells in mud. Its length can be up to approximately 22 cm. It inhabits the deeper regions of the North Sea, mostly below a depth of 40 m, and does not occur therefore in the southern North Sea, except when individual animals stray south during the cold season. Norway lobster live mainly along the east coast of Britain, near Brittany, in the Kattegat and Skagerrak, and on the coasts of Norway and western Sweden. According to Rühmer (1954) the catches of 1937 were 510 metric tons for Denmark, 475 tons for Sweden, and 153 tons for Great Britain.

Death, disintegration and burial. Young lobsters and Norwegian lobsters are particularly exposed to various dangers. The more rugged their habitat and the more hiding places it provides, the better are the chances of survival for the young lobsters. The young Norway lobsters usually bury themselves deeply in mud. Little is known about the natural causes of their death. Disintegration of the armour, mainly at the claws, is known to occur in old lobsters (and also in *Cancer pagurus* of the brachyurans) and frequently leads to death. Certain areas of the claws turn dark brown and later bluish-black, and the covering skin becomes thin and decalcified. Similar blackening of the armour can be caused by injuries due to fights with members of the same species. In such cases the blackened areas of the armour are circular and are caused by bites or claw points of adversaries. A cross-section of such a damaged area reveals how the epithelium which builds the armour produces a single layer of chitin (at a time other than at moulting) below the damaged zone. The injury is thus temporarily closed off (Schäfer 1954a). With the next moulting a healthy, three-layered armour is formed. If larger areas are damaged, a progressively larger region blackens, and the epithelium proper becomes affected and is incapable of closing the wound. Finally, the armour collapses over extended areas, and the sea water can enter freely into the interior cavity of the claws or the body and washes directly over ligaments, muscles, and nerves. Such animals generally die of bacterial and fungal infections. This type of breaking up of the armour seems on occasion to start spontaneously without an original external lesion. Widespread damage to the carapace prevents moulting, and the animals die, at the latest, when they are due to moult.

The preservation of an animal depends partly on the nature of the carcass itself and partly on its biotope and the medium of burial. While the armour of the common lobster is harder and larger, the Norwegian lobster has the more favourable environment. Soft mud is a good medium for burial. Furthermore there is little current in the deep mud areas. The carcass remains intact because there is no transportation. Complete fossil crustaceans are almost invariably preserved in fine-grained sediment. The biotope of the common lobster is altogether different; boulders below submarine cliffs are rolled and moved by the surf whenever there is some wind. Even the hardest parts of any skeleton are soon broken and what fragments remain are intensely abraded and rounded. If they remain among the boulders they are eventually completely destroyed. However, the surf of heavy gales also removes some of the remnants and transports them to sandy areas at greater depths. In a zone surrounding a scree-covered area, light but hard parts that are easily transported can accumulate in the sand (breccias of lamellibranch shells, gastropod shells, sea urchin tests, and rhizoids of *Laminaria* seaweeds, still attached to pebbles). Hard remnants may be closely packed into narrow gaps in solid rock and may thus be permanently preserved.

Only the hardest and thickest parts of the common lobster can persist among boulders and pebbles and they are often fractured. The hardest parts are the rostrum with parts of the frontal section and the eye depressions; the rostral part of the sternum, the mandibles; of the abdomen merely the thorned ventral plates of individual segments and the basipodites of the two uropods; of the appendages with claws, the short ischium, the mobile dactylus (in particular that from the stronger, breaking claw), some teeth of the claws and spines of the chelae; of the other thoracic limbs, mainly the dactyli and also the coxae.

The number of well-preserved skeletons in sand and mud would be considerably greater, if the carcass were not at once attacked by scavengers, which destroy most effectively segmented skeletons (particularly crustaceans, echinoderms, and vertebrates). Evidently, the best conditions for fossil preservation of animals with segmented skeletons are in those deep parts of the sea hostile to life, although carcasses that reach such places must have been transported—mostly floating—over a considerable distance. Only a small fraction of a population can be transported or actively move into such regions. Unsegmented hard parts, such as skeletons of sea urchins, lamellibranch valves, and gastropod shells can be preserved even in very agitated areas. In the North Sea decapod crustaceans are destroyed and fragmented by polychaete worms, scavenging gastropods (*Buccinum*, *Neptunea*), starfishes, some fishes (*Anguilla*, *Conger*, *Onos*), and mainly, by crustaceans (Isopoda, Decapoda). The activities of the crustaceans are most devastating because they tear and break with their claws, maxillas, maxillipeds, and limbs. A dead lobster is rapidly reduced to pieces because numerous individuals are attracted by the smell of the carcass and take part in the destruction; the parts and fragments of the skeleton are carried away and scattered.

The following hard parts occur frequently as a result of the destruction of Norwegian lobsters by scavengers: whole carapaces together with rostrum but without antennae and eye stalks; complete segment rings of the abdomen; individual segments of the uropods of the tail fan; chelae of the claws, individual dactyli and carpi of the claws, and all coxal segments of the pereiopods, and complete claws of the pereiopods.

(iii) *Anomura*

The Anomura, another group of decapod crustaceans, show some morphological similarities with the Brachyura, although phylogenetically they are probably not very closely related to them.

Structure and life habits. The Thalassinidae (*Callianassa* and *Upogebia*, Fig. 185) are elongated, dorso-ventrally flattened, and have a well-developed abdomen with tail fin; the Galatheidae and especially the

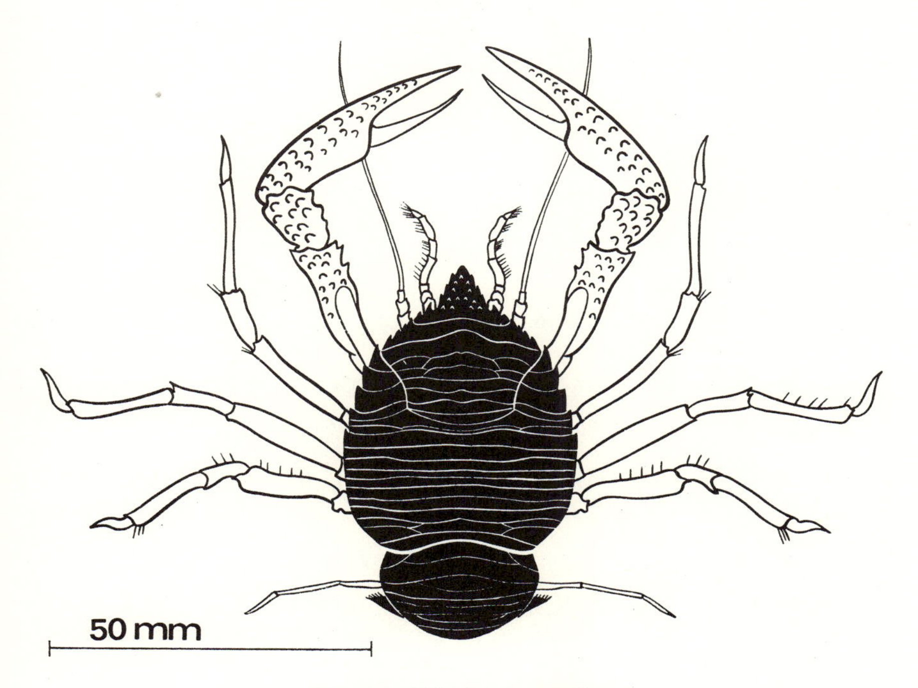

FIG. 75. *Galathea squamifera*

Porcellanidae can fold the abdomen temporarily or permanently under the thorax (Fig. 75). The best known Anomura are the Paguridae (hermit crabs) whose large abdomen is asymmetrical, covered by soft skin, and carried in a gastropod shell (Beurlen 1932). The claws are also asymmetric and differ in size (Fig. 76). The lithodids are similar to the brachyurans as the abdomen is permanently folded; however, the broad abdomen and the claws are asymmetric (Fig. 77). The Anomura crawl on the sea floor or dig in the substrate.

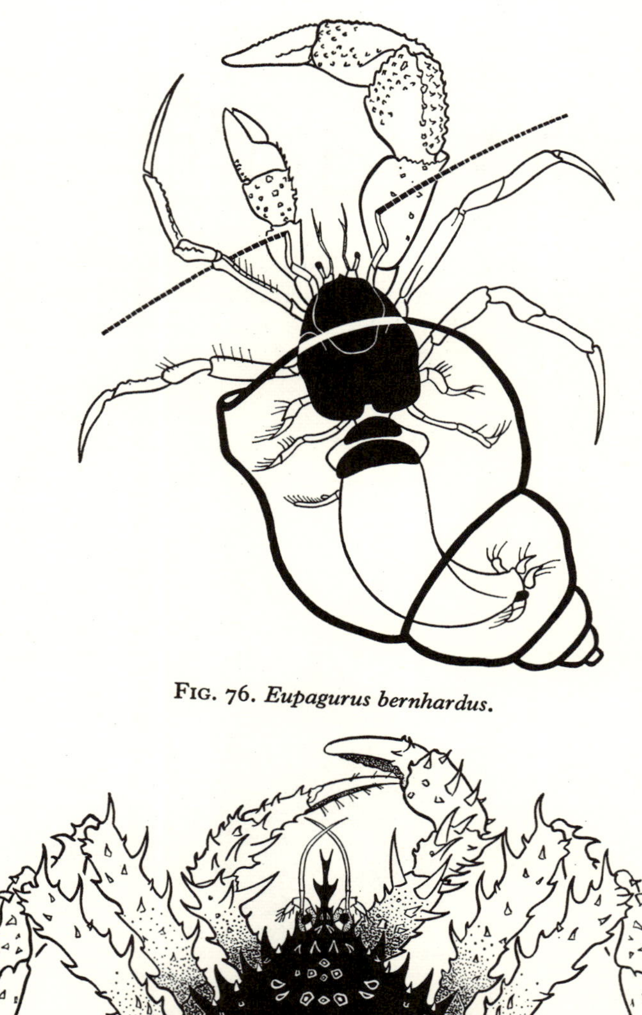

FIG. 76. *Eupagurus bernhardus*.

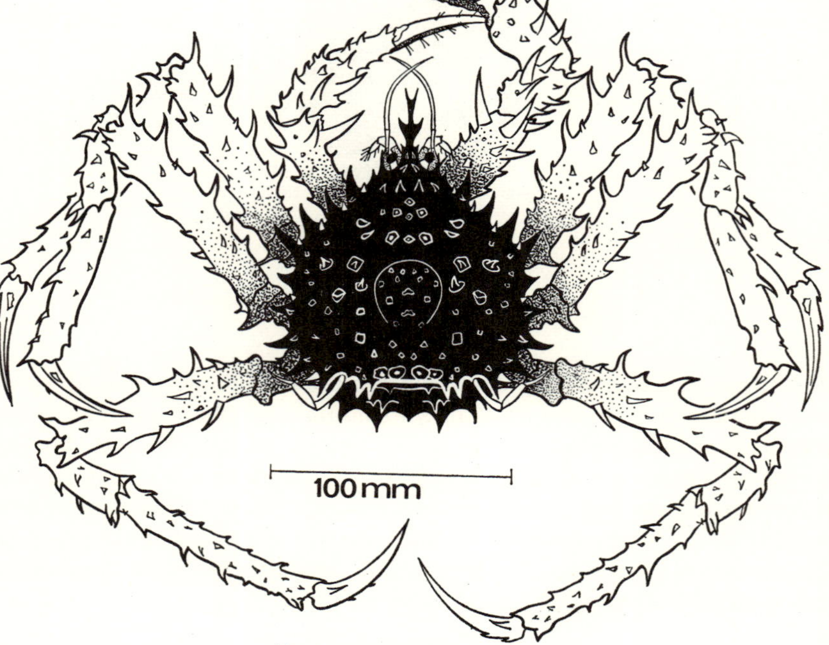

FIG. 77. *Lithodes maja*

Death, disintegration and burial. Some forms of the anomuran crustaceans live in burrows that they dig themselves or that were dug by other animals (Thalassinidae); other forms use the protection of gastropod shells (Paguridae). Both groups have sheltered dwellings which would seem to be suitable for fossilisation; however, these crustaceans leave their shelters in response to noxious chemical stimuli, and (like most marine cave-dwellers) retreat into their burrows only in response to an adverse mechanical stimulus. With approaching death and diminishing activity, the oxygen supply in the burrow or gastropod shell becomes insufficient, and therefore, as a rule, the animals leave their home before they die. The carcasses of these animals are thus generally found outside their shelters. If recent and fossil *Callianassa* claws are frequently found in the burrows of the animals, they are remnants of exuviae and not of carcasses.

Few groups of anomuran crustaceans are preserved. The majority of the Thalassinidae are small animals with soft skins; the Galatheidae are also small, and many of them live in biotopes with pebbles and in shell breccias; only some of the Paguridae have hard armours. Only the Lithodidae, and of the Callianassidae only *Gebiopsis*, are large and have hard, completely closed armour. Conditions for preservation are similar to those for the Brachyura.

The conditions for the preservation of the hermit crab are particularly interesting because they are well known as fossils, and, at the same time, they are the anomuran family which today has the greatest number of individuals. They are first known in the Eocene. The thickness of the exoskeleton and its calcification increase toward the rostrum. The massive claws are asymmetrical and sturdy. The right, larger one serves (besides its uses for breaking, grasping, and locomotion) to close the aperture when the animal retreats into the gastropod shell. The outer side of the chela of the claw is particularly resistant.

This right claw limb is frequently found by itself. Claws, or whole limbs with the claws, of *Eupagurus bernhardus*, the most common hermit crab of the North Sea, are frequently found in the swash marks of the open, sandy beaches, although this hermit crab never leaves the water and lives below the low-water line. These pieces must therefore have been transported a considerable distance. The large claw limb separates particularly easily from the rest of the body because it is heavy compared with the remaining part and because the connection at the cephalothorax is rather narrow and thinly covered by skin. In the carcass, the abdomen and the posterior part of the cephalothorax disintegrate quite rapidly because the covering skin is soft. Thus the harder sections of the cephalothorax are also loosened at an early stage, and only the locomotory limbs and especially their hard claws remain.

(iv) *Brachyura*

The brachyuran crustaceans vary in shape and life habits. They become increasingly important from the Mesozoic onward. Although some authors question the phylogenetic unity of the group (Ortmann 1901, Beurlen 1929 and 1933) we will continue to use it for the purpose of classification.

Structure and life habits. The following structural features are common to this varied group—they are decapod crustaceans whose abdomen is poorly developed, does not have a tail fin and is folded into a depression

FIG. 78. Segments of *Brachyura* (after Schäfer, 1951b). Top: carapace; bottom: ten cephalothorax segments and seven abdominal segments. (Schematic drawing)

on the underside of the cephalothorax. The cephalothorax is shortened, and the ganglia are concentrated in it. The antennae are short, the stalked eyes lie in cavities (orbitae). The maxillipeds have broad, flattened segments. The males have two pairs of abdominal limbs, the females four. The first pair of pereiopods carries claws, the others are locomotory limbs.

The cephalon, the thorax, and the abdomen become one functional unit, and the different segments of the body appear telescoped together; on the disintegration of the carcass they partly separate again. If we consider the stalks of the eyes as modified limbs, they represent, together with the first and second antennae, the first three pairs of limbs. The exoskeleton of the corresponding segments of the body is completely fused and forms the frontal edge of the animal, but also extends far backward and forms the main part of the carapace (Fig. 78). The mandibles are the fourth pair of limbs. The corresponding fourth body segment in the exoskeleton is represented in the lateral parts of the carapace. The appendages of the fifth to ninth segment are auxiliary appendages for grasping and chewing food (first and second maxilla, and first, second, and third maxilliped); although they are a functional unit, they belong partly to the head and partly to the thorax. The function of the next, more posterior, first pereiopod (with claw) varies from species to species; it may serve mainly for feeding and defence or mainly for locomotion. The remaining four pereiopods serve as locomotory organs (for running, walking, climbing, and swimming). The body segments that correspond to the maxillas, maxillipeds, and pereiopods lie closely together with their exoskeletal elements acting as supporting organs; together, they form a cone under a shield-like roof of the carapace and support it without being joined to it by chitin.

There are numerous species of Brachyura in all oceans and in the southern North Sea. Balss (1926) listed 37 species for the North Sea: Leucosiidae with the genus *Ebalia*; Majidae with the genera *Inachus*, *Macropodia*, *Hyas*, *Eurynome*, *Pisa*, and *Maja*; Corystidae with the genus *Corystes*; Portunidae with the genera *Polybius*, *Portumnus*, and *Portunus*; Atelecyclidae with the genera *Pirimela*, *Carcinides*, and *Cancer*; Xanthidae with the genera *Geryon*, *Xantho*, and *Pilumnus*; Pinnoteridae with the genus *Pinnoteres*; and Grapsidae with the genus *Eriocheir*.

The relative abundance of the various species is unknown but it is clear that they differ widely.

In the absence of frequency measurements, catches convey an idea of the density of populations. This number is multiplied if one considers the numerous exuviae discarded (see also D II (*c*) (i) about moulting).

Eriocheir sinensis (see Peters & Panning 1933) invades brackish waters and river estuaries and occurs in particularly large numbers (thousands of individuals crowded into small areas); *Carcinides maenas*, an animal living in the tidal zone, can occur in numerous individuals (up to 250 can be caught in a single sweep of the net). The distribution of *Cancer pagurus* is much wider but its density of population is moderate. South of Heligoland a drag-net brings approximately four to eight animals with one sweep over an area of approximately 20 by 1500 m. *Portunus holsatus* is caught with nets in large tidal channels. Sometimes, a single catch brings 200 animals, but usually only six to eight. *Hyas araneus* and *H. coarctatus* are less common (five to twenty animals in one catch); both prefer a rough sea floor with vegetation. *Macropodia rostrata* lives only in seaweed and colonies of hydroid polyps, is always caught individually and not frequently. *Maja squinado* lives in the central North Sea in depths of 20 to 50 m and is also caught individually. *Corystes cassivelaunus* lives in the southern North Sea between the depths of 15 and 30 m; if it is caught at all, several individuals are normally captured together. In large parts of its habitat it is rare.

There are three parts of the brachyuran body which on occasion are found individually in marine sediments and whose specific shapes are significant with regard to the animal's relations to its environment: the carapace, the pereiopods or their segments, and the claws. They may be hard and resistant, such as the claws and the pereiopod segments, or particularly easily transported, such as the wide, large carapace. Another resistant assemblage is the exoskeleton of the thorax which consists mainly of the conically shaped, chitinous covering of the musculature of the pereiopods, and of the sternal plates with the segmented inner crista; this assemblage is not specific. On the other hand, from the morphology of the three most frequently found parts of the skeleton, carapace, claws, and pereiopods, it is possible to interpret function and therefore behaviour and environment of the animal. In order to do so we arrange the Brachyura in four groups, according to the different modes of locomotion which best reflect their relations to their environment (Schäfer 1954a). They are animals that (1) crawl and run on the sediment surface, (2) dig in the sediment, (3) climb on a rough substrate, and (4) swim actively.

All four groups exist in the southern North Sea. They differ morphologically and functionally in the following features:

(1) Crawling and running Brachyura to which the following species belong: *Pirimela denticulata*, *Carcinides maenas*, *Cancer pagurus*, *Geryon tridens*, *Xantho* animals move about in search of food, or else, at certain times of the day or year, they travel over large distances. The shallower the water in which the animals live, the stronger generally are the currents. Certain features are therefore required for shallow-water crawlers. The body is compact and has no protrusions and therefore offers little resistance to flowing water. The body is dorso-ventrally flattened, has a wide anterior edge and no rostral structures; the largest extension of the cephalothorax is frequently at right angles to the plane of symmetry. The mouth appendages are shaped to be recessed into the rounded outline of the body. The eye stalks can be folded back into the orbitae, first and second antennae are *hydrophilus*, *Eriocheir sinensis*, and *Pilumnus hirtellus*. For most of their life these short, the claws can be tucked away, their curvature fitting their resting place at the anterior margin of the cephalothorax. The pereiopods are strong and often long, locomotory limbs. There is little room for the coxae of the pereiopods along the short, telescoped body; they therefore sit at right angles to the plane of symmetry of the animal. The almost parallel locomotory limbs point to both sides of the body and are flattened (sideways running is explained in D I (*a*) (ix)).

This general structure is slightly modified in the various species in adaptation to specific conditions. There are two subgroups, slow-walking animals with heavy armour and heavily toothed breaking claws and fast-running animals with thin armour and catching and grasping claws. *Cancer pagurus* is slow and has short, sturdy appendages which cannot lift the heavy, large body far above the substrate. The clumsy claws help in the slow walking locomotion. The animals never travel over long distances but can bury themselves in the surface of the sediment. *Carcinides maenas* (Fig. 79) is fast and has long, sturdy limbs which lift the light body above the surface of the sediment. When the animal is walking, the small claws are held tightly against the anterior margin of the cephalothorax and do not take part in locomotion. The animals run swiftly over the tidal flats. They also dig shallowly into the sediment. *Eriocheir sinensis* resembles *Carcinides maenas* in body proportions and mode of life. It has long, closely set, flattened locomotory limbs, a small body and small claws that can be tucked close to the body. It also digs open burrows with its appendages.

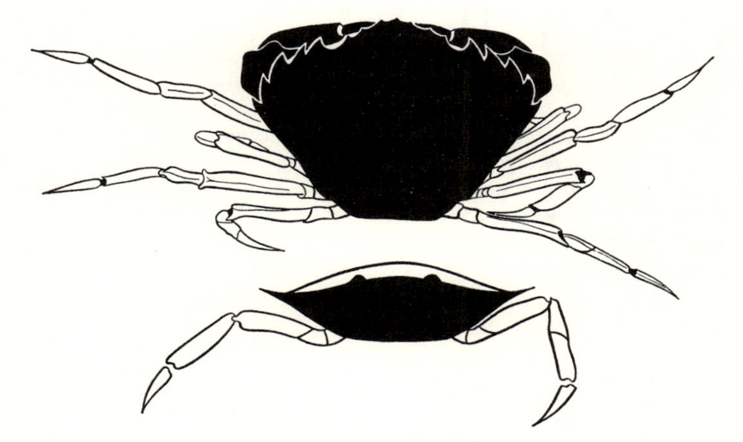

Fig. 79. *Carcinides maenas*. Top: dorsal view; bottom: cross-section. (Actual size × 0·6)

(2) Digging Brachyura: *Corystes cassivellaunus* is the only species representing this type of burrower. The cephalothorax and carapace are elongate, narrowing to a cone toward the posterior end (Fig. 80). Spines at the anterior edge point forward (i.e. away from the direction of digging). The animal has no other protrusions, and

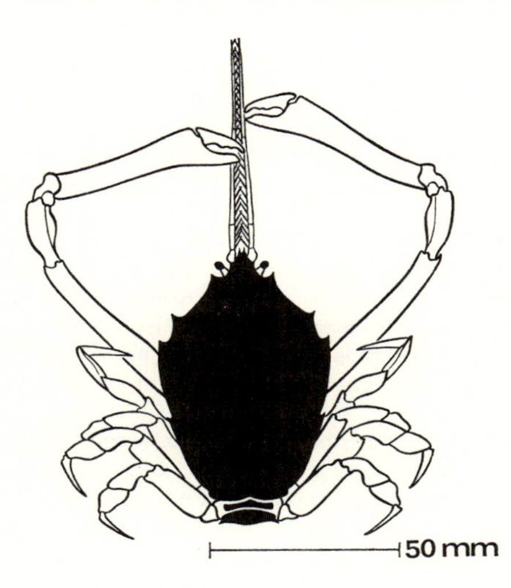

Fig. 80. *Corystes cassivelaunus*. Dorsal view

the cephalothorax is smooth. The appendages of the mouth are recessed into the curvature of the cephalothorax. The eyes have stalks, the antennae are long and can form a breathing tube. The locomotory limbs decrease in size posteriorly and carry the body in backward sloping position when running. The last pair of locomotory limbs serves for burrowing; the coxa joint makes it possible for the limbs to move backward and forward about a dorso-ventrally oriented axis. The claw-bearing limbs cannot be tucked to the cephalothorax. They are very long, the chelae of the claws being especially long, and the grasping claw is set with fine teeth. In other oceans the burrowing Brachyura are represented by many more species with a variety of other burrowing techniques (see also Schäfer 1954a).

(3) Climbing Brachyura: To this ecological group belong *Inachus dorsettensis*, *Macropodia longirostris*, *M. rostrata*, *Hyas araneus*, *H. coarctatus*, *Maja squinado*.

Locomotion on a rough substrate or through hydroid polyps between algae is necessarily slow and consists in climbing from one hold to another. The cephalothorax is generally drop-shaped and pointed toward the anterior end of the body

FIG. 81. Left: *Hyas araneus*; right: *Macropodia rostrata*. Dorsal views. (Actual size × 0·35)

formed into a pseudorostrum (Fig. 81). The carapace carries tubercles, thorns, spines, and also bristles. The eyes have stalks, the mouth appendages protrude. The locomotory limbs are long and spider-like and are set radially under the body; the individual sections are rounded. The claw-bearing appendages are long and cannot be folded to the cephalothorax. The chelae of the claws are curved, and the cutting edges of the claws have fine teeth. These animals never burrow in the sediment.

(4) Swimming Brachyura include *Polybius henslowi*, *Portumnus latipes*, *Portunus arcuatus*, *P. puber*, and *P. holsatus*. They show many resemblances with the walking Brachyura, with more pronounced characteristics. The cephalothorax is flattened dorso-ventrally; the anterior margin of the cephalothorax is broad, toothed, and has no rostrum (Fig. 82). The mouth appendages fit the cephalothorax, and the chitinous armour is generally thin but has reinforcing ridges on particularly stressed sections. The locomotory limbs are long, flattened, and are parallel to each other. The claws can be tucked against the margin of the cephalothorax and are shaped to fit it. The claws have teeth used for catching. Swimming plates are present on the dactyli of the second, third, fourth, and fifth pereiopods of the genus *Polybius* and only on the fifth pereiopod of all other genera. Locomotion is by

swimming sideways in an inclined position. The animals also burrow in the surface of the sediment but never run.

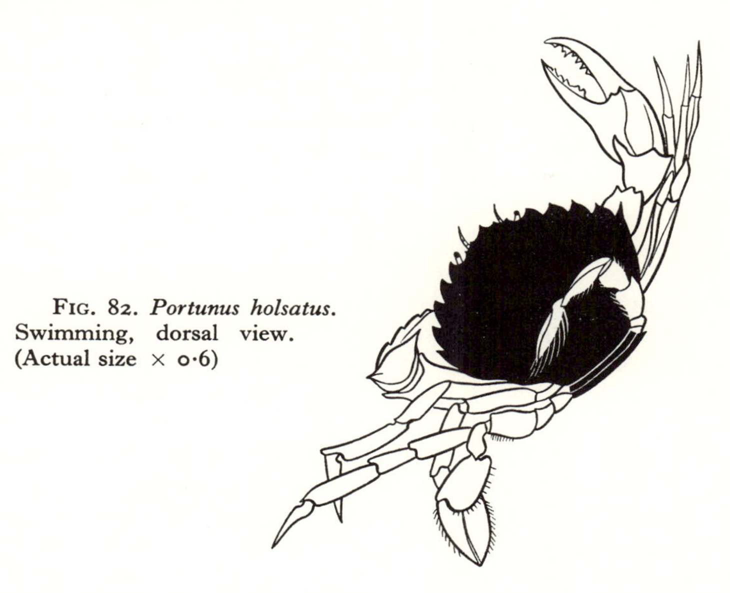

FIG. 82. *Portunus holsatus.* Swimming, dorsal view. (Actual size × 0·6)

The exoskeleton which is renewed many times during an animal's life is the only part of a Brachyurum that could ever be preserved as a fossil. This armour consists of five layers of different thickness (Fig. 84). At the surface the armour is covered by a very fine skin not of chitin. Inward follow four chitinous layers, a thin exterior, a thicker pigmented, a massive principal, and a thinner interior layer; this in turn rests on the matrix. The individual layers are composed of lamellae of various thicknesses consisting of fibrous structures oriented at right angles to the lamella. A filling material connects the lamellae (Vitzou 1882, W. J. Schmidt 1924). Calcified zones are often as wide as the lamellae and are arranged in rows; they are more frequent in places that need special reinforcement. The matrix forms the exoskeleton and determines its shape and thickness from place to place. The skeleton is especially thin at the places which serve as joints, and is highly calcified, folded and ribbed where strength is required. Such particularly calcified and folded zones occur, for example, at the anterior margin of the carapace, the teeth at the edges of the claws, thorns, spines, ridges of the sternum. Their chances of being preserved are greater than those of thinner parts of the skeleton. Other decapod crustaceans have the same kind of armour as the brachyurans. The chitinous armours of lower crustaceans have hardly been investigated but are essentially similar.

Death, disintegration and burial. The life span of different Brachyura varies widely. *Cancer pagurus* can reach an age of twenty years and a width of 25 cm. A mature animal of this size and strength has few enemies and dies only from disease or old age. *Carcinides maenas* reaches its maximum size at four years, and an adult animal was kept in an aquarium for five years. *Macropodia rostrata* seems to reach an age of five to six years. The ages of other species are not known. Most animals die through cannibalism at the time of moulting (Schäfer 1951a). Small animals are eaten by fishes, larger ones only by specialised feeders such as sea wolves (*Anarhichas lupus* and *A. minor*). The cephalopod *Octopus vulgaris* feeds on Brachyura. Frequently the parasitic cirriped *Sacculina carcini* causes the death of *Carcinides maenas* in the tidal sea. Many of those crabs that remain on the open tidal flats at low tide are affected by *Sacculina carcini*. The armour of such crustaceans is frequently yellowish-red or bright red (especially the underside and the locomotory limbs). In certain areas of the tidal channels where at times up to three metres of sediment are deposited in twenty-four hours (Schäfer 1956a), *Carcinides maenas* and *Cancer pagurus*, and smaller brachyurans even more easily, can be killed by the influx of sediment. Carcasses of buried animals show characteristic positions of their locomotory and claw-bearing limbs. The locomotory limbs are generally raised and turned over the cephalothorax, and the claws are also raised and slightly opened. It follows that the buried animals must still have been able to bring the limbs into position for an upward push but must have lacked the strength to raise themselves against the rapidly increasing weight of the sediment.

The carcasses of the Brachyura of the shallow, coastal, southern North Sea are frequently transported by currents. The organic material decomposes rapidly and completely in the oxygen-rich water which is also favoured by scavengers. Parts of skeletons, rather than complete, undamaged animals, are therefore normally found in the sediments. If a crustacean carcass lies in the water, the skin of the joints remains flexible, and the appendages can therefore be shifted and moved by currents. Carcasses, and also exuviae, thus get into positions that are impossible in life. Disintegration of the skeleton around the rotting soft parts usually begins with the separation of the carapace from the thorax. The skin between the anterior margin of the carapace and the thorax, however, remains intact for a while, a phenomenon also observed on exuviae. The carapace hangs loosely on the thorax, like a door on its hinges, and flaps back and forth in agitated water. Before this connection is destroyed, decomposition gases unroll and stretch the narrow, segmented abdomen and align it with the thorax. In this position of the three body sections, the carcass can look deceptively like living Astacura. The same position has been described for fossil trilobites, found with stretched abdomen and thorax, both ventral side downward, and the attached carapace lying on its dorsal side. This is called 'Salters' position' after its first investigator (Fig. 83). According to Richter (1937c) such a position is only possible in exuviae; however, it is just as frequent in carcasses (Schäfer 1951b). In the next stage, the carapace separates completely from the thorax. The separation reflects the division of the crustacean body into its main sections; the carapace keeps all its appendages, eye stalks, antennulae, antennae, and mandibles, after it has separated from the thorax (Fig. 83). The mandibles are particularly hard and heavily calcified and are usually buried unbroken. In swash accumulations of small fragments at sandy beaches they are often found in large numbers.

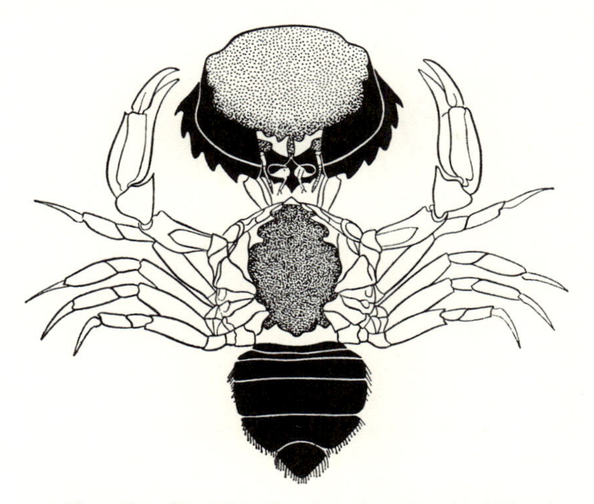

FIG. 83. Exuvia of a brachyuran in Salter's position. Top view. From top to bottom: carapace, dorsal side down; cephalothorax, ventral side down with claws and walking legs, abdomen. (Schematic drawing)

After the carapace has fallen off, the abdomen eventually separates from the thorax. Each of the three main body sections remains intact for a long time; further disintegration sets in much later. Due to this pause in the disintegration, numerous body sections, usually with all appendages, are found at the beach and on the sea floor. Because they are heaviest the chelipeds separate first from the thorax. In shell deposits and swash marks, dactyli of chelipeds are the most frequently found parts of appendages; dactyli of locomotory limbs, points of chelae (digitus fixus) and coxae of the pereiopods are moderately frequent.

If a living brachyuran becomes the prey of cannibals or other predators, disintegration occurs in a different way. Some animals that live on living or dead crustaceans do not damage the hard parts, but accelerate disintegration (*Buccinum*, *Neptunea*, *Nassa*, *Asterias*, *Solaster*, *Gammarus*, *Eurydice* etc.; see also D I (*b*) (i) about feeding methods).

In mud sediments the armour begins to lose strength after four weeks. Light pressure breaks the skeleton into small splinters. Hecht (1933) ascribes this fragility to decalcification. However, experiments show that decalcified crustacean armour does not break easily; on the contrary, its elasticity increases as with decalcified vertebrae. Wax-embedded sections of such fragile armour show that the connective tissue between the chitinous lamellae is missing (Fig. 84); the external layer of the armour still holds together but the lamellae of the principal and internal layers have disintegrated into loose, smaller or larger bundles. Embedded and

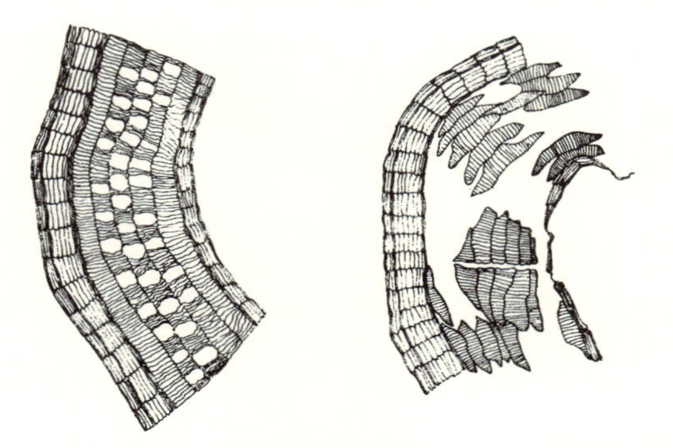

FIG. 84. Cross-section through armour of a brachyuran (after Schäfer, 1951 and 1954). Left: fresh, from the outside inward: covering, external, pigmented, main, and internal layers. Shaded—chitinous lamellae; white—calcite. Right: after two months' decay in sea water. The sheaves of chitinous lamellae of the main and internal layers have disintegrated after solution of the cement. (Schematic drawing)

sediment-filled partial or complete armours with all appendages are not further destroyed despite the lack of connective substance unless they are disturbed or subjected to increased pressure. It can be finally preserved by diagenetic hardening of the fossil and the surrounding sediment. However, most sediments with their embedded animals are destroyed in the agitated, shallow sea before they have a chance to harden.

(*e*) INSECTA

The majority of insects are land or fresh-water animals; only few species have returned to a marine life. In the North Sea there are several beetles, called 'salt beetles', whose habitat is the beach near the mean high-water line. Periodic or occasional flooding by sea water does not harm them. Lengerken (1929) lists halobiont Coleoptera that occur exclusively in the coastal areas but not in the salt areas of the mainland. In the salt marshes of the tidal coasts the following insects are particularly frequent: *Bembideon concinnum* (see also Schäfer 1941d), *B. normannum*, *Pogonus litoralis*, *Atheta varendorffiana*, and *Dychirius globosus*.

Many insects build burrows in the sediment. Some, for example *Bledius spectabilis*, care for their progeny by depositing the eggs in special chambers in their burrows (Larsen 1936), and the larvae spend their early life there (Wohlenberg 1937). All these animals are small; they live mostly in sand and their armour is soon destroyed after death. Hence these beach-dwelling insects are rarely preserved.

Flying insects that are more or less passively transported over the sea are more important as eventual fossils. After drifting for some time on the surface, they die, reach the bottom, or are stranded. There is little chance for the preservation of small animals which as 'plankton of the air' often drift over wide distances. Speyer (1937) reported that numerous aphids of various species on occasion arrived at the island of Memmert (off the East Frisian coast) and at the island of Heligoland. The distance from the mainland to Heligoland is more than 100 km. The same author reported that numerous aphids *Dilachnus picea* were found on the snow in Spitzbergen after a southeasterly gale from the direction of Kola peninsula, 1300 km away. He also mentions the young larvae of the moth *Ocneria dispar* and other caterpillars suspended on long silk threads which can be transported over many kilometres in high winds. Swarms of larger individual insects with harder exoskeletons are also occasionally carried over the sea by winds and then die in masses; a few insects fly out over the sea and eventually come down.

Most fossil insects are fairly large animals. However, they are found mainly in lacustrine, fluviatile, and aeolian deposits, or are preserved in resin. However, of the ten thousand known species of fossil insects (Schröder 1925) a fairly large number comes from marine sediments. The sediment is always fine-grained, as shale, calcareous shale, or marl, and not reworked. Insects are never preserved in sand because sand grains do not stop shifting after they are first deposited. Small and fragile insect bodies are destroyed if they are transported together with surrounding sand particles. Only dragonflies (Odonata) and beetles (Coleoptera) are frequently observed in the North Sea and its coasts. Many other orders of insects occur as fossils but their mode of preservation cannot be studied in the North Sea.

(i) *Odonata*

The four-spotted dragonfly (*Libellula quadrimaculata*) occurs each year in large numbers at the East Frisian coast. The animals do not originate there, but live as larvae in brackish beach ponds and bays of the Baltic Sea. The development of the dragonfly is essentially temperature-dependent. The time of egg-laying, transformation into the chrysalis stage and emergence of the imago are governed by the temperature. As a result, the imagos appear almost simultaneously; this is particularly so when the temperature increases suddenly in May and June after weeks of cool spring weather. Then dragonflies fly in swarms along the coast from east to west. They reach the southern North Sea after several days of active flight in swarms of hundreds of thousands or even millions. That they are not simply windborne can be seen from the fact that they travel not only during east winds but also during quite strong westerly winds. The same species develops each summer in large numbers in the marshes of north-western Holland (Leege 1935).

The migration of *Libellula quadrimaculata* takes approximately three days to pass through the area of the southern North Sea. During this time, trees and grass in the coastal area are thickly covered with the large, lustrous insects. In 1958 a particularly dense swarm obstructed road traffic west of Wilhelmshaven for several hours. Mass flights of these insects also cross the sea; invasions of dragonflies are not rare on the island of Heligoland (G. Vauk, personal communication). Many of them die during flight and drop on to tidal flats, wet sand, or the open sea. If temperatures drop suddenly or if it rains continually, as happens frequently in June, the animals' flying ability is reduced, and they die in masses. After death the carcasses float for approximately fifteen days, generally with extended wings. During this period the head falls away, and the abdomen begins to split lengthwise. Approximately after twenty days the carcass sinks to the bottom. When the thorax also splits lengthwise ventrally and dorsally and disintegrates, all four wings are dropped. They do not float.

(ii) *Coleoptera*

Far more Coleoptera are buried alive or dead in the coastal sediments than insects of other orders. Trusheim (1929b) and Schwarz (1931) discussed the causes of mass deaths of Coleoptera which occur almost every year. Both authors report that particularly large numbers of beetles rise in swarms on sunny days following rainy periods. A seaward wind can drive hundreds of thousands or millions of them over the sea. They may be pushed down on the water by gusts of wind, or drop on the water at sunset. Occasionally masses of live animals drift together in the sea or on the waves near the coast and are finally washed ashore on the swash marks of the beaches. Although cold and rigid they recover faster on the beach than Diptera, Hymenoptera, and Odonata under the same circumstances (Schwarz 1931). The author reports further that beetles stay afloat for a long time in contrast to flies and Hymenoptera which are fragile, often hairy, and more easily wetted.

Franz (1931, and in Trusheim 1929b) identified the species involved in this type of mass mortality. In 1928 they belonged to the following groups: 19 species of Coleoptera, 6 of Diptera, 6 of Hymenoptera, 8 of Hemiptera, and some unidentifiable Lepidoptera fragments. In 1931 there were 37 species of Coleoptera (belonging to the Carabidae, Scarabaeidae, Hydrophilidae, Coccinelidae, Byrrhidae, Heteroceridae, Cantharidae, Chrysomelidae, and Curculionidae), 2 species of Diptera, 3 of Hymenoptera, 1 of Hemiptera, and 1 species of Odonata.

Lochmaea suturalis was the most numerous beetle in the mass death of 1928; this small, yellowish-brown beetle lives on leaves of marsh plants, birches, and willows. According to Trusheim (1926b) approximately 13 500 beetles of this species were found in 1000 ccm of swash accumulation material spread along a 1·5 km stretch, which gives an overall number of 40 million beetles.

Carcasses of salt beetles whose habitat is in vegetation of halophytes near the mean high-water line are just as important as the inland beetles because they are more regularly incorporated into the sediment. However, artificial constructions along the beaches and estuaries have altered the original biotope of the salt beetle (Schäfer 1941d) to such an extent that the conditions for burial today are too different from the natural state to offer a valid model. Before dams were constructed, the zones of gain and loss of land were wide. Today dams separate the safe, dry land from a narrow strip of salt marshes covered by halophytes and of muddy tidal flats where in the middle ages large brackish-water zones lay between these biotopes. The brackish zones were covered with dense reeds and other vegetation and criss-crossed by open channels, silting up tidal flats with halophyte vegetation and ponds of stagnant, brackish or salt water were scattered over the area. These were the regions where the salt beetles normally lived and died. Dead beetles and other insects were much more often covered by sediment than today. The muddy coastal deposits full of plant material and roots contain especially numerous remnants of beetles.

IV. Mollusca

The molluscs are among the most conspicuous organisms of the Recent shallow sea. Sea shells are a familiar part of any beach scene. Molluscs evolved from benthonic animals, and the majority are still benthonic with the exceptions of the Pteropoda, Heteropoda, and Cephalopoda. Important as the Cephalopoda are for geologists, they cannot be thoroughly investigated from the point of view of actuopalaeontology because the forms which were common and easily preserved in the geological past are now extinct. Cephalopods still living have few hard parts that can be preserved. The Placophora and Scaphopoda can be ignored, and we will confine our discussion to gastropods, lamellibranchs and cephalopods.

(*a*) GASTROPODA

There are two orders of marine gastropods, the Prosobranchia and the Opisthobranchia. The prosobranchs are more important from our point of view. They are by far the more abundant and more numerous in species in the North Sea. All species have easily preserved hard parts. Of the few species of opisthobranchs in the North Sea, only the tectibranchs have shells.

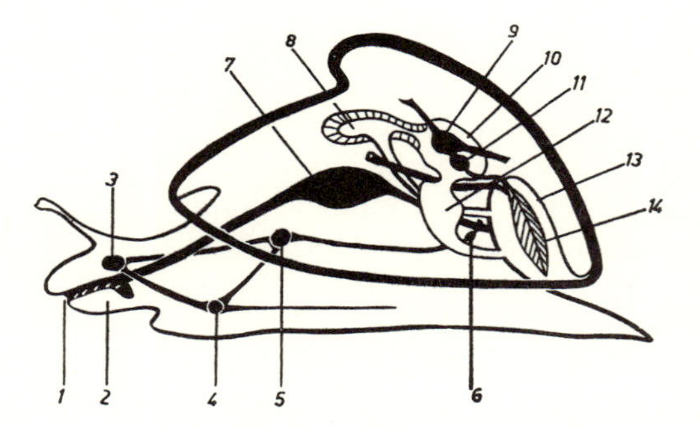

FIG. 85. Prosobranch structure, shown on generalised gastropod. (1) mouth; (2) radular cartilage with radula; (3) cerebral ganglion; (4) pedal ganglion; (5) pleural ganglion; (6) visceral ganglion; (7) stomach; (8) ovary; (9) heart ventricle; (10) pericardial cavity; (11) heart atrium; (12) excretory organ; (13) mantle cavity; (14) gill. (Schematic drawing)

Structure and life. The palaeontologist's only source of information about a gastropod's morphology, functions, and ecology is the shell. The varied forms of recent gastropod shells can be considered as having evolved from a form similar to the amphineurans.

A hypothetical, primitive gastropod is shown in Fig. 85 to demonstrate in a schematic way the general structure of the animal, the positions of the various organs, and their mutual relationships. The bilaterally symmetrical animal crawls on a foot, has an anterior head and carries its visceral mass in a dorsal hump. An inward and upward fold of the skin marks externally the borderline between the hump and the rest of the body. At the posterior side of the hump the fold widens to enclose a wide space, the mantle cavity. Two gills lie in this cavity, and the nephridia, the gonads, and the intestine open into it. The epidermis above the hump and, especially, the fold over the mantle cavity secrete a chitinous-calcareous, layered shell which arches conically over the hump. The shell grows at the rate at which the shell-secreting fold of the mantle covers the growing hump. The shell serves to protect the visceral mass against external damage (Fig. 86). Of all molluscs only the cephalopods have modified

FIG. 86. Asymmetry of prosobranchs and of their shells. Left: *Pleurotomaria* type, shell forming a plane spiral, with mantle slit, symmetric heart (diotocardian); centre: *Gibbula* type, right atrium somewhat reduced, single gill; right: *Paludina* type, right atrium rudimentary (monotocardian). (Schematic drawing)

this function of the shell concurrently with other changes of organisation. Their shell functions additionally either as a gas chamber aiding in locomotion (ammonites and *Nautilus*), or it is completely modified into an internal supporting element (belemnites and decapode cephalopods).

The uniformity of function of the various gastropod shells is reflected in the uniformity of shape. Whatever the biotope and conditions in which shell-bearing gastropods live, the coiled shell into which the soft body can partly or entirely retreat is rarely modified. Gastropod shells tell us little about habitat and life habits of the bearers, and so palaeontologists can use them rarely to interpret facies problems. Only a few species have characteristic shell structures that tell us something about the life and behaviour of their owners.

The more common gastropods of the southern North Sea fall ecologically into eight groups, but, with one exception, the shell forms give no indication of the biotope. The groups are:

(1) In the surf zone on rocks or plants (Fig. 87): *Acmaea virginea*, *Gibbula cineraria*, *Littorina littorea* (in the entire tidal zone), *L. obtusata* (near the mean low-water line), *L. saxatilis* and *L. neritoides* (both in the

spray zone), *Margarites helicinus* (on *Laminaria* seaweeds), *Nucella lapillus*, *Patella vulgata*, *Lacuna divaricata*, *Skeneopsis planorbis* (in tidal ponds of Heligoland), *Tritonalia erinacea* (on *Laminaria*). Except for *Acmaea* and *Patella*, coiled shells dominate. The main difficulty in this biotope is to anchor the animal to the substrate, and this is the task of the foot. The animals can cope with the conditions regardless of the presence or absence of whorls, or differences in whorl width.

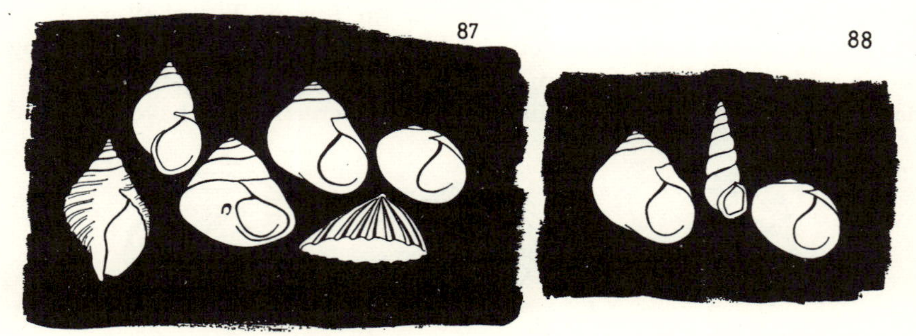

FIG. 87 (*left*). Prosobranchs in the surf zone on rocky ground. From left to right: *Nucella lapillus*, *Lacuna divaricata*, *Gibbula cineraria*, *Littorina littorea*, *Patella vulgata*, *Littorina obtusata*. (Schematic drawing)

FIG. 88 (*right*). Prosobranchs at the coast on sand. From left to right: *Littorina littorea*, *Truncatella truncata*, *Littorina obtusata*. (Schematic drawing)

(2) In the surf zone on sand (Fig. 88): *Littorina obtusata* (near the mean low-water line), *L. littorea* (in the entire tidal zone), *Truncatella truncata* (occasionally burrowing). All species of this group occur also in other biotopes.

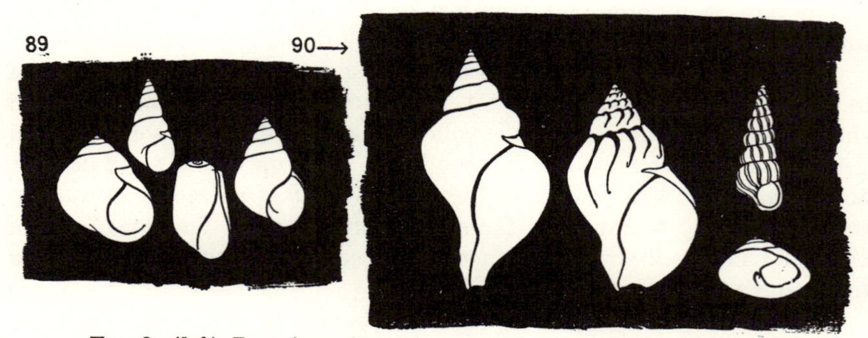

FIG. 89 (*left*). Prosobranchs at the coast on mud or silt, sheltered from surf. From left to right: the opisthobranchs *Littorina littorea*, *Hydrobia ulvae*, *Retusa obtusa*, and *Assiminea grayana*. (Schematic drawing)

FIG. 90 (*right*). Prosobranchs in the shallow sea on sand. From left to right: *Neptunea antiqua*, *Buccinum undatum*, *Scala clathrus*, *Gibbula tumida*. (Schematic drawing)

(3) At beaches protected from surf, on mud and fine-grained sand (Fig. 89): *Assiminea grayana* (between halophytes), *Hydrobia ulvae* (in open tidal flats near the mean high-water line), *H. stagnalis* (in brackish water), *Littorina littorea*, *Retusa obtusa* (an opisthobranch). Only *Assiminea* is restricted to this one biotope (Schäfer 1940, Sander 1950 and 1952); this limitation is not due to the shape of its shell.

(4) In the shallow sea on sand (Fig. 90): *Buccinum undatum* (also on muddy sand near the coast), *Gibbula tumida*, *Neptunea antiqua*, *Rissoa inconspicua*, *Scala clathrus*, *S. clathratula*.

(5) In the shallow sea, burrowing in sand (Fig. 91): *Caecum glabrum*, *Hydrobia ulvae* (also in the tidal zone), *Bela turricula*, *Lunatia nitida*, *L. catena*, *L. montagui*, *Nassa reticulata* (mainly in the tidal zone). Adaptation

FIG. 91 (*left*). Prosobranchs in the shallow sea, burrowing in sand. From left to right: *Lunatia nitida*, *Bela turricula*, *Nassa reticulata*, *Caecum glabrum*, *Hydrobia ulvae*. (Schematic drawing)

FIG. 92 (*right*). Prosobranchs crawling on and burrowing in mud. Top, from left to right: *Scala clathrus*, *Hydrobia ulvae*, *Gibbula tumida*; bottom, from left to right: *Velutina velutina*, *Aporrhais pes pelicani*, *Lamellaria perspicua*, *Turritella communis*. (Schematic drawing)

of the shells for burrowing could be expected in these animals; however, each of the species has a different shell form, all of which also occur in species with an entirely different way of life.

(6) In the shallow sea, creeping and burrowing in mud (Fig. 92): *Aporrhais pes pelicani*, *Gibbula tumida*, *Hydrobia ulvae*, *Lamellaria perspicua*, *Mangelia brachystomia*, *Nassa incrassata*, *Philbertia linearis*, *Scala clathrus*, *S. clathratula*, *Turritella communis*, *Velutina velutina*. The shell of *Aporrhais* shows signs of adaptation to the biotope and the behaviour of the animal.

(7) In the shallow sea, on rocks and plants (Fig. 93): *Bittium reticulatum* (on *Laminaria*), *Buccinum undatum*, *Gibbula cineraria*, *G. tumida*, *Helcion pellucidum* (on *Laminaria*), *Lacuna divaricata*, *L. pallidula* (on *Laminaria*), *Lamellaria perspicua*, *Rissoa parva*, *Skeneopsis planorbis*,

Trivia arctica, Urosalpinx cinerea, Trophonopsis muricatus. Some of these animals are grazers or scavengers and thus restricted to a rocky substrate. The foot is generally adapted in these species—not the shell.

FIG. 93. Prosobranchs in the shallow sea on stones or plants. Top, from left to right: *Rissoa parva, Helcion pellucidum, Gibbula tumida, Acmaea testudinalis*; centre, from left to right: *Lacuna divaricata, Bittium reticulatum, Trophonopsis muricatus, Lamellaria perspicua*; bottom from left to right: *Gibbula cineraria, Buccinum undatum, Trivia arctica, Urosalpinx cinerea*. (Schematic drawing)

(8) Parasites and commensals (Fig. 94): *Brachystomia ambigua* (on *Pecten opercularis* and *P. maximus*), *B. rissoides* (on *Mytilus*), *Crepidula fornicata* (no specific host), *Pelseneeria stylifera* (on sea urchins). Even the shells of the ectoparasitic gastropods do not reveal their biotope and way of life.

FIG. 94. Parasitic and commensal prosobranchs. From left to right: *Crepidula fornicata, Pelseneeria stylifera, Brachystomia ambigua*.

Certain features of marine gastropods are quite striking and deserve examination, although little is known about their function. Indeed in some cases it is uncertain whether they have any function at all.

(1) Siphonal canals. If prosobranch shells have a siphonal (i.e. anterior) canal, the canal encloses a muscular and generally mobile siphon from the dorsal side; this extension of the mantle protrudes beyond the shell (Fig. 242). A siphon is usually accompanied by a sensory organ (osphradium) which lies at the opening of the siphonal tube in the mantle cavity. Experiments have shown that the osphradium of the prosobranchs tests the water for smells. Whereas the existence of an anterior canal on a shell indicates that the carrier must have been a predator or a scavenger, its length does not give any information about the length of the muscular siphon. The anterior canal, however, gives some information about the use of the siphon and the environment in which it is used: if the anterior canal of the shell is short and flared at the end (as in *Buccinum*), the siphon is used to test for smells in open water. An animal buried in sediment can extend the siphon beyond the canal and use it as a connection to the surface of the sediment and to the open water (e.g. *Nassa*). If, however, the anterior canal protrudes far (usually in the direction of the shell axis), the canal surrounds the siphon entirely even when it is extended. Together with its protective calcareous cover it can be moved through the sediment, and it can search for dead organic material in the water of the interstices between sand grains (e.g. *Neptunea*).

(2) Inhalant and exhalant currents. The path of the inhalant current through the mantle cavity of prosobranchs is well known. If a siphon exists and the shell has an anterior canal, the inhalant current is through this organ. However, the exhalant current may also have a specific path as can be seen from the shape of the shell. Apparently, a specific channel is necessary for animals that crawl partly immersed in mud and have their breathing cavity mostly closed. In such cases the outer and parietal lips meet at the end of the suture in an acute angle which is often set back in a sharp fold against the rest of the outer lip (e.g. in *Conus*, *Ranella*, *Murex*, *Gastridium*, *Voluta*); or this upper angle of the lip is extended and curved upward and backward and forms a siphon-like protrusion (*Aporrhais pes pelicani*); when the animal is moving, this angle lies at the back. In the opisthobranch *Retusa truncatula* the involute shell shape seems to have a bearing on directing the water current.

(3) Teeth of the shell aperture. One or several teeth can occur on the outer or inner lip of the aperture. They seem to have some relation to the position of the shell in locomotion. According to our observations these ridges act as guides for the columella muscle when it glides over them, and prevent the muscle from slipping while the soft body moves out of the shell and brings it into an upright position. Moreover, the columella muscle is aided by the teeth in carrying the shell more securely. It is not known what biotopes and life habits require such teeth.

(4) Thickness of the shell. A familiar rule states that species living in

the surf zone have thick shell walls and those living in still water, on sheltered coasts or at depth, have thin shells. There are, however, so many exceptions to this rule that its validity is questionable. Generally, the factors which determine the thickness of the shell are the salinity and the calcium content of the sea water in which the animal has grown up; the degree of water agitation is not an influencing factor. Remane & Schlieper (1958) discussed modifications in size, shape, and structure of brackish-water organisms with decreasing salinity. What saves gastropods from being broken in the surf is not a thick shell but the ability of the foot to hold on to a solid substrate. Thick shells that are not firmly held by the foot are just as easily destroyed by surf action as if they were thin.

FIG. 95. Aperture of *Murex fortispina*. Arrow: tooth which serves for opening bivalves. After photograph (Ankel, 1938a). (Schematic drawing)

Salinity and calcium content of the water determine the thickness of the wall and the size of the shell, even within one species. If *Buccinum* gastropods live in the low salinity tidal sea, they have thin shell walls; if they live in the open sea at greater depth in still water, their shells are thick, large, and lavishly sculptured. The shells of *Littorina littorea* are also smaller if the animals live near the coast in muds with lower salinity, larger if they live on the coast of the island of Heligoland.

(5) Shell parts as tools. According to Ankel (1938b) predatory prosobranchs use certain parts of their shells as tools for opening bivalves. According to François the predatory *Murex fortispina* (absent from the North Sea) opens large bivalves using a denticle on the inside of the outer lip to prise the two valves apart (Fig. 95). In the examined specimens, this tooth shows traces of use. Thus a feature of the shell that was considered purely ornamental was shown to be significant as a feeding organ; it does not seem impossible that other unexplained features of shells may also have ecological functions.

Colton (1908) reports that *Fulgur carica* pushes the bivalve *Venus* so hard against the outer edge of its shell that the valves are forced open.

According to Ankel (1938b) *Sycotypus canaliculatus* intending to eat an oyster sits 'waiting on the upper valve' of the chosen victim until it opens; with a rapid turn, the gastropod then presses the free edge of its shell between the valves and can safely insert its proboscis into the crack.

(6) Rapid increases in diameter from one whorl to the next lead to low shells with wide apertures as in the capuliform shells. The depressed-cone shape of the shell has been interpreted as an adaptation to conditions in the surf or in strong currents. The relationship is probably different: the wider the aperture, the larger and stronger a foot it will accommodate. The large foot, however, can have different functions in the species concerned: either it serves to cling to a hard substrate (as with *Patella*, *Helcion*, *Acmaea*, *Littorina obtusata*, *Crepidula*, *Fissurella*, and *Theodoxus*) or it is used to hold the prey (e.g. *Lunatia*), or to crawl over soft ground (as with *Buccinum*, *Assiminea*, *Velutina*, *Gibbula tumida*, *Lamellaria perspicua*, *Philbertia linearis*, and *Scissurella crispata*).

Death and shell destruction. Shell-bearing marine gastropods have few enemies. The shell is an effective protection, and only specialist feeders can cope with it. Adult catfishes, for example, are able to break *Neptunea* shells (Fig. 36). Some flatfishes (especially *Scophthalmus maximus* and *Hippoglossus hippoglossus*) often eat numerous *Lunatia*, *Turritella*, and *Scala*. Sea birds (mainly *Larus argentatus*) eat *Littorina littorea* and *L. obtusata* if they live at the beach and in the tidal zone, and break the shells in their muscular stomach. The stomach region of *Astropecten irregularis* is often stuffed with shells of *Lunatia nitida*. Gastropods are rarely killed by covering sand or mud because they are mostly capable of moving upwards through thick layers of sediment. Animals that live on rocks in the surf zone have such an effective holding mechanism in the sole of the foot that they can usually cope with most storms (*Patella*, *Littorina*, *Gibbula*, *Acmaea*, *Nucella*). How this mechanism functions is not yet explained (Ankel 1936a, with data on the holding power of *Patella*). As long as the gastropods can cling to a solid substrate, they remain unhurt. They do not live in areas with coarse pebbles or boulders, where neither a thick shell nor a strong foot nor both combined would give them a chance of survival. It can be demonstrated that the effectiveness of the foot and not the thickness of the shell is the decisive adaptation for life in a rocky surf zone. Gastropods with hard, thick shells that belong to a different biotope (e.g. *Buccinum*) are rapidly broken if they stray into such a zone. Other species from biotopes at greater depths are even more sensitive (e.g. *Turritella* or *Lunatia*).

With few enemies and good resistance against the mechanical forces in their habitat, marine gastropods die frequently from old age. Thus empty shells of immature gastropods are rare compared with those of lamellibranchs, echinoderms, and crustaceans; most gastropod shells that are found in shell deposits or individually are adult forms. Little is known about the age reached by prosobranchs. Most animals are sexually mature but not yet fully grown at the age of one year. In some forms the shells are modified in old age (e.g. *Trivia monacha*) or their aperture has changed (e.g. *Aporrhais pes pelicani* or *Buccinum*). It seems certain that the life span of many prosobranch species is more than six to eight years. Free-living *Littorina littorea* were marked by us as full-grown animals and recovered seven years later. We kept *Buccinum* in captivity for five years.

There are few data concerning the density of gastropod populations. Lamellibranchs normally far outnumber the gastropods of the same biotope. *Hydrobia ulvae* is the most frequent species of the tidal sea. Its density in the Jade Bay area reaches 20 000 per square metre (Linke 1939); the population of *Littorina littorea* near groynes and around Heligoland is only 400 per square metre, of *L. obtusata* only 30 per square metre. In some areas of the open North Sea (e.g. near light ship P 12) *Buccinum undatum* is one of the most frequent species. A catch of 100 kg swept up by a drag-net can contain 40 to 70 living individuals. Ziegelmeier (1954) reports a population of 50 per square metre for *Lunatia nitida*. Six to eight nautical miles north of the island of Spiekeroog we observed a dense population of *Bela turricula*, estimated as 200 per square metre. *Nassa reticulata* lives in the North Frisian tidal sea with a density of 2 to 3 per square metre, *Scala clathrus* and *Turritella communis* with less than 1 per square metre.

Relatively few empty gastropod shells are produced by populations of such low density. Only a small fraction of the mollusc shell deposits of the coasts and tidal channels consist of gastropod shells; and only *Littorina littorea* and *Hydrobia ulvae* are frequent enough in the southern North Sea to furnish a considerable proportion of the shell deposits or to be locally concentrated to the exclusion of all other shells. The population increase of *Littorina littorea* at the mainland and island coasts must have occurred since groynes, moles and quays were built, because their habitat is on hard substrates in the tidal zone.

The shells are preserved in great quantities in mud and fine-grained sand. On Heligoland, on the other hand, where numerous *Littorina* live under natural conditions the shells are completely destroyed between surf-rolled boulders and pebbles. Shells of *Hydrobia ulvae* form beds washed up by storm tides mainly at the edges of salt marshes against the tidal flats. They are also accumulated in sand flats by the continuous reworking by *Arenicola* at the level of the U-bends of their burrows; shells can be concentrated to the exclusion of all other sediment. Swash accumulations of the frequent *Hydrobia* shells along the open beaches of the islands can rarely be preserved. Those species that live in mud or are transported there are most likely to be preserved. In sand, continuous reworking tends to sort the hard parts and leads to the accumulation of shell breccia (Ankel 1929). Before they are buried gastropod shells can be damaged by drilling organisms or abrasion, impact or pressure.

Several species of *Polydora* bore into mollusc shells; they seem to be selective and to prefer certain calcareous shells. Borings into several *Lunatia* and into *Littorina littorea* are particularly frequent; they are less frequent on *Gibbula cineraria*, *Crepidula fornicata*, *Buccinum undatum*, *Neptunea antiqua*, *Patella vulgata*; and *Scala* and *Turritella* are rarely attacked. C. Hempel (1957b) listed the molluscs that can be affected alive but his list is incomplete for gastropods. The borings by these worms run more or less horizontally in the shell walls forming U-shapes and loops. They are frequently found on the columella area which is sheltered from abrasion. In surface-dwelling gastropods (*Crepidula*) they occur also in the convex part of the whorl. Where several worms have attacked a single shell, their borings are usually clustered. If perforated shells are transported repeatedly, the walls between neighbouring perforations often break, and individual U-bends cave in. These areas are often the starting points for further destruction.

Although less clearly visible, the work of micro-organisms on the shells may be even more devastating. Wetzel (1937b), Mägdefrau (1937), and Pia (1937) described micro-organisms that sometimes attack in enormous numbers. Their activity will be discussed in the section on the destruction of lamellibranchs.

When gastropod shells are exposed to impact and pressure in coarse, frequently moved shell breccia, or between pebbles or boulders, the arched whorls and the apertures often break first. The shape of the whole shell and the sculpturing determine the course of the mechanical destruction.

Some cases are shown in Fig. 96. If the aperture is particularly hard as in adult

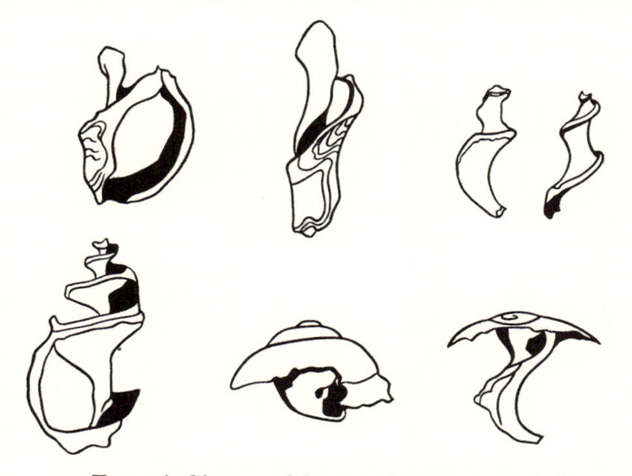

FIG. 96. Shapes of fractured gastropod shells. Top, from left to right: two fragments of *Buccinum undatum* and two of *Littorina littorea*; bottom, from left to right: *Nassa reticulata*, *Lunatia catena*, *Littorina obtusata*.

Buccinum, *Murex*, *Nucella*, and *Nassa* it remains intact for a long time. In elongated shells (e.g. *Turritella*, *Scala* etc.) the apex breaks before the columella is entirely exposed, and the aperture usually fractures at an early stage. The most resistant part of moderately long shells is the columella, particularly its lowest section. Columellas of *Buccinum* and of *Littorina* can be found even between the coarse, constantly moved pebbles on the dune beach of Heligoland. The apex with the first whorls is the hardest part of short shells as in *Lunatia* or *Littorina obtusata*. In *Gibbula cineraria* the shell breaks in two crosswise; the apex with the first whorls can frequently be found, and also the last whorls with the aperture intact. Such a cross-fracture occurs regularly in umbilicate shells because they are not supported by a columella.

The majority of gastropod shells are short-conical or roundish and therefore cannot show much preferred orientation under the influence of currents. Only the long-conical forms (e.g. *Cerithium*, *Turritella*, *Scala* etc.) show marked orientation. According to R. Richter (1922a) high-spired gastropods are oriented by their long axis, without preference for apex or aperture for the up-current direction, patelliform shells are oriented according to the rules that have been developed for lamellibranch valves but are valid for any dish-shaped body. According to Krejci-Graf (1932) high-spired shells can be oriented apex upward in quicksands. Papp (1944) observed that gastropods living in the upper sediment layers can be covered up by sedimentation while they attempt to creep upward; in this position the apex of the shell points downward.

(*b*) LAMELLIBRANCHIATA

Species of lamellibranchs in the North Sea far outnumber those of the gastropods. Individuals of most lamellibranch species are so numerous that their hard remnants are a characteristic component of North Sea sediments and of the sediments of most other recent shallow seas.

Structure and life. The relation between morphology and function of many lamellibranch valves is almost as difficult to interpret as in gastropod shells. The calcareous shell of lamellibranchs consists of two valves connected by a ligament. Their protective function does not give rise to fundamental variations of the basic, more or less oval form. However, specific adaptations to various habitats and ways of locomotion arise (Ziegelmeier 1957).

The species of the North Sea can be arranged in three groups, for each of which the relation between ecology and morphology will be considered separately.

(1) Attached to hard objects (fixed-sessile): *Anomia ephippium*, *A. squamula*, *A. patelliformis*, and *Ostrea edulis*, all belonging to the order Anisomyaria. The two valves differ from each other. *Anomia* is attached

to the substrate by the right valve, and *Ostrea* by the left. Both valves, but especially the lower one, can grow so that their shape fits the substrate exactly, even when it is very rough or consists of sessile animals or mollusc shells (Schäfer 1938b, Papp 1939). This adaptability of growth to specific needs compensates for the lack of locomotion. *Anomia* is attached by a calcified byssus which passes through a perforation under the beak of the right valve to the substrate and which becomes permanently fixed in the adult stage. For their sessile life they require water that is rich in nutrients but contains little suspended matter. *Anomia* settles frequently on drifting objects.

(2) Capable of attaching themselves temporarily to hard objects (facultatively fixed-sessile): *Musculus marmoratus*, *Mytilus edulis*, *Modiolus modiolus*, and *Dreissensia polymorpha*, all belonging to the order Anisomyaria; or living on the surface of the sediment (loose-sessile): *Cyprina islandica*, *Astarte borealis*, *A. montagui*, *A. sulcata*, *A. triangularis*, *Pecten varius*, *P. maximus*, and *P. opercularis*. The animals of the first group attach themselves to a hard object by means of the byssal threads secreted by the base of the foot. Inanimate objects and shells of living bivalves are used as substrates, especially the shells of members of the same species; all three species can form multilayered colonies. The valves sit upright so that both touch the substrate with what is functionally the underside of the animal. The undersides of the valves are more or less flattened or at least broadened; the valves close tightly. The animals have no siphons and therefore no pallial sinus. They live in places that are almost free from sedimentation. Numerous epizoans favour them therefore as a substrate (Balanidae, actinians, young *Lanice conchilega* living in their tubes, and interwoven hydroid polyps). The orientation of the epizoans is often a further clue to the living position of the bivalve serving as substrate. Seilacher (1954b) described a fossil example of this type of valve, the Triassic *Lima lineata*, together with its epizoans.

Cyprina and *Astarte* lie on the sediment surface on either one of the symmetric valves and have a well-developed periostracum. Neither has a siphon, and their foot is too weak to bury the large, heavy body in the sediment. It is barely strong enough to rotate the animal about a vertical axis, pivoting it on the point of support provided by the strongly convex shell.

(3) Burrowers (sediment-vagile): about thirty species of lamellibranchs, the majority of those occurring in the southern North Sea, live in the sediment. Life in sediment requires a constant connection with the surface of the sediment for breathing water and for food. There are two siphons, one inhalant and one exhalant. The retractors for the siphons are at the posterior end, and their size is proportional to the length of the siphons. The margin of the mantle is clear of the area where the retractors

are attached. If the retractors are large, the pallial line, where the muscles of the mantle are attached on the inside of the valve, curves around the retractor area forming a sinus (forms with a pallial sinus). Large sinuses indicate a habitat under the surface of the sediment, no sinuses generally a habitat above the sediment.

Life in sediment must necessarily be locomotory, because both the addition of sediment or its removal require the ability of the bivalve to compensate for the change in level. The muscular foot which lies in such cases at the anterior end can move the body.

FIG. 97. Opisthotruncate bivalves with short siphons, living just below the sediment surface. Top, from left to right: *Cardium edule*, *Venus gallina*, *Cardium echinatum*; bottom, from left to right: *Venerupis pullastra*, *Spisula subtruncata*, *Donax vittatus*. (Schematic drawing)

In many species, movement does not only compensate for changes in the level of the sediment surface but is also required for the search of food. If bivalves living in the sediment do not feed on plankton but graze by sucking detritus and diatoms from the surface of the sediment with their inhalant siphon, not only vertical but also horizontal movement is required. Grazing bivalves move horizontally through the sediment lying on their side with the foot forward. Both valves show little convexity, in particular in the area of the beak, and they are extremely flat. The siphons are separated from each other. *Scrobicularia plana* lives usually in muddy sediments of the tidal zone; *Abra prismatica*, *A. nitida*, *A. tenuis*, *A. alba*, *Angulus tenuis*, and *A. fabula* live preferentially in sandy sediments of the southern North Sea; *A. fabula* is especially frequent. This bivalve lies always in the sediment with its right, ribbed valve upward; hence its siphons can reach the sediment surface only by bending upward. In mature animals the valves are adapted to this position in the sediment and are bent at the tapering posterior end. G. E. & N. MacGinitie (1949) illustrated the 'bent-nosed clam' (*Macoma nasuta*) living on the American west coast with the same habits and showing the same curve of the posterior end.

Seilacher (1954b) called a certain type of valve 'opisthotruncate', giving the Triassic *Myophoria cardissoides* as an example. Recent inhabitants of the North Sea of this type are: *Spisula subtruncata, Donax vittatus, Venerupis pullastra, Venus gallina, Cardium edule, C. echinatum* (Fig. 97). The animals are oriented in the sediment in such a way that the flattened posterior end of their valves reaches near the sediment surface and is shaped to be flush with the surface, giving it a cut-off appearance. At this point the normally short siphons reach the open water. Being so short they are suitable only for feeding on plankton from the water; thus, although they live so near the sediment surface these animals do not graze and therefore are not dependent on continuous locomotion. However, the animals are frequently uncovered because the top layer of the sediment inhabited by them becomes agitated by the weakest of bottom currents. They are therefore able to burrow rapidly. The long, mobile foot is either hatchet-shaped, or the point of the foot can swell up and anchor the body for a downward pull into the sediment.

(4) Boring into hard substrates: *Petricola pholadiformis, Saxicava rugosa, Zirfaea crispata, Barnea candida, Pholas dactylus.* Some of these species are rock-boring and some peat- and wood-drilling. Some make the holes by dissolving the substrate and some drill mechanically. In all case- the valves are longer than wide, generally very long and narrow, the cross-section is circular. Boring bivalves always drill with their mouth end forward.

Chemically boring bivalves dissolve the rock by a secretion from glands at the margins of the mantle and need no specific drilling and rasping structures nor mechanisms to spread the valves by muscle action. *Saxicava* bores purely chemically (on Heligoland in limestone). Consequently, there are no morphological features of their valves that identify them as boring animals.

Species that drill mechanically with a combined mechanical and chemical technique, such as *Pholas, Petricola, Zirfaea,* or *Barnea,* have roughnesses or rows of denticles mainly at the anterior end, or filing ridges running obliquely over the curvature of the valves. In the southern North Sea *Pholas* bores in chalk; *Petricola, Zirfaea,* and *Barnea* drill mainly in peat but also in wood. In *Zirfaea, Barnea,* and *Pholas* the anterior 'closing' muscle opens the valves and lies outside them (Fig. 126). The valves of *Zirfaea* and *Pholas* are recessed at the anterior end to leave space for the protruding foot, so that the anterior part cannot be closed.

All boring bivalves are sessile living deep in a narrow tunnel. Therefore they depend on plankton for food and have siphons. In *Zirfaea* the valves curve at the posterior end to let the long, strong siphons pass through. All boring bivalves show a distinct pallial sinus.

(5) Swimmers: *Pecten maximus*, *P. varius*, and *P. opercularis* can rise from the sea floor and swim for several metres. Buddenbrock & Moller-Racke (1953) investigated the swimming of the Mediterranean *Pecten jacobaeus* and its light-sensitive organs. The curvature of the two valves differs, the strongly curved valve is usually on the underside. Opening of the valves fills the mantle cavity with water which can be ejected by rapid closing. The resulting water jet lifts the animal from the bottom, and its direction and force can be regulated by the position of the mantle. Swimming is an escape reaction from enemies, and we observed that it can be provoked instantaneously by traces of raw bivalve juice in the water. Minor adjustments of position, not connected with flight, can also be achieved by ejecting small jets of water.

(6) Boring in wood: the body and the valves of the genus *Teredo* are thoroughly modified in accordance with specific life and feeding habits (Fig. 129). The valves have lost all protective functions and serve only for rasping and drilling in wood (Kühl 1957a and b, and 1958, gives details about their distribution in the southern North Sea).

Modes of death. The three most important modes of death of lamellibranchs are wash-out, covering with sediment, and predation.

(*a*) Most bivalves live in the sediment, and in the shallow sea they die most frequently by being washed out of the sediment. Most of the lamellibranch species are able to dig deeper and to compensate to some degree for erosion by strong currents. However, such downward movement is restricted by the increasing consolidation of the sediment with depth. Mud and silt rapidly harden by compaction, subsidence, and drainage. If this process has gone too far the sediment becomes too stiff for digging by the foot. The bivalve, still in living position, is eventually eroded. If the animals live in tidal flats they die from exposure at the sediment surface where they dry out, or by lack of oxygen, or they become the prey of birds or other animals. Constantly immersed lamellibranchs die from repeated transportation or they are eaten by fishes or crustaceans. Many young bivalves are able to dig again into the sediment with their foot, and some species, especially those with short siphons, retain this ability throughout life.

(*b*) Bivalves living in sediment respond to new sedimentation by moving upward. If the rate of deposition is too high the lamellibranchs cannot follow, the weight of the new sediment keeps them from moving up, and they die in their living position. We must assume that this kind of death is particularly frequent in shallow seas with strong sedimentation. Even so, bivalves are rarely found in living position in fossil shallow-sea sediments. The reason is that primary beds with dead lamellibranchs are rarely preserved, and that in most cases the covering sediments are soon reworked together with their fossil content and then redeposited. In the process the valves generally separate, and lamellibranchs with both valves are extremely rarely preserved. This can happen only if they are finally buried near the place of death. Bivalves living on the surface of the sediment can also die by sudden sedimentation, usually caused by abnormally bad weather. Such sedimentation has catastrophic effects on species incapable of long-range locomotion (*Mytilus*, *Cyprina*, and *Modiolus*). Thus a whole colony can die at a time.

(*c*) Many bivalves are regularly or frequently consumed by fishes and crustaceans; the larger fishes are not satisfied with washed-out animals but dig extensively through superficial sediment in search of lamellibranchs. The prey consists mainly of the small species with weak valves. Fishes of the North Sea that feed regularly on bivalves are: *Raja clavata*, *Acipenser sturio*, *Gadus morrhua*, *Brosme brosme*, *Mullus surmuletus*, *Anarhichas lupus*, *A. minor*, *Hippoglossus hippoglossus*, *Pleuronectes platessa*, *Limanda limanda*, and some others. The fishes destroy the shells of their prey in their mouth and stomach, and only a fine shell grit is released into the sediment with the faeces. Many bivalves are killed by boring gastropods such as *Lunatia*.

Partial and complete destruction. Lamellibranch valves are amongst the most resistant organic hard parts in the sediments of the shallow sea. The enormous quantities of such remnants in Recent sediments are not only due to the large population but even more to their strength and uncomplicated structure. If lamellibranch valves were as easily destroyed by the first transport as the generally composite skeletons of most other marine animals, their remnants would be equally rare in the shallow sea. But even lamellibranch valves are eventually destroyed, reduced to fragments and ground into fine grit. Solution of calcium carbonate facilitates the mechanical destruction, especially in the last stages. Carbonate-boring organisms also help to destroy the shells.

Gradual destruction by sand abrasion and sudden breakage by impact play the most important roles in the mechanical destruction of valves. Pratje (1929) has described in detail the slow destruction by sand abrasion of recent gastropod shells and lamellibranch valves. When abrading sand grains travel over a mollusc shell in the same direction for a long time this results in the formation of a polished facet. The facet gradually forms a hole (Fig. 98). Strongly convex valves become faceted preferentially, especially if the edge of the shell is well anchored in the sediment. Pratje noted further that abrading sand currents of constant direction can only be created by tides, and that faceted mollusc remnants thus indicate a shallow tidal sea. He observed that such faceting does not occur in the Baltic Sea but frequently in the North Sea. However, he has never seen it on valves from depths greater than approximately thirty metres, that is below the zone where regular tidal currents touch bottom. Faceted lamellibranch valves and gastropod shells are also absent in the muds of the tidal flats. Hollmann (1968) reported grinding marks and faceting on valves of *Cardium edule* which the animal acquires when it moves through the sediment.

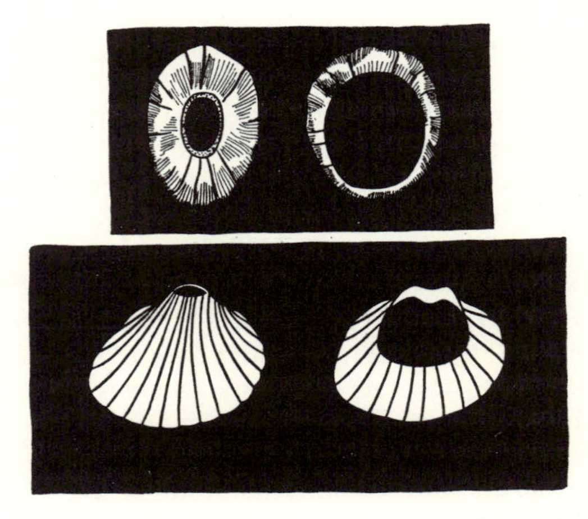

FIG. 98. Faceted gastropod shells and lamellibranch valves, originally of inverted-basin shape. Top (after Pratje, 1929), left: *Fissurella*, right: *Patella*; bottom: *Cardium edule*. (Schematic drawing)

Valves are more frequently broken by impact than destroyed by faceting. The fracture edges are irregular, but weak zones in the shell structure often determine where it breaks.

Cardium edule breaks most easily where the valve is most curved, near the beak; *C. echinatum* breaks frequently along the radial ribs; *Ostrea* breaks piecemeal, beginning at the outer edge, until only the hinge is left; *Mya arenaria*, *Scrobicularia plana*, *Solen*, and *Ensis* break along the growth lines, of the left valve of *Mya arenaria* only the hinge and the chondrophore remain, and the hinge of the right valve frequently breaks crosswise. Only the hinge remains of *Zirfaea crispata* and *Pholas dactylus*. *Mya truncata*, *Cyprina islandica*, *Mytilus*, and *Modiolus* break into large fragments. The flat parts of *Pecten opercularis* that are reinforced by ribs stay intact for a long time. *Barnea* and *Petricola pholadiformis* break crosswise, with the fracture also crossing the hinge; *Mactra subtruncata* breaks rarely, and *Donax vittatus* hardly ever.

The harder and coarser the substrate and the stronger the wave action at the bottom or the surf, the sharper is the selection. Eventually, only a few typical forms remain, but these in great numbers; thus the frequency distribution of preserved shells is entirely unrepresentative for the relative frequencies in the populations of the living species. In sandy areas, sand abrasion rounds the edges of the fragments so much that identification is difficult. Fragments in shell breccia keep sharp edges because shell fragments are not very abrasive. Considering all these facts it becomes evident that breakage is generally possible only where wave action and surf affect the bottom intensively; this means in the North Sea above a depth of 50 to 60 m. Below the zone in which wave action is effective (i.e. below the 'critical plane' of R. & E. Richter 1932), breakage is impossible in the southern North Sea (see also Schäfer 1941b). Tidal currents and exchange currents that touch bottom at greater depth are not powerful enough to break mollusc shells.

Klähn (1932 and 1936) investigated quantitatively and in detail how valves are mechanically and chemically affected, and he was the first to complement observations by experiments. Sharp sand polishes faster than rounded sand and fine-grained sand faster than coarse sand, individual shells are reduced more rapidly than shells packed in shell breccia where they have mutual protection. Pea-sized pebbles are more effective in the first nine hours of their activity than later on, because at first they polish the roughnesses and sculpturing of the shells; in sands *Cardium* is reduced to a 'minimum' within 56 days, a *Cardium* shell is dissolved in distilled water in 208 days, while it does not dissolve at all in water from the Baltic Sea which is saturated with calcium carbonate. Solution is the more intensive the smaller the weight of the original fragment (see also Pia 1933).

Calcium-carbonate-boring micro-organisms also play a considerable role in the reduction of the valves (Duerden 1902, Kylin 1935). Many investigators consider the boring thallophytes as the most important destroyers of empty, marine mollusc shells (Pia 1937). Mägdefrau (1937) came to the same conclusion for hard parts of fossil animals (for balanids, see Schäfer 1938b).

Wetzel (1937b) listed the following common calcium-carbonate-boring thallophytes: green algae (mainly *Phaeophila, Entocladia, Gomontia, Ostreobium*), blue-green algae (especially *Hyella*) and fungi (*Ostracoblabe, Lithophytum*). According to this author, shells affected by micro-organisms become chalky, their structure is loosened and made more vulnerable to chemical and mechanical influences because of the increased surface area. Only shells of dead or living animals in open water are affected, buried they are immune; but if buried shells are washed out after death, surrounded by oxygen-rich water, and regularly exposed to intensive light, they, too, are attacked. Shells that lie for a long time in the reducing zone of the bottom sediment are impregnated by iron sulphide and are made permanently proof against micro-organisms. They can be easily recognised by their bluish-black colour (see also Kessel 1936). Sulphide impregnation also hardens shells against sand abrasion. Thus an early cover maintained uninterrupted for some time gives the best protection for the shells, even if they are later repeatedly reworked.

The traces left by boring organisms are generally inconclusive. Species that may be affected by micro-organisms and the locations where they occur are listed in Table 11. Other destructive organisms are sponges (*Cliona*; see Volz 1939 for its ecology) and worms (*Polydora*, cf. gastropod shells).

TABLE 11

Lamellibranchs affected by boring micro-organisms

	Locality	Diameter of bore hole	Micro-organism
Ostrea, *Mytilus*, *Placuna*	Heligoland, Pacific coast, Chinese coast	0·013–0·024 mm	
Mytilus, *Ostrea*, *Placuna*	Baltic and North Sea	0·003–0·007 mm	? *Entocladia*
	Heligoland	0·004–0·007 mm	? *Ostreobium*
Dreissensia, *Nucula*	Fresh water, in foss. micaceous clay	0·0022 mm	*Plectonema*
Mytilus, *Anomia*, *Pecten*	North and Baltic Sea	0·0014–0·0028 mm	? *Lithophytum*
Mytilus, *Ostrea*, *Tapes*, *Pecten*, *Placuna*, *Fissurella*, *Buccinum*, *Concholepas*	Baltic, North Sea and Pacific	0·0018–0·0025 mm	? *Ostracoblabe*
Ostrea, *Pecten*, *Tapes*, *Astarte*	North Sea	0·0012–0·0014 mm	

Orientation of valves. On a vertical face an embedded fossil valve appears usually in profile. The position of the convex sides of valves can serve as a criterion for the original upward facing in folded beds. R. Richter (1922a and 1942, with bibliography) observed that a valve lifted by agitated water has a tendency to turn about a horizontal axis and to flip over so that it comes to rest on the opposite side. The facing rule states: generally, dish-shaped bodies (lamellibranch valves, certain gastropod shells, and brachiopod valves) come to a final rest with the convex side facing upward. The rule holds on most plane, water-covered sediment surfaces regardless of depth, if only the agitation or current are strong enough to flip the shells over.

Under certain circumstances and fairly infrequently, the majority of valves lie in a concave-up position; this can only happen in restricted areas and in environments that are easily eroded, so that valves with this orientation do not fossilise easily. Richter mentions the following cases causing a concave-up position: swash accumulations, current shadows downstream of obstacles (called still-water traps by Quenstedt 1927), particular morphological features of dish-shaped bodies, preservation of the living position or dying position in still water, or, possibly, orientation caused by sinking into mud.

Disk-shaped bodies remain unoriented if the water movement is too weak for tilting or if the bodies are prevented from tilting by close packing, by anchoring with byssal threads, or by being wedged between rocks or plants.

Richter made his observations on the beach, in the tidal zone, and in the top few metres at the edge of the permanently submerged area, but he generalises his rule for all depths. According to Pratje (1929) the convex-upward position occurs exclusively in the beach zone whereas at depths of 3 to 28 m off the islands of Norderney, Juist, and Borkum the concave-upward position predominates even in areas with considerable agitation of the water. He does not report how he obtained undisturbed sea floor samples on which he bases his findings. H. E. Reineck (1958b and c) sampled with an especially designed cutting box in which orientations remain undisturbed, and he found that the convex-upward position of valves predominates in the North Sea also at the depths covered by Pratje.

Valves are not only tilted but also oriented within the bedding plane; the body rotates around a more or less vertical axis. Richter calls one special case of this orientation 'swinging', the case of a valve pivoting around a fixed end, like a ship swinging when held by a single anchor. The pivot can also lie in the middle or on the side of a valve. We observed that, once a valve is turned convex-upward and has thus reached a relatively stable position, the valve becomes oriented with the heavy hinge pointing up-current. The orientation appears most clearly with nearly symmetrical valves. The more stable an orientation, the stronger the current required to dislodge it. Such an orientation can be observed in many other forms; washed-out worm tubes, for example, are found so oriented (Reineck 1960). Wherever unidirectional bottom currents exist, oriented valves can be expected; yet most observations today are made at beaches and in tidal zones or in flumes. In the future, under-water photography and television will bring more information about the orientation of valves at greater depths. Pictures of *Ensis ensis* from the English Channel (Barnes 1956) show orientation at considerable depth.

Not only single valves can be oriented but also complete shells still connected by the ligament, but gaping. The stability of such shells depends on their outline. If the edges opposite the hinges are straight and if the valves are elongated (as in *Mytilus*, *Modiolus*, *Arca*, *Venerupis*, *Petricola*, *Ensis*, *Phaxas*, or *Donax*), the valves, supporting each other in the current, can rest on the outer edges like the runners of a sledge. If the valves are rounded (as in *Cardium*, *Scrobicularia*, or *Mactra*) the shell always rests on one valve while the other points upward like the lid of an opened tin. Whichever way the valves rest on a substrate, the heaviest part, the hinge with the beak area, is always oriented pointing up-current

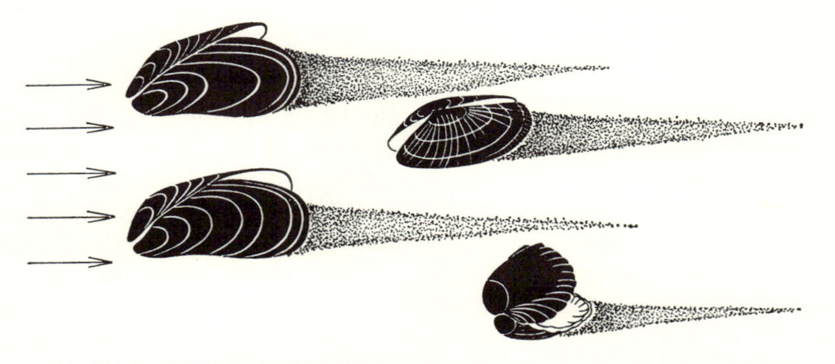

FIG. 99. Hinged valves of *Mytilus*, *Donax* and *Cardium* oriented by current. (Schematic drawing)

(Fig. 99). The valves gape only after being exposed to air where the ligament dries and shrinks and thus opens the valves.

The two lamellibranch or brachiopod valves of one species are often found in different numbers at a given locality; this has been described for both fossil and recent examples. The reason for this unequal occurrence is usually that the two differently shaped valves of one specimen are transported at different speeds and therefore over different distances. Richter (1922b and 1924a) distinguished three cases: (1) If the curvature of the right and left valves differs in a species, the degree and the kind of resistance offered to the water current is also different (e.g. *Ostrea*, *Pecten*, or *Angulus fabula*). They are thus transported different distances at different speeds from the living place. (2) If a valve has a projecting chondrophore (e.g. the left valve of *Mya arenaria*) the chondrophore acts as an anchor as soon as the valve has turned into the convex-upward position. The right valve does not project outward and is more easily transported. (3) If the two valves are of different strength as in *Anomia*, the harder valve can withstand more transportation before it is broken up than the thinner one. Thus external influences achieve some selection.

In the spatial disposition of the valves the term 'valve pavement' is used as distinct from shell accumulation. We call an aggregation of valves a pavement, if the shells rest on a bedding plane convex-upward after prolonged transportation. Jessen (1932a and b) distinguished between 'scattered pavement' and 'complete pavement'. If valves—usually of one species—have formed a complete pavement, newly arriving valves carried by the current find no hold on the smooth humps of the existing pavement and not enough space between the shells to anchor their edges in sediment until they reach the down-current side of the pavement where they continue the formation of the pavement. Usually valves accumulate where the power of the current wanes; such a place at the beach is the zone of spent waves below the swash mark, or on tidal flats near the mean high-water line. Repeated sedimentation over a pavement followed by formation of another pavement can lead to multilayered pavements. Valve pavements have been investigated in intertidal areas. There the wave action makes contact with the ground over a large surface and transports valves in the same direction and carries them to the same deposits. Therefore complete pavements can be observed mainly in muddy marginal areas of the tidal flats where fresh valves are brought in frequently but in moderate quantities. We can assume that complete pavements will also form on permanently submerged sea floors with strong bottom currents, but this has not been confirmed. However, one of Jessen's conditions for the formation of pavements—a poverty of species—is absent there. We think that scattered pavements are transitional and are in the process of being shifted from one place to another; but a very stationary, complete pavement must have started as a scattered pavement.

In shell accumulations, valves exceed sand in volume, and frequently they contain nothing but shells. All transitions exist from pavement to shell accumulation, and a pavement often forms the base of an accumulation (see also E III (*b*) for a discussion of shell breccia). Shells accumulated at swash marks usually lie in a disorderly fashion. Where the shells are subject to strong, lateral thrust or behind obstacles where the surf usually flows down, valves can be oriented in an upright position and be closely packed, but only if all the accumulated valves have approximately the same size (due to similar age) and if they are flat or only slightly convex as, for example, *Mya*, *Scrobicularia*, *Donax*, or *Pecten*. Any discordantly shaped or more convex valve disturbs the mosaic of the packed, upright valves. Upright valves can form several layers in a shell deposit (Schäfer 1941a and 1950c). Mii (1957) showed good pictures of upright valves (*Actinopecten yessoensis*) from the coast of the Japanese peninsula Natsudomari. As Häntzschel (1939b) has described for the first time, valves occasionally lie like roofing tiles on the floor of narrow tidal channels, and the direction of the current can be deduced from the sense of imbrication. It is not yet known whether such imbrication can also occur on the bottom of wide channels.

(*c*) CEPHALOPODA

According to Grimpe (1920, 1921, and 1925) and Jaeckel (1948 and 1958) these most highly developed marine invertebrates are represented in the southern North Sea by the following (dibranchiate) species: *Octopus vulgaris*, *Eledone cirrosa*, *Sepia officinalis fillouxi*, *S. elegans*, *Sepiola atlantica*, *Sepietta oweniana*, *Loligo vulgaris*, *L. forbesi*, *Alloteuthis subulata*, *A. media*, *Ommatostrephes sagittatus*, *Illex coindeti*, *Stenoteuthis bertrami*, and *S. caroli*.

Living areas and life habits. The species mentioned do not occur in the same abundance. Only a few of them spend the whole year in the southern North Sea; the majority are seasonal. However, the periodic visitors deserve special attention because they are so abundant. The only permanent inhabitants are *Eledone cirrosa*, *Sepietta oweniana*, and *Sepiola atlantica* which are uncommon and unimportant because they are small and lack hard parts. Some of the seasonal migrants are *Sepia officinalis fillouxi*, *Loligo forbesi*, *L. vulgaris*, *Alloteuthis subulata*, and *A. media*. These species usually spawn in the southern North Sea, and large quantities die there. *Ommatostrephes sagittatus* and *Octopus vulgaris* occur irregularly, but occasionally in huge swarms. The remaining species invade the southern North Sea only from time to time or are strays.

Sepia officinalis fillouxi arrives in May at the Dutch coast from the southwest; in Heligoland Bay this species appears only in the middle of June and stays to the end of September. According to Jaeckel (1948 and 1958) a few of them probably live permanently in the southern North Sea. The majority of *Loligo vulgaris* arrives in June and July in Heligoland Bay and disappears again in September. *Loligo forbesi* comes later in the year and also leaves later in October or November. *Alloteuthis subulata* invades Heligoland Bay each year in large swarms; the majority arrive in the summer, but a few appear in the Jade estuary at the beginning of April. In some years a few individuals are caught in drag-nets even in November and December. Almost every year, individual dead *Ommatostrephes* are washed ashore on the coasts of the East Frisian Islands between January and March. Catches give a distorted picture of the frequency of individuals of any cephalopod species; this holds in particular for the large and fast-swimming *Ommatostrephes* and *Loligo*. Off Spieekeroog in June we caught with the drag-net up to twelve *Loligo* in one haul, but up to five hundred of the smaller and considerably slower *Alloteuthis* (for the Dutch coast, see Tinbergen & Verwey 1945).

Since most of the species occurring in the North Sea require normal salinity they avoid, except for *Sepia*, the tidal flats; consequently, their carcasses seldom reach flats north of the East Frisian Islands and never the wide, muddy bays such as Dollart and Jade Bay. *Sepia*, however, is regularly caught during the summer in shrimp nets in the inner Jade Bay where the salinity is 29 per mil and the water temperature 18° to 20°C.

Death, disintegration and burial. Little is known about the life span of North Sea cephalopods; however, they grow rapidly and do not live long. Animals of several species die after spawning, for example *Sepia*, *Sepiola*, *Octopus*, and *Eledone*. Carcasses therefore become quite numerous at certain seasons. Captive specimens of *Octopus* whose mantles were 7·0 cm long ventrally, grew to a length of 19·0 cm within ten months. *Loligo vulgaris* grows even more rapidly. The animals reach a length of 20 cm in one year and are then sexually mature. Apparently the large species, especially the deep-sea cephalopods, require a longer period to reach their potential size; in several species unlimited growth can lead to enormous body sizes. According to Grimpe (1925) animals of the species *Alloteuthis subulata* continue to live and grow after sexual maturity. Jaeckel (1958) compiled a list of proven or probable ages of cephalopods living in the southern North Sea; none of them grow older than three or four years. It is probable that many fossil cephalopods lived much longer, for their external shells were often large, had thick walls, and were divided into chambers. Their development must have required longer periods than that of the recent decapod cephalopods which have few or no hard parts. Such hard parts suitable for preservation are the chitinous jaws, the sucker-rings and teeth on the arms, the chitinous, vestigial shells (gladius) and the calcareous cuttle-bones. The jaws look like beaks of parrots and have calcareous zones at the points.

Many individuals do not die naturally because they are a favourite food of many fish species and most marine mammals. The toothed whales feed regularly on them, some species exclusively. In many cases their much reduced dentition has small teeth, suitable only to cope with soft-bodied prey, as in *Grampus griseus*, *Globicephala melaena*, *Lagenorhynchus albirostris*, and *L. acutus*; *Phocoena phocoena*, *Delphinus delphis*, and *Tursiops truncatus* also feed almost exclusively on cephalopods during certain seasons, as do the large *Physeter*, *Kogia*, *Ziphius*, *Berardius*, *Hyperoodon*, *Mesoplodon*, and *Monodon monoceros* which do not occur in the North Sea. The movements of these marine mammals are influenced or even completely determined by the yearly or seasonal movements of the cephalopods. The hard parts of the prey, mainly the chitinous jaws, the gladii, and the calcareous cuttle-bones, accumulate in the faeces of these marine mammals. In the coprolites they form a disorderly aggregate of partly broken and split chitinous splinters and calcareous fragments. The typical layered structure of cuttle-bones allows easy identification, even of fragments.

After death the squid body floats; the musculature of the arms is completely limp. As a result, the arms and particularly the two tentacles that are normally carried in pouches droop from the head section and become considerably stretched. For several hours before death the tentacles can no longer be completely drawn in. Instances of cephalopod mass death with unknown causes have been described (see Jaeckel 1948). Thus in certain years millions of dead *Loligo pealii* which had been spawning in swarms off the east coast of North America are said to have been thrown on to the beaches. The Dutch ship *Vriedentrouw* is reported to have sailed, on 10th January 1858, for two hours through masses of squids which covered the water surface as far as could be seen.

Squid carcasses float for three or four days depending on the water temperature and disintegrate almost completely before the parts sink to the bottom. This total destruction during the period of drifting seems to be the reason for the absence of fossil remnants from decapod cephalopods. They are missing in fine-grained sediment from anaerobic depths that are otherwise favourable for preservation because the carcasses never reach them, least of all the gelatinous, tender, pelagic forms. When such animals or their moulds have been preserved in the geological past, the surrounding sediments were probably coastal muds that dried out from time to time. Only outside the water is a cephalopod sufficiently heavy to form an impression of its body with all the arms in the sediment. Jaeckel reported (after Vérany) that a large, 30 to 35 cm long *Sepia officinalis* weighs 5 kg, and *Ommatostrephes sagittatus* up to 15 kg, but on the average only 1 kg. A 145 cm long *Stenoteuthis caroli* from Juist weighed 13·1 kg. *Loligo* with a mantle length of 50 cm weighed between 1·6 and 2·0 kg. Today few complete, drifting carcasses reach the muddy sediments of certain tidal flats which favour impressions, but if they do they produce moulds. The soft body parts disintegrate completely except for the ink pouch and its duct; their skin full of iridocytes (Schäfer 1937) and the ink granules from inside the pouch can be preserved as body fossils.

When the drifting carcass of *Sepia* disintegrates, the broad, dorsal cuttle-bone separates, passes through a lengthwise gash forming in the dorsal skin, and floats away on its own. Cuttle-bones are transported to the northern coasts of Norway and they have been found at the coasts of the Shetlands and even Jan Mayen. They are suited for almost unlimited transportation as long as the lamellar calcareous layers remain undamaged and act as walls for gas-tight compartments. Occasionally, long beards of green algae are carried by cuttle-bones and are good evidence of the length of the journey. An undamaged cuttle-bone frequently ends its travels only when it is washed ashore. However, most cuttle-bones disintegrate while they are floating and are finally reduced to fine grit because either lime-boring algae settle in the spongy, loose mass, or they are hacked to pieces by sea birds and never reach shore. If a floating cuttle-bone breaks at sea, water begins to seep into the porous spaces, and the fragment eventually sinks. Large fragments float usually for less than eight days, smaller fragments less long. (According to Geissler 1938, empty shells of *Nautilus* do not float longer than four weeks.) Evidently, cuttle-bones, unlike hard ammonite shells, never reach the sea floor complete. Only the rostral end of the cuttle-bone can sink to the bottom because it is mainly chitinous and has only a thin layer of gas-conserving calcareous lamellae. Such sunken pieces, however, do not withstand repeated transport with the sediment, and preservation of fragments can only be expected in deep parts of the sea.

The number of cuttle-bones that are washed ashore give some indication of the abundance of *Sepia* in the North Sea. It can be assumed that the schools that migrate into the North Sea consist of animals of equal age that die at about the same time, because cuttle-bones arriving simultaneously at the beach are almost all the same size. There are years when cuttle-bones lie in continuous swash accumulations along the beaches or form accumulations numbering tens of thousands. At other times they are almost entirely missing in the swash marks of the sandy beaches and at the coasts of the tidal flats.

Some gladius bearers among the decapod cephalopods are washed ashore as complete carcasses; but the majority of animals dying at sea disintegrate there. In *Ommatostrephes*, *Loligo*, and *Alloteuthis* (and probably also in some others) the dorsal gladius separates from the disintegrating tissue as in *Sepia*. The gladius, being purely chitinous, sinks immediately to the bottom. Its preservation or the preservation of its mould is possible if the gladius sinks into undisturbed muds. In living decapod cephalopods, however, this shell organ is so reduced that the whole structure consists only of the thin proostracum that is reinforced by longitudinal flattened folds and whose end bears what is left of the rudimentary phragmocone. In some cases the phragmocone has completely disappeared.

Octopus has a poor chance of being preserved. Most species have no shells and many live in rocky environments.

V. Bryozoa

The bryozoans and especially the exclusively marine Gymnolaemata (also called Stelmatopoda) have certain qualities in common which generally favour their preservation as fossils and are helpful taxonomically. They live in sessile colonies called zoaria. The individual animals of a zoarium are called zooids. They rest with their soft body in a hard chitinous shell (zooecium) that is often calcified. Individual zooecia rarely reach a length of 0·1 cm; the zoaria, however, extend in some cases to 1000 square centimetres. The colony grows by asexual budding; either it extends laterally, and each individual zooecium of the colony is directly connected to the substrate, or the colony is upright and branched or grows in fronds; in this case most of the individuals are not directly connected to the substrate that carries the colony. Instead it is borne by a trunk or stem consisting of numerous individuals. Individuals which form the stem are modified and degenerated.

The soft body that rests in the chitinous-calcareous zooecium cannot be preserved. The lophophore and the tentacles that surround the mouth aperture are part of the soft body and can be retracted into the coelom cavity. The anus lies near the mouth but outside the ring of tentacles. Sexual reproduction occurs and leads to a pelagic larva which later metamorphoses.

Gislen (1930) classified all sessile, marine animals according to their external features (crustida-encrusting forms, corallida-stem or tree-shaped forms, etc. with many subdivisions). In Remane's (1943) opinion such grouping is just as important for an external characterisation of colonies as for botany the terms 'tree', 'shrub', 'liana', superficial as they may first appear (see also C. D. Müller 1966). The palaeontologist can afford even less to neglect the external features of a fossil, because in many fossils no anatomical and taxonomically important features can be seen. For the student of actuopalaeontology an understanding of the shape of sessile animals or colonies (whether encrusting, stem-shaped, or tree-shaped) is essential, for fossilisation depends mainly on the strength and nature of the connection with the substrate.

External skeleton. A thin, ectodermal epithelium that consists of a single layer of cells envelops the body; in the area of the posterior body section (metasoma) the epithelium secretes a chitinous, layered cuticle (Nitsche 1871).

The Ctenostomata exoskeleton is purely chitinous and is generally preserved only as moulds. In some of these forms small sand and shell particles can adhere to the gelatinous outer layer. These adhering particles partly conceal the structure of the individual zooecia. Purely chitinous skeletons are usually completely destroyed during fossilisation. No identifiable trace remains of these forms unless they have left a clear mould before their decay. However, Voigt (1933) believed their stolons can be preserved if they are incrusted; moreover, some forms bore or etch into their substrate, such as in the North Sea, *Lepralia edax*, *Terebripora*, and *Spathipora* according to Remane (1940). Their etchings and borings as well as their moulds may be preserved.

The Cyclostomata and Cheilostomata have a calcareous skeleton in addition to their chitinous cover; magnesium and calcium carbonates are deposited within the chitinous layer between a dense network of fine, horny fibres. Partial solution of this inner carbonate layer during fossilisation can blur or modify the sculpture of the individual zooecia. In some cases calcification is strong, and the colony as a whole becomes rigid and coral-like as in the Horneridae.

In the Cyclostomata (*Cancellata, Rectangulata*) and Cheilostomata a second skeletal wall is frequently found inside the other, especially in the basal and frontal area of the zooecium; as a result there is the primary body wall, the gymnocyst, as distinguished from a second wall, the cryptocyst; the two are separated by the hypostegal coelom. Since both walls, or either one of them can be preserved, three different fossil aspects of the same species are possible, even within a single colony. It is therefore necessary that all available zooecia of a colony are investigated in thin section, and not only part.

The box-shape and the tube-shape can be considered as the two basic external shapes of individual zooecia. The box-type is represented by zoaria that bud two-dimensionally or are crust-forming, the tube-type by the tree-like forms. Common features of the cuticular skeleton in all individuals are a frontal side, or aperture area, with the opening for the passage of the ring of tentacles, the orificium, and a basal surface.

Within a species the various external features are mainly constant; there are, however, some modifications in size, thickness, and shape, caused by external factors. Frequently, modifications are a reaction to the sessile way of life as in many other sessile epibionts (surface dwellers; exact definition, see section D I (*b*) about ethological and ecological interpretation of movements). Both in recent and fossil specimens many individuals from different biotopes should therefore be investigated for a complete diagnosis. Since the mode of budding in a colony and the spatial distribution of the zooecia vary particularly widely, depending on environmental conditions, they can therefore only be used with great caution for taxonomical purposes, and only the detailed morphological features of the zooecium are suitable for use in identification. The approximately 6000 species are based almost exclusively on these features.

The following features can be preserved and are also of taxonomic importance: shape, size, and composition of the apertural area (chitinous or calcareous), reinforced edge of the orificium, the peristome; apertural area free or protected by thorns or by frontal, broadly joining shields; size and shape of the autozooids, the normal, feeding and sexual individuals; presence and shape of kenozooids, forming tendrils, anchoring hooks, stolons, rhizoids, stem members, and diaphragm layers; distribution and shape of heterozooids, such as avicularia, vibracularia, gonozooids, and ovicells.

In the Cyclostomata and Ctenostomata the individuals are interconnected through the separating walls by scattered pores; as a result, the fluid filling the coeloms can flow between neighbouring animals. In many cyclostomes the calcareous incrusted wall bears numerous 'pseudopores' which are closed on the outside by a chitinous membrane and thus do not communicate with neighbouring individuals. They exchange gas between the external medium and the coelom. In the cheilostomes these pores are clustered into so-called rosette plates. Pores, pseudopores, and rosette plates are known in fossils but have rarely been used for taxonomical purposes.

Budding and formation of zoaria. Budding is the asexual form of reproduction in bryozoans. Each colony starts with a single, sexually reproduced, individual free-swimming larva (or cyphonautes) settling on a substrate of its choice. This larva begins to bud in the apical region of its body wall and becomes the first individual zooid, or ancestrula, of the future zoarium. The modes and places in which the budding continues and the spatial distribution of the secondary individuals, both autozooids and heterozooids, determine the shape and the growth pattern of the forming colony.

Generally, three buds spring from the ancestrula, and this can serve for its identification in the colony because the secondary individuals of incrusting forms usually produce only two buds or one. The formation of the zooecium always precedes the complete separation of the polypid. Budding is called zooecial, if autozooecia arise from other autozooecia. In some cases a single bud is later subdivided by cross-walls, resulting in several zooecia, as in *Membranipora membranacea.* On the other hand, several buds may combine to form a single kenozooecium serving as attachment of the zoarium to the substrate. In these cases autozooids no longer bud and thus are terminal members of the zoarium.

Living areas and conditions for preservation. As sessile animals, the bryozoans need a durable, more or less solid substrate. For many species it is irrelevant whether this substrate is a living animal or inanimate object. When settled on living substrates that may themselves be sessile or locomotory, the bryozoans become epizoa. But this is fortuitous, and all that most species need is a firm and, most important, durable base (Marcus 1921).

The substrates consist of three main types (see also Remane 1940 and Hincks 1880): (1) primary hard substrates such as rock kept free from sedimentation by currents or steepness, Pleistocene boulders or erratic blocks, or pebble beds; (2) secondary hard ground, such as dead organic hard structures; shells and other exoskeletons of living animals, or living plants, and (3) sand. Although the marine bryozoans of the southern North Sea are best preserved where muds and muddy sands are being deposited, these areas never are the habitat of living bryozoans.

The primary hard grounds encourage the richest development of bryozoans but are also the most unfavourable for fossil preservation. Fossils from this biotope can only be expected to be preserved where currents remove them to areas where fine-grained sediments are being rapidly deposited. The shrub, leaf, and tree-shaped varieties are carried away most easily. But the incrusting species remain in place and are therefore rarely preserved. Where hard sea floors lie below 50 m, wave action is usually too weak to transport shrub-like bryozoan colonies. The following forms live primarily on hard ground: *Alcyonidium gelatinosum*, *Caberea ellisii*, *Cellepora avicularis*, *Crisia*, *Diastopora*, *Escharella immersa*, *Farella repens*, *Flustra securifrons*, *Hornera lichenoides*, *H. violacea*, *Idmonea serpens*, *Lichenopora*, *Membranipora pilosa*, *Porella*, *Retepora beaniana*, *Schizoporella linearis*, *S. hyalina*, *Scrupocellaria*, *Smittina*, *Tubulipora*, and *Umbonula verrucosa*.

In our context the biotopes of the secondary hard grounds are more important. Organogenic, dead, hard structures like lamellibranch valves and gastropod shells furnish good places for settlement in all areas of fine-grained sand and in mud. Colonies settling in such places, however, are endangered by further sedimentation, and their life span is restricted, as shown by their generally small size. *Membranipora membranacea*, one of the most frequent incrusting bryozoans of the southern North Sea, can grow to hand size within two months. Gastropod shells, lamellibranch valves, and crustacean skeletons that lie isolated on the sediment surface are favourite settling places, for they are often the only suitable substrates over wide areas of sand or mud, and the free-swimming cyphonautes settle on them in great numbers. Sand and mud concretions play a similar role. In fossil concretions such zoogenic crusts provide a well-defined boundary layer with the surrounding sediment.

Living substrates temporarily protect the colonies from the dangers of sedimentation because host animals avoid being covered. A considerable length of time usually passes before the host dies, or, if it is a crustacean host, before it moults and discards the epizoa with the old skeleton. Bryozoan colonies on mobile animals, particularly crustaceans, are therefore often quite large. The conditions for their preservation are favourable. The following species of the North Sea settle on secondary hard grounds other than plants: *Alcyonidium gelatinosum*, *Bugula flabellata*, *B. neretina*, *Flustra foliacea*, *F. securifrons*, *Gemellaria loricata*, *Scrupocellaria*, and in brackish coastal areas *Victorella pavida*. Leaves of submerged higher plants (e.g. *Zostera*) and thalli of algae are suitable for permanent settlement of bryozoans because they are rarely covered by sediment. Incrusting forms and those with clinging stolons live most frequently on plants; shrub-like colonies are rare. Wave action can transport the plants together with their epizoans into muddy and sandy sediments where they may be buried. Because mechanical and chemical effects destroy the plants rapidly, fossil incrusting bryozoans seem to lie freely on the sediment surface and seem to have grown there. The following bryozoans live on plants in the North Sea: *Aetea*, *Alcyonidium polyoum*, *A. hirsutum*, *Bowerbankia*, *Cribrilina*, *Crisia*, *Farrella*, *Flustra hispida*, *Flustrella*, *Hippothoa*, *Lichenopora*, *Membranipora membranacea*, *Scrupocellaria reptans*, *Tubulipora*, and *Valkeria uva cuscuta*.

Only few forms are adapted to a life in pure sands. Remane (1936, see also 1938) discovered and described an individual, hardly 0·2 cm long, bryozoan living in solitude in the coastal sand which he called *Monobryozoon ambulans*. This animal can attach itself to sand grains by tubular processes provided with cement glands; the same organs allow active movement through the interstitial spaces in the sand. These processes can also bud. The solitary and locomotory way of life is probably an adaptation to life in sand. Fossil preservation is impossible in this environment.

The small, branched colonies of *Cellaria fistulosa* and *C. sinuosa* live on sand surfaces; however, both can also be found on lamellibranch valves. The thick calcification on the individual branches that are kept articulating by connecting chitinous sections seems to be an adaptation to this habitat. Fossilisation of these animals in their habitat is possible. The different species of the genus *Cryptozoon* also seem to live in sand (Cori 1937). Their stolons form dichotomous branches, the zooecia sit in the forks and are enveloped in spherical, sticky masses of sand grains into which the tentacle rings can be retracted.

VI. Polychaeta

Polychaete worms are rarely preserved as fossils in marine sediments. Polychaetes lack a supporting and movable skeleton, and unlike, say, a medusa, their body lacks a specific shape that would attract attention if it were preserved as a mould. Almost every detail known about the existence, distribution, and life habits of these worms in ancient marine sediments we owe to the fact that their life activities produce recognisable traces in the form of trails and burrows. The only hard parts that can be preserved under certain conditions are certain cuticular structures, the jaws, teeth, bristles, scales, and paragnaths.

Only the Errantia have *jaws*. These animals are creeping predators or scrape in search of detrital food particles, or graze. The Sedentaria, on the other hand, remain in one place and feed by grasping with their tentacles or by whirling water toward their mouth and therefore need no jaws.

The masticatory mechanisms of the polychaetes evolved from simple chitinous swellings of the oesophageal epidermis serving as protective, firm coverings of the oesophagus wall where it is most heavily used. From this beginning more complex chitinous structures have developed such as a toothed, hook-shaped or stiletto-shaped jaw component or individual teeth provided with an effective musculature. These jaw parts and teeth are called scolecodonts.

The shapes of the scolecodonts depend on the way of feeding and vary from one species to another. Similarities do not necessarily indicate a close phylogenetic relationship but rather analogous feeding habits. By their function and morphology two groups of jaws can be distinguished.

(1) Composite jaw mechanisms consisting of several individual teeth and bars are a single functional and often also morphological unit. Evolved from extensive, chitinous swellings they obtain their particular shape from folds and processes of the epidermis. Frequently, the masticatory mechanism is buried as a unit with blades, bars, muscle scars, and teeth. The Eunicides, *Eunice*, *Onuphis*, *Lumbriconereis*, *Ichthyotomus*, and *Ophryotrocha*, have such jaws. In these animals the jaw mechanism

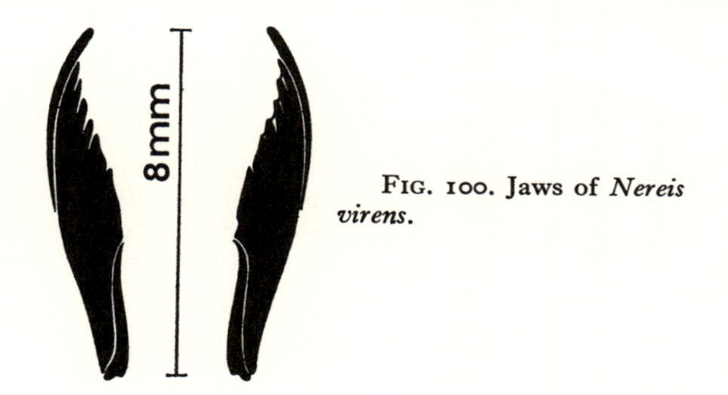

FIG. 100. Jaws of *Nereis virens*.

lies in a cul-de-sac ventrally from the oesophagus (Heider 1925). In some species the masticatory mechanism is adapted to an ectoparasitic way of life.

Usually, the individual jaws of such a composite jaw mechanism are designated by Roman numerals, from the most posterior forward; the chitinous jaws and blades showing muscle scars are called supports. This terminology is very rarely applicable in fossils because the number of individual teeth varies from one species to another and because the numerical order derived from the sequence in the oesophagus does not correspond to form or function. Neither is the numerical sequence any help in the reconstruction of a jaw mechanism from individual, available parts, because the numbers in different species do not apply to homologous structures. Another classification of the jaws into mandibles and maxillas as in arthropods is unsatisfactory because the jaws of the polychaetes have no relation to those of the arthropods which are modified limbs.

(2) Separate jaws that occur as a pair or in two pairs have an attachment plate on which the muscle is anchored, and a free, usually dark-coloured blade which is usually richer in chitin. The jaws have no chitinous connection with each other; consequently fossil jaws are usually found individually. The Nereidae (Fig. 100) and the Nephthydidae are

generally predatory animals and have two jaws. They are anchored in the musculature of the pharynx. The same holds for the Aphroditidae (*Aphrodite aculeata*, *Lepidonotus squamatus*, and *Gattyana cirrosa*), the Pisionidae (*Praegeria*) and the Glyceridae (*Glycera*), but all these animals have four jaws. They are set in a circle around the front end of an eversible pharynx, the points are directed inward and toward each other (Fig. 101). The food of these animals consists of vegetable and animal detritus and of small worms, gastropods, and lamellibranch larvae. The teeth serve to hold moving prey, but are also used to shovel detritus. *Lepidonotus* inserts its everted pharynx into the tube of *Sabellaria spinulosa* whose diameter it fits perfectly; then *Lepidonotus* grasps the *Sabellaria* with its four jaws and pulls its head out of the tube like the cork out of a bottle.

FIG. 101. Top view into the mouth of *Lepidonotus squamatus*. Four jaws (black) and radial mouth lobes. The lobes carry ciliated cells (not shown) and serve as tactile sensory organs. (Actual size × 20)

Scolecodonts are rarely found in fossil marine sediments with many trace fossils. Their absence is not due to a lack of jaw-bearing polychaetes but to the poor conditions for preservation during diagenesis. The immense number of Recent, jaw-bearing, benthonic worms gives an indication of the quantities of scolecodonts that are constantly released into the sediment. In certain muddy sediments 4000 *Nereis diversicolor*, 300 *Phylodoce lamellosa*, 80 *Nephthys hombergi*, 15 *Scoloplos armiger*, 10 *Magelona papillicornis*, and 5 *Aphrodite aculeata* live on one square metre. Accumulations of scolecodonts cannot be due to sorting by bottom currents; they may be caused by birds and fishes which feed on great quantities of these worms. Their faeces can consist almost exclusively of scolecodonts. Mr Ehlers, who was bird warden on Mellum for several years, told me that certain sandpiper species on the tidal flats of that island preferentially eat Nereidae and *Nephthys* during certain times of the year. Their faeces contain great quantities of worm jaws. In the deep sea flatfishes prefer to feed on polychaetes at certain periods, also producing considerable accumulations of the small, chitinous jaws in their faeces.

Bristles of polychaete worms are more numerous in marine sediments than scolecodonts. There is hardly a sample that is free of them. They are rarely reported, however, and the reason is that they are destroyed or washed out during the usual sampling and sieving procedures. They are especially frequent in mud pebbles with a well-preserved fabric. The number, size, and shape of bristles of living polychaetes are quite well known because these features are important for their taxonomical classification. The chitinous bristles of the living animal are either yellowish, brownish, glass-clear, or golden iridescent. According to Hempelmann (1934) the bristles consist of an outer hyaline covering layer and a centre made of fibres or slim, hexagonal prisms. Most bristles are solid; but there are some exceptions. Thus in the bristles of the Nereidae a firm

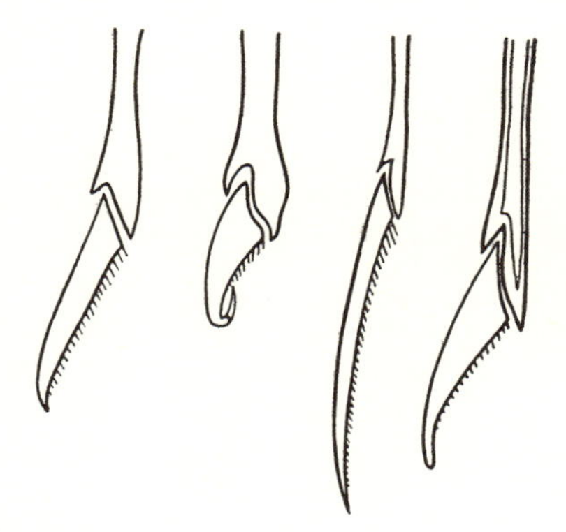

FIG. 102. Composite bristles of several *Nereis*. From left to right: *Nereis zonata*, *N. irrorata*, *N. dumerili*, *N. diversicolor*. (After Friedrich, 1938).

mantle surrounds a row of central cavities that are arranged like coins in a roll and separated by thin walls. These bristles appear horizontally striped. They easily disintegrate beyond recognition after the animals' death. Generally the bristles grow from depressions in the skin as cuticular secretions from a basal cell. This cell produces the central part of the bristle whereas the cells of the follicle wall of the bristle pouch probably build the outer covering. Muscles that can move the bristles in various ways are attached at the pouch walls. Some bristles can be replaced by new ones which form in the bristle pouch before the old ones drop out. Shed bristles can occasionally be found in the mucus lining the walls of polychaete burrows together with shell and sand particles. The bristles can bear processes of hook, arrow-head, awl, shovel, and various tooth shapes at their terminal end. Their functions are not clear. Some Polychaeta have composite bristles (Fig. 102), a feature important for their taxonomy. They consist of a shaft with a beaker-shaped end, into which the terminal section is inserted either straight or obliquely. The shaft serves only as a support for this terminal section which is often provided with teeth and hooks. The composite bristles disintegrate soon after death into their two components. A single animal body can carry many forms of bristles. Short, sturdy hooks and sickle-shaped bristles that are frequently toothed and have only short shafts are called uncini. They occur in many sedentary polychaetes and are often set in rows.

The bristles of some large Polychaeta are best preserved if the complete animal was embedded in muddy sediment. In such a case the thick bristles and the four sturdy jaws of *Aphrodite aculeata* are found in their natural positions. The strong aciculae whose function is locomotion in the sediment lie in rows of bundles in the sediment covered by a furry felt of small iridescent or greyish-white covering bristles. Even the elytra remain well preserved in their original, scale-like arrangement. However, the slightest reworking of the sediment destroys the mutual relationships and scatters and breaks the bristles.

Moulds and disintegration of carcasses. Several fossil moulds of Annelidae have been described. They always lie on the bedding plane, and in some the jaws and bristles can be identified. In our experience, moulds on the bedding plane showing a clearly segmented relief can form only on a wet sediment surface exposed to the air. Then a moist and still lifelike worm is heavy enough to impress its shape into the soft and yielding substrate, and it can dry and shrink in its mould. New sediment can then fill and cover the mould. Under water, on the other hand, a dead worm with a density hardly more than that of the water is too light to make an impression in the sediment.

Animals that live in burrows usually leave them before death if possible, including polychaete worms. It could therefore be expected that many annelids produce moulds; however, on the tidal flats of the North Sea, predatory animals, such as crustaceans, birds, and worms, soon eat the exposed animals. A polychaete worm dying in its burrow usually leaves no trace. Polychaete worms were observed dying on the surface of tidal flats at low tide. If death occurs in a large settlement of *Arenicola*, it generally affects many animals at the same time. The animals perform peristaltic movements and move about slowly without burying themselves. If the sun shines the animals die approximately after one day. Dying begins at the tail end. The muscles in the posterior region contract intensely, forcing some faeces out of the anus. Simultaneously, the most posterior six to ten segments are autotomated partially by sharp contractions with the result that faecal sediment seeps or even squirts sideways from the ruptured intestine (Fig. 103). The body section posterior to the point of autotomy soon dries in the sun, and later lies as a thin, segmented skin in the depression made earlier and now too wide for the shrunken remnants. The same happens successively to each of the six to ten segments, in forward progression. Sediment leaks out of each wound, or, occasionally, if contractions are too weak, the sediment remains in the intestine as in a string of sausages, and in this case the body dries to a thin skin retaining the sediment column which is segmented crosswise or obliquely. After several hours the worm is dead except for the anterior segments, but these and the everted pharynx continue to show contractions running anteriorly. It takes another few hours until all life has ceased.

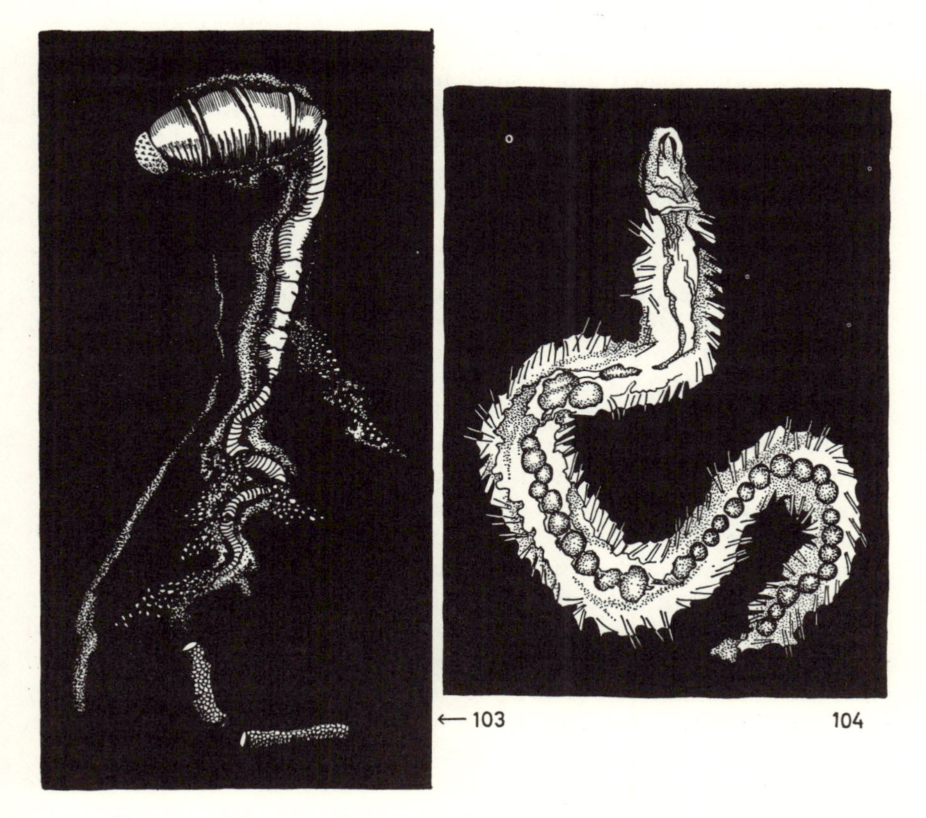

FIG. 103 (*left*). *Arenicola*, dying of exposure to air. Anterior portion with proboscis is still alive. Behind it, several portions of the body are severed by autotomy, their intestinal contents having been squirted out. (Actual size × 0·8)

FIG. 104 (*right*). Dead *Nereis* desiccated on the sediment surface. Beaded chain of faecal globules is visible through the skin. (Actual size × 0·8)

Nereis and *Nephthys* and most other Polychaeta that die on a surface exposed to the air do not usually autotomate. The body remains complete and flattens so that the paired bristles lie along the margin (Fig. 104). The position of the intestines is indicated by a string of usually sand-rich balls of faeces that lie close together. The body dries to a mummy in which the original segmentation can no longer be discerned. Jaws and proboscis are usually clearly preserved.

Agglutinated and calcareous tubes. Worm tubes can on occasion become fossilised. In another section of this book tubes are subdivided into three groups according to the building material, shell and sand particles, mud, and body secretions.

Tubes agglutinated from shell fragments or sand are sometimes preserved *in situ* and in the original vertical position. However, they can also be transported and be embedded, individually or washed together, in a horizontal position. Felted accumulations of *Lanice conchilega* tubes made up of sand or shell grit can be found washed up at the swash marks of beaches and also in deep water. R. Richter (1924d) noted that such accumulations could be the origin of certain quartzites with a 'woven' texture and of 'tendril-like' patterns on bedding planes. Tubes built of mucus masses do not withstand transportation and are only found *in situ.*

Tubes built from body secretions, mostly calcium carbonate, are especially important. Such tubes are always constructed on firm substrates where there is no danger that they might be covered by sediment. The builders of such calcium carbonate tubes spend their entire lives in them. They are thus sessile and subject to all the advantages and disadvantages of this way of life. In most cases these calcareous tubes are preserved together with the substrate (*Pomatocerus* on lamellibranch valves and gastropod shells); more rarely the substrate is destroyed before or after embedding (*Spirorbis* on thalli of algae), or the worm tubes are separated from the substrate and preserved with fragments of gastropod shells and lamellibranch valves. When the calcareous tubes are carried with their substrates on to the pebble zones of beaches, they are easily torn off. Between the surf-beaten pebbles of the Heligoland dune, calcareous worm tubes are numerous and often strongly abraded.

If only small fragments of calcareous worm tubes (Serpulidae) are found, it is often difficult to distinguish them by their external appearance from those of the Scaphopoda and Vermetidae. Microscopic methods can be used to distinguish them (W. J. Schmidt 1951 and 1955):

Cross-section of the Serpulidae: always concentric laminae.

Cross-section of the Vermetidae: the middle and usually thickest layer shows concentric laminae; the internal and external layers show laminae at right angles to the tube surface; the internal layer is sometimes missing.

Cross-section of the Scaphopoda: the middle and usually thickest layer shows laminae at right angles to the tube surface; the internal and external layers show concentric laminae.

Longitudinal section of the Serpulidae: the internal thin layer shows laminae parallel to the tube axis; the thick outer layer shows laminae that are parabola-shaped or are at more or less wide angles with the tube axis.

Longitudinal section of the Vermetidae: the middle layer, by far the thickest one, shows laminae that are parallel to the tube axis; the internal and outer layers show laminae at right angles to the tube axis, and the internal layer is sometimes absent.

Longitudinal section of the Scaphopoda: the middle, usually thickest layer shows laminae at right angles to the tube axis, the internal and outer layers show laminae parallel to the tube axis.

VII. Coelenterata

Coelenterata comprise so many types and life habits that no general statements about conditions for their preservation can be made. There are sessile and vagrant forms, benthonic and planktonic animals, inhabitants of sand and mud, others that settle on hard ground, solitary and colonial forms. The variety in appearance is increased by the frequent polymorphism involving both the configuration and the habitat of the different polymorphs of a single species. In response to a wide variety of habitats and life habits, the structure varies widely from species to species. In some the body is gelatinous without any hard parts and therefore the most perishable; in others the body is muscular and firm even if devoid of a skeleton; some have periderm structures and exoskeletons of characteristic shapes, and some form very large colonies and produce very hard and durable calcareous structures.

Of the coelenterates inhabiting the southern North Sea only a few are important for actuopalaeontology: among the Hydrozoa, the hydroids (except Trachylina and Siphonophora, not present in that part of the sea or not preservable); among the Scyphozoa, the Semaeostomeae, the Rhizostomeae, and, possibly, the small polypiform Lucernariidae. (The Carybdeidae live only in tropical waters and will not be considered.)

Many forms of the Anthozoa (Octocorallia and Hexacorallia) have hard and durable calcareous skeletons, yet the few that live in the southern North Sea are almost exclusively forms without, or with small, or with very perishable skeletal parts.

Of the Octocorallia, only the skeleton-forming Alcymacea can be investigated. The Pennatularia (sea pens) live in deeper parts of the coastal zone and only exceptionally in the southern North Sea. The Hexacorallia (Ceriantharia and Actiniaria) have no characteristic hard structures; however, their presence can be recorded in the sediment by fossilised traces of life activities.

None of the other Anthozoa with their characteristic hard structures live in the southern North Sea; possibly because the salinity is too low, or the rate of sedimentation too high, or because hard ground is so rare. Many skeleton-building forms live in areas not far from the southern North Sea (on the coasts of Scotland, Norway, Sweden, and in the Norwegian Channel) where they have a good chance of being preserved. They are especially important because they live at greater depth and lower temperatures than usual for most Anthozoa. Examples of the Hexacorallia are the Antipatharia and the Madreporaria. The Antipatharia live almost exclusively in tropical and subtropical waters; some representatives of the North Atlantic seem to occur in the northern North Sea, probably in zones with primary hard grounds and deep water. Both the solitary and the colony-building Madreporaria grow on bedrock, stony grounds, and coarse shell deposits, mainly at great depth, but some forms grow little below the low-tide line (Broch 1922, Pax 1934). The most important reef builders are *Lophohelia prolifera* and *Amphihelia oculata* which form banks and reefs and incrustations. Their compact, aragonitic skeleton is whitish. The skeletal elements of the individual polyps are interconnected. Banks of these animals are known to exist at depths of 200 to 600 m near the Atlantic coasts of Europe, the north coast of England and Scotland, the Orkney and Shetland Islands, the Norwegian coast, and the west coast of Sweden, off Koster Islands at a depth of 80 m (Pratje 1924, Remane 1940, Teichert 1958).

The Gorgonaria live in the deep, littoral zone of the North Sea (100 to 200 m), in the Norwegian Channel, in the Skagerrak and Kattegat. They are anchored by stolons in soft ground at considerable depth, or they form covering membranes on hard substrates. In most cases their skeleton is composed of isolated, disk, needle, or scale-shaped sclerites consisting of calcium carbonate (calcite, see also Schmidt 1924).

(*a*) HYDROIDEA

Hydroids are hydrozoans with alternating generations. As a rule, a sessile generation of polyps produces vagrant, sexual medusae by asexual budding which in turn produce polyps by sexual reproduction. This alternation in generations is frequently modified, and sexual individuals may remain sessile, or the entire sexual generation is suppressed, or the medusa generation is the main form while the polyp generation shows certain reductions or can be missing altogether.

The structure and appearance of the hydroids is determined by the mode of growth, by the exoskeleton, and the polymorphism of the individuals; the same factors determine the appearance of the fossil and also the mode of preservation.

Formation of the colony. Polyps do not bud from an existing polyp; instead the new buds develop from the supporting stolons or stems of the polyps. Because the various modes of budding determine the growth of a colony, a preliminary characterisation by colony-appearance (Gislen 1930) is therefore useful (Fig. 105); Gislen's classification is explained in the section on bryozoans.

FIG. 105. Several modes of budding of colonial hydrozoans. Top left: stolons; top right: rhizocauloma; bottom left: monopodium; bottom right: sympodium. (Schematic drawing)

The three modes of budding and the resulting shapes of the colonies in hydroids are as follows: a stolon creeping on the substrate sends up individual, unbranched polyps (*Clava*, *Obelia geniculata*, *Grammaria serpens*, *Laomedea johnstoni*, *Campanularia*). These are Gislen's 'crustide' forms, or Remane's (1943) 'stolonial-creeping' forms. Or the stolons rise from the substrate, support each other, grow upward, and are intimately interwoven, forming a so-called rhizocaulome. The upright stems branch irregularly, and the positions of the polyps on the interwoven stolons are equally irregular (*Lafoea dumosa*). These are Gislen's 'chitinous corallida', that is colonies on stems with a chitinous exoskeleton. Or, the colony branches in a regular fashion from the upright stem of the polyp (hydrocaulus). The mode of branching is either monopodial (*Eudendrium*) or sympodial (*Laomedea*). These are Gislen's 'corallide dendrides', that is tree-shaped colonies on stems.

Exoskeleton. The polyp body (hydranth) is generally goblet-shaped and is provided with contractile tentacles. It sits on the freely rising hydrocaulus through which it communicates with the other individuals of the colony. The polyp body has evolved from the gastrula. It is highly perishable since a supporting tissue such as a large cushion of a gelatinous and firm mesoglea is generally missing. Only those animals that secrete a firm, chitinous cover, the perisarc or periderm, can be preserved as fossils. According to the form of the periderm, the order is subdivided into two suborders, the Athecatae, whose polyp is not surrounded by a cover of periderm or theca, and the suborder Thecaphorae whose polyp does have a stiff, peridermal theca into which the polyp can retract. It is called the hydrotheca, is generally goblet or tube-shaped, and can in many cases be closed by a lid. The Thecaphorae have a more differentiated peridermal exoskeleton than the Athecatae in which only the stolon and hydrocaulus are hardened by periderm. The Milleporidae and Stylasteridae secrete aragonitic skeletons around the rhizocaulomas of their stolons. Two of them, *Stylaster gemmascens* and *S. norwegicus*, live in the North Sea, and their calcareous skeletons resemble those of certain colonial corals.

Polymorphism. In most cases the individuals of a polyp colony have a division of labour which may be expressed in modifications of their shape. Differentiation into feeding and reproductory polyps (gonophores) is common. Gonophores either become freely swimming medusae, or they may remain attached to the colony. Further specialisations are: spiralzooids with defence function in the *Hydractinia*; dactylozooids, also with defence function, in the Stylasteridae; and sarcostyles with unknown function in the Plumulariidae. These specialised individuals generally have the same chances of being preserved as other parts of the colony because as a rule, they have the same type of periderm structure (e.g. gonotheca) as the undifferentiated individuals.

Fossil preservation. Preservation of hydroids depends on especially favourable conditions of the environment and the embedding medium, and is therefore rare in spite of their frequency and wide distribution.

Fossil Athecatae can rarely be recognised as organogenic structures and even more rarely identified. The bodies and moulds of Thecaphorae, however, can become identifiable fossils.

Generally, colonies with creeping stolons have very little chance of being preserved because they are intimately attached to hard substrates. They cannot be broken off and carried to areas of fine-grained sediment where they could be buried. On the other hand, if a piece of substrate breaks off and is transported the delicate adhering colonies are rapidly crushed and abraded.

An exception are certain encrusting species, *Hydractinia echinata* and *H.* (*Podocoryne*) *carnea*, belonging to the family of the Bougainvilliidae. They have stolons covered by a thorny or spiny layer of calcareous perisarc. According to Kessel (1938a) this perisarc is produced from solutions originating in the coenosarc of the colony. Elongate, quite sturdy aragonite needles crystallise from this solution and arrange themselves into spherulites. Both species grow preferentially on gastropod shells inhabited by hermit crabs; this improves their chances of preservation. The calcareous deposits that encrust the colony are so resistant to impact, pressure, and transportation, that they are as durable as the shell itself.

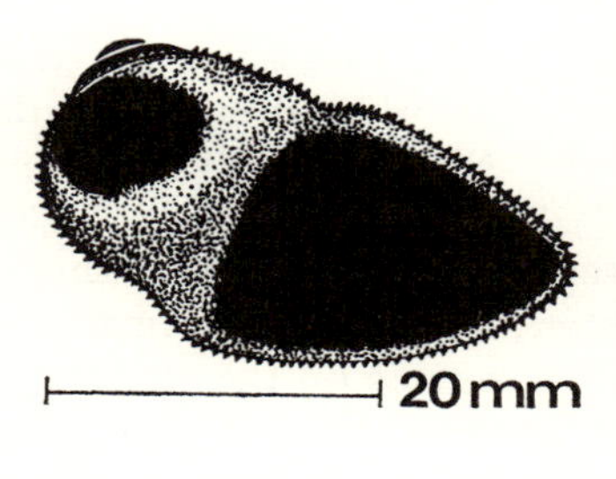

FIG. 106. *Lunatia nitida* shell, formerly inhabited by *Eupagurus* and with overgrowth of *Hydractinia echinata*. The aperture of the gastropod shell has been elongated by *Hydractinia* perisarc structures, so as to conform with the external shape of the hermit crab. On the underside of the shell the overgrowth is rubbed off

Gastropod shells (*Buccinum*, *Littorina*, *Lunatia*, *Turritella*, and *Scalaria*) covered by dead growths of *Hydractinia* and with clear marks of former habitation by hermit crabs are quite frequently found in the much transported material of coastal swash marks. The growth can be so dense that the gastropod shell is completely covered by the spiny perisarc of *Hydractinia*. The stolon network covers not only the outside of the shell but also the inner wall of its cavity. In addition it can also project from the free edge of the shell, following the shape of the hermit crab rather than the way a gastropod shell would grow. Thus the shell is enlarged and its opening becomes elongated while the hermit crab inhabits it (Fig. 106). *Hydractinia* occurs particularly frequently on the shells of *Buccinum*, *Littorina littorea*, and *Lunatia*; occasionally, however, it grows on concretions of calcareous shell fragments or on fine sand (Schäfer 1939b); this shows that *Hydractinia* does not depend on symbiosis with the hermit crab as absolutely as is generally assumed. According to Dahl (1928) *Hydractinia* (*Podocoryne*) *carnea* prefers the shells of *Nassa* to those of any other gastropod.

Colonies with a rhizocauloma or with a hydrocaulus preserve well in fine mud as body fossils or as moulds. Since the major part of a colony rises above the substrate, any small, hard object such as a gastropod shell, lamellibranch valve, living lamellibranch, worm tube, or stone can serve as substrate. Such colonies can therefore live and grow abundantly in areas with sedimentation. They can be buried when unusual weather conditions increase the rate of sedimentation, or when they are torn out and transported to some distant area of rapid sedimentation. Only in fine mud can the body or its mould be preserved. Inhabitants of the shallow sea (*Sertularia cupressina*, *Abietinaria abietina*, *Halecium halecium*, *Serturella polyzonias*, *Hydrallmania falcata*) can be transported over wide distances after storms and yet retain their shape for a very long time if they belong to the thecata forms. Due to this fact several of them have become economically important. For decades 'fishing' of sea moss has been undertaken in the tidal channels of the southern North Sea down to depths of 5 to 20 m. The dried, chitinous skeletons of the graceful, tree-like colonies are used to decorate wreaths and bouquets. Somewhat out-of-date figures on harvests indicate the frequency of these animal colonies in the southern North Sea: catch off Schleswig-Holstein in 1909, 57 000 kg; catch off East Frisia in 1910, 53 000 kg.

Not all hydroids are sessile and colony-forming. *Nemopsis* and *Margelopsis* are pelagic: their distribution is therefore much wider than that of the other species. The following species do not build colonies but are solitary: *Acaulis primarius*, *Corymorpha nutans*, *C. nana*, *Tubularia dumortieri*, *Myriothela*.

The frequency of hydroid medusae can be so high that they form a major constituent of the marine plankton; but they are so small and light, even out of water, that they rarely leave moulds on the yielding, watery mud of the tidal sea. Obviously, they leave no trace if they sink to the bottom after death. Impressions in tidal mud are usually circular depressions with distinct margins but a smooth, featureless centre in which small central depressions may occasionally be seen (the mould of the mouth tube). Häntzschel (1937) published pictures of such moulds of a swarm of stranded *Phialidium* medusae. Where drying animals were lying close together the boundary between them became a straight line. Others were distorted to an elliptical shape.

(*b*) SCYPHOZOA

Whereas only, or almost only the polyps of the hydroids have a chance of being preserved, the medusae rather than the polyps of the Scyphozoa have a better chance of being preserved as fossils, and occasionally, they produce distinct moulds.

Frequencies of species and individuals. Members of the orders Semaeostomeae and Rhizostomeae can cause impressions in the sediment because these animals are large, their bell consists of a firm, gelatinous mass, and they are heavy. The Rhizostomeae have firmer bodies and produce usually clearer traces in the sediment than the Semaeostomeae. The

Lucernariidae produce the least perfect moulds. For details about medusae of the Scyphozoa occurring in the mouth of the river Elbe, see Kühl (1964–5).

Medusae of all kinds have nearly the density of sea water. Therefore they are heavy enough to form moulds only if they are stranded on the beach or sediments of tidal flats and remain dry for some time (Nathorst 1881). Dead medusae sinking to the sea floor retain so much buoyancy that they make an insignificant impression on the sediment surface, which cannot be identified.

Four species of the Semaeostomeae can be observed in the North Sea: *Cyanea lamarcki*, *C. capillata*, *Chrysaora hysoscella*, and *Aurelia aurita*; moreover one Rhizostomea: *Rhizostoma octopus*.

In the summer months (May to September) swarms of Scyphozoa invade the southern and the southeastern North Sea from the west. They are made up of thousands of individuals belonging to one of the five listed species. Linke (1956) described the ecological conditions under which jellyfish are washed ashore. Enormous swash accumulations collect always within the short time of one or two tides. Such mass strandings of jellyfish consist almost exclusively of one species and occur usually during long periods of good weather (e.g. a Rhizostomeae swash accumulation of 10 km length and 2 to 3 m width, made up of 200 000 individuals). The formation of large swarms is not yet understood; but Linke assumed that the budding of medusae from the polyps occurs almost simultaneously in a certain area of the sea floor; thus dense swarms of young jellyfish are released into a restricted body of water of several hundred thousand cubic metres and move with it. These bodies of water retain their identity for a long time and mix only gradually. A swarm of young jellyfish is born and grows up in one and the same body of water. When the jellyfish rise to the surface during good weather they are easily driven to the beaches by weak winds and are stranded in great quantities.

Conditions for fossilisation. Schäfer (1941c) investigated the conditions for fossilisation of Scyphozoa in detail. Jellyfish washed up on open sandy beaches are likely to be thoroughly shifted and redeposited by strong surf at some time or other and have therefore no chance of being preserved as fossils. Permanent preservation is only possible in the wide bays of the tidal sea where the sediments are more muddy and where they are not much reworked. The number of jellyfish which are drifted over the much greater distance from the open sea to the tidal flats is, naturally, much smaller than that of animals driven to open beaches.

Jellyfish can be stranded lying on their oral or their aboral end, or on their side; the latter happens mainly to Rhizostomeae that have sturdy oral arms. The bells of all Scyphozoa become soft and pliable after they have drifted back and forth in the beach area for a long time and have often been washed back into the water after their first stranding. Then the bell can fold and easily land on its side. On sediment temporarily exposed to air, we can therefore expect two types of moulds: the bell appears as if seen from above or below, with the moulds of organs appearing as radial grooves and ribs and depressions, or it appears as if seen from the side showing a clear outline but with indistinct moulds of individual organs which overlap and disturb each other or are altogether absent.

Jellyfish can only be preserved as moulds and not as body fossils. They contain no hard parts, and the water content of the tissues is so high that only a thin organic skin remains after dehydration. Therefore, the mould of the body and organs forms in the first few hours after the body of the jellyfish has settled on the sediment. At that time the parts and the whole body are still heavy enough to produce visible and sometimes identifiable moulds (soup-plate-sized jellyfish weigh up to 1 kg). During fossilisation the thin organic pellicle reinforces the depressions and elevations of the mould and protects them to a certain degree from destruction. In this way the outline of the bell and the specific shapes of its lobate rim, the large circular muscle, the radial muscles in the lobes, the genital cavities, the tube-shaped gastro-vascular cavity, and the tentacles hanging down from it can all be identified in moulds; however, any single mould will never show all these organs. Whether one or another organ is predominant in a mould depends on the consistency, the grain size and water content of the sediment, the humidity of the air, insolation, rain, and temperature. The rates at which the sediment dries and at which the gelatinous tissue decomposes are also influencing factors; others are the state of disintegration in the stranded animal body, its size, and its weight. Whether the body has been thrown on to the sediment by the surf, or whether it has settled slowly while the supporting water has gradually run off are factors to be taken into account. An additional difficulty in the interpretation of moulds arises from the fact that the decomposition gases often collect in bubbles which also leave their impressions, often more distinct than those of some of the organs. Even experimentally it is impossible to obtain two equal moulds from animals of the same species. A fossil mould of a jellyfish thus shows us more about its process of fossilisation than taxonomically informative features. The sculpted mould surface is finally preserved only after another layer of sediment has been gently spread over the organic pellicle that has protected it so far.

Distinct jellyfish moulds with traces of organs can generally originate only when the animals are exposed to air; however, on rare occasions animals in the coastal zone get buried under marine sediment during storms and are covered so rapidly that their buoyancy does not free them. In such cases circular compression zones form in the sediment as many repeated embedding experiments have shown. Sediment layers that originally settled on the dome of the bell later subside when the gelatinous material of the bell decomposes; this happens first above the gastro-vascular cavity. As the bell collapses further and its surface area is reduced, the sediment layers above it are crumpled. If the disk of the jellyfish ends up in a vertical or oblique position in the sediment subsidence also occurs upon decomposition. However, the sediment layers become disturbed in such a disorderly way that the cause of the disturbance could not be guessed if one found fossil sediments that had slumped over such decaying jellyfish. Hertweck (1966) has repeated such experiments on the fossilisation of medusae. He thinks that in certain cases moulds of medusae can be produced in dead calm water on the sea floor. However, such conditions hardly ever occur in a shallow sea.

The gastro-vascular cavity, the circular canals branching from it, and the genital pouches are frequently filled with sediment when a jellyfish is stranded. The result can be a globular internal mould with several radial projections. Living jellyfish that get close to the bottom in shallow water try desperately to get away by pumping movements of their bell, and inadvertently ingest fine mud, detritus and suspended particles, peat shreds, finest shell grit, and sea urchin spines. The more violent the pumping movements, the more detritus is sucked up and the heavier become the animals until, eventually, they become immobile and remain lying on the bottom. Whereas the jellyfish ingest mainly fine detritus and mud particles their clayey moulds are found mainly in pure sand because open, surf-exposed beaches are always made up of sand. The body cavities of many jellyfish picked up on a beach are filled to capacity with detritus. 'Stomach stones' of jellyfish led Walther (1919 and 1930) to set up the taxonomic group of the 'skleromedusae' whose detritus-filled gastro-vascular cavities were assumed to be the consequence of a creeping way of life. Details in the shape of cavity fillings should not be given too much taxonomical importance (Rüger & Haas 1925, distinguished between gastro-vascular cavities with a central mouth and so-called basal gastro-vascular cavities). We observed in recent animals that the extent to which a cavity becomes completely or partially filled is the accidental result of events shortly before the jellyfish is stranded (see also Lörcher 1931).

Embedding experiments are especially useful with gelatinous and other marine animals without skeletons to supplement observations. We made experiments with medusae of Scyphozoa under conditions similar to those in nature; the sediment and sea water were those of the animals' habitat, and animals were used shortly after their natural death. Two investigators used animals fixed in formalin (Maas 1911, for deep-sea medusae; Peyer 1957, for *Salpa maxima* and *S. fusiformis*). Such a procedure defeats the very purpose of an experiment, namely to observe the chemical and mechanical effects of the decay of an easily perishable, water-rich invertebrate carcass. Fixation hardens the tissues and preserves their structure by clotting the proteins.

(*c*) ALCYONARIA

The Octocorallia comprise several hundred species with world-wide distribution. They appear as lumpy, coarsely branched masses, fine branches, or the mushroom-shaped or conical bodies. They carry numerous, small polyps that can be retracted into a matrix (coenenchyma). These animal colonies are sessile, most of them attached to hard sub-

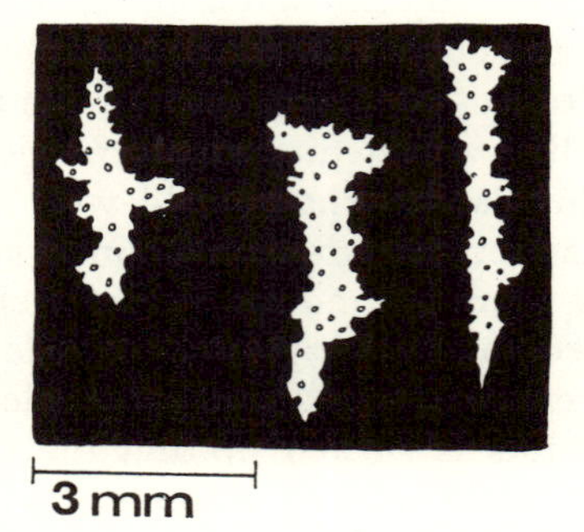

FIG. 107. Sclerites of *Alcyonium digitatum*, up to 120 microns long

strates on the sea floor, but some are able to live in loose sand or mud. All have a leathery consistency. In most cases the skeleton consists of isolated, knobbly bodies, hardly 1 to 10 cm long, called sclerites. They are secreted from ectodermal scleroblasts and consist of calcium carbonate. In some species the sclerites form firm bundles. Only these sclerites can be preserved; after the animal colony has completely disintegrated (approximately twelve days after death) they sink to the bottom and become mixed into the sediment of their former habitat. Fossil sclerites have been found.

Only the species *Alcyonium digitatum* occurs in the southern North Sea (Fig. 107). There its occurrence is dependent on gastropod shells and lamellibranch valves lying isolated in the sand. It grows (as lumpy, yellow, brownish, or white masses the size of a hand) on *Neptunea* and *Buccinum*, on the valves of oysters, *Cyprina* and *Modiolus*. Since the time of the steamship the population of *Alcyonium* has doubtlessly increased because the sea floor below the main shipping line has become a habitat where innumerable slags discarded by the ships provide a substrate. With one haul of the shrimp net, approximately twenty to thirty *Alcyonium* colonies are usually brought up in this zone; everywhere else a catch brings three to four colonies. The short and compact sclerites of this species are tiny, hardly 120 μ long according to Pax (1943). A large colony of *Alcyonium* provides the sediment with barely 3 g of sclerite material. We do not believe that sclerites can be concentrated on the sea floor of the North Sea.

VIII. Spongiae

Numerous species of Spongiae live in the southern North Sea. According to Arndt (1934) all five orders are represented; but the conditions for the preservation of these animals are extremely poor in this area. There is therefore not much point in discussing the animal group in detail. No skeletal elements, whether consisting of calcium carbonate, silica, or of spongine, can be preserved in their original structure and position under the environmental conditions of the shallow, storm-churned tidal sea.

Individual skeletal needles are occasionally found in the slime-lined tubes of worms burrowing in mud, or in the intestines of crustaceans and echinoderms; faeces of bottom-dwelling fishes, of gastropods and echinoderms may also contain these needles.

PART D

Traces of life habits preserved in the sediment

JUST as body fossils and moulds can tell us something about the former presence of animal organisms belonging to a fauna that is not itself preserved, so also can traces of a variety of life activities in marine sediment. In such traces the 'writing' enables us to draw conclusions about particular qualities of the 'writer'. Only in the last few decades have the traces of former life activities been thoroughly studied; palichnology, as this study was named by Seilacher (1953a), could bring results only after knowledge of fossils themselves had been supplemented by a better understanding of the life activities of recent animals and the way in which they influence sediment during its formation. To date this study is still in its beginning, and few recent faunas have been examined. We will therefore go into considerable detail in the discussion of life activities and of their traces.

Behaviour manifests itself in movement (Tinbergen 1952) which in turn is caused by the combined effect of the contraction of many individual muscles in the case of Metazoa, or of myonemes in the case of Protozoa such as *Stentor*, or by the work of plasma currents in the case of Protozoa such as *Amoeba*. This is true, whether these movements are simple reflex motions, more complex instinctive actions, or performances that require learning and experience.

Of all the various motions of which animals are capable, locomotion is in this context doubtlessly the most important. But not all animals, particularly many marine animals, have the power of locomotion. However, motions of metabolism, such as ingestion and digestion, and of reproduction are basic functions and therefore present in all animals. Most of these motions, or consequences of motions, are performed within the body and are therefore not projected into the preserving and surrounding sediment; yet the sediments can show indirect effects: the consequences of such motions can be either material, as in the manifestations of defecation, moulting, tooth change, and spawning, or they can be markings on extraneous substances, as in the manifestations of eating. All movements that can produce traces thus require the same attention as does locomotion.

I. Locomotion

Locomotion of a marine animal that lives on the sea floor, or occasionally makes contact with it, causes traces. They can be recognised by later impressed alterations of the original texture of the sediment due only to inorganic agencies.

Locomotion in marine animals can be considered from different points of view: if the potential sequences of movements in an animal body are considered, a physiological order of movement will result. If we try to interpret the movement, the influence of the external conditions on the sequence of movement, and the position of the animal in its environment, the answers will be ethological and ecological. If we determine which movements one particular species can put to use we can compile an inventory of its activities.

(*a*) THE FUNCTION OF LOCOMOTION

The body movements that result in locomotion in or on the sediment can be reduced to a number of mechanisms. These, in turn, are the results of certain, well-co-ordinated activities of muscles and groups of muscles, or in Protozoa, of similar processes in organelles. For an interpretation of the visible traces of movement on the sediment, the various mechanisms of locomotion must be known. Comparative analysis will show that different animal groups have developed identical or similar mechanisms. Therefore, all animal groups have to be considered. Regrettably, these mechanisms of movement and the co-ordinated work of the muscles concerned have not yet been described exhaustively, in particular not for the lower animals, although this seems to be such basic work; for our special purposes knowledge about them is most important. Every time he finds a trace the palaeontologist is confronted with the effects of a co-ordinated muscle activity and with its particular ethological and ecological relationship.

The course of this study will lead us to analogies in structure and function of the locomotory systems in various marine animals that live in the sediment, and the analogies will be applicable to the corresponding ethological and ecological aspects. From these analogies we conclude that similar or even identical traces of movement can be produced by different species.

(i) *Amoeboid locomotion, flowing of the body*

Only protozoans without shell or rigid pellicle use this mode of locomotion: local alterations of surface tension in the outer skin of the protoplasm, ectoplasm, or plasma gel cause pseudopodia to be formed spontaneously or in response to external stimuli; the pseudopodia are often directed. Apparently, flow of protoplasm is possible even in species with a highly viscous consistency of protoplasm; for during the process of locomotion the viscosity can be lowered temporarily by a reversible gel-sol process. W. J. Schmidt (1942) demonstrated that the birefringence of the protoplasm is subject to local fluctuations which indicate that processes at the surface are linked to metabolic processes in the cell. Thus the movement is performed without specific contractile structures, such as muscles or myonemes; the direction of movement is not predetermined.

The naked forms (Amoebozoa) change their shape completely during locomotion. This mode of locomotion has been termed 'flowing' of the body. The whole cell is elongated in the direction of movement, forming one or several lobes; the lobes and advancing tongues of protoplasm (lobopodia) increase in volume at the expense of the trailing cell body, until it is finally completely transferred into the former lobopodium (Fig. 108). The endoplasm at the tip of the lobopodium soon reverts to firmer

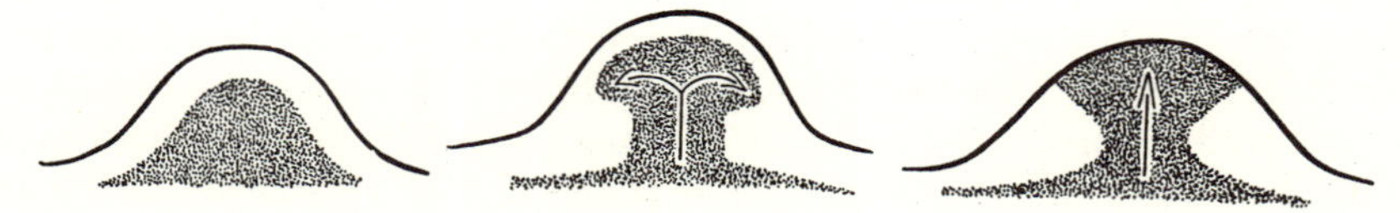

FIG. 108. Schematic sketch of three stages in the formation of an amoeboid lobopodium (after Grell, 1956)

ectoplasm. Sustained locomotion gives a streamlined shape to the cell body. Frequent and suddenly altered movements produce more or less radially directed tongues and threads. Animals whose cell body is enveloped by a firmer and denser pellicle can no longer execute a purely flowing motion. They push and shove themselves forward by means of mobile pseudopodia, and in some cases their action can be compared with climbing and pacing (Doflein-Reichenow 1953, Grell 1956).

Traces produced by amoeboid protozoans are very small, yet their traces in fine mud are more conspicuous than the originators themselves. Where amoeboid protozoans occur they can be found in large numbers, mainly in diatom-rich muds at the brackish margins of the tidal sea (Rhumbler 1928).

The locomotion of the Foraminifera can be discussed here only with partial justification. This almost exclusively marine order of the Rhizopoda possesses plasma pseudopodia which, among other functions, serve for locomotion in the sediment. The plasma body of a foraminifer is enclosed by a rigid calcareous shell or one of cemented sand. Such a body cannot flow as a whole, and plasma filaments, the rhizopodia, act as locomotory organs, outside the shell (Rhumbler 1928). These rhizopodia are either independent, unbranched strands (or with only few branches), or the rhizopodia form a more or less dense, anastomosing network which envelops the shell completely and is always capable of regeneration. The endoplasm of the rhizopodia is in an almost continuous flow which becomes observable by the movement of protein granules carried both in a centrifugal and centripetal direction along the firmer internal axis, the stereoplasm, of the rhizopodium. This current transports food particles to the cell body and undigestible remnants to the outside.

Locomotion of Foraminifera in the sand or mud of the sea floor is not yet well known in detail; the flow of plasma along the rhizopodium as a whole, and its sticky surface, seem to be important factors in promoting locomotion on the sea floor. According to Grell (1956), local contractions of the internal stereoplasm axis seem to cause the movements of the rhizopodium. Grell showed a gamete of *Iridia* in several microphotographs: it extends a broad 'leading rhizopodium' in the direction of movement while the others are trailed behind the body, forming a more or less elongate tail. In the case described by him the movement is performed on the smooth surface of a glass dish, and probably is achieved mostly by flowing plasm, possibly facilitated by the adhesive surface. In sand and mud, on the other hand, the rhizopodium itself can perform pushing and climbing movements, supported by the transporting effect of the plasma current in the outer layer of the rhizopodium. The plasma current seems capable of removing foreign particles that bar the way, often at several rhizopodia at once. As with Amoebae, the trace of a foraminifer in mud is a deep and quite narrow zigzag furrow. In sand the crawling movement leaves no trace.

(ii) *Motions of beating cilia*

In the Metazoa, ciliated epithelia are quite common; they serve to transport food in the digestive tract, to distribute mucus, to move sexual products in the oviduct, and to remove foreign particles from the trachea. In these cases the epithelium is stationary and transports material by beating with the cilia. In unicellular animals the ciliated ectoplasm becomes the locomotory organ for movement of the whole body. Only small animals with a low specific weight can use this mode of locomotion because of its low power. According to Jessen (after Hesse-Doflein 1935) the cilia of *Paramecium aurelia* can propel nine times as much weight as they themselves weigh, all weights counted after subtracting the buoyancy in the surrounding water.

A cilium projects beyond the surface of the cell membrane and consists of a central bundle of fibrils lying in a gelatinous or liquid matrix and surrounded by a membrane. The cilium is rooted in a swelling in the ectoplasm, called the basal granule, where it is probably formed. The cilia of protozoans which constitute in many cases a complete covering of cilia are arranged on the body in parallel, longitudinal rows, often in elongate spirals. Generally, the beat is a movement within a single plane. The individual beats of one row of cilia are in phase and co-ordinated in direction both of which can be altered (Fig. 109). How the co-ordination is achieved is not quite clear (Grell, 1956). In general, the forward beat of the cilium is faster than the backward movement. During the forward beat the cilium bends like a rigid hair so that the basal part is bent sharpest. In contrast, the cilium is erected in such a way that the sharp curve at the base of the cilium wanders upward along the cilium, lifting its tip gradually from the surface of the ectoplasm. The higher speed and

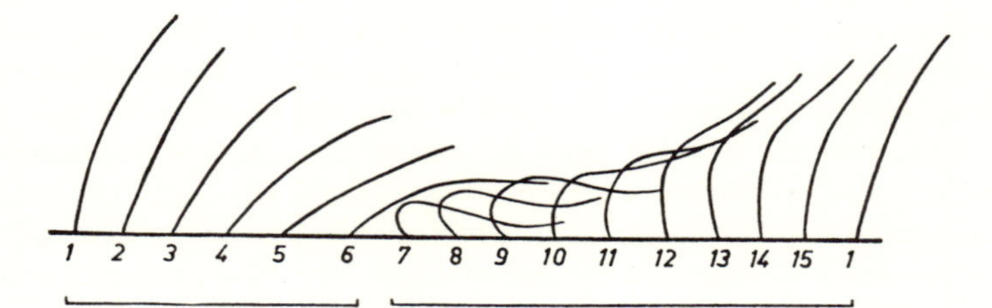

FIG. 109. Sequence of shapes of a single cilium during forward (stages 1 to 6) and return (stages 7 to 1) beats (after Grell, 1956)

intensity of the forward beat has a unilateral effect and causes locomotion in one direction. The bending of the cilium seems to be caused by differential contraction of the different individual fibres in the axial bundle.

In some cases the effect of the cilia is modified by their grouping and fusion into membranelles enabling the animal to move in special ways. Such membranelles fused from several cilia serve frequently for feeding and for the reception of stimuli. Where specialisations of this kind occur the animals are often less completely covered by cilia. Many ciliates are provided with additional locomotory organs, the myonemes, which can temporarily alter the shape of the cell by their contractions. Such deformations of the body frequently serve locomotion.

In open water cilia are used for swimming. On sandy sea floors and in the upper layers of the sand the cilia are capable of crawling or walking as on stilts over the surface of sand grains, or the animals swim and crawl in the interstitial water of the sand. Hypotrichs (ciliated on one side) occur abundantly in such biotopes. Many of the sand-inhabiting forms have contractile bodies (according to Kahl 1933: *Trachelocera*, *Heminotus*, and *Geleia*).

Locomotion in mud is particularly important. If the mud is well compacted the animals can crawl by means of the cilia; if the mud is newly deposited, is flaky, soft, and rich in water, ciliates, like the Amoebae, sink through the surface into deeper layers of mud; there the locomotion resembles digging with all the cilia in action. As a result, deeply incised traces with sharp rims can be observed. If new mud is deposited the ciliates can survive and reach the sediment surface unless the rate of sedimentation is very rapid. It is not yet known whether these movements toward the sediment surface are due to phototactical or negatively geotactical responses (possibly, inclusions in the vacuoles might serve as light or gravity sensors).

Muds are inhabited by ciliates only if they are rich in sapropel; sterile muds without organic substances do not seem to offer favourable conditions, either because of nutrient deficiency, or due to the presence of numerous mud-eating nematodes (Kahl 1933, after Remane).

A detailed inventory of faunas according to biotopes has barely been started (Kahl 1933). Certainly, the dependence of the animals on a certain kind of sea floor is only of secondary importance; Kahl has found *Discocephalus rotatorius* in the sandy bottom of Kiel Bay and also on coral reefs in the Bay of Suez. The distribution of the ciliates is probably determined rather by differences in acidity and in the concentrations of oxygen and hydrogen sulphide, than by the grain sizes of the sediment. However, it is occasionally possible to distinguish between sand faunas and sapropel faunas. This dependence on either sand or sapropel is due to the food supply. Thus decomposing sapropel sediments that produce much hydrogen sulphide are populated by many forms. Such biotopes are also the most likely to furnish traces of ciliates.

Not only protozoans use cilia as locomotory organs, but also several species of gastropods, rotatorians, gastrotrichans, and kinorhynchs, crawl over a substrate using cilia rather than the more usual undulatory movements of their sole. However, these are small forms; they weigh little in water, and their locomotory power need not be great. Götze (1938) has described the crawling of *Caecum glabrum*, a prosobranch from the 'Deep Trench' of Heligoland. Plate (1896) measured the crawling speed of *Caecum auriculatum* and found it to be 25 mm/min. Probably, the number of prosobranchs that move by means of cilia is much greater than known to date (other known cases are: *Alectrion trivitata* and *A. obsoleta*, Copeland 1919, and *Skeneopsis planorbis*, Gersch 1934). Ankel (1936a) assumed that some young animals of other species continue to crawl by means of cilia for some time after metamorphosis. All these forms with ciliated soles of the foot do not live in mud areas and do not produce locomotory traces. They live on the thalli of algae, on stones with attached plants, or in coarse sand or fine shell breccia crawling on the grain surfaces in the interstitial water system.

(iii) *Undulatory movement*

Many animals move by undulating motions; animals living in the sediment, on subaqueous or on wet sediment, in open water or on dry sand employ this kind of locomotion. The following marine animals undulate: coelenterates, nematodes, nemertines, errant annelids, isopods, acranians, and fishes. We will limit our discussion to forms which move in or on the sediment and impress their traces into it.

The body of undulators is generally elongated, and sine-shaped waves travel along the body, generally in a single plane. If the wave travels from head to tail the animal moves forward. The obliquely backward-facing parts of the body exert a pressure against the medium and thus give a forward impetus to the body. Where the medium has a certain stiffness, as in plastic mud, it has to be deformed only once to provide subsequent parts of the body with a suitably oriented surface against which pressure can be applied. This minimises the work of deformation and consequently the expenditure of energy for locomotion. The direction of travel of the undulations can be reversed to one from tail to head, allowing backward locomotion. The dependence of undulatory movement on counteraction from the medium can be confirmed by the observation of nematodes trying to move on a smooth surface such as glass. Nematodes have only longitudinal muscles and therefore depend entirely on the undulatory mode of locomotion. Unable to anchor themselves on the smooth surface, they make squirming undulatory movements of the body but no net forward transportation results.

Animals of the different phyla use different mechanisms to bend a certain length of their body. However, one feature is common to all of them: the active contraction of one side of the body causes the passive stretching of the opposite side. In coelenterates and worms the longitudinal muscles of the ectodermal muscle system contract on one side, a section at a time, in response to nervous stimulation, which affects equal lengths of muscles, alternately right and left. The wave length is constant at any given time, but it can change with time. In some cases the head and the following body section do not undulate with the rest of the body; thus in the polychaete *Nereis* the head and the anterior part of the body are kept straight. This is necessary to make another mode of locomotion effective, of which *Nereis* and *Nephthys* are capable. This is the so-called 'bolting' (see below). Both animals can shoot their proboscis forward when they are moving through sediment. For this their head must be directed forward and kept in this position. Then the animal can 'bolt' when the posterior part of the animal is still undulating and moving forward. When the same animals swim by undulation in open water they use the whole body because 'bolting' is ineffective in water.

The ectodermal muscles can contract but do not, in the absence of a skeleton, provide enough firmness to translate one-sided contraction into bending. However, constant tonus of all muscles produces hydrostatic pressure in the body fluids. This pressure acts on the elastic skin and gives the whole body the necessary elastic firmness.

All longitudinal muscles, upon a tactile stimulus, contract suddenly in a single reflex motion and not in a travelling wave of contractions. Thus external stimuli have a different effect from the internal ones that co-ordinate the orderly progression of contractions. If the ventral nerve

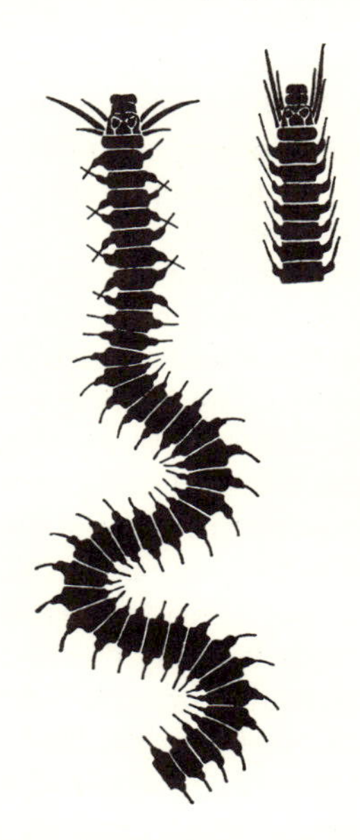

FIG. 110. Left: *Nereis* undulates but keeps the anterior part of the body straight, movements of parapodia supplement the effect of undulation; right: attitude of parapodia during sudden retraction of the worm's anterior portion. (Schematic drawing)

cord is severed the co-ordination between the severed sections ceases. The separated parts swim, crawl, and rest independently from each other (Schlieper 1955).

The nereids and many other polychaete worms have additional locomotory organs, strong and large parapodia with bristles. During locomotion their movements are co-ordinated with the undulating body movement, beating a back stroke on the convex side, a forward stroke on the concave side. They thus supplement the other locomotory efforts. If the anterior section with the head advances straight forward the pairs of parapodia of each segment perform walking movements; while one parapodium is thrown forward, the opposite one presses backward (Fig. 110). If the head section receives a tactile stimulus both parapodia of a segment fold forward; this reaction may be understood as being effective against a pull by a frontal attacker.

Annelids rarely move by undulation alone; generally, they supplement it with peristaltic movements, achieved by contractions of the circular musculature. Circular muscles are often rather weak compared with the longitudinal muscles (Nilsson 1912).

Some annelids are capable of three-dimensional undulations. Each section contains four longitudinal muscle strands which are contracted sequentially in a circle around the body. Combined with the usual longitudinal sequence the result is an undulation wave which travels in a spiral along the body, propelling it screw-fashion.

In acranians and fishes undulation is achieved by contraction of the segmented lateral musculature. A segment is called a myomere, and the number of myomeres in a muscle varies. Each myomere is in turn composed of several (at least three as in *Branchiostoma*) sections that join at an angle or in a curve and are separated by myosepta consisting of fibrous connective tissue. The myomeres are upright and fit into each other with a slight overlap. The horizontal septum laterale separates the dorso-lateral muscles from the ventral muscles. Apparently, the angles at which the myomeres meet are important for the working capacity of all dorso-lateral muscles. Narrow angles allow only weak undulations (as in *Scomber*) and wide angles permit stronger undulation (as in *Anguilla*).

Branchiostoma has a flexible chorda dorsalis, and fishes have segmented vertebral columns. These skeletal elements prevent the body from being telescoped lengthwise by contractions of the lateral muscles. This means that the effect of the contracting muscles is considerably increased over that of, say, an undulating annelid with its lack of any skeleton. Thus, animals with a chorda or vertebral column can move through sediment by undulating motions whereas annelids cannot. Nematodes, on the other hand, have a very thick cuticula serving the same function, and they can undulate effectively through sediment. Marcus (1934) pointed out that vertebrates without limbs and therefore dependent on undulating locomotion have more vertebrae than other vertebrates (e.g. Amphibia and Reptilia). He also observed (in detail on *Hypogeophus*) that in motion their vertebral column gets more curved than the outside of the body; the vertebral column lies eccentrically. The entrails must be free to be displaced because they are incompressible except for the lungs. The organs are moved in the direction of the longitudinal body axis and require reserve space in the body and a suspension by the mesenterium which is adapted to this particular requirement, as e.g. the suspension of the liver of *Gymnophiona*. Marcus emphasised that *Hypogeophus* can propel itself below the ground by undulations. Probably, the particularly large amplitude of the undulations of the acranians and fishes is an adaptation to life in or on the sediment; all animals that undulate with large amplitude also dig in the sediment or live on its surface. This is equally true for

Branchiostoma which undulates. This is the primitive form of locomotion of the chordates and of the eel-like teleosteans who have reverted to this type of locomotion by the repression of specialised locomotory organs.

(iv) *Peristaltic movement*

The peristaltic mode of locomotion is frequent in the lower marine animals; its study is important for the palaeontologist because it is used only by organisms without skeletons which could be preserved as fossils. Proof of their former existence can therefore come only from the traces they may leave in the sediment.

Peristaltic (or helicoidal) locomotion always calls for specific adaptations: the body must be elongated and worm-like, and its body wall must contain two layers of muscles, an internal longitudinal one, surrounded by a layer of circular muscles. The muscles of the double layer are co-ordinated; contractions of the circular musculature can be observed as zones where the body becomes thinner and longer, contractions of the longitudinal musculature cause the body to shorten and thicken. Contraction of one layer is always coupled with relaxation of the other. Periods of thickening and telescoping alternate with periods of thinning and lengthening in any part of the body.

In the simplest case the longitudinal or the circular muscles of all segments of the body contract simultaneously, thickening or lengthening the body alternately. This alteration of shape alone cannot propel the body without the help of supplementary, anchoring organs. If either end of the body is fixed at the time of greatest elongation, the animal can pull the free end toward it and achieve locomotion toward the fixed end. Usually, the anchors are bristles, chitinous hooks, or parapodia which attach themselves to a substrate or drill into it once they are in a certain position.

More frequently the animals stretch and telescope section by section, a section usually consisting of a group of segments. Thus thin and thick sections can be observed simultaneously, and the thickening and thinning runs in waves from the posterior end to the anterior end if the animal is moving forward. The wave direction and the sense of locomotion are reversible. The enlarged sections with relaxed circular musculature are kept firm by the pressure of the body fluid which is maintained by the circular muscles of the contracted zones. Even in rest the body fluid is under some pressure due to a constant muscle tonus and keeps the body firm. Muscle tonus is generally regulated by a central nervous system.

Sections of the body can be anchored simply by hooked bristles or groups of bristles, or the body can possess parapodia, special anchoring organs with stiff bristles, as do the marine polychaetes. Special muscle groups can adjust the angle of each parapodium to the body according to needs; moreover, in many cases bristles of the parapodia can be protruded or retracted. As a result, many kinds of movement are possible. Individual sections of the body can move on their own, or can combine to move the body as a whole. This allows animals to live in narrow burrows and to burrow and to drill into sediments, activities that depend on the ability of independent movement of parts of the body, of their retraction, or advance, or anchorage when and where needed.

Movements of individual organs and body sections must be coordinated to be effective. Control is exerted by a central nervous system consisting of the cerebral ganglion, the pharyngal junction, and the ventral nerve chord, and a peripheral plexus serving the ectodermal muscular system. Stimulation and suppression of activity are triggered by an interplay of external and internal factors, producing both reactive and spontaneous behaviour. The peristaltic mode of locomotion varies from species to species.

Purely peristaltic movements are found mainly in burrowing marine annelids. Or peristaltic movements can be combined with undulation by animals that are not restricted to life in the sediment, or they are supplemented by 'bolting' with a proboscis.

Purely peristaltic movements disturb the texture of the sediment only very slightly because the body moves without the help of laterally protruding organs. Local thickening of the body produces sufficient pressure on the sides of the burrow walls to anchor the widened section temporarily and thus allow effective locomotion. This can produce observable concentric, circular grooves in the burrow walls. Frequently, used burrows show groove next to groove. This sculpture should not be taken as a mould of the segmented inhabitant.

(v) *Glide-crawling*

Gastropods can frequently be observed crawling, and one can easily get them to do it in the laboratory, yet until now the mechanism of this mode of locomotion has not been satisfactorily understood. All the same, glide-crawling of the gastropod sole and the resulting motions are especially important in our context because the marine benthonic gastropods played an important role in the geological past and their traces of locomotion are frequently preserved.

The gastropod foot has long been recognised as a hollow, muscular organ, analogous to the ectodermal muscles of worms. It consists of a network of numerous longitudinal, diagonal, and transverse fibres and bundles of fibres whose units are anatomically separated and work in conjunction or antagonistically. Nerves control the motions of the individual units. The foot has no skeletal elements, and cavities filled with blood, or sea water under pressure, give it the necessary stability; the cavities lie between the muscle fibres and are connected with the main body cavity (Schiemenz 1884, Simroth 1896). Moreover, the muscle fibres themselves have a certain tonus (Biedermann 1905, Jordan 1918, Herter 1931).

The sole of the foot moves over the substrate by co-ordinated motions of its parts. Vlès (1908) systematically compiled the possible modes of locomotion in gastropods that crawl on their sole.

Direct wave motion. Rhythmic, alternating contractions of the longitudinal and transverse muscles in narrow zones transverse to the soles create waves that run from the posterior end forward, in the direc-

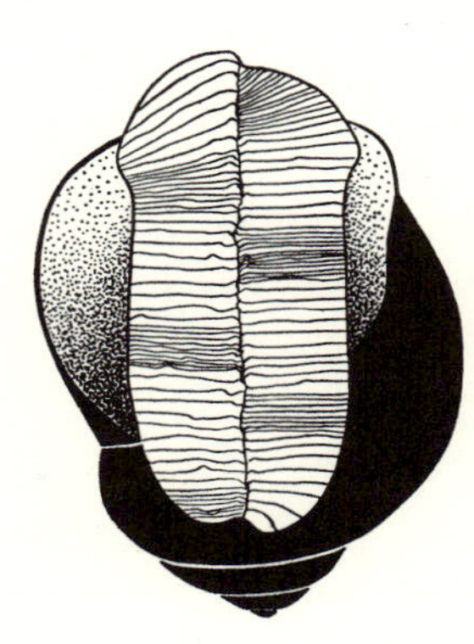

FIG. 111. Sole of *Littorina littorea*, showing one stage of contraction in parts of the sole's muscles. Example of ditaxic type of glide-crawling. (Schematic drawing)

tion of movement. A front of contractions lifts the sole just barely from the substrate, and the part of the sole that is out of touch with the substrate is eased forward. The resulting numerous little 'paces' appear as an evenly gliding forward movement.

(*a*) Monotaxical type. Successive waves of contraction pass through the whole width of the sole (*Cerithium*).

(*b*) Ditaxical type. The waves of contraction pass only through half the width of the sole. The two halves are anatomically separated by a longitudinal membrane of connective tissue (Ankel 1936a). The waves of one half are somewhat offset relative to those of the other (Fig. 111). This way of locomotion with its functional lengthwise separation is analogous to pacing; as a matter of fact, a small advancing tongue can be observed alternately at the right and left front of the foot. Similarly, the trailing end of the sole can be observed to advance alternately right and left (*Trochus turbinatus*, *Haliotis*; according to Simroth 1896–1907, *Cyclostoma*, *Littorina obtusata*, and *L. littorea*; according to Gersch 1934, *Helcion pellucidum* and *Gibbula cineraria*).

(*c*) Tetrataxical type. The waves of contraction pass through only a quarter width of the sole; therefore, four parallel waves run through the sole, each out of phase with the others (several species of small *Littorina* according to Vlès 1908).

In addition to these types of locomotion Vlès mentions also the backward-running wave; this means that waves run from the front backward along the sole, although the sole as a whole, and thus the animal, moves forward. Krumbach (1918) and von Buddenbrock (1930) maintained that backward movement is impossible. In experiments Gersch (1934) succeeded in provoking one-sided and also two-sided backward movements by reversed waves through the sole in *Gibbula cineraria*. These observations, however, have not been confirmed, and do not deserve a special place in the systematics of glide-crawling. If more observations should support the hypothesis of reversibility of wave motion our concepts of the mechanism and the nerve control of glide-crawling would have to be revised.

Only in the rarest cases is a substrate sufficiently plastic to mould the wavy shape of the sole of a crawling snail. Ankel (1936a) noted that *Nucella lapillus*, a ditaxically crawling prosobranch, can leave a twofold divided trace on a 'soft medium'.

(vi) *Push- and pull-crawling*

In pushing locomotion one or several organs are pressed against a supporting object and then become elongated so that they push and shove the body forward from behind. The organ for pulling, on the other hand, is at the front of the body. The organ first stretches forward, then anchors its front end, and finally contracts again, pulling the body forward.

Each of these mechanisms can move the body in only one direction. Such restricted movement is inadequate for life in many biotopes. If the body needs to be moved forward and backward in opposite directions, either one locomotory organ must be able to move both by pushing and pulling, or both ends of the animal must be provided with one of the mechanisms. This explains why we discuss the two mechanisms together although they act in opposite senses.

Lamellibranchiata. The lamellibranch foot is a typical example for pushing and pulling locomotion. Lying in the plane of symmetry of the animal the foot can be projected between the open valves. Presumably, the musculature of the foot stem served originally only to protrude or to retract the foot. If one takes the crawling foot of the protobranchs to be the primitive prototype in the development of the lamellibranch foot, then the broadened sole of the foot originally had its own musculature serving to propel the lamellibranch, somewhat in gastropod fashion. Such a mode of locomotion is only possible on a suitable substrate such as the sediment surface or a hard substrate. Locomotion within the sediment requires a modification of the foot: the broad foot of the crawling sole must be reduced, its lower end must then turn into an often sharp-edged, muscular rim capable of penetrating the sediment by undulating and digging motions. Also the musculature of the entire foot must be strengthened and altered into a hollow, tube-like structure capable of pushing and pulling.

The strong, longitudinal musculature that corresponds to the columella muscle of the gastropods runs in two strands through the entire length of the foot and often ends high on the inside walls of the valves. Circular and diagonal musculatures work antagonistically to the longitudinal muscles and enable the animal to change the shape of the foot in many different ways, most effectively the distal parts. Pushing movements achieve horizontal locomotion in the surface layers of the sediment and also vertical locomotion through sediment from below upward. Frequently the pushing motions have to be performed with the foot in a sharply bent position. Fig. 112 shows the musculature that performs the pushing movement and restores the starting position. In contraction, the circular musculature (5 in Fig. 112) forms a tight belt enclosing the longitudinal and the diagonal muscles; thus it prevents deformation of the foot at its bend while the other muscles contract. If the longitudinal bundles of muscles (1 and 2) contract they stretch the curved foot. If the diagonal muscles (first the one labelled 3 and subsequently 4) contract they curve the foot and thus restore its original position (see also Hollmann 1968 on the movement of *Cardium edule*).

Such mechanisms occur mainly in lamellibranchs which move horizontally in the upper sediment layers and on the surface of the sediment (*Cardium*, *Yoldia*). Lamellibranchs sitting deep in the sediment, employ pushing movements mostly to compensate for the sedimentation above and move differently; they have broad lobes at the distal end of the foot, and the lobes beat alternately right and left but not backward and forward as do the animals in the other biotope. In the starting position the foot is folded sideways and frequently also bent upward. When it stretches it pushes and presses the body of the lamellibranch upward. Then the foot is retracted and pushed sideways into the starting position at the next higher level, ready to stretch; the sides alternate (Fig. 113).

In many cases the same locomotory organs and the same muscles can also pull. While pushing leads to an upward movement of the lamellibranch in the sediment, pulling causes downward movement of the animal which is dragged behind by the locomotory organ. In this case the active phase of the movement is the folding of the bent foot beyond an angle of 90° which is achieved by the diagonal muscles 3 and 4 of Fig. 112. The foot is subsequently retracted into the mantle cavity, then pushed downward in an outstretched position and folded again; thus the lamellibranch can move downward step by step.

In many cases (*Solen*, *Scrobicularia*) the lamellibranch is equipped with a dilatable organ at the tip of its foot which anchors the outstretched foot in the sediment (Fig. 114). In these species the foot is often long. *Solen* also can move sideways in the sediment if it bends the foot toward the side and anchors it there.

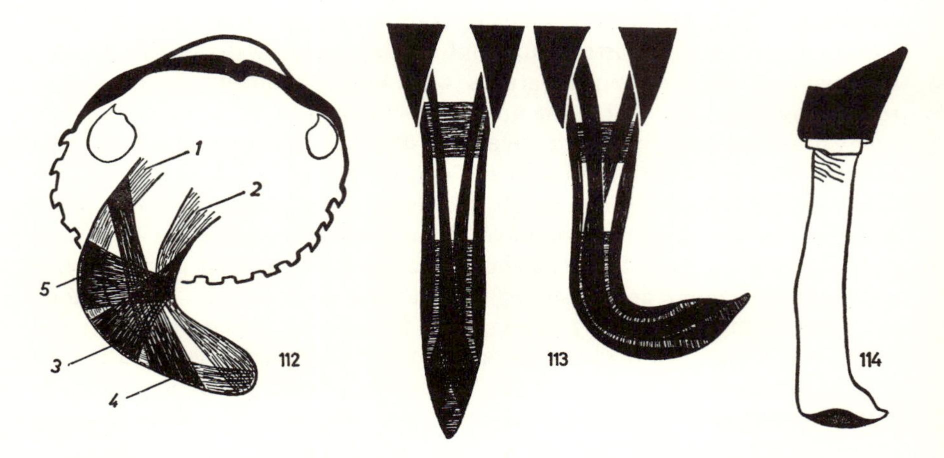

FIG. 112 (*left*). Muscles in the foot of *Cardium*, cut longitudinally through its plane of symmetry. (1) and (2) longitudinal muscles, serving to straighten the foot; (3) and (4) diagonal muscles, serving to bend the foot; (5) annular muscle, preventing a simple shortening effect of the action of muscles (1) to (4). All movements occur in the plane of symmetry. (Schematic drawing)

FIG. 113 (*centre*). Muscles in the foot of *Mya arenaria*, cut normal to the plane of symmetry. Two longitudinal muscles straighten the foot, two diagonal muscles bend it, and two transverse muscles prevent shortening of the foot by the action of the other muscles. (Schematic drawing)

FIG. 114 (*right*). Club-shaped end of the foot of *Solen*, capable of swelling. (Schematic drawing)

If the foot cannot be folded (as in *Cardium* and others) and is not provided with a dilatable organ (as in *Solen* and others) downward-pulling motions are laborious and slow (as in *Mya*). The leaf-shaped foot is pushed downward by undulating motions of its sharp edge; then the muscles of one side contract, bending the tip of the foot sideways and upward in a curve, and pull the body down. Only animals with a long and strong foot can move in this inefficient way. Young *Mya arenaria* with a valve up to 10 cm long can escape covering sediment by climbing upward foot first, the siphon directed downward (see section on escape reactions). The course of such a burrow describes a more or less stretched spiral because the animals frequently climb by repeating asymmetrical extensions of the foot toward the same side (Fig. 222). The proportion between length and strength of the foot of *Mya*, and the size and width of the rest of its body are favourable for this type of locomotion only in the adolescent stage of growth. Young animals still succeed to pull themselves downward. Later in life they are no longer strong enough because the foot ceases to grow before the valves and the soft body do (Schäfer 1956a). This explains why *Mya arenaria* are so frequently exposed to the open water in living position after sediment has been eroded, and why the animals are unable to counteract the loss of sediment cover by descending to greater depth. Young animals, on the other hand, who still can burrow are never found dead in such a position.

Scaphopoda. The scaphopod families Dentaliidae and Siphonodentaliidae are in a similar ecological situation. Their capability to move by pushing and pulling is an insurance against the consequences of rapid sedimentation or loss of sediment cover. Again, both pulling and pushing are performed by the same organ, a fleshy foot.

Fig. 115 shows the organisation of a scaphopod. The soft body sits in a tubular calcareous shell that is open at both ends and resembles an elephant tusk in shape. The animals sit in the sediment head downward. Immediately beside the head (on the ventral side) is the foot. It is circular in cross-section and can be retracted into the shell by a strong, longitudinal central muscle. This muscle is analogous to a double columella muscle and is attached to the dorsal inner wall of the shell at the distal end of the soft body. The foot carries a cone-shaped swelling at its end. In the Dentaliidae this is surrounded by two lateral lobes consisting of connective tissue and muscles, and in the Siphonodentaliidae it carries a round disk. All along the foot the longitudinal muscle is surrounded by circular muscles. The Dentaliidae have additional radial musculature.

The soft body with its calcareous shell upward is lifted by pushing movements: first the foot retracts to the shell opening, then it spreads the end organ (the two lobes of the Dentaliidae or the disk of the Siphonodentaliidae), finally the circular musculature contracts and thus stretches the foot which is also firmed up by an inflow of blood into a cavernous system. Lengthening of the well-anchored foot pushes the soft body with the shell upward. Pulling motions of the foot, on the other hand, can move the animal head first in the opposite direction. First the foot stretches downward to its full length, performing burrowing motions in the sediment with lobes or disk folded together at the tip. Then the lobes or disk spread out and anchor the foot in the sediment. Finally, the longitudinal muscles contract and shorten the foot, thus pulling the soft body with its shell downward. Both movements can be repeated until the animal reaches an ecologically favourable position in the sediment.

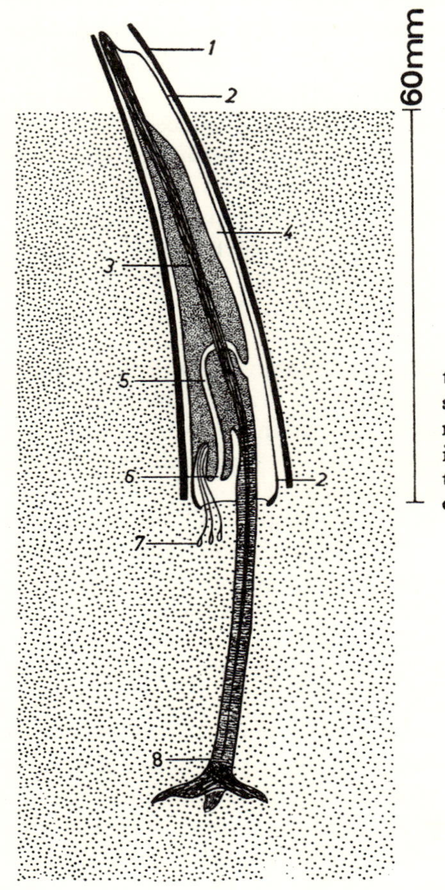

FIG. 115. Internal structure of a scaphopod. (1) tubular shell; (2) mantle; (3) foot muscle; (4) mantle cavity; (5) intestine; (6) mouth; (7) tentacles; (8) foot with anchoring organ

Gastropoda. Some of the prosobranch marine gastropods burrow in the sediment. They are either scavengers or predators. As we have seen, the lamellibranchs and scaphopods require only short-distance locomotion, but this in two opposite directions to compensate for perpetual alterations in the level of the sediment surface; they achieve this by pulling and pushing. In contrast, the gastropods that live in the sediment are so mobile that they can easily change direction during forward movement. Pulling is their only mode of locomotion in the sediment. They do it with their foot or rather with its propodium.

The prosobranch *Nassa reticulata* has a narrow and elongate foot which ends on the forward side with two lateral lobes. According to

Ankel (1936b) the foot is pushed into the sediment by muscle contraction alone, then it broadens its horizontal lobes and anchors them in the sediment and pulls the shell-covered body forward. Repeated pulling advances the animal through the sediment in stages. On the sediment surface *Nassa* moves like many other gastropods by undulating movements of the sole of its foot.

Members of another prosobranch family, the Naticidae *Natica clausa*, *Lunatia catena*, *L. nitida*, *L. montagui*, and *L. pallida*, are even better adapted to locomotion in the sediment. The front section of their foot is

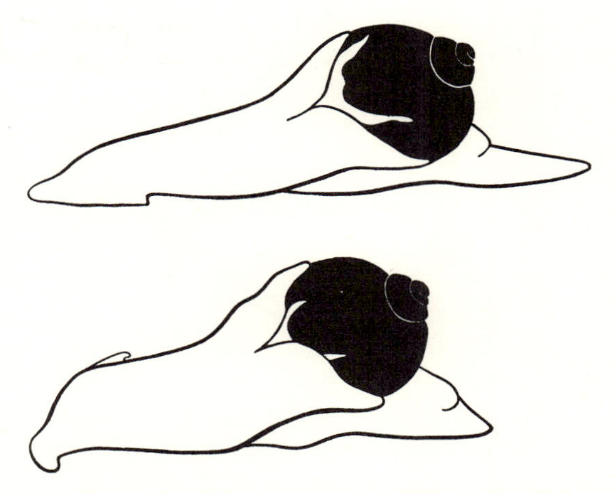

FIG. 116. Changes in shape of the propodium and metapodium of a crawling *Lunatia nitida*. (Schematic drawing)

changed into a very large propodium that covers the head entirely. The propodium can vary its shape widely, particularly its distal parts. The longitudinal, circular, and diagonal muscles whose specific functions in altering the shape are not yet clearly understood, can push the propodium forward and broaden its tip. As in *Nassa*, the outstretched foot is thus anchored in the sediment and pulls the animal and shell forward upon contraction. In forms that frequently burrow deeply into the sediment we often find an additional pushing mechanism to help in transportation of the heavy and bulky shell against the resistance of the enclosing sediment (Weber 1925, for *Natica josephina* and *N. millepunctata* from the Mediterranean Sea; and Ankel 1936a). The broad metapodium carries a pillow-like welt of connective tissue on its back which is in contact with the shell from below and behind. The columella muscle begins under the operculum and extends into the metapodium. Thus its contraction pulls the metapodium forward. The welt on its dorsal side then presses against the shell and shoves it forward. Ankel assumes that the propodium and the metapodium are enlarged and reduced in size as a cavernous system in these parts of the foot is either filled with fluid or emptied. The sequence of movements of burrowing Naticidae is quite complex. First the propodium extends and anchors by broadening its tip, then the propodium contracts, finally, the columella muscle also contracts bringing up the metapodium and thus easing the shell forward on its dorsal welt (Fig. 116).

Ophiuroidea. Locomotion of the brittle stars consists in most cases of pushing and pulling movements. They are entirely performed by the long arms; the tube feet have only a subordinate function in locomotion.

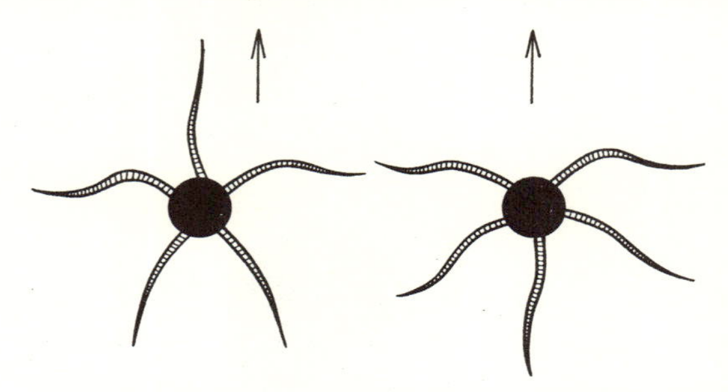

FIG. 117. Moving brittle stars. Left: odd arm in the direction of movement; right: odd arm trailing

The modes of locomotion of brittle stars have been vividly described by von Uexküll (1905 and 1909; see also Mangold 1908). Each arm is internally supported by a row of calcareous, vertebra-like bodies connected by joints and decreasing in size outward. Four strong, longitudinal muscles press the calcareous bodies together and thus achieve joint-like, mobile connections between them. Externally the arm is protected by rows of small, calcareous platelets set in a roof tile pattern which can easily slide over each other when the arm moves. Tube feet protrude from the underside of the arm and are arranged in pairs, thus forming a double row; a groove for the radial nerves also runs along the underside of the arm. The internal skeleton of the body consists of a calcareous ring on the oral side. The mouth opens downward and is the only opening of the central, sac-shaped body.

Whip-like movements of an arm are produced by contraction of some of the four longitudinal muscle strands. During locomotion two arms work symmetrically together as a pair. One odd arm of the five is always left without a partner and does not contribute to locomotion. It can be the one in front or behind (Fig. 117). When the single arm is at the back the brittle star beats the two front arms simultaneously upward and forward so that the angle subtended by them is reduced. They are then laid down in the substrate and anchored along their full length in the sediment with the tube feet and the side rows of platelets. While remaining anchored at the tip, both arms bend and form a low arch; they thus pull the body forward in a sudden jerk. This places the two backward arms further back. Now they begin a similar movement and beat in turn as far forward as the space permits. These arms bend as soon as they touch ground, lift the body and the inactive arm and move them forward. This brings the body between the two front arms and enlarges the angle

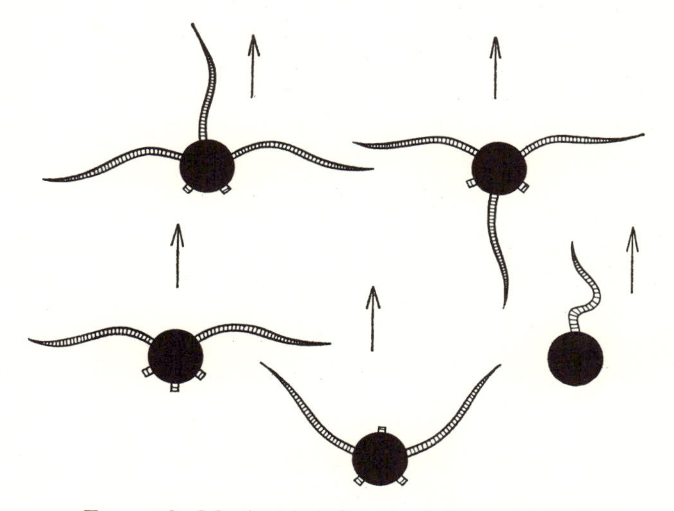

FIG. 118. Moving brittle stars with several arms missing due to autotomy

they subtend. Now the front arms lie again beside and slightly behind the body, ready to beat forward. Usually, the amplitude of the beat of the front arms is larger than that of the rear arms. The front arms thus pull and the rear arms push and lift. The synchronous action of the pairs could also be visualized as a slow-motion gallop with its alternating advance of both front and both hind legs. If the odd, inactive arm points forward the movements of the other arms are the same. But due to a less favourable weight distribution with the extra weight at the front end of the body, the individual jumps are shorter and net locomotion is slower.

Any arm can be left as the odd arm out. The direction of movement can be altered by changing the partnerships of arms. If the animal turns right or left and the odd arm is at the front, it becomes an active partner of its right or left neighbour; there is a corresponding change in the other partnership, and one of the formerly active rear arms remains unpaired.

Brittle stars are capable of autotomy. It thus happens frequently that one or several arms are missing or shortened in an animal still fit for life. In such a case the described mode of locomotion of the brittle star must be altered (Fig. 118). If one arm is missing the animal moves in the regular way with all four arms active but with no odd arm. A stump of an arm can lie either in front or at the back during locomotion. If two neighbouring arms are missing the two stumps always point backward. If two arms are missing and the stumps are not side by side, only two arms move the body while the third is inactive and always points backward. If three neighbouring arms are missing the stumps always trail while the two complete arms work by pulling. If three arms are missing and two stumps are side by side, the single stump points forward between the active arms. If only one arm is left it always points forward beating to the right and left and pulling the body on.

(vii) *Multiple, circular shovelling*

Some animals burrowing in sand use 'multiple, circular shovelling' for locomotion. This mode of locomotion occurs in animals whose body is so large and immobile or rigidly armoured that the trunk musculature of the body is incapable of moving the animal; then this function may be taken over by numerous special locomotory organs equipped with their own musculature.

Echinocardium cordatum. The irregular sea urchin *Echinocardium cordatum* is a typical example of this mode of locomotion. It lives almost exclusively in fine sand where it can burrow to a depth of 20 cm. J. von Uexküll (1921) described this activity of the animal. The calcareous armour consists of a thin test that carries a dense fur of fine spines. They sit on a ball joint set obliquely on the test. In rest the spines are therefore also in an oblique position, lying almost on the sides of the body. All the spines of a body area point in the same direction; thus the dense covering of spines has a brushed appearance. Each individual spine is slightly curved and bent backward. Its end broadens to a spoon shape. These spines can perform circular movements; those set deeper down and at the front of the body first start to move. Gradually, more spines join in the circular movement until all spines circle. Individual rows move successively, so that waves seem to run over the armour from the front and below rearward and upward, somewhat like over a wheat field in the wind.

When the circular movement of an individual spine begins the concave side of the spoon points in the direction of the beat of the spine and remains so for half a revolution. This is the active phase of the beat; during the second half of the beat the spoon moves with its convex side forward and then shifts sideways into the starting position. The spines of a row are set so densely that one spoon is immediately beside the next during the circular motion, with the effect that a slightly curved surface consisting of many spoons held in the same direction lifts the sand upward. The muscles are radially set around the foot of a spine and contract successively. Those producing the active beat of the spoon are strongest. A combination of moving and locking muscles in the musculature of the spines, and a rhythmic flow of stimuli to both types of muscles, guarantee that the spines can cope with the weight of the sand, even if it is somewhat increased over the normal.

The spines on the underside of the test work differently. They are more curved and stronger and end in much wider spoons. They transport sand from below the sea urchin to the sides, away from the median line and the mouth, until it is passed to the ordinary spines on the flanks of the animal.

Aphrodite aculeata. The large polychaete *Aphrodit .culeata* uses the same kind of locomotion. It lives preferentially in the surface layers of muddy sediment, eating detritus and scavenging (Jordan 1904).

Aphrodite looks like a somewhat spindle-shaped sack, is densely hirsute, and can be up to 12 cm long. A zone along the underside is hairless and reveals that the body is segmented. The back is covered with two rows of scales, or elytres, overlapping in a roof tile pattern. They cover a groove along the back. Strong, pointed spine-bristles run in double rows, one on top of the other, along the sides of the body. These bristles can be moved individually. The rest of the hair cover consists of tufts of dense, hairlike, iridescent bristles. The parapodia of *Aphrodite* are broad, blunt protrusions set with bristles that point obliquely downward and sideways. These parapodia (Schäfer 1956a) are moved by an especially stout, central bristle (Fig. 119). Outward it extends beyond the tip of the parapodium and inward it reaches far into the body cavity. Four bundles of muscles move the bristle. They spring from the side of the interior body wall where they are set in a circle and run to the internal end of the bristle. If one muscle bundle is contracted the bristle is inclined toward the side of the pull. Since the bristle is pivoted where it passes through the body wall the pull causes the outer tip of the bristle to move in the opposite direction and to take the entire parapodium with it The big bristle thus functions as a skeletal element and allows exact steering of the parapodium from within. Successive contractions of the four muscles make bristle and parapodium execute a circular sweep which goes from rearward to upward, forward, and downward. The sweeps always begin at the parapodia of the posterior region and are propagated wave-like toward the front along both rows of parapodia. Alternating right and left, new waves begin at the posterior end before the preceding wave has reached the head.

These circular movements of the parapodia can propel the worm on sediment and on hard rock, and even on smooth surfaces in the laboratory. If the animal wants to penetrate into the sediment the parapodia describe larger circles. Thus they shovel sediment from under the body upward and move the body downward.

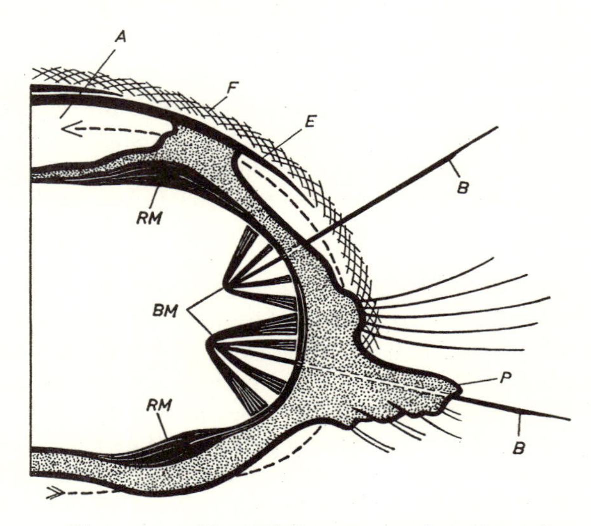

FIG. 119. Simplified cross-section through half a segment of *Aphrodite*. (*A*) respiratory cavity; (*B*) large bristle; (*BM*) muscles moving the bristles; (*E*) elytre; (*F*) felt formed by small bristles; (*P*) parapodium; (*RM*) circular muscles. Broken arrow—flow of respiratory water (from Schäfer, 1956a). (Schematic drawing)

Once the animal is halfway in the sediment an additional mechanism comes into action. The double rows of bristles above the parapodia also begin to perform circular sweeps in the same way and by the same mechanism as the bristles of the parapodia. The sediment load lifted by the parapodia is now taken over by the shovelling upper bristles which transport it further upward and backward, and the worm disappears rapidly and completely in the sediment, moving on below the surface.

Each bristle is flattened at its free end like an oar. And like an oar it can be rotated through 90° about its longitudinal axis. Thus the active phase of a sweep from the front backward can be performed so that the narrow side of the bristle cuts through the medium achieving a one-sided, rowing effect.

The muscles can move the bristle in small or wide circles, and the direction of movement can also be reversed; therefore, *Aphrodite* can move through the sediment in any desired direction, forward, backward, downward, or upward.

(viii) *Bolting*

This mode of locomotion is only effective within the sediment and is therefore found in endobiontic animals. A frontal organ is ejected forward like a bolt or slowly pressed into the sediment forming a cavity in it. The 'bolt' is an organ of the type of an eversible proboscis and consists of muscular or firm connective tissue. It is enclosed in a tough epidermis, often transformed into a cuticle. The organ can be returned to its original position by muscle action. The 'bolting' is caused partly by direct muscle action, partly by hydraulic action through contraction of the body sac. All muscles of the ectodermal system contract simultaneously, decrease the volume of the body sac, and cause pressure in the coelum fluid which pushes the proboscis forward and everses it.

'Bolting' can serve locomotion in two ways; either the worm-shaped body crawls by peristaltic movements into the prepared cavity after the proboscis has been retracted, as in *Nephthys*; or the projected proboscis anchors itself with epidermal welts in the sediment or uses the sediment at the end of the cavity for anchoring and then pulls the body sac forward, like a glove, until the proboscis is completely turned inward, as does *Sipunculus*. Both modes of bolting are related to peristaltic movement.

The locomotory mechanism of *Sipunculus* is simpler. The worm consists of two tubes, one in front of the other, with a common body cavity. The body sac proper continues forward as a thinner and shorter proboscis (Fig. 120). The proboscis carries slim lobes at its tip around the central mouth which can be completely closed by a sphincter muscle. The proboscis is a thin epidermal tube, covered on the outside with numerous, small, cuticular teeth whose tips point backward or sideways and backward. The teeth are not arranged in rows and are absent in a circular zone immediately below the skin lobes. This zone has particularly thin walls and can be blown up by pressure to form a collar-like welt. The body sac, but not the proboscis, is completely enclosed by strong muscles, circular ones surrounding concentrically a layer of longitudinal muscles. Both end abruptly at the border line of the proboscis. They are strongest in a zone making up the third quarter of the body sac. This is also the place where contractional waves of the longitudinal and circular muscles set in, they spread rapidly backward to the closed rear end of the body and also spread forward over the whole body sac (von Buddenbrock 1928). The circular and the longitudinal musculature contract simultaneously and act synergistically rather than antagonistically.

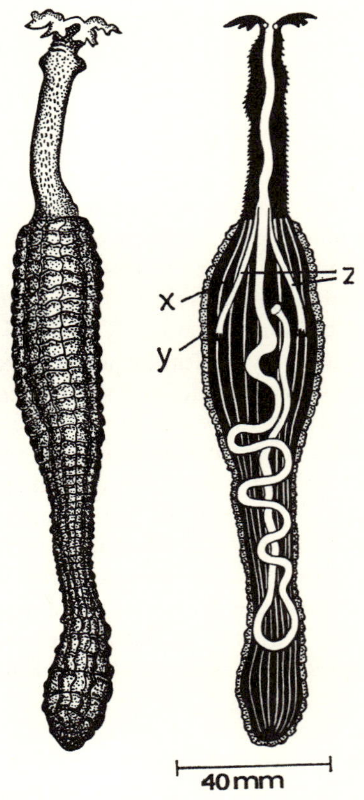

FIG. 120. *Sipunculus nudus* with erect proboscis. Left: external view; right: longitudinal section, (x) longitudinal muscles, (y) circular muscles, (z) muscles attached to the oesophagus

The wall of the proboscis has no muscles; however, grouped around the upper part of the digestive tract which hangs freely in the proboscis are four longitudinal muscle strands that are firmly attached to the oesophagus tube in the proboscis. They lose this connection with the digestive tract where they enter the body sac and are anchored to the body wall just below the anus. They serve to retract the proboscis. If, however, the proboscis is firmly anchored in the sediment by means of its cuticular teeth, inflated collar, and mouth lobes, contraction of these muscles pulls the anterior part of the body sac forward and over the proboscis and reverses it. The proboscis is eversed and protruded hydraulically by contraction of the longitudinal and circular musculature of the body sac and the resulting pressure of the coelom fluid.

The movements of the muscles follow a regular, rhythmic pattern which remains effective in the retractors and in the musculature of the body sac even after every direct connection between the muscles is artifically severed; only their connection with the central nervous system must be preserved for proper functioning. The physiology of the nervous system has been clarified mainly by von Uexküll (1933).

The locomotory processes function even when the entire digestive tract is filled with coarse sand of grain sizes up to 4 mm. The sand is

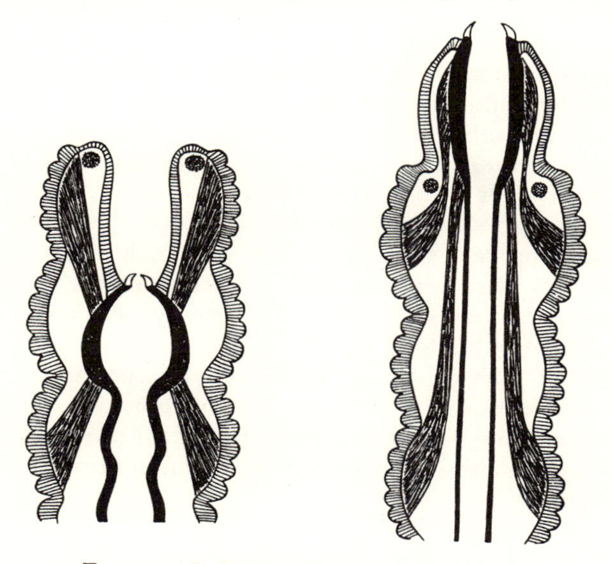

FIG. 121. Bolting action of *Nephthys*. (Schematic drawing)

ingested together with small food organisms. The loose attachment and the numerous curves of the intestines help prevent damage or tearing of the heavy, sand-filled intestines by the motions of the proboscis and of the ectodermal muscle system.

The polychaete *Nephthys* serves as an example of the second mechanism of bolting locomotion. The bolting mechanism is a powerful proboscis (Fig. 121). When it is retracted a spacious buccal cavity forms a vestibule in front of the pharynx cavity. The internal wall of the buccal cavity consists of a tough ectoderm. The pharynx also has thick walls and is provided with chitinous jaws. The pharynx is held in position by a bundle of muscles which runs from an attachment area encircling the pharynx back to the third segment. A second bundle of muscles is also attached to the wall of the pharynx and runs forward to the first segment; it is relaxed in the starting position. For bolting these muscles contract, and those attached to the third segment relax; the pharynx is pulled forward and glides into the buccal cavity, eversing its wall and turning its cuticular internal wall outward. Pressure of the coelom fluid is probably also increased by sudden contraction of the circular musculature and helps to protrude the pharynx. It does not remain in this position but is immediately retracted by contraction of the musculature of the third segment which restores the starting position. Such bolting movements can be continued in rhythmic succession for some time. Finally, they cease, and the worm moves into the gap in the sediment by peristaltic movements of its body.

Generally, this mechanism is only used in the sediment, not when the animals swim by undulating in open water or when they crawl by peristaltic motions on a substrate. However, interference with the crawling or undulating locomotion, for instance by gently pinching the body or holding it, can trigger bolting immediately.

The chitinous jaws are moved like hinged flaps by the musculature in the wall of the pharynx; they grasp the food which may be living prey, dead animals, or plant remains. The retraction of the proboscis then serves to carry the food into the buccal cavity which can be greatly widened to accept bulky food.

The effect of 'bolting' and its ecological interpretation will be discussed below in section D I (*b*) (ii) on mucous burrows.

G. P. Wells (1951a) discussed the structure of the proboscis mechanism of *Arenicola marina* and its modes of locomotion in great detail and with clear illustrations.

(ix) *Pacing*

Mobile supports are necessary for pacing movements in which the body is carried above and out of touch with the substrate. This contrasts with crawling in which the body drags along the substrate, the limbs push but do not carry it forward, and the whole body often furthers the movement by undulating. Pacing is a slow kind of walking. If it speeds up into a rapid succession of paces it turns into walking or running. The sequence in which the limbs touch the ground in running is not necessarily the same as in pacing. Limbs capable of pacing are arranged in pairs; a right and a left limb must be symmetrical in shape, size, strength, and position on the body. Limbs that have changed their function, for example from pacing to grasping, often show increasing asymmetry of shape with increasing use other than for pacing (Schäfer 1954a). Pacing requires at least one pair of limbs but such an arrangement is not very stable; where it occurs it is the result of advanced phylogenetical specialisation. In most cases there is more than one pair of limbs.

More land animals pace than dwellers on the bottoms of water bodies. In the ocean, almost only the higher crustaceans move by pacing. The lower crustaceans usually crawl, or they swim by whirling their limbs or by treading water.

The rhythms of limb movements and the nervous stimuli necessary for them are the same for freely pacing and crawling-pacing locomotion; the distinction is pointless from the point of view of the neurophysiologist. However, for the interpretation of preservable traces in the sediment, distinction is important.

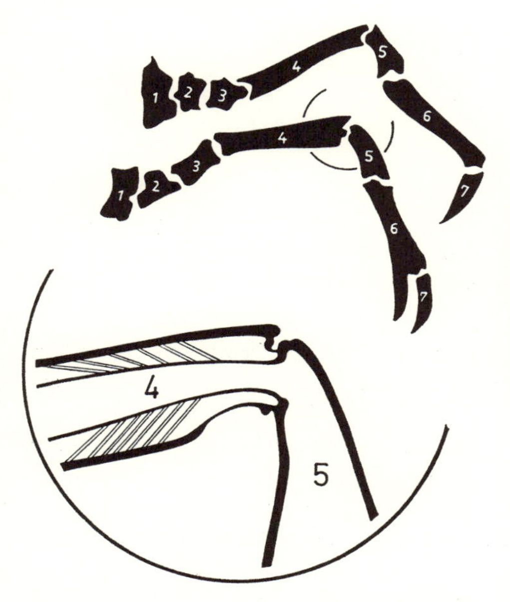

FIG. 122. Limbs of decapod crustaceans. Top: walking limb; (1) coxa, (2) basis, (3) ischium, (4) merus, (5) carpus, (6) propodus, (7) dactylus; middle: claw-bearing limb; bottom: enlargement of a joint, showing the muscles that activate it. (Schematic drawing)

Macruran crustaceans pace on the four most posterior pairs of legs of the thorax; in rest and in movement, they hold the body so high that it does not touch the ground, neither with the head and its appendages, including the heavy claws in front, nor with the abdomen and its tail appendages. Each walking limb is composed of seven tubular parts that are connected by joints: coxa, basis, ischium, merus, carpus, propodus, and dactylus (Fig. 122). The musculature that moves these parts lies inside the tube. In the third and fourth pair of walking limbs the dactylus rests on the substrate. In the first and second pair of walking limbs the lengthened propodus touches the ground because the dactylus has turned into the mobile branch of a claw which in rest lies parallel to the claw process of the propodus.

The pivoting axes of the joints are not parallel and are so arranged that the leg moves from obliquely backward to obliquely forward when the animal paces; the body is suspended between the legs and is carried straight forward during this movement (Fig. 123). The greatest muscular effort of a pacing macruran goes into the movement of the coxa and basis

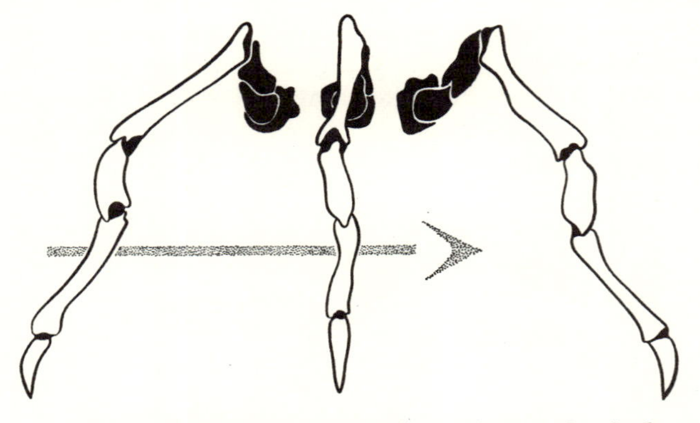

FIG. 123. Three stages of a forward pace of a single dextral walking limb of a macruran crustacean, from extreme backward (left) to extreme forward position (right)

of the thoracic limb. When these two parts rotate from the back forward and back again they carry the entire distal section of the leg in the same direction. The angles that are included by ischium and merus, merus and carpus, carpus and propodus, propodus and dactylus remain unchanged while the animal paces on an even substrate. These joints change their position only to compensate for differences in elevation and to overcome obstacles (Fig. 124). This is different from the brachyurans (see below)

FIG. 124. Cross-section through a macruran crustacean walking over an irregularity of the sea floor. The left leg and one of the right legs (black) touch level ground, a more posterior right leg compensates for an elevation without tilting the body

who move their bodies by bending and stretching the very joints that stay immobile with macrurans, and by holding their coxa and basis immobile. This is due to the sideways walking mode of these animals and is only part of the total phylogenetic reconstruction of their body.

The pacing movement of macrurans is of the same primitive type as in many polychaetes that carry parapodia. The right and left legs of one segment alternate in turning forward and backward, and neighbouring legs of the same side also alternate. These leg movements carry or push the body evenly forward. The pacing limbs are closely spaced in the

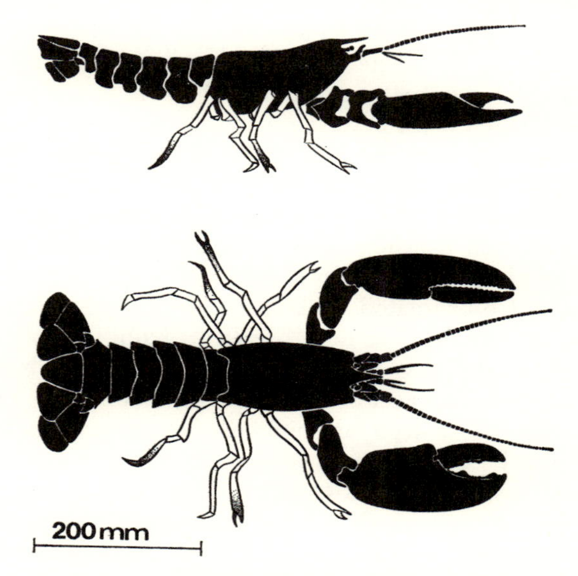

FIG. 125. Pacing *Homarus* seen from the side and top. Shaded limb extremities touch the sea floor

central part of a normally elongate body. The four individual legs of one side can move freely and alternate without mutual interference because they vary in length, and their distal ends point in different directions. The distal parts of the first and second legs point forward, the second being shorter than the first. Thus the second can pass forward underneath the first while this is placed obliquely backward during its resting period when it only supports the body. The distal parts of the third and fourth legs point backward, the fourth being shorter than the third and capable of passing underneath it. To permit these movements, the longer legs, first and third, must also in their resting position form a high enough arch for the smaller second and fourth legs to pass below (Fig. 125). The long third leg, however, cannot pass beneath the short second. Interference is avoided by the backward bend of the third leg. Thus the body rests constantly on four legs; when it rests on legs 2 and 4 of the right side and on legs 1 and 3 of the left side, legs 2 and 4 of the left side and legs 1 and 3 of the right side are moved ahead.

According to von Buddenbrock (1921 and 1953) stumps of legs work normally and move like healthy legs if they reach a substrate. However, if they do not, the movements of the remaining healthy legs change and are co-ordinated differently. This shows that the co-ordination depends on tactile stimuli from stress on the legs when they support the weight of the body. Bethe (1930) called this adaptability 'plasticity of the nervous system' because the central nervous system can adjust automatically to given internal and external conditions. This is important since the crustaceans have a tendency to autotomy of their limbs. Thus there must be a mechanism to replace existing correlations between organs by new ones when the need arises; the walking style can change depending on the number of legs that take part. Bethe regards this ability as an elementary quality of the central nervous system. The spontaneous change in the walking rhythm of crustaceans is not the work of the cerebrum but of the thoracic ganglia. Co-ordinated walking continues after the cerebrum is removed (Jordan 1910a and b; Ten Cate 1930; Kühl 1932).

The co-ordinated forward movements of the eight walking limbs can be reversed and the crustacean moves straight backward. Turns toward the sides are achieved by stretching and bending movements of the merus-carpus joint and the propodus-dactylus joint of the first or the first two pairs of thoracic legs. The last two pairs of legs continue to move in the regular way during the turn. If, on the other hand, the last two pairs of legs use their merus-carpus joints and their propodus-dactylus joints in the opposite direction relative to the first two pairs, the body of the animal turns around on the same place. If, in a rare manœuvre amongst macrurans, all four pairs of legs bend and stretch the merus-carpus joints and the propodus-dactylus joints in co-ordinated movements, the entire body moves sideways at a right angle to the body axis and to the normal direction of movement. The pivoting axes of these two joints are at right angles with those of the other joints of the leg.

This method of locomotion is normal among the brachyurans. They walk sideways at a right angle with their body axis. With the change in the direction of walking the entire animal body has been reconstructed. The most important modification is that the body is foreshortened longitudinally. The abdomen is bent and folded forward under the cephalothorax instead of being stretched back. The cephalothorax is also shortened, in many species so much that it is shorter than it is wide. This telescoping of the cephalothorax reduces the space for the coxae of the thoracic limbs, and they cannot point alternately forward and backward; instead, they lie more or less parallel to each other and are turned straight outward; this is true particularly for the extremely broadened forms (Schäfer 1954a). Thus there is no room for movements of the coxal and basal parts in the horizontal plane. However, at right angles with the coxal and basal joints are the merus-carpus joints and the propodus-dactylus joints; they remain unrestricted, and their activity is now most important in the sideways movement when the legs work parallel to each other. The more distal merus-carpus joints and the propodus-dactylus joints have taken over the original role of the coxal-basal joints. The muscles moving these joints are co-ordinated by stimuli on alternating legs which do not exist in the more primitive macrurans.

This is reflected in the subordination of these muscles. The ability to walk sideways is lost if the two commissures of the pharynx are interrupted; the ability to move forward, on the other hand, remains intact. The forward movement which is phylogenetically older is controlled by the ventral cord while sideways movement is connected with the cerebrum (Bethe 1930 and 1952). This shows that the ability for forward movement persists although it is repressed by the new ability to move sideways.

Whereas the macrurans move forward by simple pacing the brachyurans walk sideways by pacing and by pulling with the legs of one side and by pushing with those of the other. The flexors, i.e. those muscles which pull the leg toward the body, are most active in the leading, the extensors in the trailing legs. The combined pulling and pushing results in a very strong and effective sideways locomotion. The animals can move equally well to either side because the legs of both sides have flexors and extensors.

Like many other epibionts that are forced to move in yielding, soft mud, the prosobranch *Aporrhais pes pelicani* moves by pacing, or, more properly, by 'swinging forward' (Weber 1925). If the animal rests on the outermost, convex whirl of its shell on one side, and on the broad 'pelican foot' extension of the shell on the other, the aperture from which the soft body can protrude forms a tunnel-like recession (Fig. 163). It lies some distance above the substrate and runs from the front to the back between the two supporting shell parts. Thus the mobile and contractile foot, hanging suspended from its stem, has enough space to move back and forth in the tunnel without touching the substrate. When the animal moves forward it stretches the foot forward, sets its broad sole down on the substrate and pushes down; it thereby lifts the shell slightly and swings it forward. After the shell has settled down the action is repeated. This way of locomotion resembles a somewhat clumsy walk on crutches. However, only adult animals have the pelican foot. Young animals have not yet grown the pelican foot extension of the outer rim of their shell and therefore cannot walk.

(x) *Drilling*

In contrast to the so-called burrowing locomotory movements in loose sediments ('Wühlen', Richter 1937b and 1952; Schäfer 1956a) which produce 'burrowing fabrics' in the sediment, locomotory movements

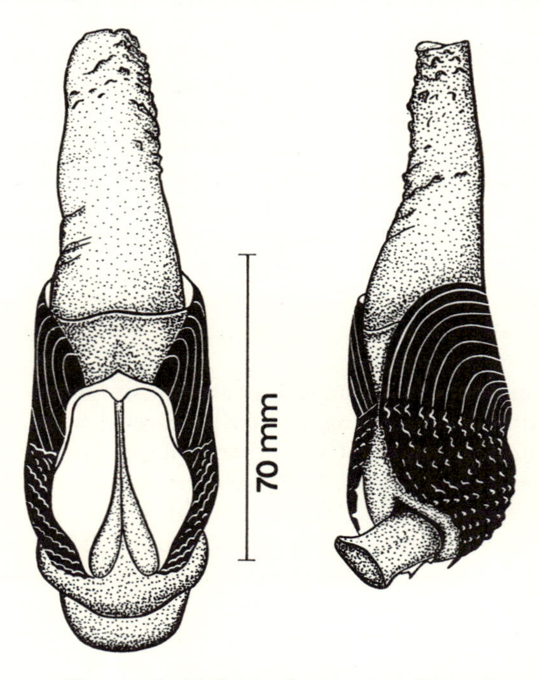

FIG. 126. *Zirfaea crispata* seen from the hinged side (left) and from obliquely below (right), showing the foot

through hard substances such as rock, peat, wood, consolidated mud, and biogenic carbonate, are called boring or drilling. This activity produces no alterations in the fabric of the material. The boring organisms either scrape minute particles off the medium with special tools, or they bore tubes and cavities by chemical means (Kühnelt 1930 and 1933).

Zirfaea crispata is a mechanical driller, with the shape of the valves and the connection between them adapted to this activity (Fig. 126). The anterior muscle which serves as a closing muscle for other bivalves is attached to the outer edges of the valves somewhat below the much reduced hinge. Two accessory valve parts protect the muscle toward the outside. When the muscle contracts the valves open. In opening, the two hump-shaped beaks of the valves roll on each other and serve as mutual supports. The weaker posterior closing muscle lies inside the valves; when it contracts it moves the two valves toward each other. The posterior end of the valves curves outward so that a wide opening remains even when the closing muscle is completely contracted. The peg-shaped foot protrudes through this opening.

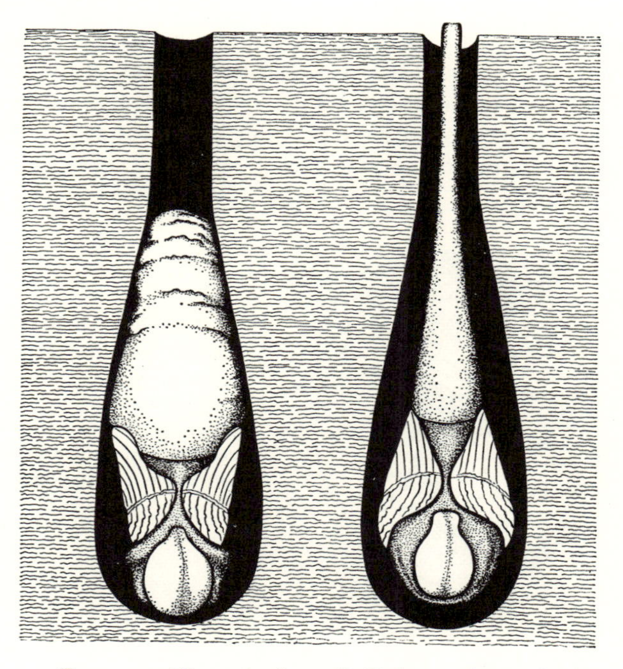

FIG. 127. Ventral view of *Zirfaea crispata* in its borehole. Left: siphon retracted; right: siphon extended. Valves have different positions in the two attitudes. (Schematic drawing)

The valves are thick and strong and can be up to 6 cm long. The growth rings form wavy, toothed welts on the outside of the anterior halves of the valves. There is a distinct boundary between the smooth posterior half and the toothed anterior half of the valve; they are separated by a transverse groove.

To drill the animal contracts the anterior, outer muscle, opens the valves and thus presses them both against the walls of the hole. The foot presses sideways lower down against the wall of the hole, attaches itself there, and then, by contracting slowly one of its retractors, it pulls the toothed valves sideways along the wall. The whole animal thus rotates a few degrees about its longitudinal axis. After this rotation the pressure of the valves against the walls of the borehole is relaxed, the foot moves sideways and reattaches, and the rotational movement is repeated. It takes the animal thirty such small jerks to complete one full turn.

In the North Sea area *Zirfaea crispata* bores in wood, submarine peat, consolidated clay, and also in the chalk of Heligoland. The shavings of wood, peat, hard clay or rock that are produced by the boring seem to be removed from the drilling hole during the operation by pumping motions of the two valves and by subsequent sudden contractions of the thick double siphon. Fig. 127 shows two possible positions of the valves in the borehole. The edges of the valves which are folded outward in a hump in the area of the hinge constitute the pivoting point about which the valves rotate during the pumping action.

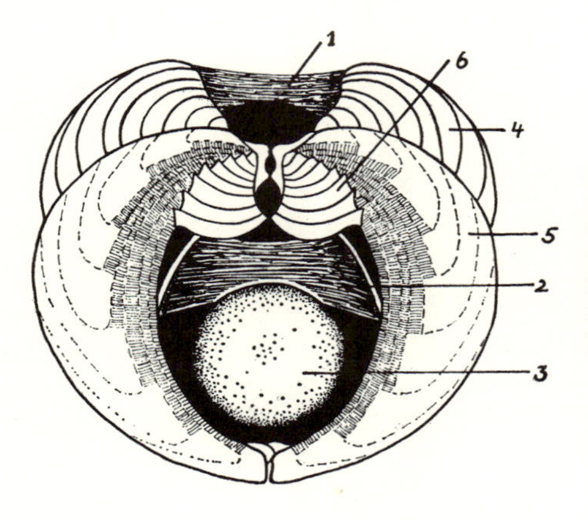

FIG. 128. Frontal view of *Teredo navalis*. (1) posterior closing muscle; (2) anterior closing muscle; (3) sole of the foot; (4), (5) and (6) posterior, central and anterior portions of left valve. (Actual size × 8)

Pholas dactylus and *Barnea candida* are two other boring lamellibranchs that live in the North Sea. In principle their boring mechanism functions in the same way. *Petricola pholadiformis* occurs in the North Sea since approximately fifty years; it opens its valves by pushing movements of the foot and not by the contraction of the anterior muscle.

Teredo navalis bores in wood. Its two valves function exclusively as boring tools and do not enclose the soft body. Fig. 128 gives a front view of the two valves. Three sections can be discerned on each valve: a frontal section which looks somewhat like a bird's beak; a protruding middle section with rows of double pointed teeth, and at the back a smooth end part in the shape of a short plate. The two valves join at two points where each has two button-shaped protrusions, and they are not connected by a ligament. The point of contact lies at the ventral edge of the shell, the homologous place with the hinge of other lamellibranchs. The buttons serve as pivoting points around which the valves move in such a way that the frontal sections are moved toward and away from each other. The anterior closing muscle which is attached to the internal wall of the frontal sections moves these toward each other on contraction; the posterior closing muscle is attached to the internal walls of the end plates. When it contracts it moves the frontal sections apart. The dimensions and proportions of the frontal and central sections are such that file-like grooves on the frontal part scrape away wood shavings at the end of the borehole and deepen the hole during the pivoting movements, while the two middle sections widen the borehole at the sides.

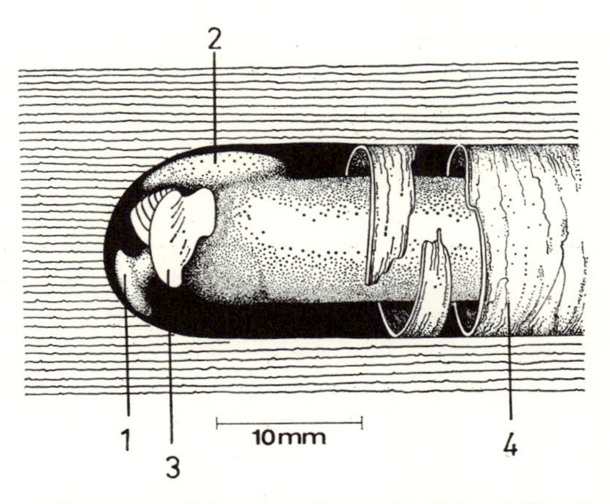

FIG. 129. Side view from the left of *Teredo navalis* in its borehole. (1) foot; (2) cushion formed by the mantle; (3) left valve; (4) calcareous lining of the borehole

The foot extends forward and downward on the ventral side; it attaches itself to the wall of the borehole by suction and thus supports the valves while they move (Fig. 129). Above, the valves are supported by the musculature of the protruding, broad, frontal part of the mantle which can swell up between the valves. While the valves repeat the scraping movements the foot presses sideways partly rotating the anterior part of the lamellibranch about its longitudinal axis; thus the scraping edges can act all around the front of the borehole, and the hole becomes cylindrical. The soft body and the siphon are protected by a solid calcareous tube which is secreted by the mantle.

Teredo drills almost exclusively with the grain of the wood, the tube runs transversely to the grain only in the oldest part where the inner diameter is smallest. The tube is thus generally bent through a right angle. Two accessory shell parts, the so-called palettes, can close off the small opening in the wood after the siphons have been retracted into the calcareous tube. Generally the boreholes serve as protective caves for the boring lamellibranchs, and their food consists of plankton which is ingested with the inhalant water. *Teredo*, however, uses the wood flour as food, and the digestive system is adapted to digest cellulose. The underside of the foot is provided with cilia and transports the drilling flour into the ventral mouth. Plankton which is ingested through the siphon is consumed in addition to the wood flour (Lasker & Lane 1953). See Schütz (1961) and Nair (1959) on the geographic distribution of *Teredo*.

Chemically drilling lamellibranchs do not occur frequently in the North Sea (Kühnelt 1930; Schäfer 1939a). *Lithodomus lithophagus* and *Gastrochaena dubia* which are both chemical borers occur rarely in the North Sea, perhaps due to the lack of suitable calcareous substrates. They live mainly at the Atlantic coast and in the Mediterranean Sea. They seem to bore with a secretion from the frontal lobe of the mantle. They are attached to the wall of the borehole by a network of byssal fibres. To bore they press the lobes of the mantle against the wall of the cavity by retracting the byssal muscles. Rotating movements of the entire animal around its longitudinal axis widen the borehole uniformly.

The sponge *Cliona cellata* also drills chemically, and preferentially in the valves of *Ostrea edulis*. The mechanism of boring is not known. Ziegelmeier (1954) showed that the boring prosobranchs *Lunatia catena* and *L. nitida* bore mechanically. They drill a small, circular hole into the valve of the lamellibranch and use the proboscis to eat the soft body. Another boring animal is the polychaete worm *Polydora ciliata* which drills its U-shaped cavities in calcareous valves, mainly of *Ostrea edulis*, *Cardium edule*, and *Mytilus edulis* (see D I (*b*) (ii) on dwelling boreholes).

(xi) *Chimney climbing*

Animals that move up and down in vertical tubes with hard walls move by chimney climbing. We find therefore this mode of locomotion mainly in those animals that build their dwelling tubes with mucus or 'cemented' walls. But there are also organisms which use tubes built by other species to live in them or to hunt the builders.

For chimney climbing it is, obviously, necessary that the diameter of the body is smaller than the inner diameter of the tube. On the other hand, the body must be large enough to push effectively against the walls. When the animal builds its own walls with the appendages of the head the tube must be constructed with exactly the right inner diameter for locomotion by chimney climbing. If the body fits exactly into the tube chimney climbing is impossible. In such a case, the body moves usually up and down the tube by peristaltic contractions of the musculature.

The polychaete worm *Sabellaria spinulosa* is a good example. *Sabel-*

laria constructs sand tubes. Usually, the tubes are glued on to valves and other hard objects, and the construction continues from there. The worms settle in colonies; the individual tubes touch each other along their entire length. Where thousands of worms settle together the so-called sand-coral reefs arise which can cover an area of many square metres between a little below low-water mark and a depth of 10 to 60 m (Richter 1920a, b, and 1927a; Schäfer 1949).

The body of *Sabellaria* has heteronomous segments: they differ in head, thoracic, and two distinct abdominal sections. The head section with the mouth carries cirri, and tentacles, the thoracic bristled parapodia, and the first abdominal section dorsal parapodia with gill functions; the terminal section has no parapodia. It is folded forward and upward and can perform peristaltic movements.

The body is freely suspended in the tube, held only by three pairs of thoracic parapodia. Their bristles protrude sideways and hold on to the tube walls by pricking into them. Water can flow through the space between the tube and the suspended body. The ciliated gills produce a water current that flows down the dorsal side, the used breathing water leaves the tube on the ventral side and carries the faecal pellets and the sexual products with it upward and out of the tube.

Fig. 130 shows the position of the bristles. Each of the three thoracic segments carries on each side a horizontal parapodium with tufts of bristles. The tuft of the first segment consists of eighteen bristles, those of the second and third segments contain twenty bristles each. They are always arranged in two horizontal rows. The bristles of the three thoracic segments and their parapodia hold the body suspended in the tube or function as locomotory organs.

The parapodium can move as a whole, but each bristle can also move individually (Fig. 131). It can project out of the parapodium or it can be retracted. Muscles are attached to the internal end of each bristle. When these protractor muscles contract they can push up to one-third of the bristle to the outside. The bristle rests partly in a muscular pillow which also contains those muscle strands that retract the bristle into the parapodium and enable the animal to spread the bristles.

Every single bristle of a parapodium is provided with its own protractor muscles so that each bristle can be protruded and retracted independently of its neighbours. Thus the individual bristle of a horizontal row can adjust to fit the internal curvature of the tube. A pair of muscles called abductor and adductor aciculi is also attached to the internal end of the bristle and serves to pivot the bristles about an axis parallel to the body axis.

As in other polychaetes each parapodium which carries a row of eighteen or twenty bristles can be lifted and lowered as a whole. It is

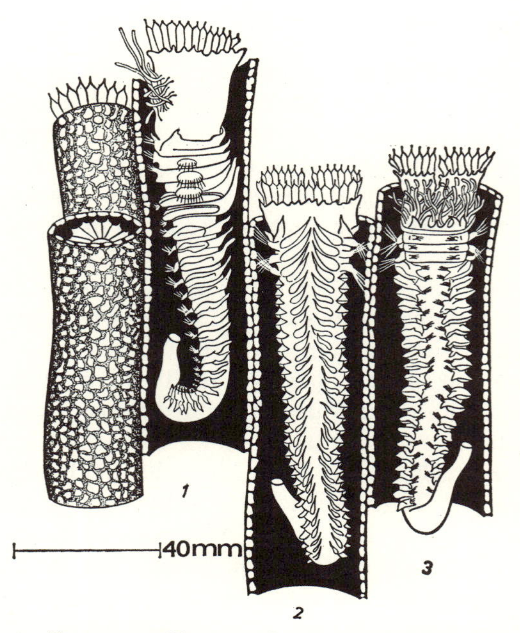

FIG. 130. Cluster of inhabited tubes of *Sabellaria*. Left: external view of two tubes, open (top) and closed (bottom) paleae. (1) worm seen from the left; (2) dorsal view of a worm, showing how it suspends itself with bristles of three segments; (3) ventral view of a worm

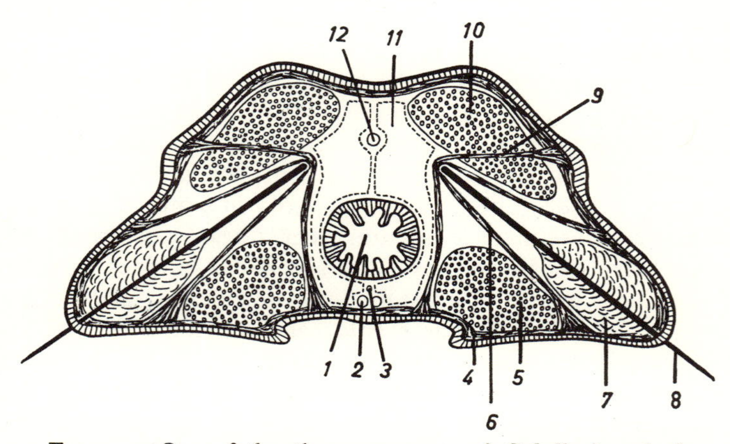

FIG. 131. One of the three segments of *Sabellaria spinulosa* which carry large bristles, in cross-section. (1) oesophagus; (2) ventral nerve; (3) ventral vessel; (4) circular muscle; (5) ventral longitudinal muscle; (6) protractor aciculae muscle; (7) muscle cushion; (8) acicula; (9) abductor aciculae muscle; (10) dorsal longitudinal muscle; (11) coelomal cavity; (12) dorsal vessel. (Schematic drawing)

moved by two pairs of muscles, a pair of adductors and a pair of abductors parapodii (they are not shown in Fig. 131). These muscles lie closely under the epidermis, they are attached to the pillow of the bristles and to the longitudinal musculature of the trunk.

Once the anatomical details of the worm are known the mode of locomotion by chimney climbing is easily understood. To move the body upward, each of the six bristled parapodia performs walking movements, parapodia on opposite sides of a segment move alternately. With each upward step of a parapodium the bristles are retracted to prevent them from dragging along the wall of the tube. After the parapodium has completed its upward movement the bristles protrude, prick into the wall of the tube and thus anchor the parapodium. The steps can be reversed to allow the worm to move slowly down the vertical tube. However, *Sabellaria* responds to sudden, strong stimuli with very rapid movements; all six locomotory parapodia beat forward simultaneously and push the worm down. This reaction is observed in many tube-dwelling polychaete worms. We have described it in *Nereis*. *Sabellaria* is so highly specialised in chimney climbing that the animal cannot move outside the tube. Once outside, it lies helpless and almost motionless on the substrate.

(xii) *Jump-swimming, jump-running and jump-burrowing*

Only in arthropods do we find any jumping movements. They may serve as swimming or burrowing movements in the sediment, or locomotion on the surface. The exoskeleton of arthropods is suited for such movements which can only be effective if individual skeletal parts are quite rigid and form a solid skeleton. The abdomen is the jumping organ and beats vertically downward with the help of its internal musculature. Three types of jumpers can be distinguished according to anatomical, functional, and behavioural differences:

(1) *Crangon* type. The abdomen contains two muscle groups (flexors), an inner ventral and an outer lateral group (Fig. 132). The strong, inner strands of each segment are shaped like a symmetrical basin. They are attached at two points in the following segment. The convex side of each strand lies in the concave side of the following muscle basin. The most anterior of these muscles connects the cephalothorax with the first abdominal segment and is flatter than the others. Plates of muscles that are arranged in the opposite direction lie above the first strands on both sides (white in Fig. 132). The inner and outer groups of muscles thus cross over each other somewhat like scissors. Their effects are additive; they connect each segment with its anterior and posterior neighbour, making a functional unit of the three.

This large number and volume of muscles works during the active phase of beating and bends the abdomen. Fewer muscles are available for extending it and restoring the starting position. Spreading of the tail fan increases the effect of the sudden bending movement. This tail fan is formed by the two leaf-shaped limbs of the sixth, or last but one, abdominal segment. If *Crangon* is lying on the sediment sudden bending of the abdomen pushes the crustacean upward and backward, and the animal springs from the substrate. In open water the abdominal beats carry the body in the same direction. If the tail fan is kept spread out during the beats, the result is an uneven zigzag movement through the water. Up to eight abdominal beats can be performed successively. *Crangon* always uses

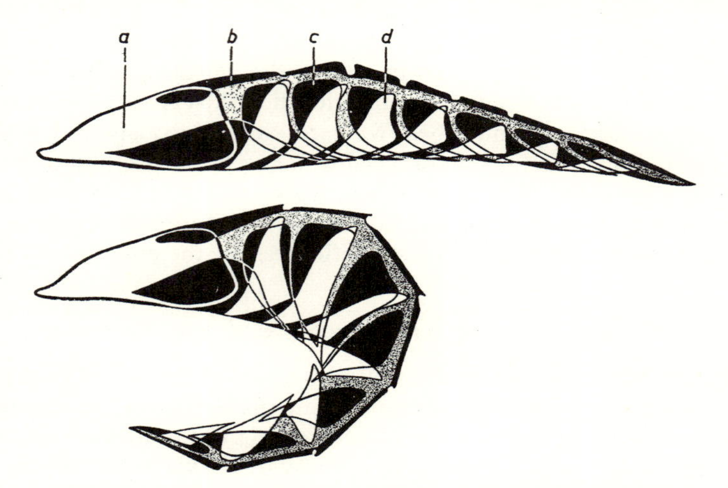

FIG. 132. Longitudinal sections through abdomen of *Crangon*. (a) skeletal frame of cephalothorax; (b) dorsal, longitudinal muscle continuous through all segments; (c) internal, ventral muscles, two per segment; (d) external, lateral muscles, also two per segment. (Schematic drawing)

them in escape from a hunting enemy who has some difficulty in following a prey that moves on such an irregular course. For normal propulsion in open water *Crangon* swims by beating with the limbs of the outstretched abdomen. On sediment the animal walks on the four posterior pairs of thoracic limbs.

Pandalus, *Leander*, and *Palaeomonetes*, like most higher crustaceans, use the same abdominal movements for the same purposes; i.e. to flee from the bottom or in open water. *Homarus grammarus*, *Nephrops norvegicus*, and *Galathea* can drive the swimming body backward over great distances by regular abdominal beats because in these animals the abdomen and the tail fan work in strict co-ordination; the tail fan spreads when the abdomen beats down and is folded when the abdomen is extended. Antennae and claw-bearing limbs are stretched forward. Again, this mode of locomotion, both on sediment and in open water, is used only in flight from enemies.

(2) *Gammarus* type. The last thoracic segment, pleon and urosom are used in jumping; they contain three longitudinal groups of muscles (Fig. 133). Two metameric double muscle strands are on the dorsal side, and

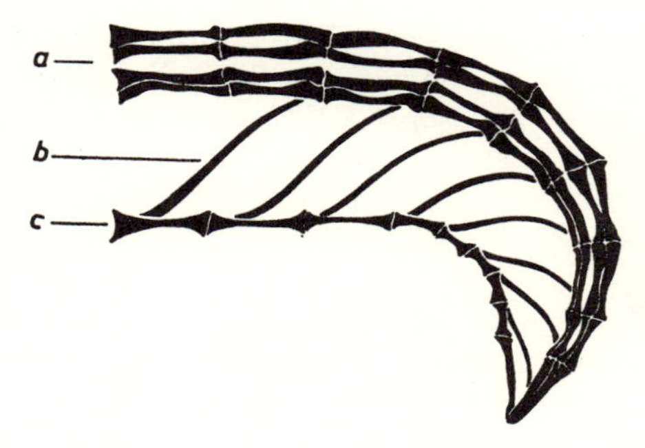

FIG. 133. Longitudinal section through abdomen of *Gammarus*. (a) dorsal longitudinal muscle; (b) diagonal muscles joining two neighbouring segments; (c) ventral longitudinal muscle. (Schematic drawing)

one is on the ventral side. This distribution indicates that the dorsal muscles do the greater part of the work. In fact, in contrast to *Crangon* the abdomen is extended in the active phase of the beat and bent to restore the starting position. The bent abdomen pushes the body off the ground and forward with the rigid uropod of the urosom by stretching it (Fig. 134). The same movements serve in open water for swimming. Two

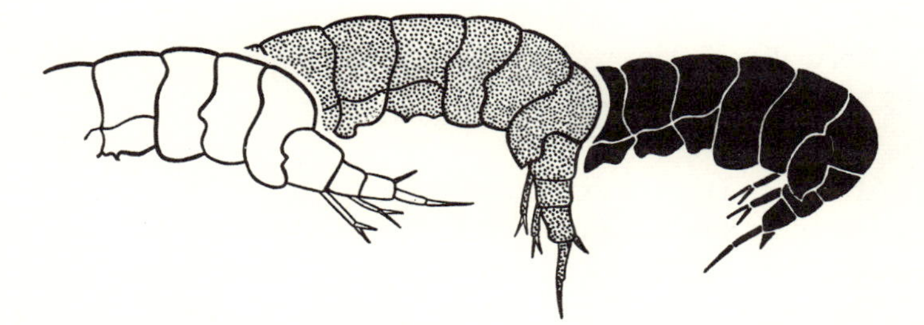

FIG. 134. Three stages of jump-swimming of *Gammarus*. Black—initial position urosoma recurved; dotted and white—progressive stages of straightening. (Schematic drawing)

groups of muscles serve to bend the abdomen, the metameric strand of longitudinal muscles on the ventral side and the diagonal muscles that run from one segment to the next (Fig. 133).

Whereas in *Crangon* and all macrurans the jerky, abdominal movement causes backward jumping used only in escape, *Gammarus* and many amphipods jump forward. They are the regular swimming movements in open water and are also used on the sediment and for burrowing. While *Crangon* can manage only a few successive beats *Gammarus* can continue many times to stretch and to bend the abdomen. The locomotion of jump-swimming and jump-running is furthered by co-ordinated movements of the pleopods which are modified into two-branched limbs (for anatomical details, see Reibisch 1927).

(3) *Cumacea* type. In these animals the entire abdomen can be moved against the cephalothorax in the dorsal and ventral direction by flexing in its two first segments (Fig. 135) as contrasted with the flexibility of mainly the two last abdominal segments of *Crangon* and *Gammarus*. Cumacea

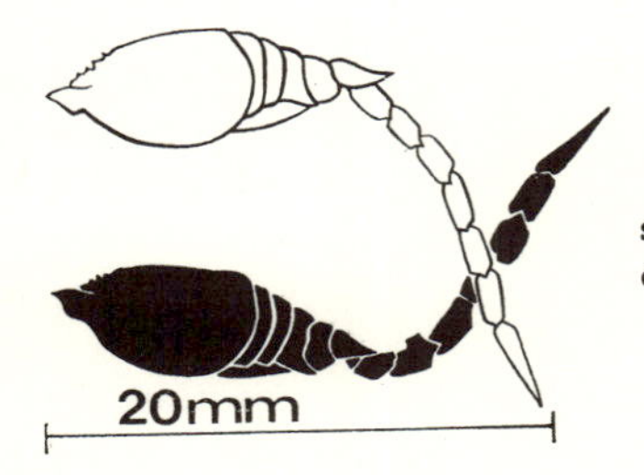

FIG. 135. Two phases of jump-swimming of Cumacean by ventro-dorsal beats of the abdomen.

keeps the abdominal segments more or less straight during the ventral and dorsal movements. Unfortunately, the muscles that produce these movements have never been described in detail. The beating movements of the abdomen produce jumps which are mainly used to escape from the sediment. Sometimes, abdominal beats are used for swimming in open water. Normally, however, the animals swim with stretched abdomen by beating with the exopodites.

(*b*) ETHOLOGICAL AND ECOLOGICAL INTERPRETATION OF MOVEMENTS; IMPRESSION OF LOCOMOTORY TRACES IN SEDIMENT

In all, twelve different modes of locomotion have been discussed. Many require special mechanisms. Certainly, more locomotory types could be added; but the types that have been described are the most important and give an impression of the great variety of locomotion in organisms. As we have seen the different modes of locomotion are not bound to definite taxonomical units; on the contrary, closely related forms can differ considerably in their mode of locomotion and in the use of specific organs while, on the other hand, quite different groups of animals can have similar or identical modes of locomotion.

It is, however, unsatisfactory to consider only the function and interplay of individual mechanisms involved in locomotion. This study must be supplemented by an investigation of the ethological and ecological relationships which require locomotion of animals. A better understanding of these relationships will help us to interpret traces of activities where nothing else is preserved. We will be concerned with actions serving only a few, but basic functions—search for food and construction of a dwelling, flight and rest, sexual and social contact.

Actively swimming, purely nektonic animals that never touch the sea floor during life can be omitted, so can the passively drifting plankton. All benthonic forms, however, are important whether they live permanently on the sea floor or whether they touch it occasionally. For our purposes the benthonic organisms can be subdivided according to their relationship to the sea floor and to the nature of its sediment cover.

Freely mobile species that live generally on the sea floor are designated as 'vagile epibionts'. Sessile species attached to the sea floor and without locomotion are called 'sessile epibionts'. Species that live mostly in one and the same place on the sediment but are capable of locomotion over short distances if certain conditions require it are 'conditionally vagile epibionts'. Freely mobile species living within the sediment are 'vagile endobionts'. Species that live in the sediment but move only under certain conditions are 'conditionally vagile endobionts'. Species that live mostly on the sediment but occasionally retreat into the top layer of the sediment are 'hemi-endobionts'. Species that generally move freely in water as nekton but occasionally touch the sediment surface, either running or crawling or brushing it, are 'hemi-nektonts'. These groupings of animals express the spatial relation between a marine organism and the sediment.

Vagile epibionts, conditionally vagile epibionts, and hemi-nektonts usually produce surface traces; vagile endobionts, conditionally vagile endobionts, and hemi-endobionts usually leave internal traces. However, all epibionts can temporarily turn into 'forced endobionts' when, as happens occasionally, they are covered by thick sediment. Then they can produce internal traces; the same holds for hemi-nektonts which frequently burrow in the sediment.

The seven spatial relationships to the sediment also involve ecological differences and furthermore morphological or ethological features necessary for settling in specific habitats, as varied as sandy, muddy, or hard surfaces of the sea floor, or the interior of sandy, muddy, or hard sediments. This leads to various 'analogies of specialisation' in structure and behaviour (Remane 1943, 1952, and 1956). These analogous aspects are often still traceable in the impressions of the organisms in the sediment and are therefore relevant for actuopalaeontological studies.

The fauna of the southern North Sea does not have examples of all types of locomotory functions that occur in benthonic animals, yet there is enough variation of biotopes to demonstrate the most important kinds of locomotory behaviour.

About four hundred benthonic species are treated in the following sections, three hundred and sixty of them produce traces. The behaviour of an animal depends on the co-ordinated work of the senses, nerves, and muscles; it determines the course that an animal takes when it searches for food, when it is fleeing, or when it is travelling for sexual purposes. To understand behaviour one must know the locomotory process, and know also what a particular movement means in the life of an animal. Both aspects are equally important for the ecological interpretation of traces of life activities.

Although a later chapter is reserved for a full treatment of yet another classification of traces or bioturbate structures of benthonic life, the biostratinomic one, a short list of its main subdivisions shall be given here. It is not so much concerned with the living animal and its actions as with the traces themselves, as they appear in a stratigraphic unit.

(1) Burrowing texture, or perturbed texture caused by burrowing (fossitextura deformativa). The bedding is irregularly deformed by burrowing organisms.

(2) Filled structures, or shaped burrowing textures (fossitextura figurativa). (*a*) sunk structures (endogenic), and (*b*) elevated structures (exogenic). Burrowing animals alter the bedding locally; either they form 'sunk structures' which turn into 'filled structures' when filled with sediment, or they build 'elevated structures', agglutinated or secreted tubes, which also become 'filled structures' when the sediment level has risen to the level of their opening and the cavities are filled with sediment.

(3) Half-structures. Locomotory animals produce grooves on the surface of the sediment ('superficial traces'), or on a bedding plane inside the sediment ('internal traces').

(4) Mobile structures. The builders of mobile structures live in agglutinated tubes and can move with them through or over the surface of the sediment.

(i) *Search for food*

Searching is the first in a series of successive functions related to nutrition; the others are recognising the food and moving towards it, grasping it, reducing it to suitable size, swallowing, absorbing, discarding the substances that have not been absorbed and the organic waste products. Only the first and last of this chain of life activities produce traces that can be preserved, search for food and defecation. The search for food can produce a trace of the movement, such as a crawling trace, the last can be manifest as a substantial remnant. Many marine animals get their food without locomotion.

Animals move actively to feed on other locomotory animals or to scavenge or graze; or they ingest the substrate on which or in which they live together with all the organisms and organic remnants contained in it.

Sessile or slowly moving animals frequently drive food contained in sea water toward them by special mechanisms. The plentiful supply of food in the water explains why so many marine animals are sessile. The prey sometimes approaches the hungry animal actively but accidentally and is grasped by arms; or 'nets' are placed into the current, and the food is caught in them; or, most frequently, ciliated zones inside the body produce an inhalant current, and food particles or suspended or swimming organisms in it are swallowed.

Depending on whether the animals live and feed either on the sediment surface, in the water above it, or inside the sediment, marine animals are epibionts or endobionts, and they use different methods for their search for food. The half-sessile and vagile forms of both groups are more liable than others to leave permanent traces on the sediment as a result of their locomotory efforts in the search for food and to modify the original bedding texture.

Hunting and scavenging

Some predators and scavengers have efficient locomotory organs, roam over wide areas of the sea floor and locate food by chemoreceptors or by tactile sensory organs. Others make little use of locomotion but hide instead in suitable places and wait for their prey. Vagile predators and scavengers make better traces and are therefore more important in this context.

Protozoa. Locomotion of protozoans is in many cases a search for food unless, of course, it is a simple response to external conditions. Autotrophic protozoans, if they move at all, obviously do not need locomotion to go out for food. Heterotrophic protozoans may feed on shaped food, plants or animals, many of them themselves capable of locomotion, or they may, like many multicellular animals, ingest inorganic and organic particles without distinction. Some rhizopods, for example, eat mud and consume the organic substances that are accidentally ingested with the mud. However, few protozoans show so little discrimination for their food; most of them do choose, and some are very particular food specialists and are then usually provided with very effective locomotion.

Unfortunately, it is unknown exactly what triggers the food-searching activity, which stimuli guide protozoans in the search, whether active searches occur at all, whether specific organelles receive stimuli, whether prey and predator meet by accident or whether there is a directional response to directed food stimuli. The fact that in some cases several or many unicellular animals come together within a short time at a food source, especially if it consists of decomposing organic material, shows that positive chemotaxis or topotaxis does exist.

Protozoans often move on a straight course over the surface of the sediment and cover considerable distances. Amoeboid movement of Amoebozoa and Foraminifera, or locomotion by cilia of ciliates produce deeply cut traces which are more distinct than their originators themselves that are often invisible to the naked eye. Fine, freshly deposited muds in a Petri-dish often show after few hours a fine network of deeply and sharply incised lines that deviate only slightly from straight lines in zigzag pattern. Amoebae, Foraminifera, and ciliates that are caught with the mud push themselves to the sediment surface and start moving on it. The traces prove that Amoebae do not simply crawl by 'body-flowing' over the substrate but push themselves forward by something like a peristaltic movement through yielding mud that squishes up around them. The animals can penetrate rapidly through moderate thicknesses of up to 10 mm of newly deposited mud. Nothing is known about their spatial orientation. The trace of locomotion is visible only in fine mud, and no traces are found in fine sand, because their small size permits the animals to move in the interstitial spaces of the sand, and the crawling motions hardly shift the sand grains.

Biostratinomically speaking, the bioturbate textures that protozoans produce in their search for food are generally superficial half-structure traces and only rarely internal traces or burrowing textures.

Polychaeta. Most marine polychaete worms are endobiontic, but very few are completely vagile predators which burrow in the upper sediment layers and have no home; most construct dwellings which will be discussed in more detail in the section on dwelling structures. They can be simple soft mucus structures, or agglutinated tubes of a firmer consistency, or tubes of lime, which are safe and lasting dwellings. In addition to its obvious protective function a structure with a rigid wall enclosing an open space also serves to maintain access to oxygenated water for breathing in an environment that may be reducing and therefore hostile to life. The respiratory organs, gill appendages at the parapodia, are outside of the body, and the worm sits in its tube, in a water bath, as it were. By active movements the animal can replace spent breathing water. Tube-building polychaetes are of great interest to the palaeontologist because either they can be bodily preserved, or they can be identified by the shape and material of the structures they build. This has been demonstrated by many samples taken with a sample box. The free-living polychaetes move through the sediment only to hunt. The disturbances of the original bedding fabric which mark their trail through the sediment are inconspicuous, simple traces of locomotion, sag trails, and inflections of bedding planes. They are identical with the trails produced by all animals that hunt in the upper sediment layers but do not live in them.

Eunice norvegica is a typical predator feeding on living animals in the sediment; it has parapodia that are specialised for burrowing. Other polychaetes that spend a considerable part of their life hunting in the sediment are the aphroditids *Harmothoe impar*, *Gattyana cirrosa*, *Malmgrenia castanea*, *Pholoë minuta*, *Polynoë scolopendrina*, and *Aphrodite aculeata*. Their body is covered by scales (elytres) below which water can circulate that can serve for breathing (Schäfer 1956a). *Nephthys hombergi* penetrates 5 to 15 cm deep into the sea floor. According to Linke (1939) it hunts for worms, small crustaceans, young lamellibranchs, and gastropods.

In their search for food vagile polychaete worms produce bioturbate textures which can be classified as half-structures and burrowing textures.

Priapulids. *Priapulus* lives in muddy sediment; it is a predator and eats small animals. Its locomotion resembles that of *Sipunculus* (see section on 'bolting'). The body of the priapulids consists mainly of two thick, short tubes, one in front of the other (for anatomical details, see Apel 1885, Baltzer 1931). The anterior smaller proboscis can be retracted by muscles into the longer trunk-tube. When the musculature of the trunk is contracted, pressure of the body fluid everses and protrudes the proboscis again. Rows of strong teeth with backward-directed points sit on the outside of the proboscis and anchor it in the sediment. In addition, the proboscis can swell up in individual sections to reinforce the anchorage. As the proboscis is slowly retracted the entire trunk is pulled forward. The movements are violent, strong, and repeated (Friedrich & Langeloh 1936).

Priapulus caudatus (Fig. 136) is up to 8 cm long and eats small animals living in muddy sediment, such as polychaetes, small starfish, brittle stars, and even members of its own species (Halton 1930, Lang 1949, Phillips 1853, Rauschenplat 1901, Schulz 1931, Wesenberg-Lund 1929). Several concentric rows of teeth arranged in circles sit in the anterior end of the proboscis. When it is eversed the teeth of the rows move outward and can grasp the prey. The retracting proboscis then pulls the prey into the oesophagus. These large animals leave quite distinct sag trails and perturbations in the soft, layered coastal mud sediment down to a sea floor depth of 500 m. The animals penetrate 8 to 12 cm deep into the mud when they move about. They do not construct dwellings. Caspers (1938) has found them regularly in the Channel of Heligoland, but according to him they occur rarely elsewhere in the southern North Sea.

The bioturbate textures due to *Priapulus* are burrowing textures.

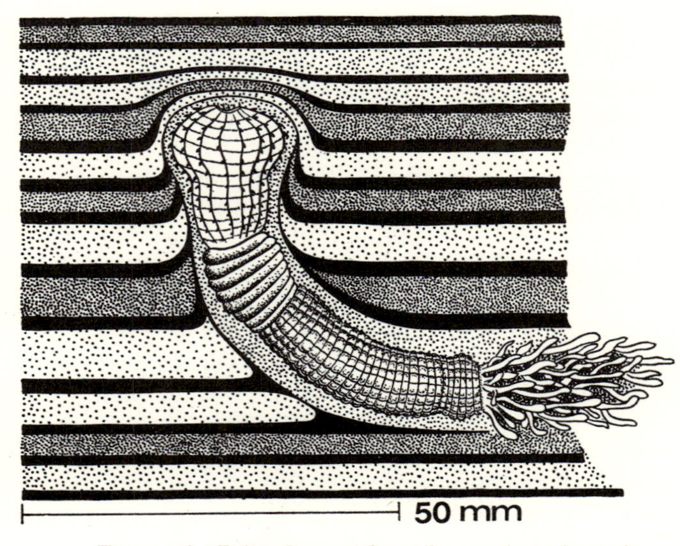

FIG. 136. *Priapulus caudatus* burrowing through bedded sediment

Gastropoda. A good many prosobranch species are predatory and scavenging. Some of them move on the surface of the sea floor; these are pure epibionts and produce surface traces. Other species penetrate deeper into the sediment and disturb its texture permanently. Purely superficial traces are unlikely to be preserved, particularly in the heavily reworked sediments of the shallow sea and shall therefore be discussed only briefly.

Buccinum undatum is the most common and the most striking scavenger among the gastropods of the southern North Sea. It is seen alive above the low-water line only when it has been thrown up by the surf. If not injured, the animal immediately tries to regain the deeper, permanently covered sea floor. *Buccinum undatum* has a depth range from the low-water line to a depth of 1200 m and lives in tidal channels as well as in the open North Sea or the deep North Atlantic. It feeds mainly on dead organic material. A chemoreceptive sensory organ, the osphradium, helps in locating the food (Brock 1933, 1936). It lies in the mantle cavity into which the breathing water enters through the siphon. *Buccinum* can also follow the drag trail of a dead animal by palpating the sea floor with its tentacles. The crawling traces produced by an adult *Buccinum* are very shallow in spite of the great weight of the animal. While *Buccinum* crawls the mucus glands at the front and side edges of the foot secrete a wide stream of mucus over which the animal crawls. This is probably an adaptation favourable for crawling over soft, muddy substrate, preventing, together with the great width of foot, almost completely any sinking into the mud. In agitated water the mucus band floats up from the trail and twists into a braid. Mucus braids with adhering sand grains may have a better chance of preservation than the crawling traces themselves (Schäfer 1943). The shell is occasionally dragged behind the sole and then draws long furrows through the sediment.

Neptunea antiqua occurs frequently in the central and northern North Sea. It is larger than *Buccinum*, with the shell up to 17 cm long. Generally,

FIG. 137. *Neptunea antiqua* ploughing through sediment with the siphonal spur of the aperture rim. (Actual size × 0·2)

it holds its shell with the apex up so that it can never drag the apex or other parts of the shell over the ground. *Neptunea*, like *Buccinum* and *Murex*, moves almost exclusively for the purpose of scavenging. It crawls usually with the broad foot 10 mm deep in the sediment. It tips its shell forward and downward and pushes the elongate siphonal process into the substrate forward of the front edge of its foot. In this way the sediment is ploughed up and pushed sideways (Fig. 137). The shell with the siphonal process is moved slowly to and fro while the animal crawls. During the activity the siphon lies protected in the siphonal process and is never extended as far as in a moving *Buccinum*. The animal ploughs through the sediment in order to find dead organic material. When the animal crawls about in its search it disturbs the upper sediment layers considerably down to a depth of 15 mm.

Nassa reticulata penetrates even deeper into the sediment than *Neptunea* (Ankel 1936a and b). *Nassa*, however, does not dig for food but only hides in the sediment; through its long siphon that reaches to the sediment surface the animal pumps respiratory water into the mantle cavity where it also checks it for chemical stimuli. The trace in the sediment is therefore a resting trace. If a chemical signal is received by the osphradium the gastropod comes to the surface and moves toward the prey following the concentration gradient of the stimulating chemical message. In flowing water the siphon of the resting gastropod is directed up-current, and currents also steer the animal when it is crawling on the sediment surface (positive rheotaxis). *Nassa* can also find dead organic detritus that lies in the sediment if it is not buried too deeply and the open water carries its scent. *Nassa* uses its long proboscis to reach buried prey. It has been observed that *Nassa* attacks animals while they are still alive if their mobility is inhibited. It propels its body by rhythmic pulling movements of the propodium and produces distinct sag trails in the upper sediment layers but hardly any deep impressions on the surface of the sediment.

The members of the genus *Lunatia*, *Lunatia nitida*, *L. catena*, *L. pallida*, *L. montagui*, and their life habits are particularly important. *L. nitida* has been investigated in detail by Ziegelmeier (1954, 1958, 1960) and by G. Richter (1962). All lunatias approach lamellibranchs in the sediment, bore through one of its valves and eat the soft body of the living lamellibranch with the proboscis. According to Ankel (1938a) *Lunatia nitida* can also be cannibalistic. In every swash mark along the coasts of the open North Sea numerous valves with circular borings can be found, borings produced by this predatory gastropod. Its special method of feeding forces the gastropod to concentrate on lamellibranchs that are usually buried in the sediment. Ziegelmeier has demonstrated by experiments that the gastropod pulls the prey into the sediment before it starts boring, if it happens to catch them on top. The front edge of the propodium functions as a tactile receptor with which the eyeless *Lunatia nitida* locates its prey. Its burrowing range is to a depth of 7 cm and more; it carries oxygenated water in its mantle cavity. It therefore does not need a siphon as many other endobionts do, and is not restricted to a certain maximum depth prescribed by the length of its siphon. The breathing water enters the mantle cavity through a cone-shaped fold in the upper edge of the propodium which slightly overlaps the shell. During the long stay in the sediment the animal replaces the breathing water partly from the interstitial system of the sand. If oxygen is lacking in the sediment the animals must rise after some time to the surface of the sea floor. The duration of its possible stay in the sediment increases with the grain size of the sediment; so do the length and depth of the burrowing traces that the animal produces. In muddy sediment it must occasionally return to the surface. However, it is not restricted to certain properties of the sediment, and it occurs in pure sands as well as in sandy muds.

Its mode of locomotion through sediment has been discussed in the section on crawling by pushing and pulling. The animal is well adapted to this locomotion and performs it rapidly and with apparently little effort. It seems therefore odd that it is frequently eaten by the less mobile *Astropecten irregularis*. When the gastropods are hungry they travel untiringly through wide areas of the sea floor. Usually, they move so near the surface of the sea floor that shallow, furrow-shaped depressions can be observed from above that give away the position of the gastropod. According to Ziegelmeier *Lunatia* often drags a captured lamellibranch around for an hour, holding on to it with the rear part of the foot while it crawls through the sediment in search of a favourable place for boring. It circles a tracked-down prey several times to find an appropriate place for boring through the shell.

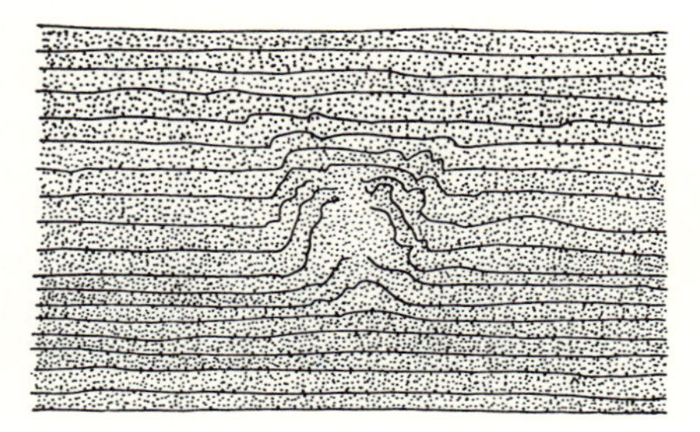

FIG. 138. Cross-section through laminated sand perturbed by a burrowing *Lunatia nitida*. (Schematic drawing)

These observations show how effectively *Lunatia* is adapted to a vagile life in the sediment, and also how intensive and widespread the disturbances must be that a single gastropod produces in its hunting ground. On its crawl the animal constantly secretes mucus which makes the body slippery and also seals the slit between the propodium and the shell, preventing sediment particles from entering the mantle cavity. The mucus does not consolidate the disturbed sediment (Fig. 138).

In 1951 Ziegelmeier found three to five animals in 0·1 m^3 grab samples northwest of Heligoland. The diameter of the near-spherical shells determines the diameter of the perturbation trails in the sediment. *L. nitida* has a shell diameter of approximately 17 mm, *L. catena* of 30 mm, *L. pallida* of approximately 20 mm, and *L. montagui* of approximately 10 mm.

The prosobranchs *Scala clathrus*, *S. clathratula*, *Mangelia*, *Typhlomangelia*, *Lora* and *Philbertia* all dig through the sediment. *Scala* and *Mangelia nebula* penetrate 5 cm deep into the sediment by means of their muscular foot. *Mangelia*, *Typhlomangelia*, *Lora* and *Philbertia* are mostly scavengers but they have poison glands which indicate that they are also predators. *Nucella lapillus*, *Urosalpinx cinerea*, and *Tritonalia erinacea* bore into lamellibranch and gastropods like *Natica*, and also into balanids (Ankel 1936a; Orton 1930; Pelsener 1925).

Among the opisthobranchs the tectibranchs *Retusa truncatula*, *R. umbilicata*, *R. nitidula*, *R. obtusa*, and *Scaphander lignarius* are predators. Several *Cylichna* species, *Acera bullata*, *Pleurobranchus plumula* and the genus *Philine* feed on dead organic material and small living animals by burrowing through muddy sediment. All these forms must either dig through the sediment or crawl over its surface on their soles to find their dead or living organic food. If they dig through the sediment their traces may well be preserved but they are unspecific.

The bioturbate textures produced by scavenging and predatory gastropods are either half-structures or perturbed burrowing textures.

Scaphopoda. The scaphopods hunt small animals living in the interstitial water in deep sand. Their particular way of feeding explains the shape of their shell and of their soft body. Structure and behaviour of these animals and their effect on the sediments in which they live have been described by Schäfer (1956a). In the families Dentaliidae and Siphonodentaliidae the shape of the shell resembles that of a hollow elephant tusk open on both ends. The shells are roundish in cross-section, ivory-smooth or occasionally fluted lengthwise but always shaped so that they glide easily through sediment. The soft body adheres to the shell by the dorsal side of its mantle. In the wide end of the shell there is room for head and foot, and the worm-shaped visceral sac extends to the other, narrow end (for anatomy, see Thiele 1926, and Lang 1888–94).

The animals are buried more or less vertically, preferably in muddy sediment, and only the thin end of the shell protrudes with an opening that can be closed by skin lobes at the back end of the mantle. Water enters through this opening to the mantle whose epithelium presumably functions as respiratory organ together with the 'water lungs' at the distal end of the intestine. There are no gills. Movements of the soft body push the used water out at the back; the anus discards faeces into the mantle cavity from where they are expelled with the water. The mouth points downward at the broad, lower shell opening. Long, very mobile, worm-shaped tentacles transport the food to the mouth.

Beside the mouth lies the muscular foot which is round in cross-section and serves for downward burrowing and locomotion in the sediment. Both head and foot can be completely retracted into the shell by a double columella muscle which runs through the entire body. The Dentaliidae have a foot with a cone-shaped and pointed end; at the base of the cone are two side lobes that can be spread. The Siphonodentaliidae have a foot with a disk-shaped end plate which can be spread or contracted into a bell shape fitting tightly over the end of the foot. An inner longitudinal muscle and outer circular musculature move the foot (the Dentaliidae have additional radial bundles of muscles, Lang 1888–94). The foot stretches when the circular muscles contract and blood flows simultaneously into the lacunae (van Benthem-Jutting 1931). The foot becomes elongated and worm-shaped and penetrates deep into the sediment while the side lobes or the disk are folded close to the foot. After this digging motion is completed the foot lobes or disk spread and anchor the foot in the sediment. This locomotion thus follows a principle applied by many animals that live in the sediment. The surface of the foot carries a ciliated epithelium with unicellular glands. It is not known how these organs take part in the digging movement. Certainly, they are not used to construct a cemented sediment tube which would mean almost sessile life habits. Probably, they function to consolidate and glue together sediment flakes or grains that immediately surround the foot. The way of feeding and the kind of food exclude a near-sessile life: the animals collect small organisms, Foraminifera, diatoms, and lamellibranch spat from their immediate surrounding by means of their ciliated mouth tentacles (captacula) which are sticky from glandular secretions. After the food at one locality has been harvested the animals must move horizontally. Each advance is preceded by a movement of the foot, and a small cavity is formed in front of the mouth into which the tender tentacles can then be stretched for feeding.

Little is known about the individual movements in the sediment. According to Schäfer the animal crawls horizontally with the concave side of the shell forward by retracting the foot, then pushing it obliquely downward, anchoring it by spreading the lobes, and finally contracting the longitudinal muscles of the foot, pulling the shell horizontally and also downward, somewhat deeper into the sediment. To compensate for the downward component the shell must be pushed up to bring it to the former level.

The foot also serves to move the animal upward if it is getting covered by sedimentation and needs to rise to the surface (Fig. 139). In this case the foot lobes, or the disk, prevent penetration of the sediment by the foot when it stretches and presses downward, pushing shell and body upward. During this upward movement the narrow shell opening lies in front, and the upper end of the mantle closes the aperture to keep the sediment out; during downward movement the frontal lobes of the mantle play a similar role at the wide, lower shell opening. Jaeckel (1953) lists localities in the North Sea where scaphopods are found.

Upward movements of scaphopods through the sediment have the best chance of being recorded as a fossil trace because downward locomotion is always triggered by erosion. Rapid sedimentation, however, drives the animals up until the narrow back end of the shell is again exposed. Episodic rapid sedimentation is likely in the habitat of those scaphopods that live in the coastal areas. Fossil scaphopod trails have not been recognised. The animals leave distinct traces consisting of local drags, thickenings or complete perturbations of beds but they are not sufficiently characteristic to be positively identified.

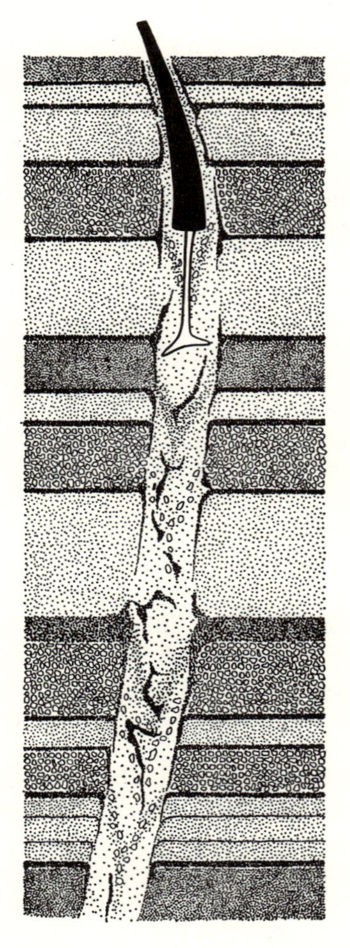

FIG. 139. Perturbed trail created by a scaphopod ascending to the new surface after being buried by rapid sedimentation. (Schematic drawing)

The polychaete worm *Pectinaria coreni* (see D I (*b*) (ii) on agglutinated dwelling tubes) has similar life habits to the dentaliids. The worm lives in a mobile tube which it has constructed. This tube plays the role of the calcareous shell of the scaphopods. The shells of the two animals are carried in the sediment in the same way. They have the same general shape, the locomotory organs lie at the lower, wide aperture of the shell, the foot of the scaphopods, or the bristles of the head of *Pectinaria*.

Adjacent to the locomotory organs are the organs for grasping food, in both cases flexible, thin, worm-like and contractile tentacles. Both animals whirl breathing water into the shell aperture which projects above the sediment surface. Locomotion in sediment is possible because the soft bodies of both animals are highly flexible, contractile and attached high in the shell; contraction is achieved by the columella muscle of scaphopods and by the lateral trunk musculature of *Pectinaria* which holds on to the upper end of its tube by claw-bristles. Because the two animals have the same type of life habits their locomotory traces in the sediment correspond perfectly.

The bioturbate textures produced by scaphopods are perturbed burrowing textures.

Crustacea decapoda. Almost all decapod crustaceans of our area are carnivorous, either predators or scavengers, and many cover wide distances on the sea floor to find dead or living prey. With their articulated limbs they crawl, walk, run, pull, push, or burrow. In D I (*a*) (ix) on pacing locomotion, the mechanism and rhythm of decapod limb movement is described in detail. In most cases the limbs touch the ground with the pointed dactylus. Occasionally, the walking limbs are bent so that they touch the ground with the outer edge of the dactylus, giving the crustacean a larger area of support and keeping it from sinking into soft ground. This occurs mainly in deep-sea forms which have to walk on soft substrates most of the time. In macruran crustaceans the two anterior pereiopods carry small claws which are usually placed half-open on the ground. Even when the dactylus is set straight down on to the substrate, obliquely downward-pointing tufts of stiff bristles at the tip of the dactylus prevent it from sinking too deeply. These bristles have therefore not only a tactile receptory function. Such aids reduce somewhat the depth to which the animals may sink, but nevertheless, dactyli penetrate enough to make traces consisting of clear depressions in the surface. The dactyli of an adult lobster can penetrate soft sediment to a depth of 22 mm; *Cancer pagurus* reaches 18 mm, *Carcinides* 15 mm, and *Eriocheir sinensis* 18 mm. Traces that are caused by such deep impressions are more than simple surface phenomena and can be considered as internal traces. This makes them more important because internal traces have a better chance of being preserved than surface traces.

The forward-moving walking limb approaches the substrate with the tip of its dactylus; and also, before being lifted completely the tip frequently drags for a certain distance when it is withdrawn from the sediment. The dactylus impressions in the sediment rarely appear as clear-cut punctures but are drawn out both forward and backward or in one of these directions (Fig. 140). A stab with a broad, heavy, pointed dactylus disturbs sediment layers even below the depth of penetration. Within an extremely small area the layers are pushed downward without being torn, producing a clear mark in otherwise undisturbed series of beds. This happens in both sandy and loose muddy sediment. The traces caused by punctures lie somewhat obliquely in the sediment corresponding to the angle at which the dactyli enter. The points where the moving limbs of large macrurans touch the ground lie far apart because with different lengths of the legs the distances between the body and the dactyli of various limbs are also different. Therefore, the dactyli

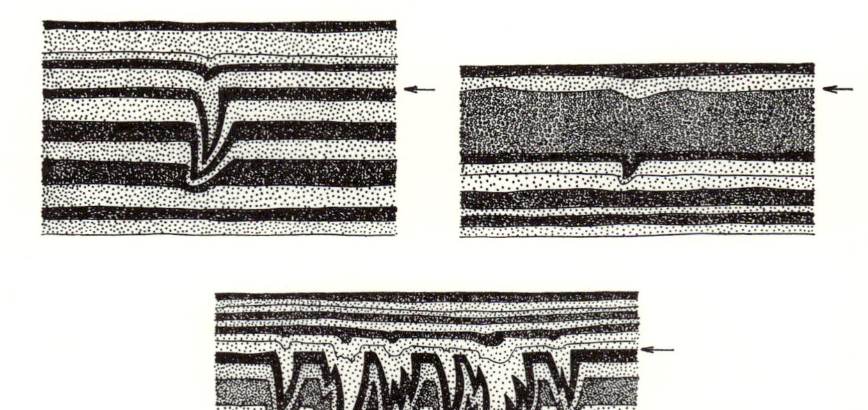

FIG. 140. Cross-section through bedded sediments marked by the incisions of dactyli of walking brachyuran crustaceans. Arrows—direction of walking. Top left: single incision by *Eriocheir*; top right: same, but not visible in a layer of coarse sand; bottom: partly overlapping incisions made by several limbs. (Schematic drawing)

of the third and fourth pairs of limbs do not destroy the traces of the first and second pairs. Brachyurans, on the other hand, often blot out the traces of the four legs of one side with the impressions of those of the other side which follow on the same course. The eight separate impressions are left only if the brachyuran moves in a curve, and the trailing leg impressions are slightly shifted with respect to impressions of the four leading legs.

Sediment transport easily destroys the traces near the surface. The deeper parts of the impressions, however, are often better preserved. Firm mud layers underlying loose sand can receive dactylus punctures below the surface and preserve them even if the rest of the trace is lost in the surface layer (Fig. 141).

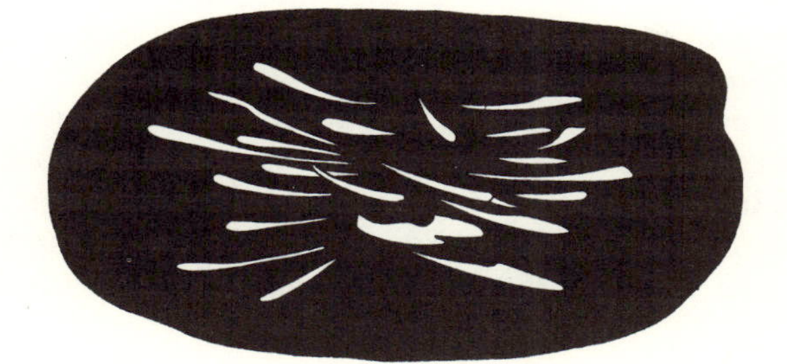

FIG. 141. Internal trace on the top of a mud layer overlain by sand into which a *Carcinides maenas* has burrowed. (Actual size)

Homarus gammarus is the largest and heaviest crustacean living in the southern North Sea. The length of thorax and abdomen reaches 55 cm, the clawed limbs become 35 cm long and have a maximum circumference of 18 cm. The lobster lives mostly on rocky ground and cliffs. In the southern North Sea area it is therefore restricted to the environs of Heligoland and to a few places with artificial solid structures. No traces are formed in those rocky areas. Quite frequently, however, fishing vessels pick up lobsters in drag-nets in the sandy and muddy zones between Heligoland and the German coast to the south of it. These are probably migrating individuals on the long journey over soft sediments between Heligoland, the East Frisian Islands, and the groynes at the Minsen Oldeoog in the mouth of the Jade river. Lobsters have settled in these places over the last decades, as catches from these areas indicate, and there may be an exchange between these populations through such migrating individuals.

Such wanderings are done exclusively by walking. The crustacean moves its pereiopods step by step covering many nautical miles and impressing its characteristic footprints into the sandy and muddy substrate. Only in escape does the animal use its supple and efficient backward-swimming technique of rapidly flapping with the abdomen and at the same time spreading the tail fan.

The smaller and more fragile long-tail crustacean *Nephrops norvegicus*, on the other hand, lives only on muddy ground. Its walking traces can therefore be preserved as internal traces; the conditions are particularly good because the animal lives mainly at depths between 30 m and 700 m in calm water. It occurs in the Skagerrak and northern Kattegat, but also in soft, muddy areas of the English Channel and around the Dogger Bank. There the populations are dense, indicated by commercial catches in 1937, 510 tons in Denmark, 153 tons in England, and 124 tons in Germany. They are caught around the year with trawls and other drag-nets. According to Rühmer (1954) the young animals bury themselves deeply in the mud. Numerous surface and internal traces must be produced by *Nephrops*.

The shrimp *Crangon vulgaris* populates the tidal seas in enormous quantities. This animal runs frequently on the sea floor against the current. The thoracic limbs carry the body, and in deep water, their impressions alone make up the walking trace (Fig. 142). If, with the receding tide, the animal gets into shallow water on the tidal flats, and the body is not

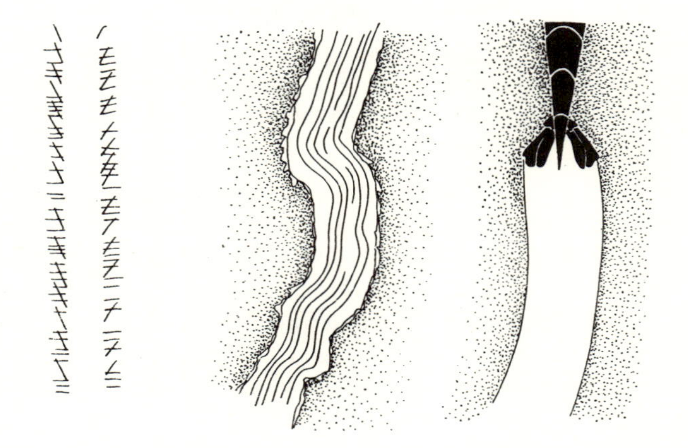

FIG. 142. Surface traces produced by *Crangon*. Left: impressions of the limbs, formed in deep water; centre: dragging trail of the whole body in extremely shallow water; right: dragging trail of the tail on moist surface of emerged tidal flat. (Actual size)

completely supported by water it becomes too heavy for the thoracic legs. Then the trailing abdominal limbs touch the ground and draw several parallel-running furrows. If *Crangon* walks on wet mud the tail fan rests also on the ground, is dragged along by the pulling limbs and produces a broad, unspecific trail that blots out all impressions produced by the limbs. Such a trace cannot be distinguished from the crawling trace of a gastropod.

The brachyurans are particularly adapted to ranging wide over sandy or muddy sea floors. Many of them cover large distances every day. The beach crabs *Carcinides maenas* which inhabit the tidal zones walk several kilometres with each tide in search of food (Schäfer 1954a). They spend the periods of low tide in the channels of the large rivers and in other permanently water-filled areas of the tidal sea. But with the incoming tide they run landward, first following the tidal channels, and at high tide, they walk on to the wide tidal flats. When the low tide sets in they leave these areas just as rapidly. Hardly a crab can be encountered during the periods when the tidal flats lie dry. Those left are usually sick animals (e.g. infected by *Sacculina carcini*). On the open sand flats of the outer coastal zone journeys of over 5 km have been measured within eleven hours. The dorso-ventrally flattened cephalothorax, claws that can be folded to the thorax, and eight additional long and very strong pereiopods are evidence of perfect adaptation to environment and mode of living (Fig. 79).

The crab *Eriocheir sinensis* has similar body proportions. It is equally adapted to life in the coastal zones. These crabs travel all through the tidal areas and far upward into the lower parts of the rivers as is well known and has often been described (Peters & Panning 1933). Occasionally, they accumulate by the hundreds of thousands on certain of their travel routes in currents of brackish water. Sediments that have been crossed by such crowds show mark beside mark of the dactyli over distances of many kilometres, amounting to a complete reworking of the fine bedding of the top layers. Such traces can be preserved because the large and long-legged animals can produce impressions up to 18 mm deep.

In contrast to *Carcinides maenas*, *Eriocheir* continues its landward and upriver migration independently of, and often against the tidal and river currents. *Eriocheir* produces not only walking traces but also burrows that run for up to 40 cm horizontally into the vertical or overhanging banks of meandering tidal channels. It burrows not only with its claws but by digging and scraping with all eight remaining lateral pereiopods (Schäfer 1954a).

Cancer pagurus lives mainly on sandy ground in deep water far from the coast, but occasionally advances into coastal areas. In contrast to *Carcinides* it has no need for speed and perseverance of walking in its habitat. The walking legs are short and slender compared with the massive but dorso-ventrally flattened cephalothorax against which the clumsy, clawed anterior limbs can be folded to reduce the resistance against strong water currents that are prevalent where the animal lives. The clawed limbs must help to carry the body in its sideways walk and their movements are therefore co-ordinated with those of the walking legs. In running, the clawed limbs touch the substrate with the outer edges of the palmae and produce rather shallow, basin-shaped impressions. In cross-section they appear as local elongations and thrusts of the fine bedding. The clawed limbs are used efficiently for burrowing into the sediment when the animal seeks shelter. While the short walking legs dig and loosen up the sediment the broad, strong claws alternately move forward and heap up a wall of sediment in front of the animal. It can then lie for many days on the same spot, more or less deeply buried. It walks only to find dead or sick animals for food; fast-moving animals never become its prey.

Eupagurus bernhardus is much more mobile. This hermit crab populates roughly the same areas as *Cancer pagurus*; it lives on sandy or muddy sediment but also on stony and rocky ground and uses the claws together with two pairs of pereiopods as walking limbs. The left claw is small, the right one large; this asymmetry compensates for the oblique position of the body caused by the gastropod shell that is carried by the animal and could not be supported symmetrically by symmetric limbs. Locomotion consists of pushing and pulling rather than walking. The animal can move forward, backward, and sideways. During these movements the gastropod shell which the abdomen carries, is not simply dragged over the ground, but lifted by the abdominal musculature and thrown in the direction in which the animal advances. Thus the abdominal musculature of the crab plays the role of the strong columella muscle of the gastropod. Such skilful handling of the shell enables the animal not only to move over mud and sand, but also to climb over uneven substrates like rubble and rocky cliffs. It can also rise to the sediment surface together with its gastropod shell when it is covered up by sediment.

Tuft-like tactile sensory organs on the carapace, mainly on its posterior part, take care of the delicate contact with the shell (Luther 1930). With these organs the crab can constantly check whether the shell projects far enough over the posterior body sections. As soon as the tuft organs lose contact the abdominal muscles contract in jerks and pull the somewhat displaced shell up again. The locomotory trace is not a continuous drag trail of the shell but a regular series of individual impressions produced by the periodic forward throw of the shell. In soft muds *Eupagurus* usually moves backward because in this way, all locomotory limbs can push the animal simultaneously. This requires less energy than alternating pulling movements of the legs either forward or sideways. The pointed dactyli of the pereiopods leave marks in the bed that resemble those produced by macrurans and brachyurans.

The constant leg work of higher crustaceans not only moves the body over the largest possible distances but also allows chemoreceptory sense organs in the skin of the dactyli of the walking limbs to trace food on the ground. Luther (1930) has recognised as such organs the 'funnel-canals' which sit on the legs and claws and on the outsides of the first maxillipeds. He has found them in *Carcinides maenas*, *Hyas araneus*, *Eriocheir sinensis*, *Portunus holsatus*, and *Eupagurus bernhardus*. To find food in the sediment the crab must pierce the surface innumerable times with the points of the dactyli. This also explains why the dactyli do not have broad soles but pointed tips that sink into the ground with each step somewhat contrary to their primary function as organs of locomotion over sand and yielding mud. They act as chemical probes that examine the sediment as deeply as is compatible with efficient walking.

The needle-sharp inflections of beds that we observe in cross-section are therefore traces of the search for food as well as of locomotion. Both fast runners and slow walkers search for their food in this way; the slow ones are restricted to a smaller territory which they exploit more thoroughly by shovelling more deeply through the sediment with their legs and claws. But all of them rework the sediment, sometimes to the point of destroying all evidence of bedding. Further examples of this kind of activity are *Hyas araneus*, *H. coarctatus*, *Ebalia tuberosa*, and the anomurans *Lithodes maja*, and those species of the genera *Galathea* and *Munopsis* that live in sand and mud.

The bioturbate textures that decapod crustaceans produce while they search for food are superficial and internal traces of the half-structure type and perturbed burrowing textures.

Isopoda. Many isopods are predators consuming molluscs, annelids, and crustaceans, others are ectoparasites on fishes, and many are scavengers. Except for the parasites the isopods are very mobile and swim, run

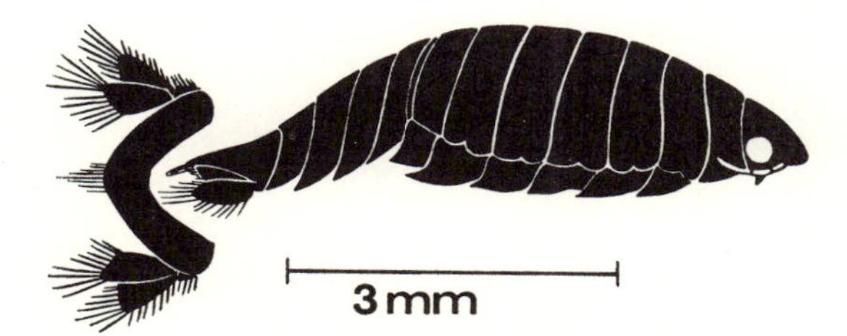

FIG. 143. *Eurydice pulchra*. The animal swims by beating its posterior segments and the uropods. Side view of the whole animal, and (left) top view of tail

and crawl, or dig in the sediment. Some who climb on algae, sponges and hydroid polyps are unimportant for our investigation. Burrowing forms produce distinct and preservable disturbances of the bedding, and all make surface traces that are exposed to destructive forces as all such traces.

Eurydice pulchra is a predator. It grows to only 5 mm long and populates the sand bottoms from near the beaches down to the central part of bays of the southern North Sea coast such as Jade Bay. These isopods occur often in large numbers in flooded puddles when the sun is shining on the sandy, sunny beaches of the East Frisian Islands; but they can also be found in great numbers in deeper sandy zones where they race about in curves and circles searching for food. They accumulate near dying or dead animals and bite into them with their mouth mechanisms. Even the foot of a wading human is not safe from them. They swim by beating rapidly with their last abdominal segments and their uropods which widen into bristled fins (Fig. 143). The five pairs of plate-shaped abdominal legs beat alternately to support the whirling activity of the last segment. These animals crawl over sandy substrates with seven pairs of thoracic legs and produce surface traces that appear as shallow and indistinct grooves, resembling those of *Hydrobia*. They use the same movements to search below the surface of sandy sediments for covered,

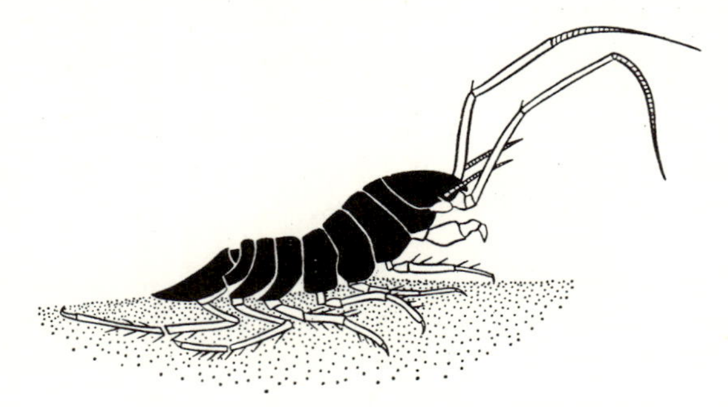

FIG. 144. Isopod *Munna boecki* resting on sediment surface. (Actual size)

dead organic material. The interstices of the upper sandy sediment layers provide enough oxygen for breathing. These isopods seem to avoid muds because of the unfavourable conditions for breathing.

Numerous other isopods live on muddy ground; some of these forms have extremely elongated thoracic limbs to prevent them from sinking into the yielding mud cover, an adaptation also achieved by some decapods. The surface traces they make consist of curved loops and semicircles, drawn by the long and often bristled limbs; in soft mud the

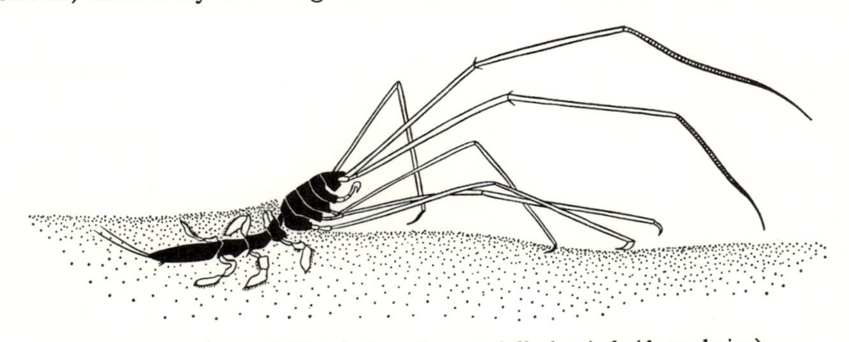

FIG. 145. Isopod *Munopsis typica* partially buried. (Actual size)

abdomen produces an additional broad and often multiple trace. These mud dwellers live mostly at some depth and down to 1000 m where even surface traces have a chance to be preserved because the water is absolutely still. Such isopods are *Munna boecki* (Fig. 144), *M. limicola*, *Munopsis typica* (Fig. 145), *Llyarachne hirticeps*, *Pseudarachne hirsuta*. *Eurycope producta* and *E. cornuta* bury themselves in muddy sediments. Efficient diggers are *Calathura brachiata* (Fig. 146), *Cyathura carinata*, *Anthura gracilis*, and the blind *Leptanthura tenuis*. Several features they

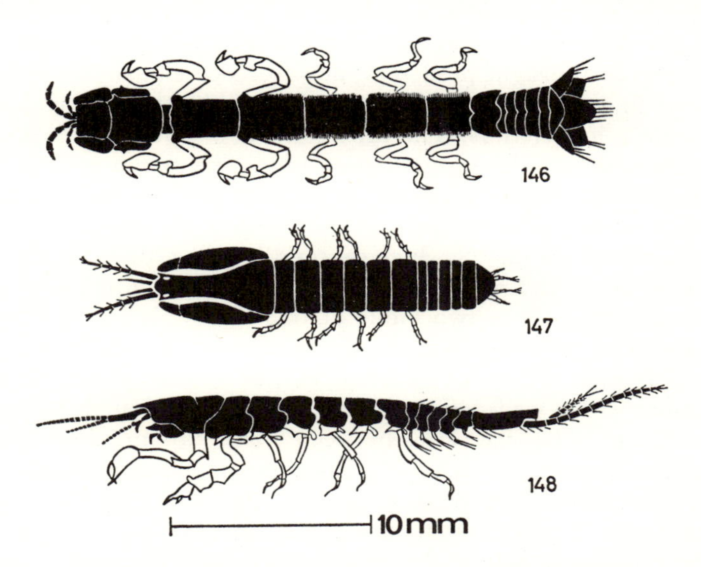

FIG. 146 (*top*). Dorsal view of the isopod *Calathura brachiata*

FIG. 147 (*centre*). Dorsal view of the tanaid *Heterotanais oerstedi*

FIG. 148 (*bottom*). Side view of the tanaid *Apseudes talpa*

have in common are adaptations to life in the sediment: effective statocyst organs in the pleotelson, elongated body, and strong claws for burrowing. Probably, the rosette-shaped posterior end of the abdomen formed by the telson and uropods has some respiratory function.

The bioturbate textures caused by isopods are surface traces and perturbed burrowing textures.

Tanaidae. Many anisopods (or tanaids) are endobionts. Their body is narrow and elongated and usually flattened dorso-ventrally. The trunk of the animal is very mobile because six of the eight thoracic segments are free and not covered by the short carapace. The small coxal plates also leave room for a fairly great mobility of the trunk. The abdomen is short and compact. The second thoracopod has a claw function and is built as cheliped (or gnathopod). Its palma and the whole limb are very strong. Thus the animals are well adapted to burrowing. They feed on small crustaceans and worms which they catch by digging in muddy sediment. Several species probably eat mud. They do not breathe through gills; instead, breathing seems to take place in the walls of the cephalosoma and carapace that enclose a breathing cavity through which epipodial appendages of the exopodites of the maxillipeds pump the water.

The animals are vagile predators; however, occasionally they construct a small dwelling tube with open space which enables them to remain in the sediment for a longer time. The tube consists of a mucous coating with small pebbles and mud lumps adhering to it. The mucus is produced by quite large glands which lie laterally in the third, fourth, and fifth thoracic segments, continue into the limbs on those segments, and have ducts that open at the tips of the dactyli. Such mucous dwelling tubes are abandoned after a short time, and new ones are constructed when necessary.

The tanaids produce two kinds of traces: simple burrowing traces which appear as narrow zones of perturbation of the bedding structure that run irregularly through layered, muddy sediment; the others are tube sections with a well-defined boundary, sitting obliquely or vertically in the sediment. The size varies between 1 and 20 mm; the tubes differ therefore in diameter and in length. The animals populate areas from the coastal region down to depths of 4000 m, but probably they are most frequent at great depth (Nierstrasz & Schuurmans-Stekhoven 1930). The Tanaidae *Paratanais batei* and *Heterotanais oerstedi* (Fig. 147) have very strong chelipeds which lie in depressions in the carapace; *Strongylura cylindrata* has an elongated body; *Cryptocope abbreviata*, on the other hand, is small, compact, and has a pointed head. The Apseudidae do not seem to build mucous tubes. *Apseudes talpa* (Fig. 148) is approximately 6 to 7 mm long and *A. latreilei* is 7 to 9 mm long.

Tanaids produce bioturbate textures that are perturbed burrowing textures or shaped burrowing structures.

Echinodermata. The majority of the echinoderms are predatory or scavenging; a few are herbivores and graze (e.g. *Echinus esculentus* and *E. acutus*). The predatory and scavenging echinoderms live mostly on the sediment surface (with the exception of *Astropecten*). Those forms that consume sediment and small animals usually burrow in the sediment; their traces are therefore more likely to be preserved than those of the predators.

Asterias rubens is one of the most common, largest and most voracious predators among the asteroids of the southern North Sea. The starfish moves by stepping with the tube feet, keeping the arms spread. In soft, yielding mud this function of the tube feet is replaced by pushing and pulling with the highly flexible arms. This results in quite deep drag trails and curved traces. Locomotion is slow but steady and can be maintained over great distances. Frequently, starfish congregate by the tens of thousands at places where food is abundant as on a patch of lamellibranchs that have been washed out by erosion. The traces at the surface of the sea floor are easily destroyed. However, starfish that get covered by large amounts of sediment can rise to the surface by pushing and pulling. Then they leave conspicuous burrowing and sag trails in the newly deposited layers. The trails are so obvious because the animals are large and numerous. *Asterias glacialis* takes the place of *Asterias rubens* in the northern parts of the North Sea. *A. mülleri* is considerably smaller and inhabits an area which has its southern border near Heligoland.

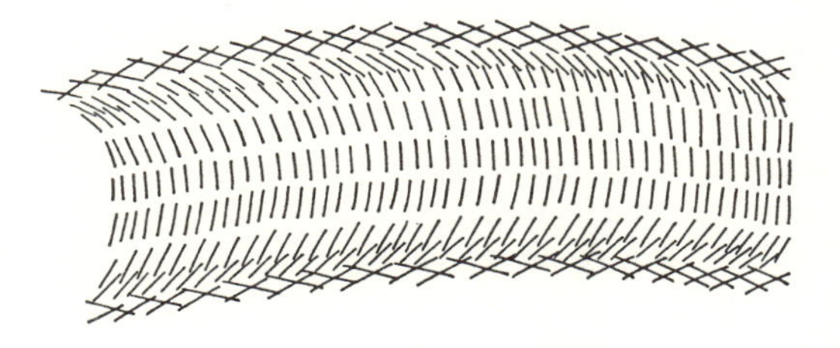

FIG. 149. Crawling trace of *Psammechinus miliaris*. (Actual size)

Psammechinus miliaris is the only important sea urchin living on soft ground. It lives mainly in 5 to 18 m deep tidal channels and scavenges or grazes on animal colonies, e.g. of *Botryllus* or sponges. It particularly enjoys opening egg capsules of *Buccinum* and eating the embryo. The surface trace of *Psammechinus* is a wide, delicately patterned band. When crawling in mud the back spines beat transversely to the direction of movement whereas the lateral spines scrape from the front backward. The crawling trace retains only the picture produced by the hindmost spines; it consists of fine, transverse grooves (Fig. 149). The tube feet do not make traces.

Locomotion of the ophiuroids has already been described. *Ophiothrix fragilis*, *Ophiopholis aculeata*, *Acrocnida brachiata*, *Ophiura texturata*, *O. albida*, *O. affinis*, and *O. sarsi* occur in the North Sea. In their trace, the individual steps of the pushing arms can be well distinguished. They are connected by the drag trail of the preceding or following arm (Fig. 150). If the muds are too soft the long arms push the body disk forward by undulating motions. The brittle stars *Amphiura chiajei* and *A. filiformis* usually crawl on the sediment surface by pushing, bending and stretching their very long arms. If they scent animal food their movements become

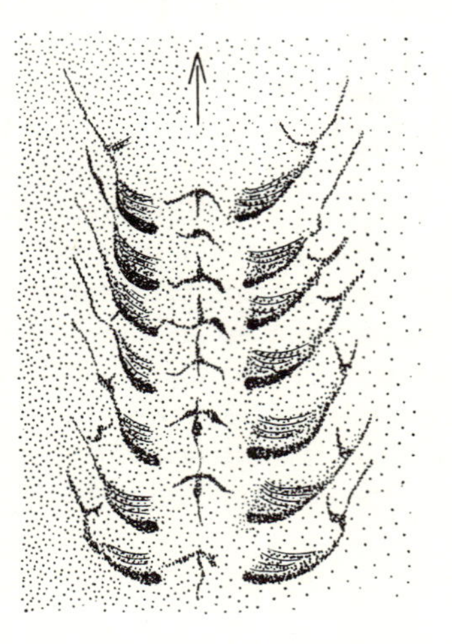

FIG. 150. Crawling trace of a brittle star. (Actual size)

very lively, and soon many accumulate near where the food lies. Where they are very crowded their traces overlap and form a pattern of fine grooves in which the impressions of a single animal can no longer be identified (Fig. 151).

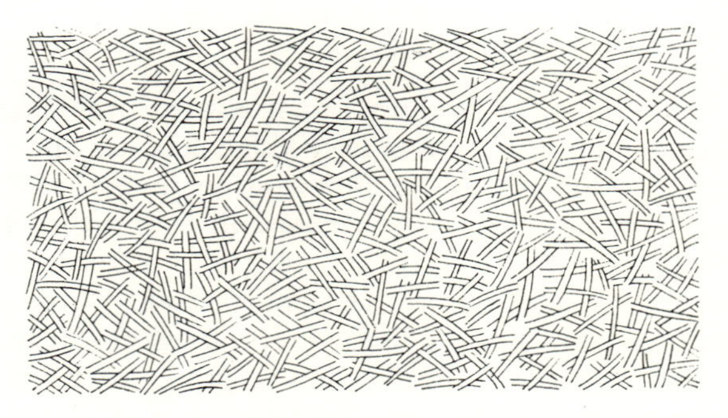

FIG. 151. Mud surface marked by numerous brittle stars crawling at random. (Actual size)

The bioturbate textures that predatory and scavenging echinoderms produce are surface traces and perturbed burrowing textures.

Pisces. Fishes feeding on the sea floor produce traces. Some fishes pick up the prey from the sediment surface, others dig and push into the sediment while they search the ground with their eyes and palpating tactile organs. In both cases they produce swimming traces. Bottom-dwelling fish move either by undulating horizontally or vertically, or by beating with the pectoral fins, or by walking on limbs.

There are many examples of the first group. The animals have an elongated body, and in many cases the dorsal and anal fins extend the body considerably in the vertical. The more common species are *Scyliorhinus caniculus*, *Galeorhinus galeus*, *Anguilla anguilla*, *Conger conger*, *Gaidropsarus mustela*, *Pholis gunnellus*, *Zoarces viviparus*, *Anarhichas lupus*, and *Liparis liparis*. The locomotory traces of all these species are similar. When the swimming fishes stay constantly in touch with the ground the resulting trace is continuous and undulating. If the fish raises the body slightly from the bottom only the lowermost parts of the body touch the sea floor and produce semi-circles and sections of curves. If the fishes swim very fast incisions by the tail obliterate all other traces that may have been created by anterior parts of the body. If an undulating fish is pursued it moves as close as possible to the bottom, frequently alters its direction and makes U-turns. Turns tend to be made where the bottom offers points of resistance. Traces in the form of arcuate grooves through 180° or less are produced in escape, and they are irregularly distributed over a small area.

Vertical undulation is the mode of locomotion of flatfishes; numerous examples in the North Sea are *Scophthalmus rhombus*, *S. maximus*, *Hippoglossus hippoglossus*, *Hippoglossoides platessoides*, *Pleuronectes platessa*, *P. microcephalus*, *Limanda limanda*, and *Solea solea*. Strictly speaking, flatfishes move by undulating laterally; but by lying on their side they undulate in the vertical plane. The outline of the body is a more or less elongated oval. The dorsal and anal fins form a lateral fringe which can undulate either independently or together with the trunk. The undulating movements of the body continue into the spatula-shaped tail fin. The majority of the flatfishes are bottom-dwelling; they take their food from the sea floor, or they push their mouth into the ground and tear out their prey such as lamellibranchs, gastropods, small crustaceans, or worms. Therefore, these fishes frequently swim and search closely above the sea floor; probably rhythmic beating with the edges of their fins on the bottom startles the prey and chases them out of their hiding places in the ground. If flatfishes are fleeing from enemies they beat violently on the ground with the edges of the fins. The raised clouds of sand and mud divert and blind the pursuer.

When the vertically undulating body is swimming parallel to the substrate rhythmic impressions of the anterior part of the fin-fringe are overprinted by the posterior part of the fringe. The arcuate impressions cross each other (Fig. 152). These impressions tend to obscure the direction of movement. But the impressions of the tail fin are concave forward and thus indicate the direction of movement. However, if the animal swims in an upward-inclined position, only the backward beats of the fins produce an impression in the sediment. The tail fin does not always cause a drag trail.

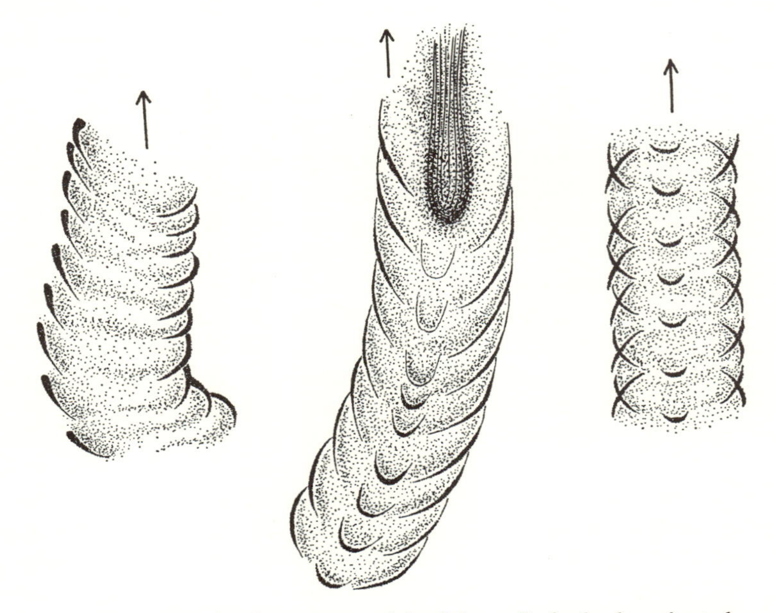

FIG. 152. Surface traces of flatfishes. Left: body oriented slightly tail downward; during downward beats the fins of the dorsal and ventral edges of the rear part of the fish touch the sea floor; centre: body oriented more strongly tail downward, the dorsal and ventral fins and also the tail fin touch the sea floor during downward beats; right: body horizontal, anterior and posterior portions of dorsal and ventral fins and the tail fin touch the sea floor during downward beats. (Schematic drawing)

Few fish in the North Sea swim by beating with the pectoral fins: *Agonus cataphractus*, *Gobius microps* (and several species of the same genus), *Callionymus lyra*, and *Myoxocephalus scorpius*. All are poor swimmers and therefore stay close to the bottom.

Agonus is the poorest swimmer, it is heavily armoured and usually moves in discrete, clumsy jumps. The pectoral and ventral fins simply beat backward. The short ventral fins scrape through the sediment in the process and produce half-moon-shaped traces in a regular row (Fig. 153). When the fish swims at a somewhat higher level the tail and the tail fin take part in swimming by undulating weakly and ineffectively. The body lies obliquely in the water. During each interval between beats of the pectoral fins the body sinks back a little, and the tail fin strikes the ground producing a short and narrow trace. The strokes do not form a straight line but are offset by the slight tail undulations. Presumably, regularly arranged strokes from the tail fin in fossil swimming traces are also caused by similarly poor swimmers beating their pectoral fins. They do not glide

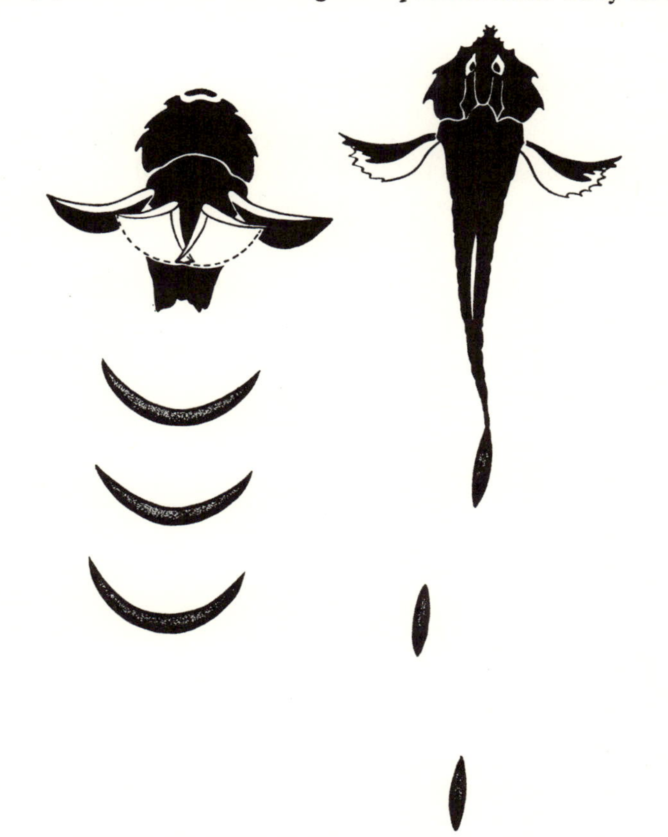

FIG. 153. Swimming traces of *Agonus cataphractus*. Top left: ventral view of anterior portion with pectoral fins extended sideways and brought together ventrally (the two positions connected by dotted line); bottom left: trace produced on sea floor by the beats of pectoral fins; top right: dorsal view; bottom right: trace produced by the tail fin. (Schematic drawing)

forward continually but can only 'jump' clumsily. Continual gliding would produce continuous furrows rather than separate strokes.

Gobius has a similar mode of swimming. It moves in jumps along the sea floor mainly by the beat of its large pectoral fins. After each jump the animal settles down on a suction cup formed by its modified ventral fin and produces small, regularly spaced circular depressions; in this mode of swimming the tail fins make no marks. When the fish swims fast it throws the tail violently to and fro and produces regularly spaced, narrow strokes in the sediment.

Callionymus has perfected swimming with the pectoral fins to such a degree that the body can glide continually. The pectoral fins are held backward close to the body and perform a whirling vibration. The rest of the body is kept straight and motionless, the large ventral fins are spread and function as gliding wings in the water or they touch bottom. The body touches the ground so lightly that, apart from almost imperceptible drag trails, the sediment stays undisturbed.

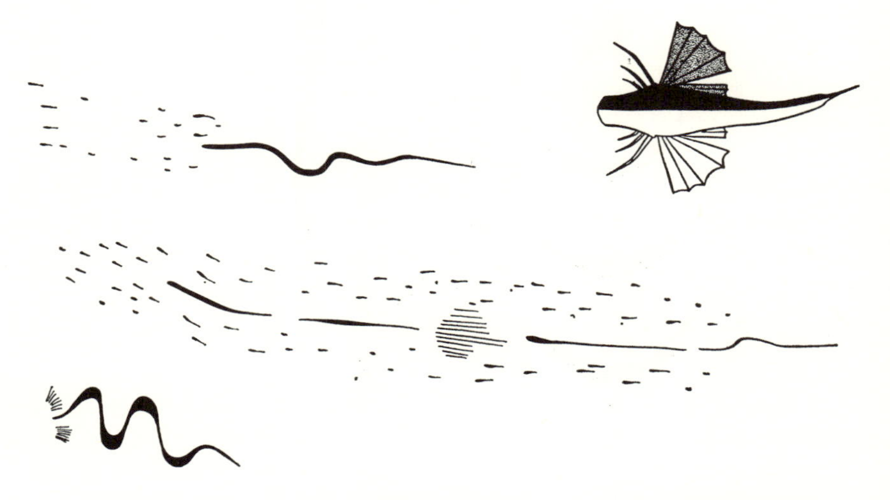

FIG. 154. Swimming traces of *Trigla*. Top right: dorsal view of the fish; top left: trace produced by landing on the sea floor; centre: trace produced by walking on pectoral fins, aided by undulations of the tail fin; (ruled lines) braking trace produced by pectoral fins; bottom: undulating trace of the tail fin, braking fins; bottom: undulating trace of the tail fin, braking trace produced by pectoral fins. (Schematic drawing)

Myoxocephalus also uses the large pectoral fins for swimming, but a strong musculature in the tail also enables it to undulate. It can therefore rise rapidly in the water. The animal hunts by perching on an observation post on the sea floor and suddenly advancing toward the prey. Swimming traces are therefore rarely produced. Moreover, this fish lives preferably on stony ground.

The only fishes of the North Sea to walk with limbs are *Trigla corax*,

T. gurnardus, *T. pini*, and *T. lineata*. The three ventral rays of the pectoral fins function as limbs. They are detached, have only small fin membranes, are articulated and can be moved independently. It has often been described how these fish walk by working their six limbs alternately and pulling themselves forward, resting lightly on the bottom. The dragging tail fin produces a straight trail. To the right and to the left of this trace are the impressions of the walking fin rays. These fin rays differ in length, and therefore their impressions do not obliterate each other (Fig. 154). The three free rays of the pectoral fins incidentally also serve as palpating organs to trace prey that may be hidden in the sediment. *Trigla* is a good swimmer; in open water it swims by undulating the broad tail fin. When the fish returns to the sea floor and changes from swimming to walking, it therefore produces undulating traces of the tail fin in addition to braking traces from the ventral fins.

The bioturbate textures produced by bottom-dwelling fishes during their search for food are surface traces and perturbed burrowing textures.

Grazing

One uses the term grazing to mean eating from a more or less dense plant cover while moving slowly about. Whenever the food consists only of a thin film each bite of the grazing animal removes all the food at that spot. To get new food, grazing animals must necessarily move and must also keep eating while they move, in contrast to predators, scavengers, whirlers, filterers, or net-catching animals who can afford to be stationary while they eat.

The term 'grazing traces' has been used for a long time in the palaeontological literature. Seilacher (1953b) used 'pascichnia' to designate traces of biting and of the concurrent locomotion, leaving it undetermined whether the grazing trace consists of the sequence of the individual bite traces, or of the traces which the locomotory organs of the animal produce while it moves about. Zoologists usually distinguish one case clearly from the other (Ankel 1936a, 1937c, 1938b; Eigenbrodt 1941) and restrict the term to the series of biting traces. In the case of a gastropod grazing on a thin coating of algae the term is used for the gaps in the algae, not for the trace of the moving foot. The situation is different for the palaeontologist: in most cases fossil grazing traces are traces of locomotion for the purpose of grazing; individual biting traces are preserved only rarely and under special conditions. According to Seilacher (1953a, p. 434) fossil grazing traces are 'caused by vagile sediment eaters, feeding by crawling and digging'. Thus he also leaves it open whether the term is applied to impressions produced by locomotory or feeding organs. Seilacher's Fig. 10 (1953a) is a sequence of clear illustrations showing transitions from one extreme type of grazing trace to the other, from pure biting traces to pure locomotory traces. Only the traces of bites or patterns of multiple bites are a definite proof of a former grazing activity; the criteria for the grazing function are always uncertain. However, certain spatial features of the trace are indicative of grazing, such as obvious avoidance of already grazed areas or meandering; but they are inconclusive, animals may have had other reasons for such crawling patterns.

By definition only herbivores can graze. But grazing is rare among the invertebrates, apart from some gastropods. Most ingest large amounts of sediment together with plant food like diatoms or green algae. Thus there is no sharp distinction between grazing animals and sediment eaters. Neither does Seilacher draw a sharp line in his definition (1953a, p. 434), 'grazing traces are produced by sediment eaters that feed while they crawl or dig'. To avoid ambiguities the term grazing is here used to exclude any 'grazing in sediment' which consists mainly of sediment eating. The term shall be restricted to the activity of animals moving on the sediment surface and ingesting pure or almost pure plant food. Grazing traces must therefore always be half-structures in the biostratinomic nomenclature. Even this sharpened definition leaves the activities of certain small forms undetermined; they live within sandy sediments and are called 'sand lickers' (Zimmer 1932, 1933a; see also Remane 1940). They ingest individual sand grains, remove the algae and detritus from them and spit out the clean grains of sand. The sand-licking activities of these animals are of no interest to the palaeontologist. Only few groups of lower marine animals are true grazers and leave grazing traces; gastropods, lamellibranchs, and polychaetes will be discussed in detail.

Prosobranchia. The prosobranch marine gastropods are typical grazing animals. The best known example is the beach gastropod *Littorina littorea*. It inhabits muddy and sandy tidal flats, areas with shell breccia, stony beach zones, and artificial masonry. It has been demonstrated that *Littorina* does not reject animal food (Schäfer 1950a); generally, however, it feeds predominantly on the diatomean coatings in the tidal flats and the fine cover of green algae on stones. Ankel (1936c, 1937c) has described in detail its mechanism of feeding, how the highly differentiated radula-apparatus functions, and how the individual teeth of a row work during feeding. Each individual working phase of the radula against the substrate represents a 'bite' of many teeth and rows of teeth into the algae cover; scraping traces of the individual teeth are engraved into the coat of algae as a result (Fig. 155). The material of algae that is left after the first bite would certainly be sufficient for several additional scraping movements of the radula on almost the same spot, but this does not happen in grazing.

Instead, the eating movements of the radula are coupled with turning motions of the whole head and mouth that carries the radula.

Thus bites are not superimposed on each other but lie beside one another on a trail which is curved according to the turning motion of the head. The composite feeding traces follow a more or less meandering course because the sole of the foot carries the body forward while the head is turned to and fro. If the gastropod advances slowly the resulting feeding trace is a narrow, meandering biting trace; if the animal crawls

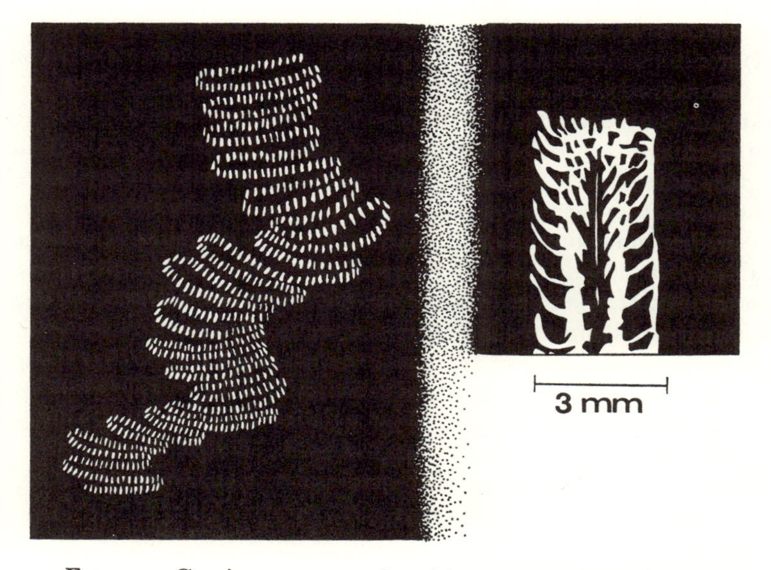

FIG. 155. Grazing traces produced by gastropod radula. Left: *Helcion*; right: *Littorina* (after Ankel, 1936c)

faster the individual loops become wider. Many not closely related gastropods graze in essentially the same way. Meandering traces are known of (Ankel 1938b, Eigenbrodt 1941, Schäfer 1950a) *Littorina obtusata*, *L. saxatilis*, *L. neritoides*, *Helcion pellucidum*, *Gibbula cineraria*, *Lacuna divaricata*, *L. pallidula*, and also of the opisthobranch *Aeolis papillosa* and of the pulmonate *Helisoma nigricans*. The time-relation between movement of the radula, turning of the head, and crawling forward suggests that these movements may be coupled. This is not the case in *Littorina* (Schäfer 1950a) as can be shown by the fact that they do not always graze. When they eat animals (e.g. *Balanus*) they do so at a standstill and without any of the typical grazing motions. The only gastropods known to combine both feeding methods are *Littorina littorea* and *Aeolis papillosa* (Schäfer 1950b).

Meandering feeding traces are rarely preserved whether they are impressions in diatom lawns of the sandy and muddy tidal flats, or in algae lawns on lamellibranch valves, stones or rock faces. Since they are engraved into living substance the biting traces perish sooner or later together with the living plants. All that can remain of the grazing activity are the impressions of the advancing gastropod foot into the substrate. The problem is, essentially, to determine whether the path of a crawling trace on the sediment surface can tell us something about the ecology. Was the crawling in a certain case simple locomotion, steered, perhaps, by the direction of the sun rays, or by the consistency of the substrate, or did it serve the special purpose of grazing? Before we concentrate on this question we will discuss the crawling trace of gastropods.

Gräf (1956) investigated the various superficial crawling traces of *Littorina littorea* on different sediments. These observations indicate that the grain size does not greatly influence the shape of the trace; the definition of its internal sculpturing, however, increases with decreasing grain size. A more or less distinctly sculptured median band (trace bed) can always be distinguished from the two bordering raised edges. The width of the trace bed corresponds to that of the gastropod foot. According to Gräf the changing intensity of the undulating movements of the foot modifies the sculpturing of the median band. The sideways pendulating movements of the head and anterior part of the gastropod body produce the sculpturing of the bordering raised edges. The head pendulates even when the radula is inactive.

However, the undulating movements of the musculature of the foot sole (D I (*a*) (v)) are very rarely transferred directly to the sediment. Thus the coarse, transverse divisions of the median band which Häntzschel (1938c) has illustrated have a different cause. Weber (1925) first discussed this problem for several prosobranchs from the Mediterranean Sea. Along with several other cases of 'arhythmic forward movement' he mentions *Conus mediterranus* which moves by undulating the sole monotaxically (see section on glide-crawling); the foot crawls forward below the stationary, heavy visceral mass until the columella muscle contracts quite suddenly and pulls the visceral mass and the shell forward. As they shift, they press the frontal section of the foot somewhat deeper into the substrate. As a result, transverse divisions appear in the trail as described by Häntzschel (1938c). On soft, pliable ground *Littorina littorea* adopts a similar mode of locomotion and, as a result, produces traces with transverse divisions. On firm substrates the median band remains smooth. The narrow crawling traces of *Hydrobia ulvae* also show transverse divisions; the phenomenon is more common than has generally been assumed. Transverse divisions in traces of amphipods (*Gammarus*, *Corophium*) have other causes (see D I (*a*) (xii) on jump-swimming, jump-running, and jump-digging). Transverse ridges caused by sudden weight shifts are curved forward in the direction of locomotion. However, on occasion the back edge of the gastropod foot produces rhythmic impressions, creating backward-curved transverse ridges.

The question remains whether it is possible to interpret the purpose of the locomotion from the course of the trail: was the trace produced by an animal fleeing, searching for a sexual partner, hunting or scavenging, travelling on a daily or seasonal migration, or grazing? Seilacher considers unbranched and heavily looped traces as typical grazing traces; more generally, he interprets all the traces as grazing traces which are distributed over a given area, forming a 'planar ornament'. Certainly, these are features of grazing traces, but they can also occur elsewhere. Even Recent traces produced by known animals cannot be identified with certainty as grazing traces. Chemoreceptors in tentacles, mouth, and edge of the foot enable gastropods to recognise and avoid areas that have been grazed over before. Old traces are either crossed nearly at right angles or avoided altogether. *Littorina* is often guided by an old trace and crawls closely along its raised edge for some distance (Schäfer 1950a).

The bioturbate textures that prosobranchs produce while they are grazing are surface traces.

Lamellibranchiata. Lamellibranchs that live on the surface of the sea floor consume plankton drawn from the surrounding water, have no siphons and show little movement; often they are provided with byssal glands whose fibrous secretions serve to attach the body. Locomotion, if it exists, serves to correct the orientation of the body or to free the animal from covering sediment.

Many plankton-eating lamellibranchs live permanently in the sediment. The length of their two siphons, one inhalant the other exhalant, corresponds to the depth of the permanent dwelling place. Both are contractile, have roughly the same length, and are frequently joined along half their length or their entire length (Schäfer 1950c and d). By means of strong locomotory organs the animals can free themselves from covering sediment. If they are washed out of the sediment they are capable of burying themselves again, at least while they are still young.

In addition to these two groups of filter feeders there are grazers. They, too, live in the sediment and have two siphons. However, the inhalant siphon also serves for feeding; it can be stretched far over the sediment surface to suck off the diatom coat. The exhalant siphon is shorter and just projects above the surface. With the used water the animal also discards faecal pellets.

Macoma baltica is a good example for this form of life. This flat lamellibranch lives in the coastal areas of the North Sea to a water depth of approximately 15 m. Fig. 156 shows the superficial grazing trace of this lamellibranch. The short and the long siphons are separated by a loose mud hill marked by tension gashes. This hill is caused by pushing movements of the lamellibranch in the sediment and by water ejected upward from the mantle cavities. There is a pile of discarded faecal pellets at the opening of the short exhalant siphon. The long inhalant siphon is very flexible. Contractions of its longitudinal muscle bundles

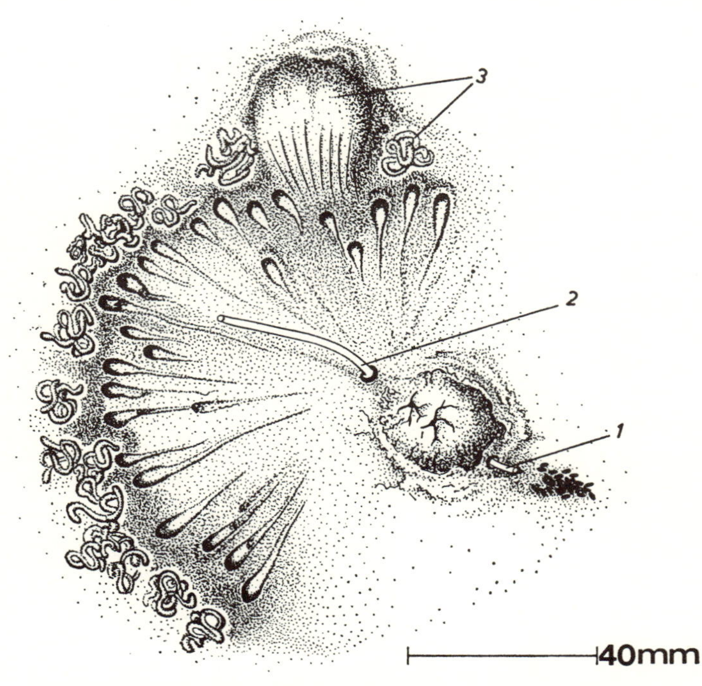

FIG. 156. Surface grazing trace of the bivalve *Macoma baltica*. (1) exhalant siphon, faeces nearby; (2) inhalant siphon, producing radial scrape marks; (3) mound of semi-fluid sediment extruded from the mantle cavity by the inhalant siphon

cause circling, undulating and bending movements of the siphon. Individual suction marks are produced by the siphon and form a radial pattern with drag trails. While the animal sucks off the diatom lawn on the sediment surface it also ingests much mud and fine sand. Before the inorganic material can reach the mouth it is collected and discarded at intervals to the outside through the exhalant siphon. For this purpose the siphon stretches as far as possible to keep the grazing area free and clean. Depending on the consistency of the discarded sediment mass, either long, narrow, spreading bands or vermiform faecal cylinders form upon ejection.

Like all grazing animals *Macoma* must move after it has depleted the immediate environment of the tip of its siphon of diatom covering. It achieves locomotion by pushing the body sideways through the sediment at a rate of up to 12 cm in six hours. The combination of surface harvesting and sub-surface locomotion produces two kinds of grazing traces: radially arranged sucking trails at the sediment surface and travel trails

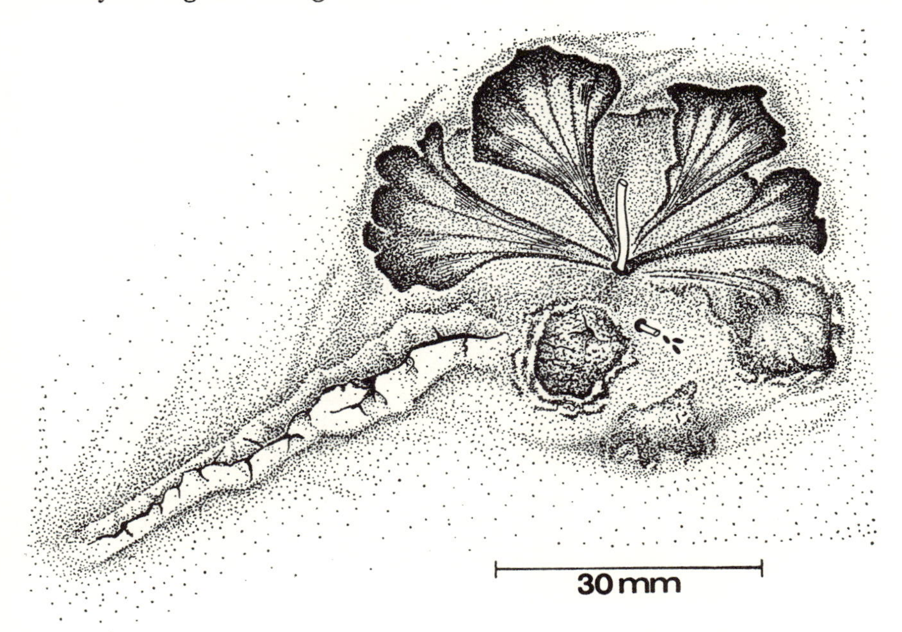

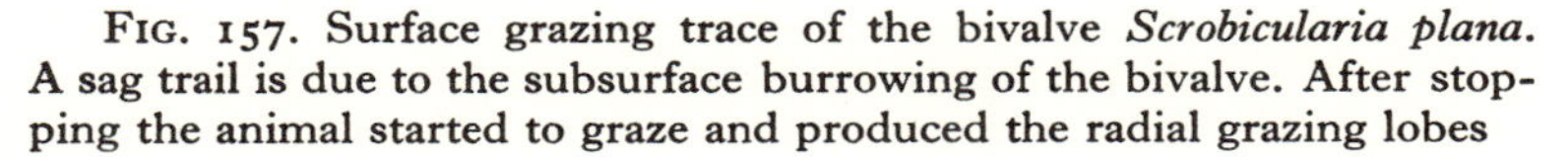

Fig. 157. Surface grazing trace of the bivalve *Scrobicularia plana*. A sag trail is due to the subsurface burrowing of the bivalve. After stopping the animal started to graze and produced the radial grazing lobes

inside the sediment. Disturbances of the bedding are quite considerable, especially if the animals are densely settled. Up to thirty animals per square metre have been found in some tidal channels of the outer Jade Bay. *Macoma*-made bioturbate textures inside the sediment are thus grazing traces.

Scrobicularia plana lives exclusively in tidal mud flats and has life habits similar to those of *Macoma*. It also grazes over the mud surface, but for algae. When it wanders through the sediment it seems to rise into higher sediment layers than those in which it remains stationary while grazing. Occasionally, more or less straight sag trails can be observed at the surface connecting the grazing traces proper (Fig. 157). These are somewhat star-shaped and have often been shown in illustrations because they are easily accessible on the tidal flats (Häntzschel 1934, Linke 1939, Schäfer 1957). *Syndosmya alba* is another lamellibranch with similar life habits, and so is *Yoldia* that grazes over the sediment surface by means of its whip-like mouth lobes. It is a shallow burier and maintains connection with the water by short strong siphons that grow together. The locomotory organ is, again, a strong foot with a sole.

The bioturbate textures which grazing lamellibranchs produce are surface traces and perturbed burrowing textures.

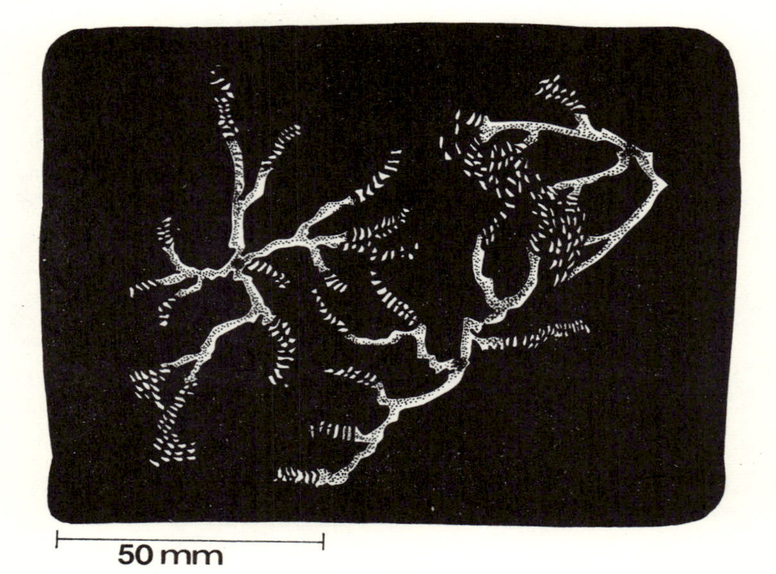

FIG. 158. Surface grazing traces of *Nereis diversicolor*. Near the openings of the several dwelling tubes the grazing trails are continuous. In their continuation lie separate, irregularly spaced incisions of the jaws

Polychaeta. Illustrations of star-shaped grazing fields of *Nereis diversicolor* have been published by Linke (1939). It grazes by removing a diatom cover from the sediment surface with its jaws. The individual bites can be impressed into the coating of diatoms, and as a result, the grazing trace appears transversely divided. These worms inhabit a mucous dwelling tube, and the branching grazing traces are often grouped around one point, the opening of the dwelling (Fig. 158). Most grazing worms create traces branching like antlers: the animals tend to reverse part of the way in their old trails, then readvance in a new direction. Branching of a trace is therefore not a criterion against its being a grazing trace. The spiraling grazing traces of another polychaete, *Paraonis fulgens*, are dug within the sediment and are a special case mentioned in D I (*b*) (ii).

The bioturbate textures produced by grazing polychaetes (except *Paraonis*) are in most cases surface traces.

Sediment eating

A common mode of feeding among lower marine animals is to ingest sediment with the food and to digest the organic material. This kind of feeding includes careful grazing over muddy sediment surfaces covered with a diatom lawn but allowing detritus to enter the digestive tract with the diatoms; and it includes ingestion of coarse sand, a habit shared by *Arenicola** and *Sipunculus*. They pass large quantities of fine and coarse quartz sands through their body in order to digest the organic substances between and on the grains (Krüger 1958 and 1959). The great majority of the sediment feeders, however, live in mud and eat fine sediment rich in detritus; often, only the topmost detritus layers are chosen after careful palpating with tentacles. But generally, organisms that ingest the sediment itself are less strictly limited to the surface than the grazers proper; they can penetrate downward, construct feeding burrows, or more frequently, feed on the sediment at the walls or the opening of the dwelling burrow. Usually faeces are produced in large quantities. The continuous ingestion and egestion as faeces of sediment-eating animals rearrange and rework the sediment and modify its textures. Some authors assume that the digestive processes free, or 'defoul', the sediment from decaying substances (Richter 1924c, also see bibliography there).

Polychaeta. The large and mobile forms capable of marking the sediment permanently when they are feeding and moving are relevant here. *Arenicola marina* is very common near the coast and is the best known sediment eater among the polychaete worms. It lives in self-made feeding burrows that are not permanent dwellings, and so do the Capitellidae. *Notomastus latericeus*, *Capitella capitata*, and *Capitellides giardi* are fairly large forms, widespread in the soft muds of the North Sea; *Heteromastus filiformis* is particularly frequent at the coasts and in the tidal sea. Their life habits are still unknown, except for *Heteromastus*. It favours the muddy tidal flats, and its population reaches a density of 7000 per square metre (according to Linke 1939, for Jade Bay). The worm is 1 mm thick and up to 15 cm long. It lives in very fine mucous tubes with fairly resistant walls in which it probably remains for a considerable time.

* In sand areas of the tidal zone (e.g. Arngast sand, Jade Bay) *Arenicola* lives in coarse sand and fine gravel. Pebbles of more than 6 mm diameter have been observed in *Arenicola* faeces.

There is never a halo of oxidation around the tube. 10 cm long pieces of washed-out tubes can be seen drifting with the currents. Occasionally, mud particles and fine sand grains stick to the viscous mucous coat and make the wall somewhat rigid. When they are washed out these mucous tubes are always collapsed. The thin tubes run almost vertically from the surface to depths of 20 to 25, or even 30 cm. They are called faecal tubes because the worm moves in them whenever it wants to discharge its faecal pellets at the surface of the sediment. The 0·5 to 0·6 mm long

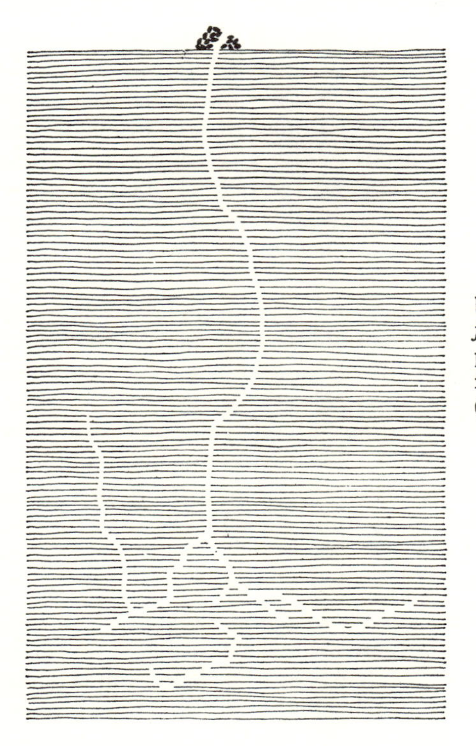

FIG. 159. Internal feeding burrows of *Heteromastus filiformis*. A small pile of faecal pellets is deposited at the surface opening of the burrow. (Actual size × 0·3)

pellets are oval and enclosed in a thin mucous cover. Thus the piles of faecal pellets on the surface of the tidal flats indicate clearly the locations of *Heteromastus* burrows (Schäfer 1952b). The worm rises backward to the sediment surface when it wants to discharge the faeces. With depth the tubes are less distinct and harder to identify; they branch out, some branches running downward, others horizontally. These are feeding tubes, changed constantly to find more sediment. The food sediment is black, belonging to the reduction zone, and therefore the faecal pellets are also black. The main burrow is used for some considerable time, and the faecal pellets discarded at the surface can build up to large piles. Fig. 159

shows one faecal and several feeding burrows of *Heteromastus* in cross-section. Almost nothing beside fine destructions of the individual beds indicates the course of the various burrows as thin traces. The tubes are so narrow that they are not filled by sediment after the burrow is abandoned or the occupant has died. Nevertheless they can be distinctly observed in fine muds. *Aricia kupfferi*, *A. cuvieri*, and *Scalibregma inflatum* have similar habits. The last, a vividly red worm, is widely distributed over the muddy bottoms of the North Sea (according to Caspers 1950a, 400 animals per square metre live on the oyster bank of Heligoland, during mass development in the summer this number can rise to 2200). Their burrows seem to be little consolidated.

There is also *Maldane sarsi*; the opheliids *Ammotrypame aulogaster* and *A. cylindricauda* live in soft muds. *Ophelia limacina* and the genus *Polydora* are widespread on sand.

Certain animals do not consume sediment wholesale but are selective and sweep and palpate the detritus-rich top sediment with tentacles or lip appendages. Some live in more or less permanent burrows in the sediment.

Poecilochaetus serpens palpates the sediment surface from the opening of its small U-shaped burrow; *Scoloplos armiger* roams through the sand ground of tidal flats. It does not have a permanent dwelling tube and rarely rises to the surface. While burrowing the worm constantly secretes a mucus which does not consolidate the burrow walls. It thus creates digging trails which cause only minor disturbances of the beds. The worm ingests small organisms and fine detritus together with sand. Its faeces are not cohesive and thus have no definite shape. *Pygospio elegans* lives in the tidal sea floor in tubes that are consolidated with sand grains. It feeds on detritus lying on the sea floor, its tentacles palpate the neighbourhood of the tube opening. The food particles are transported along a tentacle groove in a current created by cilia and enter the mouth. A few detritus-eating polychaetes live on the sediment surface and produce crawling traces in the soft muds. They are *Harmothoe sarsi*, *Sphaerodorum claparèdi*, *Gastalia punctata* and the genus *Sphaerosyllis*, and they all have dorso-ventrally flattened bodies.

The bioturbate textures that sediment-eating polychaetes produce are either perturbed or mucus-consolidated, shaped burrowing textures.

Sipunculidae. The sipunculids are a small family of burrowing worms with varying life habits. Formerly they were combined with the Echiuridae and Priapulidae as the Gephyreae, a term that is now obsolete. Their body is unsegmented and they have neither bristles like the annelids nor a head lobe. The intestines loop forward and run to the anterior third of the body. These features distinguish sipunculids from annelids (Fig. 120).

Sipunculus nudus can grow to a length of 25 cm. The body is elongate-cylindrical and with approximately thirty longitudinal grooves. Strong longitudinal muscle bundles enclose the body sac. The mouth is surrounded by a lobed tentacle membrane; it is the end of a strong, eversible and retractable proboscis, covered with papillae. (For anatomy and physiology of the sipunculids, see Andreae 1882, Lang 1888, Metalnikoff 1900, Fischer 1928, von Uexküll 1921 and 1933). Its locomotion has been discussed in the section on locomotion by bolting, D I (*a*) (viii).

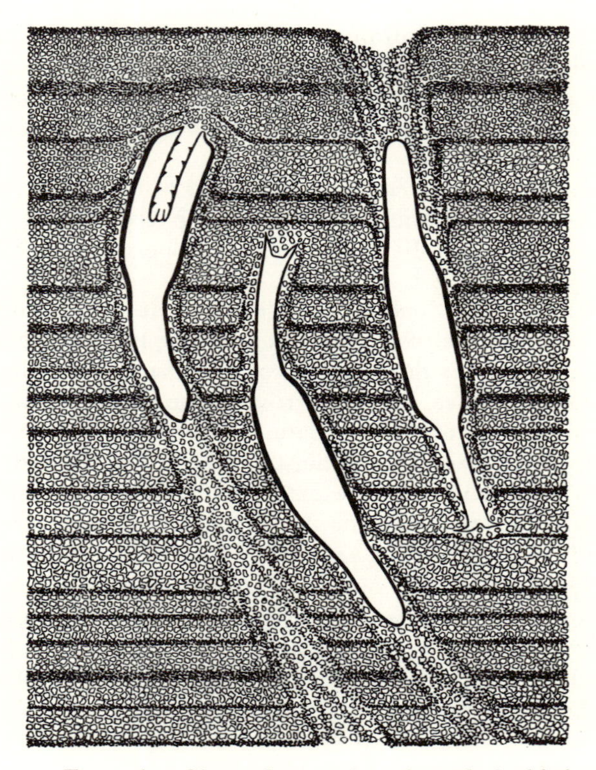

FIG. 160. *Sipunculus* moving through bedded sand. Beds are dragged in the direction of the worm's advance. (Actual size × 0·3)

All sipunculids feed on sediment and detritus. With their extensible tentacle membrane they enclose and grasp sediment lying in front of the mouth. Peristaltic movements transport it from there into the digestive tract. The animals ingest almost sterile sands as well as muddy fine sands and muds. The intestines are always filled with considerable quantities of sediment with grain sizes up to 4 mm. In contrast to other species and genera, *Sipunculus* has no home burrow and moves frequently, a mode of life to which it is adapted by its special locomotion, by breathing sediment water through the skin, and by its independence of any specific type of food.

The size of the animals and their mobility are responsible for striking and widespread perturbations of the bedding. Consolidating mucus, however, is rarely secreted. Occasionally, minor consolidations of burrows are supposed to occur. But as far as is known, true burrow holes are never constructed.

Sipunculus glides smoothly with its elongated body through the sediment layers which it perforates (Fig. 160); as a result, the sediment retains its position after the animal has passed. The cut-off beds do not sag due to gravity as they do with most other burrowers. Instead, the cut-off beds always point in the direction in which *Sipunculus* moved, regardless of whether the animal burrowed upward, obliquely, or downward. Several other sipunculids inhabit the North Sea (according to Fischer 1928 and Caspers 1938). *Phascolosoma elongatum*, an animal up to 8 cm long with unknown life habits, lives at the Belgian coast in mud at depths down to 29 m and also in the Skagerrak. *P. margaritaceum* is up to 19 cm long, lives in mud, in the deep Channel of Heligoland and in the Gullmar Fjord; *P. minutum* is hardly 1·5 cm long and yellowish-green, its life habits are unknown. It is found near Heligoland and in the Skagerrak in sand rich in shell breccia, at depths from 15 to 18 m. *P. vulgare* up to 15 cm long, also of unknown life habits, lives at the Belgian, Dutch, and English coasts, and also near Bergen and in the Skagerrak. Finally, *Sipunculus nudus* whose life habits have been described is found at the Belgian and Dutch coasts in muddy and pure, frequently coarse, sands at shallow depths.

The bioturbate textures which sediment-eating sipunculids produce are perturbed burrowing textures.

Kinorhyncha. These purely marine worms belonging to the Nemathelminthes, grow to a length of barely 1 mm. In size they are thus at the lower limit of significance as trace producers. Those that live in mud are capable of producing distinct locomotory trails on the sediment surface and also in the finely layered muds. The animals are so numerous in certain coastal muds of the North Sea that the shiny mud surfaces look scarred and segmented into small sections by their crawling traces. The upper layers of sediment, if sufficiently fine-bedded, are also distinctly marked by the vertical, oblique, and horizontal digging trails.

The compact, worm-shaped body of the kinorhynchs is composed of thirteen segments whose first segment constitutes the voluminous head section, and the second segment the neck section. The ventral side is flattened in the central and posterior body region. The trunk segments carry an armour of two ventral plates, a dorsal plate, and a varying number of hard spines. The last segment bears one or more tail spines.

Locomotion on or in the sediment is achieved in the following way (Remane 1928): the head section is perfectly retractable; in this position it is protected toward the front by a closing mechanism formed by the second segment. As the head section shoots forward, it spreads its several rings of hard bristles, or scalids, which anchor the animal in the sediment. During this movement the whole body stretches lengthwise, and the tail spines and those of the last few segments press into the sediment for

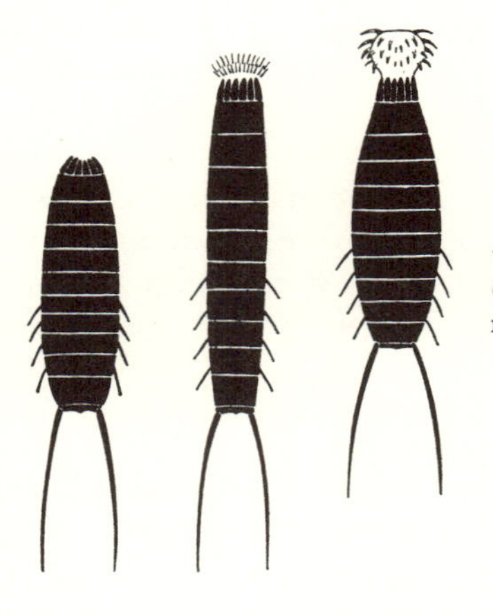

FIG. 161. Kinorhyncha in three stages of locomotion (after Remane, 1928). (Schematic drawing)

support. Once the head is anchored in the sediment the trunk contracts and is thus pulled forward as a whole (Fig. 161). The movements of the animals in mud are quite lively. Within an extremely short time these animals can pass through newly deposited sediment layers if the sedimentation threatens to bury them too deeply. They are vagile and do not construct mucous burrows. Remane observed *Echinoderes dujardini*, *E. setigera*, and *E. borealis* in the North Sea.

The bioturbate textures that sediment-eating kinorhynchs produce are surface traces and perturbed burrowing textures.

Echinodermata. Amphiura filiformis moves by pushing and undulating movements of the long arms and by rotating movements of the body disk, often in deep mud, but also in fine sand. There this brittle star ingests small animals together with plenty of detritus. The stomach cavity therefore often contains much sediment, besides parts of small animals, such as polychaete bristles or shell parts of small crustaceans. Frequently, however, they are more selective and search for small animals, particularly in sand. *A. chiajei* seems to feed in the same way; it lives at greater depth (20 to 200 m). *A. filiformis*, on the other hand, prefers shallow water 1 to 40 m deep. The animals are quite mobile and change their location constantly. Their activities therefore create extensive bioturbated zones in the upper sediment layers.

Ophiura albida and *O. texturata* live on the sediment surface. Predatory as well as detritus-eating, they travel constantly and produce locomotory traces on the sediment surface that are quite regularly arranged.

The Holothuria *Stichopus tremulus* and *Leptosynapta inhaerens* feed on detritus and small animals that are partly or wholly buried. Their digestive tract usually contains sediment. *Leptosynapta minuta* lives on mud and fills the intestine completely with detritus and muddy sediment. The irregular sea urchins *Echinocardium cordatum* and *Spatangus purpureus* live in sandy sediment and eat sediment exclusively.

Sediment-eating echinoderms produce bioturbate textures that consist of surface traces, internal traces, and perturbed and shaped burrowing textures.

Whirling, tentacle feeding, and filtering

A great many marine animals gather planktonic animals and plants and also suspended detritus particles from the surrounding water for their food. Large animals need a great quantity of these small particles for food and can therefore not hunt for them individually. Various mechanisms of different shape and function serve to accumulate planktonic food. Most frequently a water current is created by cilia and driven either through the body or along certain of its surfaces where the food is made to stick to mucous fields or collected in sieves or filter mechanisms. This mode of food capture may be called 'water whirling', and the animals who employ it, 'whirlers'. Frequently, cilia are arranged in grooves or on certain areas and beat in specific directions, thus accumulating the caught particles and steering them toward the mouth (Willemsen 1952, Theede 1963, Dral 1967, Coughlan 1969). Tentacle feeders wait in the water for their prey, stretching tentacles provided with sticky secretions or with stinging mechanisms into the water. An animal that touches the tentacles is captured (Hertling 1953, calls this 'to sediment'). To catch plankton the filterers rapidly pull a net through the water which is often shaped like a basket formed by bristled limbs. Or they spread such a net in the water current.

Whirlers, tentacle feeders, and filterers can be divided into three groups according to their mode of life: (1) Purely sessile forms. Whirlers among them are sponges, bryozoans, and sessile tunicates. Tentacle feeders are the sedentary coelenterates and some sedentary worms.

Filterers are the lepadids and balanids. In our context, they are insignificant because they are sessile, and, moreover, live in areas that are free from sedimentation. (2) Plankton feeding on plankton. Whirlers among them are the larvae of some gastropods, lamellibranchs, and polychaetes, some copepods and phyllopods, mysidaceans and appendicularians. Tentacle feeders are the planktonic coelenterates (Hydromedusae, Scyphomedusae, and Ctenophores). The life activities of animals of this group cannot be preserved. (3) Epibionts and endobionts with sufficient locomotory potential to cope with shifting and newly deposited sediments.

Only this last group is important, for its members leave preservable traces which are caused by locomotion in the search for food. Many of the epibiontic and endobiontic lamellibranchs belong to this group but also many other forms (see also D I (*b*) (iii) on escape).

Branchiostoma. *B. lanceolatum* is a typical whirler (Webb 1969). It feeds with its whole body buried and only the mouth above the surface of the sediment. Its behaviour has been described by Schäfer (1957). It lives mostly in sand which may be coarse and rich in shells or shell breccia (M. M. Wells 1926). Position and function of the breathing organs determine the position of any sediment-dwelling animal and also its choice of sediment type such as coarse sediment with a wide interstitial system. Most aquatic, sediment-dwelling vertebrates have to hold the head far enough above the sediment surface so that both mouth and gill slits are free for the water to enter and to leave. In contrast to terrestrial vertebrates, the breathing current is unidirectional, and the mouth does not serve for both intake and outflow. As long as both mouth and gill opening are kept free most aquatic vertebrates can remain in all kinds of sediment, even in those with finest grain. Not so *Branchiostoma* whose exhalant opening for breathing water, the porous branchialis or atriporus, lies at the end of the long peribranchial cavity in the posterior half of the body (Fig. 162). Therefore the used water can be ejected only if the sediment is coarse enough to let it flow through the interstices. The separation of the gill outlet and of the entire peribranchial cavity from the outside is not an original feature of primitive vertebrates but a specific adaptation of *Branchiostoma* to life in the sediment. It protects the gills against injury and pollution. The habitat, however, becomes limited to loose and relatively coarse sediment, or to somewhat consolidated fine muds with enough cohesion to allow establishment of a space between animal and enclosing sediment for circulation of the used water. Generally *Branchiostoma* lies in the sediment with its ventral side up; this facilitates the whirling in of water and food (van Weel 1937) and also the discharge of the used breathing water (Fig. 162). Rows of cirri at the mouth which carry sensory organs at their outside serve as a sieve against pollution from the front and above; they also can close the mouth completely. The water current in the body is created mainly by a single, large flagellum in the mouth and supported by cilia on the epibranchial and hypobranchial groove (Grassé 1948).

In open water and in sediment the animal moves by undulating laterally; backward movement is possible over a small distance. To get into the appropriate permanent breathing position the animal either digs

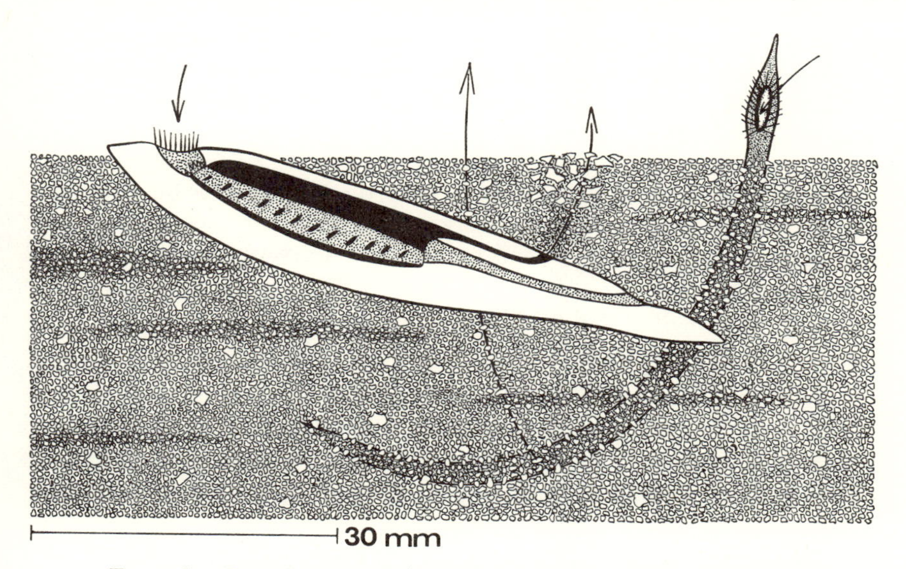

FIG. 162. *Branchiostoma lanceolatum* lying in sediment dorsal side downward. (Arrows) direction of respiratory water current; (black) peribranchial cavity, opening outward through the porus branchialis, with coarse sediment above; (background) another individual in ventral view, the animal lies in the sediment laterally curved (after Schäfer, 1956a)

down head-first by shaking the body and then arching the head up, or it pushes the body backward into the sediment by shaking movements until nothing but the head peeks out. The rostral and caudal fins form a sharp edge around the body and are used for both movements. They penetrate into the sediment like pointed knives. Either the rostral or the caudal fin can penetrate first, depending on the direction of movement.

What has been said in section D I (*a*) (iii) on locomotion by undulation holds for *Branchiostoma* with its laterally compressed body. There it was explained that, the larger the amplitude of the undulations, the more energy has to be expended deforming the medium, especially if it is stiff and does not yield easily. *Branchiostoma* is able to vary the amplitude; in loose sands it moves undulating in large curves, and in denser sediments that offer more resistance it advances by small undulations. In the extreme case, undulating can be reduced to trembling and shaking. According to Böker (1935) this is the original mode of locomotion of the limbless vertebrates, as in the cyclostomes, to which different vertebrates sometimes revert (fishes, Amphibia, Reptilia).

The skin is coated with mucus to facilitate smooth gliding of the flanks of the body when they push against the granular medium during undulation and at the same time are displaced; the mucus also protects the epidermis against scraping injuries that could occur while the animal moves. However, the mucus does not penetrate into the sediment and does not make it more cohesive than it was. The undulating movements cause local buckling and inflections of the bedding. Large amplitude undulations also shift the layers either to one side or the other. Small undulations hardly show alternating deformations in the burrow, and shaking movements even less so; only narrow zones of dislocation of the beds mark the trail. In fine shell breccia the burrows of *Branchiostoma* cannot be seen at all.

Branchiostoma leaves the sediment and escapes by swimming when there is sudden sedimentation. Probably the sensory organs on the cirri of the mouth receive stimuli from moving sediment that warn it of the impending danger. Hagmeier & Hinrichs (1931) have described observations on captive *Branchiostoma*. Animals sitting in sand can cause local shifts in the sediments by the pressure of the exhalant water. Coarse sediment particles, pebbles and shell fragments, are moved upward while the fine sand sifts down. Undulating movements of the body accelerate the process.

The bioturbate textures that *Branchiostoma* produces are perturbed burrowing textures.

Prosobranchia. Whereas epibiontic and endobiontic lamellibranchs frequently feed by whirling, gastropods rarely do. The greater mobility of the gastropods may be the reason for this fact. The few forms that are provided with whirling mechanisms are almost sessile species, as *Crepidula fornicata*, or species that burrow, as *Aporrhais pes pelicani* and *Turritella communis*. In all cases the breathing and feeding functions are closely connected as in lamellibranchs, *Branchiostoma*, brachiopods and ascidians. For our problems, *Aporrhais* and *Turritella* are important because both species whirl in the sediment.

Weber (1925) clarified the mode of locomotion of *Aporrhais* on the surface of the sediment. The animal walks clumsily 'on crutches' by steps (D I (*a*) (ix)). Yonge (1936–7) described the behaviour of the animal in the sediment and its mode of burrowing. All the movements of *Aporrhais* are slow but powerful. First the animal pushes the pelican foot, the finger-like, broadened lobe of the shell at the aperture, into the sediment, obliquely towards the right. In the sediment the prosobranch continues to move in the same way as on the surface by alternately pushing the shell forward and swinging the narrow, muscular foot forward. The animal does not advance on a straight course but in a slight curve to the right because the shovel of the pelican foot offers greater resistance than the rest of the shell when the foot pushes up. Evidently, this kind of walking in the sediment can only succeed as long as the animal moves in superficial layers that are not yet consolidated by compaction, such as soft muds full of water, or occasionally layers of sand and gravel. If rapid sedimentation sets in *Aporrhais* leaves the sediment immediately and tries to avoid being covered by being constantly on the move. This is not always successful, and very fast sedimentation can cover and immobilize the animal.

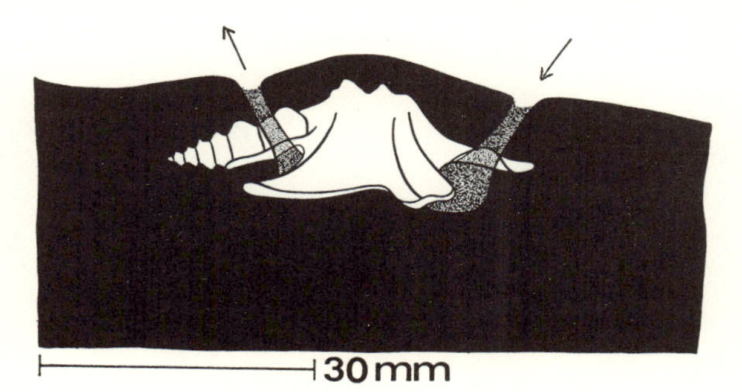

FIG. 163. *Aporrhais pes pelicani* in the sediment with channels for aspiration and ejection of breathing water

When the animal has penetrated into the sediment to a certain depth it stops and begins to construct two ducts to the surface. It begins the work for the exhalant duct at the back by lifting slowly the broad pelican foot; thus it creates a water-filled space under the aperture of the shell. It then lets the shell drop suddenly and thus pumps the water out of this cavity. In a quick but strong jet the water pushes outward and upward between the end of the foot and the posterior embayment of the pelican foot. After this pumping movement has been repeated a number of times, the mud is disrupted, forming an open canal to the surface (Fig. 163). For building the inhalant duct at the front the highly flexible proboscis and the narrow, long tip of the foot dig an opening through the mud. Glands at the proboscis and at the edge of the foot seem to provide mucus that helps to consolidate the walls of the inhalant duct. Stomping movements of the foot widen the canal in front and under the shell, and also the exhalant duct, and consolidate the walls of these cavities. If any of the ducts become clogged they are immediately unblocked by the same technique. Only the two small, round holes in the sediment cover give away the presence of an *Aporrhais* sitting in the sediment. The water current that enters through the inhalant duct is created by unidirectionally beating cilia which cover the mantle cavity and particularly the gill filaments. Fig. 164 shows the flow of the water. The position and shape of the mantle cavity of prosobranchs cause the water current to leave the shell near the head somewhat to the right. The spent water, laden with

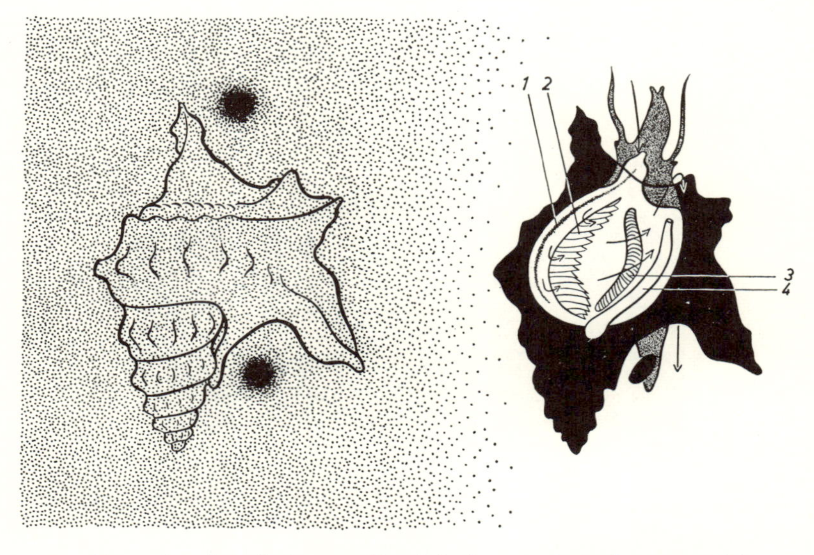

FIG. 164. *Aporrhais pes pelicani.* Left: top view of gastropod in the sediment with aperture of ingestive and egestive channels; right: view into the mantle cavity; (arrows) current direction, (1) accumulation of planktonic food, (2) gill, (3) hypobranchial gland, (4) terminal portion of intestine. (Actual size)

faeces and chemical excretions, collects therefore near the lower end of the inhalant duct, and must be transported from there toward the back in individual batches by vertical pumping movements of the foot, to be ejected through the exhalant duct. The small siphon immediately posterior to the head whirls plankton together with the water into the breathing cavity. Plankton sticks to mucus from the hypobranchial gland in the anterior cavity and is accumulated on it until large quantities of food are collected; they are then ingested by the mobile proboscis. Yonge believes that *Aporrhais* feeds exclusively on phytoplankton.

Turritella feeds also in the sediment by whirling plankton towards itself. The animal sits so deep in the sediment that only the apex of its narrow, tall shell projects above the surface of the sea floor; the aperture is directed downward (Hunt 1925). The mode in which the animals whirl water and food downward is not yet known, nor is the course of the water current. *Turritella* moves on muddy surfaces by looping; the elongate foot is advanced in two successive stages, first forming a short arch, then anchoring the rear end and pushing forward by stretching the arch. The long shell drags behind along the ground (Ankel 1936a).

Prosobranchs that feed by whirling food toward them produce bioturbate textures that are perturbed burrowing textures.

(ii) *Living in a burrow*

The sand and mud areas of the shallow sea appear to be almost barren of life over wide distances, and on underwater photographs they often look like desert regions. However, a dredging sample of the same sea floor proves otherwise: these sands and muds are full of abundant and varied animal life under the surface. Some of the inhabitants are true 'vagile endobionts', free predators that wander through the sediment and mark their digging courses in the beds. The majority of the bottom-dwelling animals, however, are conditionally vagile endobionts. They move only if special circumstances call for locomotion, e.g. if they are pursued by enemies, or if new sedimentation or erosion force them to abandon their place, or if lack of food makes it necessary to move. Generally, they live undisturbed at the same place in self-made burrows for considerable times.

In most cases the ecological reason for this behaviour is probably a need for protection. It has been said that the most important aspect is protection against visual detection by enemies, and that dwelling burrows must therefore be restricted to well-illuminated, shallow depths, and hence that bioturbate textures from such burrows in sedimentary rocks indicate shallow-water deposits. However, such a far-reaching conclusion with all its important consequences for palaeontology and geology must be considered premature as long as the dark depths and their sediments have not been investigated in greater biological detail than has been the case so far.

Oxygen-consuming organisms can live in sediment only if they have a way of maintaining a supply of oxygenated breathing water. There are several ways of solving this problem. The respiratory epithelia of perfectly vagile endobionts frequently lie in body cavities which serve as reservoirs of oxygenated water. Their mobility is thus little restricted, as in *Lunatia*, although the breathing water has to be replaced occasionally at the surface of the sediment. Or some organisms are provided with breathing tubes, e.g. siphons, which connect the animal permanently (lamellibranchs) or temporarily (*Aphrodite*) with the open water. The use of siphons restricts the mobility of the animal. The respiratory epithelia of certain other animals lie on appendages outside the body. If vagile, such animals can only live in coarse sands where water circulates freely and can thus reach the gills in sufficient quantities. The proportion of body size to the volume of the interstitial system of the sand is especially favourable for small forms, and the majority of animals with this kind of breathing apparatus are small (Remane 1952, and publications of several of his students).

Exclusively vagile life habits in the sediment are the exception. In most cases endobiontic animals can breathe only if dwelling tubes with permanent walls are burrowed into the sediment, and if a water current driven by movements of the inhabitant circulates inside the dwelling tube. Particularly in muddy and clayey sediments burrows so constructed may be a necessity. In many sediments the interstitial water either does not move at all or so slowly that it cannot supply enough oxygen for most endobionts. Such animals can leave their dwelling tube and forage into the sediment. The situation is aggravated where the concentration of hydrogen sulphide is high. In experiments with organisms living in sea floor sediments, Theede, Ponat, Hirobi & Schlieper (1969) investigated their resistance to lack of oxygen and to hydrogen sulphide.

Endobiontic animals must be able to return to a place where they have communication with the sediment surface and the oxygen-rich water above. The dwelling burrows with self-supporting walls are thus vitally important for many endobionts. Animals become as dependent on them as if they were internal organs.

If a dwelling burrow becomes covered by new sediment—and this happens frequently in the shallow sea—the inhabitant is forced to leave and to strive toward the new surface. The action is triggered by lack of oxygen in the dwelling tube. Lack of oxygen causes many animals to abandon their dwellings. This holds for all tube-dwelling worms, for crustaceans living in gastropod shells, for fishes and crustaceans living in mud, and for many others. They all abandon their home, even if there is no more oxygen outside than inside. This means that an action is induced by a specific external stimulus. Peters (1948) called this type of reaction a 'Wirkschema' (scheme of effects), an instinctive mechanism which determines a certain action or series of well-co-ordinated actions as a response to a specific situation.

An endobiontic animal responds to different stimuli when it wants to enter the sediment from the surface. From surroundings with abundant oxygen it moves to oxygen-poor regions deeper in the sediment. Sometimes, intensive light may trigger such an action; more probable and more generally effective, however, seems to be a need of the animal to be exposed to constant touch-stimuli on as large a surface of the body as possible. This urge is especially strong in animals that live in dwelling tubes; they immediately accept artificial tubes and hollow structures of any kind made of transparent material even in a well-lit environment. In an aquarium without overhead lighting in which the glass walls are the most brightly illuminated areas, tube-building worms crowd to the walls and start constructing semi-tubes, using the glass as part of the enclosure.

Endobiontic animals that are experimentally brought to the surface soon escape. They come to rest only when their body is in touch, as completely as possible, with the walls of a structure produced by them. Often, the resolution of this urge requires complicated activity from the animal. Respiration has to be taken care of, but the animal also requires to keep 'in touch'. The endobiontic animal must treat the sediment through which it digs in such a way that it changes the simple digging trail into a dwelling tube which makes breathing permanently possible; for the sediment is poor in oxygen and thus hostile to life. Peters calls this type of concerted action a 'Werkschema' (working scheme) with which the animal is endowed from birth and which is always ready for action. In our case, this 'Werkschema' causes a dwelling tube to be constructed by a certain technique as long as the need for touch from all sides is effective and acts as an impulse, and as long as other, inhibitory factors are absent that prevent the execution of this specific series of actions. When the animal has finished constructing its burrow it stops further construction.

This particular behavioural sequence which is directed toward the dwelling as a goal, does not apply to the so-called feeding burrows (Fodinichnia, according to Seilacher 1953a) which are caused by eating sediment. According to their nature these burrows are never finished or in any way completed. They lack home character and all features connected with permanence.

As mentioned above, dwelling in a burrow implies safeguarding of breathing. The burrow is so oriented that respiratory movements of the inhabitant can drive a current of oxygen-rich water from the surface into the burrow and discharge the used water to the surface along the same or another path. A kind of building hygiene is necessary for a successful burrow design. Frequently, undulating movements that run along the whole body of the animal achieve the transport of fresh water into a dwelling tube with self-supporting walls. These rhythmic body pulsations are stationary and have no locomotory effect. In polychaete worms (e.g. *Nereis*) breathing undulations occur in the dorso-ventral plane in contrast to the lateral undulations for locomotion. Bohn (1904) first described the breathing movements in *Arenicola*; he calls them annular. After the posterior body section is first thinned, several segments form a thickened annular body zone by relaxing the circular musculature of that part. Thus the body wall is pressed on all sides against the wall of the burrow, forming a tightly closing plug. This thickened zone closes off a volume of oxygenated water in the upper section of the tube. As the thickened zone is propagated to the next segments and so on down the whole body, the trapped, fresh water is pressed downward toward the gills. These waves of thickened zones can run in either direction along the body and can therefore push the water back or forth, and either take in fresh water or discharge used water. The gills can be rhythmically protruded into the free space inside the tube and retracted into the body wall so that the blood is rhythmically recharged. These ventilation movements occur for periods of 10 to 15 minutes and are interrupted by intervals lasting 50 minutes (van Dam 1937). The gill bundles of *Pectinaria coreni* show the same rhythmic movements.

In another frequent mode of respiration, certain areas are covered with unidirectionally beating cilia, mostly on the gills but also on the exterior skin. The current produced by the cilia renews the water inside the tube. In the polychaete worm *Sabellaria* such a current created by cilia is so strong that it forms a vertical jet above the dwelling tube which keeps dancing sand grains in suspension.

With reduced oxygen concentration the speed of the water flow through the dwelling tube is generally increased, in *Arenicola* up to eight times the normal speed (van Dam 1937). If after an interruption oxygen-rich water again enters the tube, the breathing frequency is temporarily increased to ten times normal (see also Wells 1966).

The construction of a dwelling burrow necessary for easy breathing ties the animals of many species to a certain place. They tend to lose their mobility and become settled. If they are still capable of moving, locomotion is restricted to an area surrounding the dwelling from the moment it is completed. From then on the living space is circumscribed although a variety of actions can still be performed. It becomes a territory, surrounding a place of maximum safety, and this link with the territory remains as long as the same burrow is inhabited. Hediger (1946 and 1956) analysed these relationships in higher animals. He says that there exist 'relationships between the internal organisation of an animal and that of its territory; physiological, behavioural, and territorial organisation form one single, organised unit'. He cites also C. G. Carus (1886) who defines instinct as an organisation that reaches beyond the body and into the surrounding space.

In some cases mobility is completely retained despite the link to a home, because effective locomotory organs permit long-distance excursions. Then an animal can construct a dwelling tube and yet be a predator or grazing animal. However, where the tie with the burrow is so strong that its builder can only move within the tube, and because of its highly specialised locomotory organs (see D I (*a*) (xi) on chimney climbing), the mode of feeding must be equally specialised, being either angling, filtering, net-fishing, or detritus-eating.

In the following pages several patterns of behaviour leading to the construction of dwelling burrows shall be explained by means of examples. An understanding of the relationship between behaviour and burrow is important for the palaeontologist because, in most cases, he can see the burrow but not the burrower. The trace fossils are his only evidence of former life.

The mucous burrow

The simplest way to transform a digging trail through the sediment into an open burrow with self-supporting walls is to coat the sediment particles with mucus secreted by cells that are either scattered over the body surface or are concentrated in certain places. Animals with scattered mucous glands simply coat the burrow walls while the body glides back and forth in the sediment. The mucus penetrates into the spaces between sediment particles and cements them as soon as it hardens. The method is analogous to the sedimentologist's preparation of a peeling film which also aims at consolidation of sediment particles in their original position by pressing a rapidly hardening liquid into the interstitial system. While the sediment layer is soaked with secretion but is still yielding, the animal smoothes and adjusts the inner space to fit the cross-section of the body. Finally, additional films of secretion cover the internal walls of the burrow in a more or less thick coating and smooth them perfectly. Where the secreting cells are limited to certain regions of the body of an endobiontic animal the construction becomes a concerted action, performed by means of a tool, a specialised organ.

The course of the burrow and its special orientation are determined by the direction in which the animal moves in the sediment. The dwelling tube represents a certain section of an often much longer trail through the sediment since the secretion of mucus does not occur constantly but can be started or interrupted at will. Mucus is produced only if the part of the animal which produces it is either at the sediment surface or in an open tube connected with it. The animal starts coating at the surface and works downward so that a dwelling tube can be constructed only with an open connection to the oxygen-rich water above the surface. The simple mucous burrow is the most primitive form of all burrows. It can be constructed quickly but it is equally easily destroyed during movements of the sediment, and the owner must abandon the burrow. Many of the builders of such burrows are animals that do not entirely depend on a dwelling and are capable of moving safely outside the burrow.

Mucous burrows are preserved as shaped burrowing textures.

Ceriantharia. The Ceriantharia and Hexacorallia do not occur frequently in the sediment of the North Sea. Where they occur, however, they can be found in great numbers. According to Pax (1925, 1928, and 1934) they inhabit preferentially muddy sand and fine shell breccia. They avoid pure muds. *Cerianthus lloydii* lives at the Belgian coast 24 to 39 m deep, in the Gullmar Fjord 18 to 20 m deep, and in the Skagerrak at a depth of 720 m (see also Caspers 1938).

The body of *Cerianthus* is elongated, worm-like, and up to 30 cm long. The aboral pole is pointed and, in many species, provided with a terminal pore; it does not bear a foot disk. A strong, ectodermal, longitudinal musculature enables the animal to undulate the body slowly in all planes; the ectodermal musculature is covered by an epithelium, full of glands (Fig. 165). Like many other actinians *Cerianthus* rests in the sediment with the elongated body and spreads its wreath of tentacles over the surface. In contrast to endobiontic actinians, however, it constructs the vertical tube in the sediment with the mucous walls somewhat wider than the diameter of the body; when the animals are disturbed they quickly retreat completely into their tube.

The construction of the burrow starts by digging downward. The undulating movements of the body widen the cavity, mucus is copiously produced; it glues the walls of the digging trail which consolidate as the mucus hardens. Shell fragments, mud particles, and sand of all grain sizes are cemented to a tough-walled cylinder; a homogeneous mucous layer coats the wall on the inside and smoothes it. In this way, *Cerianthus* continues to build downward until it can completely disappear in the dwelling tube. An open tube is necessary for living permanently in the sediment because this animal must breathe with its entire body surface. The animal probably moves up and down in the wide tube by undulating the body and supporting it against the walls during movement. A *Cerianthus* that is lying flat on the surface begins to dig downward into the sediment bending its worm-shaped body into a right angle (for orientation in space, see Loeb 1891). Doing so it pushes the edges of perforated sediment layers distinctly downward. Once the animal has completed the dwelling tube it does not abandon its home without external cause. It compensates for slow, new sedimentation by continuing the construction upward by steps. In this case the newly deposited layers form a right angle with the walls of the tube without any inflection. Sediment that has dropped into the dwelling tube settles in layers at the bottom of it and protects the uninhabited lower section of the burrow from deformation. In longitudinal section, this part of the tube is distinguished by internal, crescent-shaped layers, independent of those outside, indicating that the filling is secondary. Where such slow sedimentation continues over a

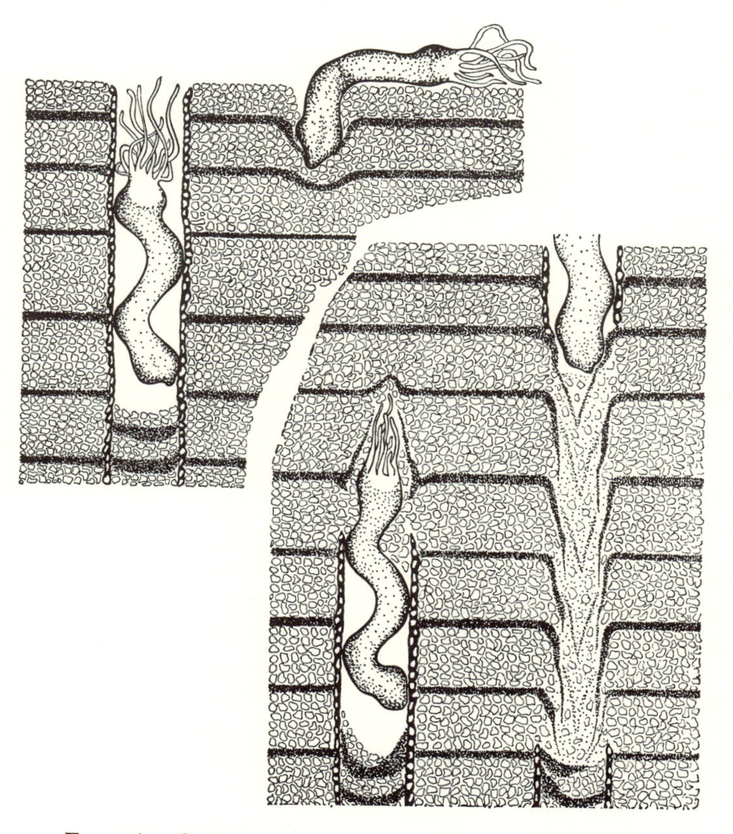

FIG. 165. *Cerianthus.* Left: dwelling burrow with mucus-covered walls and sediment filling the bottom, second animal starting construction of a burrow from the surface; right: the sea anemone has been covered by sediment, leaves its burrow and tunnels up toward the surface. (Schematic drawing)

long time the dwelling burrow can grow very long (according to Remane 1940, up to 1 m).

Cerianthus abandons the burrow if sedimentation is so rapid that several new layers cover the animal and its dwelling tube. The animal pushes its way vertically upward on an escape trail until it reaches the surface of the sediment. While the animal digs upward it folds the labial tentacles of the mouth and the marginal tentacles together to a point; it does not retract the tentacles into the oral opening. During escape burrowing the epithelial mucous glands do not secrete; the sediment does not become cemented or consolidated, and the burrow is marked only as a vertical sag trail in the new layers. Cross-sections of dwelling tubes can thus be clearly distinguished from those of digging trails.

Polychaeta. Most polychaete marine worms are endobiontic and stay at the surface of the sea floor only for short intervals. The majority of them construct temporary or permanent mucous burrows; only a few are strictly vagile and build dwelling burrows.

The mucous burrow of *Nereis diversicolor* is a good example of polychaete structures. In the North Sea the genus *Nereis* is represented by approximately eleven species. Few species are well known ecologically. Their life habits may differ considerably in spite of great morphological similarity. *Nereis diversicolor* moves through the sediment in two ways: 'bolting' with the proboscis opens a cavity in the sediment in front of the animal; it then crawls into the newly created cavity by pushing with the parapods and by peristaltic movements of the body. (The worm undulates in wide, large bends only when it swims in open water.) This mode of burrowing by bolting and peristaltic movements enables *Nereis* to pass through almost any kind of sediment, coarse and fine sand, mixtures of sand and mud, pure mud, and even layers of consolidated clay. This ability permits a wide distribution of this species. However, the animal prefers soft mud. There the animals settle densely, up to 500 adults or 4000 young animals per square metre.

Nereis starts to burrow from the sediment surface downward head-first; after it has penetrated 8 to 10 cm it turns and digs upward again toward the surface with the head up. In this position it secretes mucus while it moves up and down by peristaltic and gently undulating movements. The mucous glands are distributed over the entire surface of the epidermis. Sideways movements push the mucus-soaked walls of the tube back and consolidate them until the upward branch of the digging trail has turned into a 4 to 6 mm wide mucous burrow with self-supporting walls and an open connection to the surface. Oxidation of ferrous solutions in a halo around the tube can create rusty-brown iron hydroxide which helps to consolidate the walls. Movements of the parapodia and dorso-ventral undulating movements of the body whirl breathing water down the open tube.

Soon after completion of the dwelling tube *Nereis* begins to build branches to the burrow that run obliquely upward; as a result, the burrow finally consists of an entire system of tubes with many connections to the surface. Probably, the animal acts in this way in response to an urge to increase the current of water. Branch tubes are so connected that the current of breathing water is unidirectional throughout the entire burrow. Occasionally, a large number of individual *Nereis* construct their burrows preferentially in a certain layer of sediment so that a great network of burrows results in that plane (Fig. 166). Possibly the layer is especially favourable for digging, or the layers below are so consolidated that the worms dig sideways instead of downward. During the summer, burrows in the tidal sea do not go as deep as they do in winter when there is danger of freezing near the top; maximum depth in summer is 30 cm, in winter it is 40 cm.

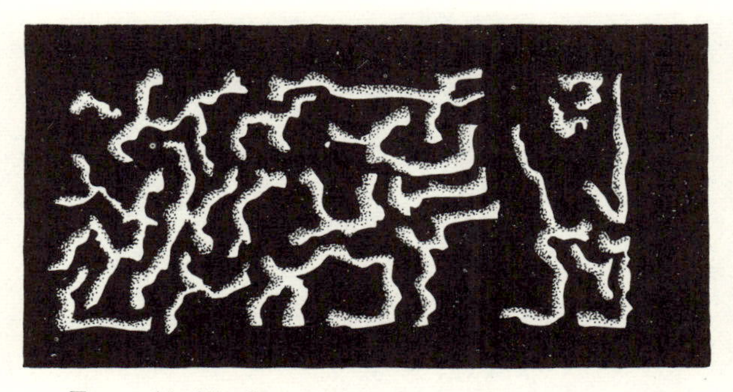

FIG. 166. Horizontal burrows produced by *Nereis* at depth because it burrowed preferentially in a single bed. (Actual size)

The gradually constructed, branched system of communicating pathways is not restricted to the interior of the sediment but is frequently extended to the surface where it takes the form of crawling trails. They lead from one opening of the dwelling tube to another, are just as well kept and cleaned as the underground burrows, and have mucus-cemented walls. Similar crawling paths may also run radially from one or several openings of the main dwelling tube and connect them with more distant auxiliary burrows. On the crawling trails the worm crawls between two raised mucous strips as if between railings. Occasionally, the crawling trail becomes a complete mucous tunnel which has not only bottom and walls but also a durable mucous roof (Schäfer 1952b and Fig. 167).

These surface structures are also guide lines always leading to a tube opening, and are particularly useful for quick retreat (Herter 1926). They play the role of the mythological Ariadne thread, making it impossible for the animal to miss its burrow. This is another good example of Hediger's (1956) relationship between the internal organisation of an animal and that to which it subjects its territory. Like a spider net, the system of mucous pathways within and above the sediment is the product of secreting glands encompassing a territory in space. But whereas the spider net serves mainly for feeding, the burrow of *Nereis* serves mainly for protection. Even the star-shaped feeding traces link the animal to its dwelling tubes while it grazes for diatoms at the sediment surface, and thus show the same subordination to the protective function. Once a large system of dwelling burrows with its subterranean and surface sections has been completed, generally no more burrows and mucous paths are built, and the system is frequently inhabited for many weeks if there are no significant disturbances from the outside.

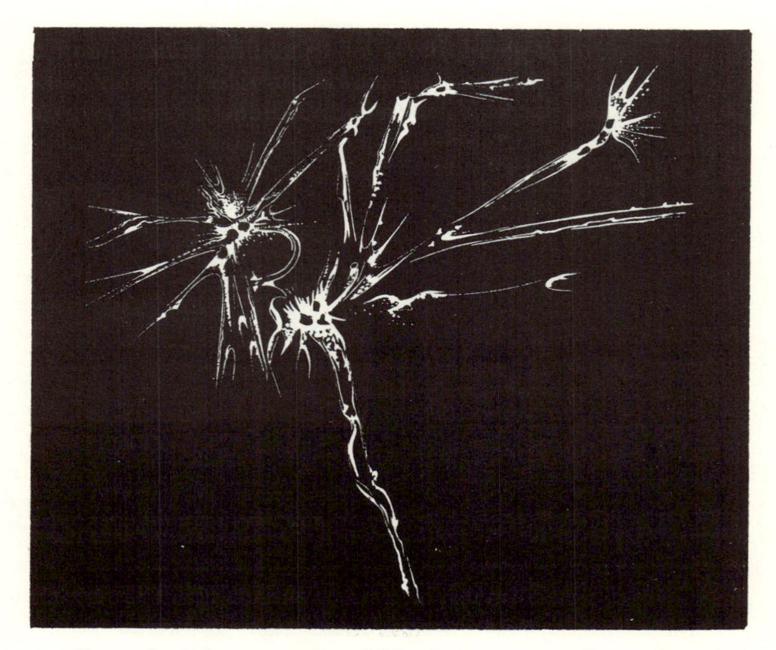

FIG. 167. Mucous trails of *Nereis* connect at the surface the openings of several internal burrows. (Actual size × 0·5)

If the density of settlements is high the structures of neighbouring individuals can be linked. Linke (1939) found that some burrows are used by two individuals. This, however, is probably an exception, for the animals defend their burrows. Considering the close tie of *Nereis* to its territory this is only to be expected. Wherever animals are closely linked to a territory and defend it, population density cannot rise above a certain level. This prevents excessive crowding, even where living conditions are optimal. Each individual requires a certain minimum area for its existence, and some individuals must die if a population becomes too dense (Schäfer 1956d).

Animals whose link to their dwelling is strong defend it not only against animal intruders but also against other hostile influences such as detritus that may penetrate the burrow. A certain amount of sediment or detritus necessarily enters into the openings of the burrow during slow sedimentation or due to slow transport of detritus above the settled area. Reineck (1958a) found in nature, and also in thin sections through sediment samples hardened in the laboratory, that *Nereis* presses sediment

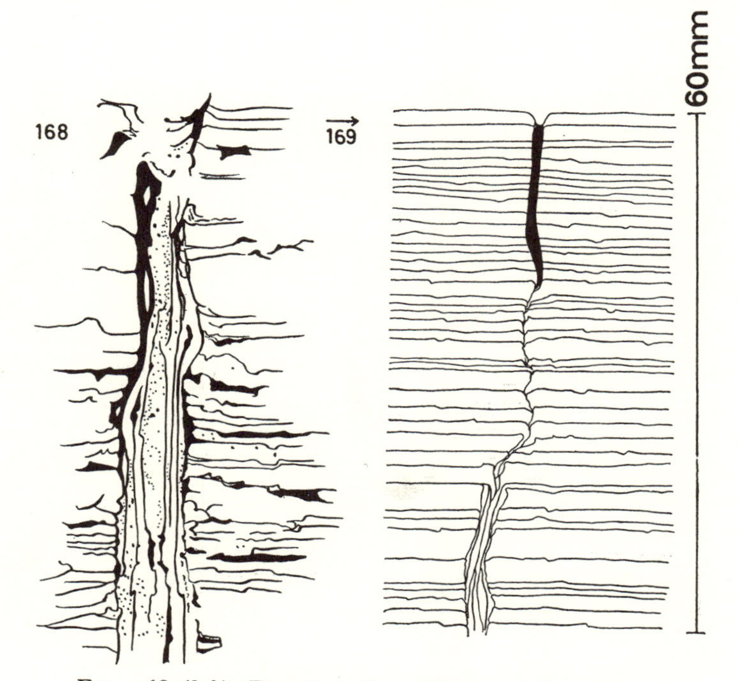

FIG. 168 (*left*). Drawing of a sediment thin-section (by H.-E. Reineck) with a dwelling burrow of *Nereis*. The vertical shaft is lined with several layers of sediment flattened against the wall and consolidated by the worm

FIG. 169 (*right*). Dwelling burrow of *Nereis*, showing from bottom to top: lining of walls, flight trail made after the original burrow was suddenly covered with sediment and new open burrow (black) connected to the surface.

that has recently entered into the tube against the walls by peristaltic and undulating movements and then restores the original diameter of the tube by locally swelling the body. After the sediment is thus pressed aside (Reineck calls this 'Räumauskleidung', literally meaning reaming and lining) it is mixed with mucus and rolled against the walls, coating them thinly with wallpaper, as it were (Fig. 168). Each time new sediment enters through the opening of the tube it is transformed into another 'wall-paper' covering an extensive part of the original burrow wall. Even faecal pellets that have entered the burrow are treated in the same way. At places where the burrow is more or less horizontal, intruding sediment collects only on the bottom wall and is cemented to it. With time the burrow thus migrates upward. We call this activity stowing as does Seilacher (1957); however, as Reineck emphasises, it is in this case not due to food turn-over but to one-sided reaming and lining. Dwelling burrows lined on all sides or unilaterally indicate moderate sedimentation which has not prevented continuous use of the habitation.

Heavy sedimentation to the point that *Nereis* is incapable of clearing the burrow forces it to abandon the dwelling. It must ascend to the surface of the sediment. The fugitive does not secrete mucus, and the trail is marked only by a more or less distinct sag trail through the penetrated layers (Fig. 169).

Regrettably, our ethological knowledge of other marine polychaetes is still patchy. Building of mucous tubes in muddy, soft ground and in sand must be quite common because a purely vagile way of life is only possible if the body is especially adapted to it. Many endobiontically living polychaetes do not build burrows according to a definite pattern; they only need burrows with one or more connections with the surface to establish a water current through the inhabited tubes. If sedimentation sets in the animals usually rise toward the new surface and build a new network of mucous tubes (Schäfer 1952b).

Other species build burrows in patterns specific to the species; in some cases the geometry of the burrow is unique to the producer, in others the burrows are the result of behaviour common to several species. Most of these burrows are more or less vertical. The dwelling burrows in the sediment can be clearly distinguished from surface traces and from traces laid out on a bedding plane that was not at the surface at the time. These forms will be discussed first.

The mucous tubes of the polychaete *Paraonis fulgens* are produced as horizontal internal traces in a bedding plane, and are therefore particularly important. Gripp (1927) was the first to describe these digging trails, without being able to identify their producers. Gripp found the trails while digging at the sand beach of Niendorf at the western Baltic Sea during very low water caused by a strong seaward wind. Breaking up the sand with the spade he noticed spiral-shaped, perfectly regular, hardly $\frac{3}{4}$ mm thick, trace patterns with a diameter of 4 to 5 cm at different depths. Since then such spiral-shaped traces of *Paraonis* have been observed on many beaches. Seilacher (1949) first found them at the coast of the southern North Sea on Wangerooge, and the author has found them on the sand beach of the island of Mellum. Their distribution is quite extensive as we know today. However, it is difficult to see them because straight, smooth cuts of the spade destroy the traces; they only remain preserved on bedding planes exposed when sand clods break open. On Mellum Seilacher and Reineck have observed them on sand surfaces exposed by planar erosion.

The *Paraonis* traces found by Gripp were of some significance at the time because Richter (1924b, 1927b) had introduced the term 'guided meander' for helminthoids in Flysch sediments without being able to give Recent examples (Fig. 170). He so designates crawling traces arranged

FIG. 170. Guided meanders. Helminthoids, Eocene Flysch at Blomberg, near Tölz, Bavaria, Germany (after R. Richter, 1927b). (Actual size)

in meanders that can be explained by automatic reaction to an external tactile stimulus. Such guided meanders, he explains, are evidence of a behaviour which is biologically significant because it enables an animal to use a superficial nutritious layer to good advantage. Without going into the question whether 'guided meanders' of the helminthoids are surface or internal traces it is now known that Recent traces of this type can be either. Those of *Paraonis* are doubtlessly produced inside the sediment, but the author (1950a) demonstrated and illustrated also short but perfect 'guided meanders', produced at the surface by a *Littorina littorea* which obviously recognised its own old trace and was guided by it while it produced another one. Apart from the 'guided' meander of *Littorina* we know of spiral tracks also in rivers and the deep sea. Gastropods have probably produced spiral tracks on the algae-covered surface of a mud-bank on the Rhine (Schäfer 1965). Identical tracks on the bottom of the Pacific are made by an enteropneust (Bourne & Heegen 1965, Hülsemann 1966).

Paraonis spirals investigated by the author have a diameter of approximately 8 cm; the individual, thread-thin, complete turns are 3 to

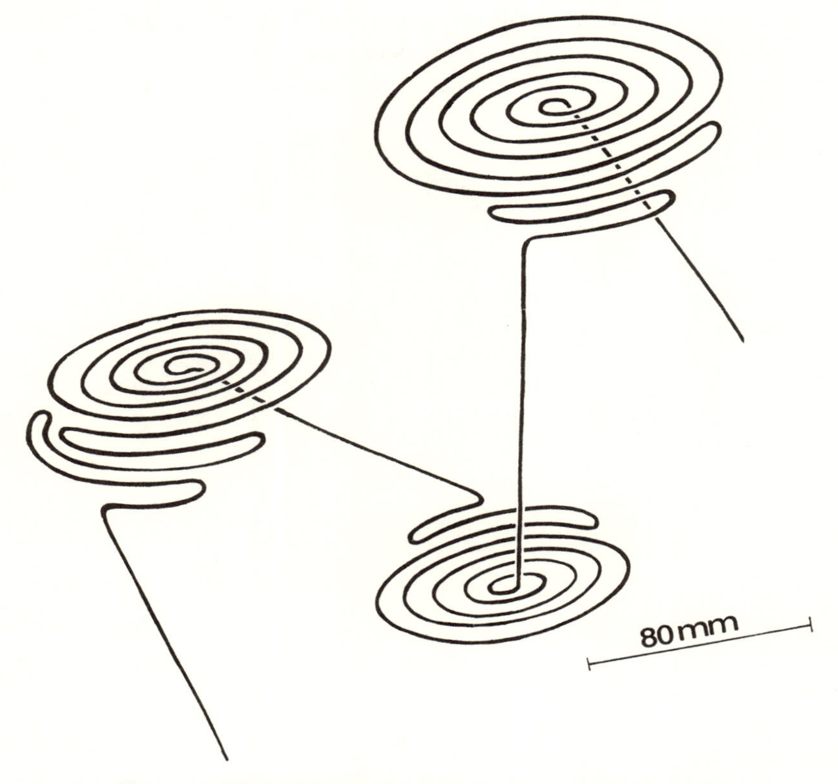

FIG. 171. Plane-spiral mucous burrows of *Paraonis fulgens*

5 mm apart (Fig. 171). The spiral-shaped traces are always horizontal, in most cases in a single bedding plane. The digging trails contain so much mucus that in broken spirals the mucous covers occasionally project beyond the fracture plane and appear as fine tubes. The spirals are connected with other spirals of a higher or lower level by more or less vertical, ascending or descending digging trails. A particularly interesting feature is an occasional 180° turn in the crawling trail near the outer edge of a spiral, resulting in a truly meandering trail that follows the older bend.

The question remains whether *Paraonis* is guided on its spiral course by a tactile stimulus that is caused by the previously completed trail and leads the worm on its path in regular loops. Any thigmotactic, guiding stimulus would have to be caused by slight compaction of sediment near the old digging trail. If this is assumed, sand compaction would have to be registered by the worm at a distance of 6 to 10 times the diameter of the trail, since *Paraonis* is somewhat less than 0·5 mm thick and the spirals lie 3 to 5 mm apart. This appears as the weakest point in this interpretation of spirals with turns so precisely equidistant, almost as if made by a precision instrument. They do not give the impression of being laid out by piecemeal palpating for small density differences in sediment that could not have been uniformly dense to start with. One could hypothesise that lateral, longitudinal muscle strands are initially contracted during spiral-digging and relax slowly, evenly, and regularly, as the animal advances; such relaxation could gradually stretch the body of the worm and force it automatically to follow a spiral-shaped course, while the animal crawls forward by peristaltic movements of its circular musculature which is quite independent of the state of contraction in the longitudinal muscles. (The worm would crawl in a circle if the contracted, longitudinal musculature were kept in uniform tension.) A combination of reactive and spontaneous behaviour would doubly ensure a regular progress of the instinctive action, and it is a frequent combination. There is an additional requirement; an organ of equilibrium must maintain all digging trails in the horizontal plane because thigmotaxis in itself would permit one more degree of freedom, and chaos would result rather than a plane spiral.

Digging in meanders at the edge of the spiral may indicate that the effect of the impulse is fading which had initially started the animal in automatic spiral digging. The animal continues to dig in meanders outside the spiral until the planar steering ceases also, the worm breaks out of the plane of the spiral and digs upward or downward. But soon the worm starts another horizontal spiral. It remains an open question which particular momentary interplay of stimuli triggers the worm to turn from the vertical to the horizontal direction, and which starts the digging of a spiral in this plane. Presumably chemical stimuli play a role; for in areas that are richer in micro-organisms and decomposing organic material, spirals are more frequent than in other regions.

This leads to the question of what may be the ecological significance of digging in spirals or in meanders. Gripp was the first to express the thought that the patterns reflect the animals' careful search for prey in the populated top 30 cm of sediment. Spiral structures that lie in different planes above each other form an efficient three-dimensional pattern of non-overlapping search paths for small mobile prey.

If this were the right interpretation the spiral-shaped burrow of *Paraonis* would not be a permanent dwelling structure but rather a system of burrows for searching, systematically built by a purely vagile animal. However, because the spiral-shaped burrows are quite durable mucous burrows they were discussed in the context of permanent dwelling tubes.

Heteromastus filiformis, a capitellid worm, builds a widely branched network of burrows whose walls are covered with a coating of mucus. These worms occur in large numbers in the tidal flats of the North Sea coast where the ground is sufficiently muddy. They are 8 to 18 cm long but, as the name implies, thread-like; the diameter of the burrow is barely 1 mm. There is no oxidation halo around the burrows, yet they are clearly visible due to their mucous coats. When tubes are washed out, as happens occasionally, the mucous wall of the tube holds together for some time, even if it is repeatedly transported and redeposited. Linke (1939) assumes that the lobe-shaped, enlarged posterior parapodia of this worm have a breathing function and that the worm which lives head-downward in its tube ascends to the surface not only for discharging faeces but also for breathing. *Notomastus latericeus*, another capitellid, is widely distributed in the North Sea and builds small, spiral-shaped burrows in muddy sand (Hertweck & Reineck 1966).

Ephesia gracilis belongs to the Sphaerodoridae. It is a shortish worm, provided with numerous epidermal papillae, and it secretes abundant mucus. The mucous burrows that the animal builds have a wide diameter. *Ephesia* lives in soft grounds that are interspersed with small shell fragments. *Lumbriconereis impaticus* and *Glycera alba* choose similar places. The mucous tubes of *Glycera* are often encrusted with shell fragments and have a firm consistency. *Glycera* seems to occur more frequently in Heligoland Bay than *Ephesia*.

Several species of the family of the Nephthydidae build mucous burrows. They are predatory polychaetes with strong jaws on a muscular, conical, eversible proboscis that is used for 'bolting' in the sediment: *Nephthys malmgreni*, *N. cirrosa*, *N. hombergi*, *N. ciliata*, *N. caeca*, and *N. longesetosa*. Most live in mud, *N. incisa* and *N. ciliata* exclusively so, but some prefer sand, e.g. *N. hombergi*, *N. caeca* (according to Remane this large species occurs frequently in the eulittoral zone), and probably also *N. longesetosa*. Common features of all these species are burrows that are built with a moderate amount of mucus, and to which new branches and ducts to the surface are added constantly. Their predatory way of life seems to be incompatible with a prolonged stay in a certain section of the burrow. Only *N. hombergi* is well known. It is common at certain localities on the sandy tidal flats near the coast where the burrows reach depths of 20 cm; surface paths in the form of open grooves are also found occasionally. Linke observed that *hombergi* swims excellently by undulating laterally with its strong, longitudinal muscles.

Scoloplos armiger occurs quite frequently on the tidal flats and prefers sand to muds. It can be found from the tidal flats to a water depth of 20 m. This worm roams widely like the nephthydids and does not build permanent dwelling burrows. However, it is not a predator but feeds on sediment. Its burrows are cemented with more mucus than those of the nephthydids and are therefore more clearly visible in bedded sediments, particularly if they are soft.

Exclusive sand inhabitants are the spionids *Scolecolepis squamata* and *S. bonnieri*. *S. squamata* inhabits the sandy surf zones whereas *S. bonnieri* lives in muddy sands under deeper water. *S. squamata* is very thin and lives in a vertical or near-vertical tube that is well coated with mucus; its long tentacles extend above the opening of the tube and palpate the surrounding for food. The animal can be easily recognised by the scraping circles of approximately 1 cm diameter made by the tentacles on the surface of the sand (Seilacher 1953a, Schäfer 1956a).

In muddy ground the mucous burrows of *Notomastus latericeus* and *Diplocirrus glaucus* can be found. *Scalibregma inflatum* (see Hertweck & Reineck 1966), *Nerine ciliata*, *N. fuliginosa*, and *Ampharete grubei* live in both mud and silt.

All mucous tubes of polychaetes have one feature in common, the sediment layers cut off by the burrows are little or not at all inflected and meet vertical tubes at right angles. Where the inside of the tube is not too narrow the burrow is frequently filled with layered material. Burrows that do not follow a special pattern such as U-shaped or spiral burrows are unspecific.

Mucous burrows produced by polychaetes are shaped burrowing textures.

U-shaped burrows. They are constructed in the sediment or on the sea floor by members of several animal groups. Compared with the large number of known fossil U-shaped burrows, however, there are few Recent species that build such structures; they are amphipods, echiurids, enteropneusts, holothurians, and polychaetes. In the ideal case, these burrows consist of two vertically descending tubes connected by a downward-arched tunnel deep in the sediment.

The construction of a burrow of whatever kind requires a working scheme (Werkschema), a plan of action residing in the nervous system ensuring similar constructions. It is not clear which stimuli effectively guide the construction of often very regularly shaped structures. These structures are normally built in water-covered, marine sediments, often with circulating sea water in the interstitial spaces. Marine animals with a density very near that of sea water do not feel gravity in most of their body. However, special organs can serve to sense the field of gravity, and most of them are probably like statocysts. The same or an even greater difficulty exists for animals which live and move in the sediment. Not only are they surrounded by a non-fluid medium in which they cannot feel positive or negative buoyancy, but they are also cut off from light. Some sediment dwellers and burrowers have no statocysts. Possibly, such animals orient themselves crudely by the gradients of temperature or oxygen concentration. Such an imperfect means of orientation does not

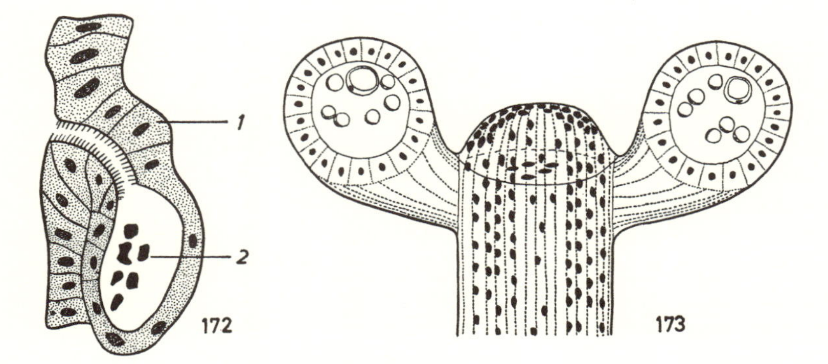

Fig. 172 (*left*). Statocyst of *Laonome*. (1) epithelial layer forming an inward bulge; (2) statoliths (after Erenkamp, 1931). (Schematic drawing)

Fig. 173 (*right*). Statocyst of *Leptosynapta* (after Becher, 1913). (Schematic drawing)

explain the equal perfection of small and large U-shaped burrows, produced by individuals of the same species, but of different size.

Where statocysts have been found, for example in polychaetes (Fig. 172; see also von Buddenbrock 1912, 1913), in crustaceans and holothurians (Fig. 173) these sensory organs probably play a significant role in their orientation in the sediment. However, the statocyst only indicates inclination. Therefore, statocyst-bearers are dependent on additional stimuli which enable them to determine depth.

Von Buddenbrock (1913) noted that *Arenicola grubei* does not continue digging downward if the 'resistance of the sand which the worm has to overcome exceeds a certain value'. However, this author does not agree with this interpretation because the types and textures of the sediment through which *Arenicola* can dig, differ greatly, and yet the burrows of individuals of the same size have everywhere the same depth and proportions. It could be imagined that tactile sensory organs determine the lengths of the individual straight or curved segments of the burrow. For it is notable that for many worms the depth of the U is roughly equal to the length of the body. This fits with an observation of von Buddenbrock (1913) that *Synapta digitata* stops digging downward as soon as it has completely disappeared in the sand. Finally, in addition to effective sensory organs, the animals require a series of automatic responses since after a certain, and always the same, time of downward digging, the direction is changed by 180°.

A series can be set up of increasing perfection of the tube from animals that simply lie in the sediment through others that dig irregular dwelling tubes to those that make regular U-shaped burrows. *Cucumaria* just lies in the sediment, bent to a U-shape; in order to get into this position it needs tactile sensory organs in the skin, supported by organs for equilibrium. *Nereis*, *Callianassa*, *Calocaris*, and *Gebiopsis* build several burrows that are connected with each other, and all end somewhere at the sediment surface. In these cases, the oxygen content of the water in the newly dug burrow seems to help the animal in orienting itself in the sediment, and particularly, in determining the depth of the burrows. The animals always respond to a deficiency of oxygen in the tube by crawling upward while the statocyst indicates the vertical direction. *Nereis* also responds to temperature changes with depth, and burrows deeper during winter. In some cases, two of several dug burrows are lined with mucus, they are constantly inhabited, and the flow of water is brought mainly through these two tubes. The other burrows disintegrate gradually, or become clogged with faeces. *Polydora ciliata* frequently digs several burrows with several openings in loose sediment but then limits use to two neighbouring, subterraneously connected tubes. These U-loops are quite irregular, the transverse connection of the two ascending tubes is asymmetrical, and the ascending tubes themselves are often not vertical. They are unlike the strikingly regular U-shaped burrows; the clear working scheme is missing. The polychaete *Poecilochaetus serpens* digs U-shaped burrows with non-vertical ascending tubes; it appears to be a step closer to the regular shape because it has no side branches.

Up to this point in the sequence of burrowers, the animals are predators, eat suspended food or palpate for food. The burrow serves only as shelter, built so that oxygen-rich water can be pumped along its entire length if it is to be permanently inhabitable. Little else is required, and a regular U-construction is not essential.

More regularly built burrows, such as those of *Arenicola* and of the enteropneust *Balanoglossus* have a clear U-shape in plan if not always in execution. (Just 1924, investigated the neurophysiology of *Arenicola*.) Both animals feed on sediment. Their U-shaped tube is built to serve as a collecting pit; the animal causes sediment mixed with organic substances to slip constantly and to drop to the point where the animal can take it up with its proboscis (Figs. 174 and 175). Thus the burrow serves not only for protection and dwelling but has additional functions, differentiated even between the two branches. Other worms turn back and forth in their tubes, but not *Arenicola* or *Balanoglossus* whose two tube exits have different functions, one serving as feeding tube, the other as faecal tube. It is to be expected that the faecal tube is more complete and better coated with mucus because it is used for a longer time to transport waste to the surface. The feeding tube, on the other hand, is less complete, its direction is often changed, it may even be temporarily missing altogether, or two or more may be serving at the same time. This explains the different

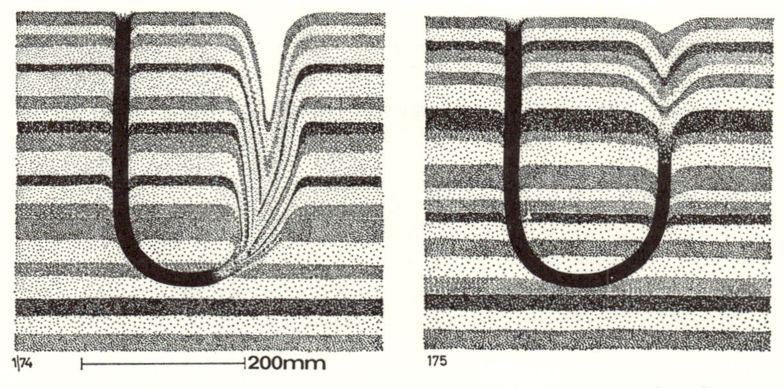

FIG. 174 (*left*). U-shaped burrow of *Arenicola marina* in soft sediment. No feeding burrow is needed

FIG. 175 (*right*). U-shaped burrow of *Arenicola marina* in firmer sediment. The feeding burrow (right) is moved toward the surface

views on the course of *Arenicola* burrows, all based on correct observations, yet seemingly contradictory (Wesenberg-Lund 1904, Richter 1924a, Thamdrup 1935, Häntzschel 1938a, all with additional bibliography). Only the structure and texture of the sediment into which the burrow is dug determines how the feeding tube is constructed. If the sediment is coarse and moves easily the ascending branch of the feeding tube needs only to be short or can be missing altogether (Fig. 174); if the sand is more compacted and/or of finer grain size, the worm often has to ascend nearly to the surface of the sediment (Fig. 175); then it digs a distinct ascending tube that often appears transversely segmented. If the sediment is sticky and does not slip, the upward branch is continued to the surface, and a new one is then constructed somewhat offset against the former one. Linke (1939) described the different forms of the burrows of *Arenicola*.

In *Arenicola* burrows the inhalant water cannot flow continually as in other tubes of burrow-living animals because this particular tube is rarely a complete U with two unclogged openings; rather, one branch is usually plugged or incomplete. As the function of the burrow has changed from a simple dwelling tube to a feeding tube, there is also a change in the ventilation. *Arenicola* climbs backward to the surface and there discards the faeces. Then fresh water enters the opening of the tube from the outside while the posterior end of the worm is elongated and narrowed. Peristaltic muscle contractions transport the water from the posterior end forward until it reaches the zone of tuft-shaped gills. The gill tufts are retracted into the body, as the peristaltic swelling zone passes along the worm, but are immediately protruded again into the newly arrived water. The breathing water is pushed back and forth several times in the gill region and is finally driven into the sands of the feeding tube and funnel.

It has been assumed that *Arenicola* feeds on organic substances that accidentally enter with the sand. We know today from the investigations of Krüger (1959) that the worm actively accumulates food in the sand of the funnel before the proboscis takes it up. The animal uses the following method: the worm introduces organic detritus through the open faecal tube, together with the breathing water, and after transporting it forward the worm presses the water and detritus in front of the head into the subsiding sand of the feeding tube. The sand acts as a filter and is thus actively enriched with organic substance before it is ingested by the proboscis. *Arenicola* is therefore a particle eater that also ingests sand because it functions as plankton filter. Krüger measured the nitrogen content of freshly eaten, plankton-enriched sand and found it to be five to ten times greater than that of samples of the surrounding sediment. If nitrogen is taken as a measure of the concentration of organic material this shows the effectiveness of the sand filter.

Locomotion in the sediment is strictly peristaltic and enables the worm to move rapidly back and forth in the dwelling tube. As contraction waves of the strong, circular musculature run along the body, the walls of the burrow are constantly reworked. They are pressed mechanically and glued with mucus at the same time; in addition, they become distinctly sculptured and show a regular transverse segmentation (Häntzschel 1938a). The transverse division in the walls of the burrow, mainly of the faecal tube, however, is not a mould of the body segments of the worm. In many cases the sculptured wall of the tube is markedly hardened by iron hydroxide. Häntzschel (1938a) and Linke (1939) illustrated iron-hydroxide hardened faecal tubes which project several centimetres above the surface of the tidal sea, exposed by planar erosion. Frequently, the transverse segmentation of the inner wall is also visible on the outside of the tube where the outside is exposed by consolidation and subsequent erosion (Fig. 176).

Reineck (1958a) observed that the faecal tubes of U-shaped burrows in tidal flats with pure sands do not show multilayered walls. If the animals live in muddy areas, on the other hand, the faecal tubes often have several walls. The individual walls are formed from muddy sediment that has dropped into the faecal tube and is then pressed against the wall. This is another case of Reineck's 'reaming and lining'.

The density in *Arenicola* settlements is extraordinarily high (up to 120 animals per square metre). The feeding funnels of the individuals lie so

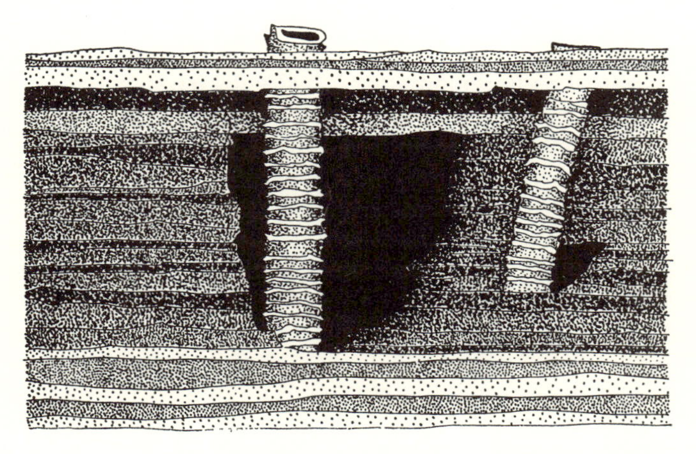

FIG. 176. Faecal burrows of *Arenicola marina*, encrusted with iron hydroxide, transversely banded, and resistant to erosion even after removal of surrounding sediment. (Actual size)

close together that they intersect. The inhabited sediments become disturbed, and the clear U-shape of the individual burrows is destroyed. Since the feeding funnels are frequently changed, less and less remains of the original beds until, finally, they are replaced by tilted layers deposited in the feeding funnels. In areas of *Arenicola* settlements extensive transportation occurs from the surface layers downward as the sediment slides toward the mouths of numerous worms. The sediments ingested by these worms are returned to the surface as faeces which are easily and quickly transported away over the surface. The worms ingest grains of sand size or finer almost exclusively, although the entire sediment, including coarse and fine shell fragments, slips down into the feeding funnels. These shell fragments thus collect in the end at the level of the U-arches of all *Arenicola* individuals of any one area where they frequently form thick, closely packed shell breccias (van Straaten 1952, 1954, 1956; Reineck 1958a). These packed shell breccias frequently consist of hydrobian shells; but fragments and whole *Cardium* shells also occur abundantly (Fig. 269). The shell breccia that collects at the level of the U-bends has the same depth below the surface over wide areas, between 30 and 40 cm, because the worms living in the same area are usually roughly of the same size and so build U-shaped burrows of the same depth. This order according to size, which also means according to age, is probably due to active migration. The youngest animals settle in the upper zone of the tidal flats near the high-water line. That is the place to which the eggs of these worms are carried, rolled along on the sea floor by tidal currents and by the surf; the eggs are denser than sea water. After some time, the young animals leave the settlements in the upper tidal zones and swim toward new living areas in the deeper tidal sea, sometimes in prolonged migrations (Newell 1948, 1949). Such wanderings are repeated later on (Werner 1954); probably these changes are influenced by temperatures and seasons. *Arenicola* swims by undulating, the movement being produced by alternate contractions of all four longitudinal muscle strands. The resulting movement follows an irregular, stretched helical path.

Like pure mucous tubes, agglutinated tubes can also be constructed in a U-shape. Seilacher (1951, 1953a) noted the burrows of the polychaetes *Chaetopterus variopedatus* and *Lanice conchilega* which, beside simple U-shaped tubes, also build W-shaped tubes by branching at the level of the bend. These W-shaped burrows become necessary for these worms (for *Chaetopterus*, see Enders 1909, Trojahn 1913) because tubes built during youth cannot be widened due to their rigid lining. When the worm grows thicker it must back out of the thin, old branch of the tube and build a new branch, resulting in a W-shape. However, not more than two of the three branches of the W are inhabited, and the thinnest is not used.

If U-shaped burrows are dug in the way described the space between the two branches cannot be used for further construction. When the worm is forced to move up or down (by erosion or sedimentation) it must therefore dig a new burrow which can either be connected with the old system of tubes or can be started independently at a new location.

In another type of U-shaped tubes, numerous formerly open, transverse bends connecting the two vertical tubes sit one above the other in the plane defined by the verticals. They indicate that the U-shaped burrow was frequently lowered while it was inhabited. Generally, both vertical branches of the U retain the same distance from each other and are gradually lengthened as they run deeper. Such U-shaped tubes with multiple bends were first observed on fossil U-shaped burrows, where biogenic textures can generally be more clearly recognised than in Recent sediments.

Clearly, the burrowing technique leading to such textures is completely different from that of *Arenicola.* It is not caused by adaptations of the burrow to changes in surface elevation due to erosion or sedimentation, as many investigators have assumed. The organism making such a burrow starts by digging an elongate cavity, about as wide as the body, and in time, the animal deepens this cavity to a flat, pouch-like shape. Then the burrower constructs a centrepiece into the pouch, leaving a channel along the outer edge of the pouch open. This channel is then the U-shaped burrow.

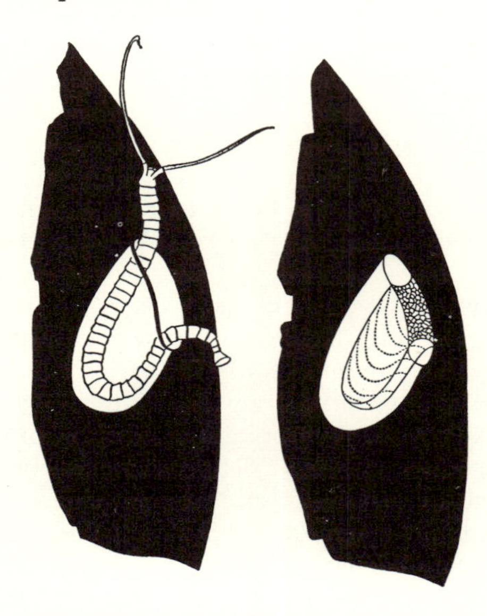

FIG. 177. Burrow of *Polydora ciliata* in a calcite shell. Left: a pouch is formed by dissolving calcite; right: the pouch is converted into a U-shaped tube by infilling with a septum made of cemented sediment grains. (Schematic drawing)

This building technique can be particularly clearly observed where the material of which the centre consists, differs from that of the substrate around the pouch. An example is the burrow of the polychaete worm *Polydora ciliata* which is sometimes built in the hard, calcareous shell of a lamellibranch and not in the sediment. In this case, its body, forming a loop, dissolves a pouch in the shell and then fills the centre of the pouch with individual, sticky grains of sediment that have been fetched from the outside. Only a U-shaped dwelling tube is left open (Fig. 177). The animal places the sand layers in the centre in such a way that the lower row of sand grains runs parallel to the curve of the pouch.

Evidently, an animal working on this kind of task needs no special guidance by sensory organs for equilibrium or by auxiliary stimuli. And these U-shaped burrows do not have to be built oriented with respect to gravity. They can be constructed vertically, horizontally, or at any angle. The course of the burrow is related to the orientation of the surface of the substrate into which it is built (Fig. 178).

The amphipod crustacean *Corophium volutator* demonstrates most clearly this independence from organs of equilibrium when a burrow is constructed. The 4 cm deep U-shaped burrows are frequently built into muddy, ragged escarpments that form on the steep outer bank of a curved tidal channel. *Corophium* digs and deepens the pouches with its limbs. It deepens the burrow at the bottom and sticks new material to the ceiling on the opposite side. However, the centres are rarely made up of

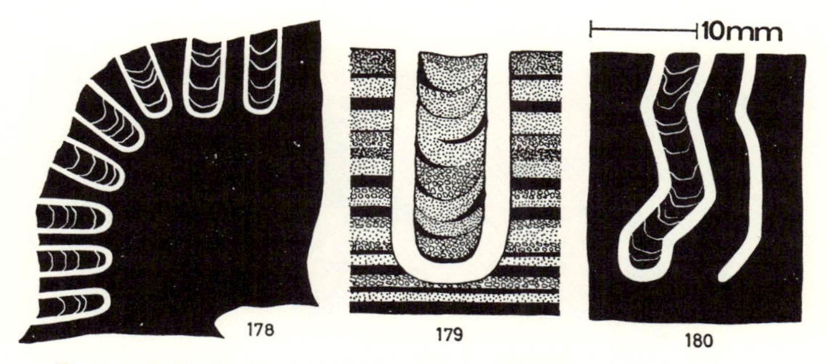

FIG. 178 (*left*). U-shaped burrows of *Corophium volutator* are built at right angles to the surface, independent of the orientation of the surface; layering of the infilling indicated

FIG. 179 (*centre*). U-shaped burrow of *Corophium volutator*; independent layering of the infilling shows that it is not a remnant of the surrounding bedding but actively constructed

FIG. 180 (*right*). Frontal and side views of angularly deflected burrow of *Corophium volutator*; the bends of the parallel sections lie at the same elevation, this is proof of the origin of the burrow as a pouch

clearly discernible layers or show any layered structure; they can be perceived only where more muddy and more sandy layers alternate (Fig. 179). This explains Häntzschel's opinion (1939a) that *Corophium* does not form its burrows by infilling the centre of the pouch. However, even without direct observation of the burrowing activity, this can be concluded from the deviations of many of these U-shaped tubes from a plane (Fig. 180). No other technique could explain how the two ascending tubes have kinks, bends, or plane portions at exactly the same levels. It should be added that *Corophium* does not always build U-shaped burrows (Schwarz 1929a and b, Häntzschel 1939a, Seilacher 1953a). It builds them mainly in consolidated ground, particularly in well-compacted mud deposits. In sand, on the other hand, one finds simple, sometimes branched pits. A simple pit suffices in sand because the whirled-in and used water can flow through the spaces in the sand, and a separate tube for the outflow is not necessary.

Probably U-shaped burrows are more common than we know today. Even for a well-investigated area such as the North Sea, reliable data are still incomplete. The very large U-shaped tubes of the species of the genus *Echiurus* (family Echiuridae, formerly considered as belonging to the gephyreans) may serve as a last example. Their muscular, cylindrical body can be up to 14 cm long (Fig. 181); the animal pushes through the sediment by peristaltic movements. Two hooked bristles on the anterior portion of the body, bristles on the tail, and the papillae arranged in rings

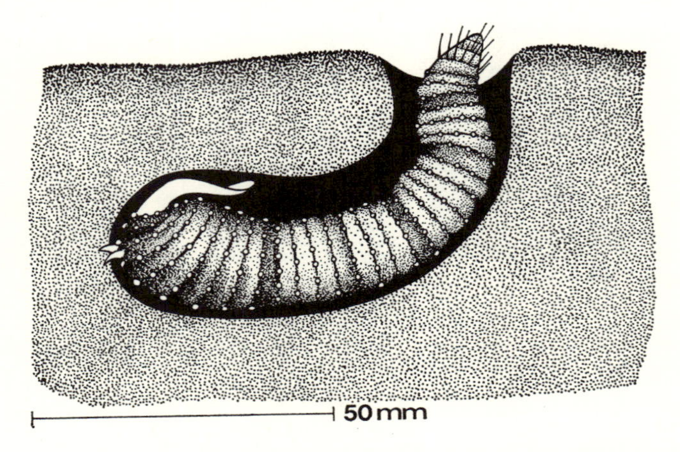

FIG. 181. *Echiurus echiurus* building a U-shaped burrow

on the skin serve as auxiliary organs for locomotion in the sediment. Digging forward, *Echiurus* folds the lobate proboscis together with the foremost part of the body backward. Thus the ventral part of the animal with the two strong, retractable, hooked bristles forms the front of the digging animal. The proboscis serves only for feeding and has no locomotory function. It can be greatly elongated and is extended from one of the openings of the U-shaped tube; with it the animal licks detritus from the surface in radial strokes. The food is carried to the mouth by means of cilia. The animal does not swallow sediment.

Echiurus digs its U-shaped tubes in fine sand to depths of 50 cm. It hardens the walls with mucus to prevent collapse. The tube openings are narrow. Because of the great depth of burial the animals are rarely caught even with deep dredges, and their distribution is probably wider than is known. In the North Sea *Echiurus echiurus* lives in shallow water and can be found in the English Channel, on the coasts of Belgium and Holland, the Frisian coast, and the coasts of Heligoland, Föhr, Sylt, and Denmark down to depths of 30 m. By means of the box sampler (see Reineck 1963) undisturbed sediment blocks with burrows can be quite easily hoisted on board. Good preparations of U-shaped burrows can be made by impregnating and hardening the sediment with plastics (Hertweck & Reineck 1966).

Echinocardium. The mechanism of shovelling has already been described in D I (*a*) (vii) for *Echinocardium cordatum*. The animal can move in this way in pure sand. It cannot move in muds, especially not if they are consolidated. Therefore, in the North Sea *Echinocardium cordatum* is limited to medium to fine sands at depths of 3 to 40 m (Remane 1940). Settlements of juvenile and adult *Echinocardium* have been observed in even the deepest zones of sandy tidal flats (north Mellum).

When *Echinocardium* lies on the surface of the sand it begins to burrow at once. The spoon-shaped spines on the underside begin with their circular movements and transport the sand from under the sea urchin sideways and upward, so that the sea urchin slowly starts to subside. Then the fine spines on the sides of the body also begin to work; and, at the same rate as sand is being moved toward them from below, they heap it in ramparts on both sides of the body (Fig. 182). Once the sea urchin has dipped below the surface it would be covered by loose sand but for a special mechanism that pushes the sand back and consolidates it. Spine-bristles are set in a paint-brush-like arrangement on top of the body; they start eccentric movements as soon as the sand starts to flow from all sides over the sea urchin and push the advancing sand outward. Simultaneously, small organs near the base of the bristle-brush secrete sticky mucus. These organs are bulbous, pigmented bodies arranged along a plate junction. While the spines of the brush move to push the sand outward they also press the mucus against the sand walls. The mucus hardens and holds the sand. As a result, a consolidated sand ring is built above the sea urchin. The deeper the sea urchin moves into the sediment by the constant, circular movements of the body bristles, the longer becomes the chimney that leads to the surface. It is kept open as a feeder tube for breathing water (Fig. 182). The sea urchin penetrates to a depth of 15 to 18 cm and forms a chamber with walls that are also made self-supporting by mucus. The breathing chimney is cleaned by the so-called brushing foot, and if necessary, the mucus-secreting organs at the tip of it repair the tube.

Echinocardium feeds on sediment, the sea urchin must therefore be mobile at depth. From the dwelling chamber it builds horizontal feeding channels. While moving forward, it scrapes the sand with the ploughshare-like scoop formed by the calcareous test and lets it flow into the mouth.

The ingested sand contains the animal's food, small organisms and detritus. The large quantities of discharged (digested) sand alone make it necessary that the animal moves constantly through the sediment. A shapeless sand slurry leaves the intestines, very unlike the consolidated and shaped faeces of epibiontically living, algae-eating sea urchins.

When shovelling through fine sand, *Echinocardium* produces trails with clean-cut edges. Its method of excavating inflects the perforated layers only slightly, and only very near the cut; in coarse sediment,

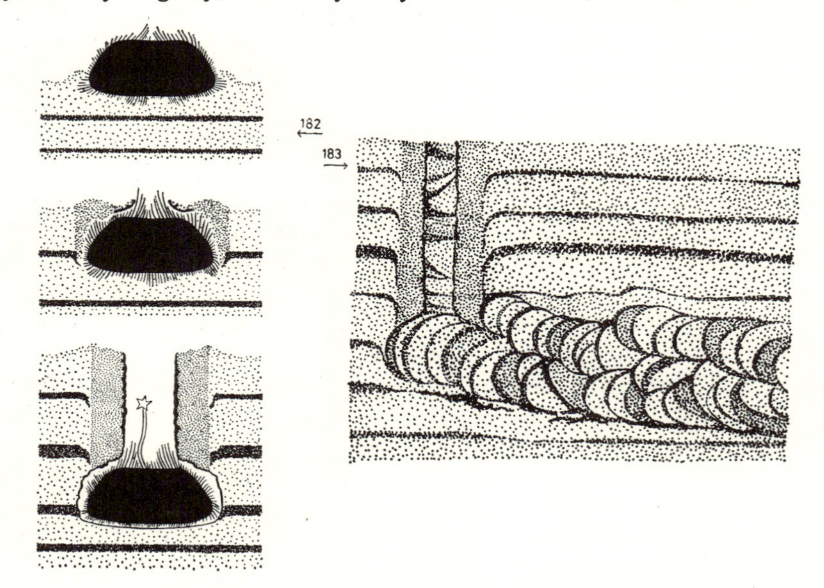

FIG. 182 (*left*). *Echinocardium cordatum* building a mucus-covered dwelling burrow. Top: beginning of digging by circular shovelling with the spines; centre and bottom: stages of subsidence of the animal and of the forming of a mucus-covered shaft. (Actual size × 0·3)

FIG. 183 (*right*). Burrowing textures produced by *Echinocardium cordatum*. Left: mucus-covered shaft; right: feeding trail with textures due to refilling. (Actual size × 0·3)

however, the cut-off beds sag somewhat more. The sediment within the digging trail becomes perfectly mixed.

Generally, *Echinocardium* settles so densely that the wide digging trails of the animals impress a new, burrowing texture on the large volumes of the sediment through which the trails are dug. The vertical chimneys that are abandoned when the animals move on, are soon filled with detritus, peat fragments, and bedded sand (Fig. 183). The sideways movements in the sediment produce specific cut-and-fill trails. The voluminous faeces, discharged backward, together with sediments that have been pushed back along the sides of the body, fill the digging trails of the sea urchin so that in cross-section they appear as crescent-shaped areas fitting into each other. Frequently, these successive crescent-textures lie in several layers above and beside one another, partly intersecting or even crossing each other; the reason for this is that *Echinocardium* often changes direction and level of crawling. Such horizontal burrowing trails are found surrounding the vertical chimney for a distance of 50 to 60 cm. This shows that the interstitial water in the sand provides the sea urchin with enough breathing water for a considerable time, and that the animal is not entirely dependent on a constantly open chimney to the surface of the sediment. Occasionally cut-and-fill trails with their crescent-shaped textures separate as complete bodies from grab samples, indicating that they are very slightly hardened by secretions (see also Reineck 1968).

Although the animals live and move deep in the sediment they are frequently washed out by waves during periods of bad weather and heavy seas. Then the animals have to begin again to burrow into the sediment. In the swash marks of the dune islands in the southern North Sea, the more or less broken, white, calcareous tests of these sea urchins collect by the thousands and form large mounds (Häntzschel 1936a).

Amphipods. Beside *Corophium* which has been discussed above there is a considerable number of other amphipod crustaceans which build tubes with self-supporting walls by hardening a digging trail with mucus. The link to a dwelling burrow results again in a more sessile way of life in these species. In all cases, the reason for constructing a mucous burrow or a mucous coat around the body is doubtlessly the need for breathing water while the animal is burrowing in muddy sediment.

Unicellular glands have been found on the middle sections of the fourth and fifth thoracic limbs of such tube-building gammarids (Nebeski 1880). Their ducts run toward the claw into a common collecting bladder from where a short duct leads to the outside (Fig. 184). The mucous wall of the tube consists of inter-woven mucous threads which have been individually excreted.

Numerous vagile amphipods live and dig mostly in sand, for example *Ampelisca spinipes*, *A. diadema*, and *A. tenuicornis*. Almost all amphipods burrow in the sediment by scraping with the walking limbs and then pushing by strong beats with the abdomen. *Cerapus crassicornis* is small and lives in short, membranous tubes that are open at both ends. Usually the tube lies open at the surface of the sediment. The first section of the first antenna closes the front end of the tube whereas the curved abdomen with the uropods closes the other end (Stephensen 1929). Many *Siphonocoetes* species live in mucous tubes covered with sand grains. They are flattened in cross-section, fitting closely to the body they house. However, the animals have also been found in tubes of *Dentalium* which have been extended by a tube section of cemented sand grains (Remane). *Siphonocoetes coletti*, on the other hand, lives in tubes of consolidated shell fragments, small pebbles, and fine spines of sea urchins. *Ericthonius* species attach their mud-built dwelling tubes on hydroid polyps and on

FIG. 184. Thoracopod of a gammarid crustacean. Single-cell mucous glands with ducts leading to a collecting bladder in the point of the claw. (Schematic drawing)

sponges (Zavattari 1920). Several species of the genus *Corophium* live in cemented mud tubes. Examples are *C. bonellii*, and in brackish water *C. lacustre*. *Amphitoë rubricata* builds a dwelling nest with mucus-cemented walls in shallow water among sea weeds (Skutch 1926). The genus *Haploops* (e.g. *H. tubicola*) also lives in self-made tubes.

Callianassa. Few higher crustaceans in the southern North Sea construct dwelling burrows. However, the genus *Callianassa* with world-wide distribution (of the Thalassinidae of the Anomura) provides good examples. Six species from the European seas have been described, and Lutze (1938) has described the life habits of two of them in some detail: *Callianassa subterranea* and *C. helgolandica*, both living in the North Sea (Fig. 185).

Callianassa uses the much enlarged and widened right or left claw of the first pair of walking limbs as the main digging tool with which it loosens the sediment and moves it backward. The other walking limbs help digging by scraping and pushing. The pleopods and the tail fan whirl and clean the burrow once it is established. Adult animals are 6 to 7 cm long. They start burrowing after their fourth moulting. In youth they live as plankton, and the first pair of pereiopods is still symmetric.

The diameter of the burrow exceeds that of the body, permitting free movement within. Young animals build burrows 1 cm wide, but as they grow the burrows become 3 cm wide. They can be 20 to 30 cm deep and

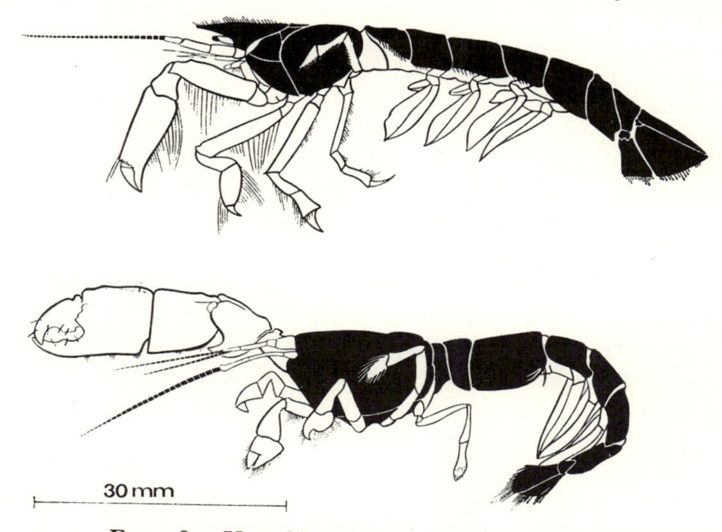

FIG. 185. *Upogebia deltaura* (top) and *Callianassa subterranea* (bottom)

are built in muddy sand or mud. From the start several communicating tubes are constructed, and construction apparently continues as long as the burrow is inhabited. The waste which is accumulated near the burrow orifices therefore becomes abundant and gives the burrows away. Below the surface a multibranched net of burrows runs through the sediment, mostly on the same level, and several connecting tubes lead to the surface (Fig. 186). At the branching some sections are widened allowing the

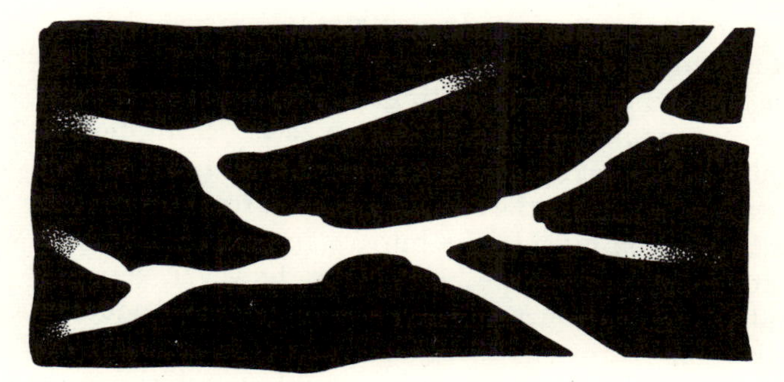

FIG. 186. Top view of mucus-covered burrows of *Callianassa*. The passages are wider at bifurcations. (Actual size × 0·3)

inhabitant to turn in the tube (see Seilacher 1954a, 1956b; MacGinitie 1934, 1949). It has not yet been clarified whether and to what degree the walls of the dwelling burrow are protected from collapsing by a coat of mucus. Lutze (1938) stated that there is no mucous lining and that the stability of the walls is achieved in a different way. According to him the burrow walls are reinforced by pressing mud lumps against them, producing a kind of 'roughcast' lining; the small plates of the third pereiopods serve as trowels. *Callianassa pestae*, burrowing in loose mixtures of clay and sand, on the other hand, is thought to coat the walls with mucus from the mucous glands of the pleopods. Schellenberg (1928) says that probably all these burrowers mix their building material to some degree with mucus, more in sandy sediment than in mud. The animals frequently leave their burrows for distant swimming excursions, especially at night. Often they do not return from such trips; the sea floor therefore abounds with abandoned burrows. The structures either collapse or they are soon filled by sediment which can be stratified in the burrows, the layers being discontinuous with those of the sediment into which the burrows were dug.

If the burrow is inhabited the owner removes penetrating sediment to the outside as long as he can; material that cannot be taken out is flattened on to the old wall, making many walls multilayered. The new coats lie on each other like scales. Bachmeyer & Papp (1951) described similar scale-like and cone-in-cone arrangements from fossil burrow walls in the French Jura and the Schlier formation of Austria. This does not mean that the described burrows were made by *Callianassa*. Reineck found similar formations which he calls 'Räumauskleidung' made by *Nereis* and many other recent burrowing animals. Many different groups of animals make wall structures from sediment that penetrates the burrows after their completion.

Only occasionally does one find fossil remnants of *Callianassa*, mainly the large palmae of the claws. In contrast to the rest of the armour they contain a fair amount of calcium carbonate. Probably all preserved parts of *Callianassa* are shed in moulting. Unlike the lobster which leads a preventive war against members of the species before moulting in order to depopulate the territory at this time (Schäfer 1951a), *Callianassa* retires for that period into the relatively safe burrow. The tender moulting exuvia is later taken outside the burrow together with other waste and disintegrates completely. The heavy claw, however, separates from the exuvia and is left in the burrow; it is too heavy and rounded to be grasped. Like many other burrow-dwelling animals *Callianassa* does not die in the burrow but leaves it when it becomes unable to whirl enough breathing water through the tubes. Thus dead animals are rarely found in the burrow, and their claws are embedded outside the dwelling, if at all.

Other burrowing thalassinids also have compact, elongated bodies, their clawed limbs have strong palmae, and the dactyli of these limbs are made for digging. The other pereiopods are wide and set with bristles. Like *Callianassa* they build large systems of burrows in the sandy mud of the sea floor. How the walls of the dwelling burrows are hardened is unknown. *Calocaris macandreae* and *C. coronatus* live in soft mud, *Upogebia litoralis* which is up to 10 cm long and *U. deltaura* live in more sandy mud. *U. deltaura* digs wide burrows, and is common in some parts of the Heligoland trough where some bottom samples contain up to five specimens (Caspers 1938). The conditions for the preservation of their frequent and extensive burrows are similar to those of *Callianassa*.

Anguilla. Many teleosts spend a large part of their life hidden in caves or tunnels. These fish are found mainly on rocky sea floors and in calcareous reefs, but not in sand or mud where a home burrow must be actively constructed and hardened if it is to persist in loose sediment. However, the European river eel *Anguilla anguilla* builds dwelling burrows. This vagile, hemi-endobiontic teleost migrates from the sea to rivers and back again during its life cycle, and is frequent near the North Sea coasts.

The eel moves exclusively by undulating, whether in the open water or in the sediment. As was mentioned in the section on this mode of locomotion (D I (*a*) (iii)) it can be considered as an especially suitable adaptation to life in or on the sediment. The elongated, widely undulating body can be pressed over its whole length against a relatively solid medium. On the other hand, the sculling action in which almost only the tail is active is most advantageous in open water. Thus this kind of movement is used by the majority of the fast-swimming teleosts. *Anguilla* reverts to undulating, the original mode of locomotion for vertebrates (e.g. *Branchiostoma*), as a secondary adaptation to life in the sediment.

The eel cannot work itself slowly into the sediment the way an organism capable of peristaltic movements can. Water offers much less resistance to an undulating body than sediment; thus an eel buried only slightly in sediment and with its rear still in water could not possibly advance. The eel 'throws' itself into sediment; it swims very fast in open water, throws the head obliquely downward and pushes it into the ground, whipping violently with the posterior part of the body. In a few moments the animal disappears completely in the sediment. If it is muddy, an obliquely downward-directed hole remains, marked at one side by the trace of the beating tail. In sand only a disturbed spot remains on an otherwise traceless surface. After a few minutes the head of the eel reappears. As a first sign, the surface of the sediment slowly breaks as the earth does above a surfacing mole; but never does the head reappear in the continuation of the line along which the eel entered, rather at approximately a right angle to this direction.

Once the eel has forcefully advanced a specific length of its body into the sediment it makes a right-angle turn with its head; then a single wave of contraction runs from the front to the back along one side of the body. In this way, the body always pushes against the same part of the burrow wall, the outward side of the sharp bend, and advances the anterior part of the body, which is held straight, through the sediment. When the tail has passed through the bend, several low-amplitude undulating movements pull the body forward and upward until the head pierces the surface. Thus the length of the digging trail does not much exceed the length of the body.

While the animal passes through the sediment, the scale-free skin constantly secretes abundant mucus which facilitates penetration. The mucus-secreting organs are goblet cells, serous glandular cells, and probably also unicellular organs which are a specialised kind of basal cell of the epidermis. Such a club-shaped cell is discarded as a whole to the outside after passing through the epidermis.

In muddy sediment a heavy coat of mucus keeps the burrow walls from collapsing for some time. A gentle but constant undulating movement performed while the animal rests in the burrow helps to consolidate the walls. Whenever the eel retracts into its burrow, the body undulates with a greater amplitude than during rest. These movements widen the burrow in several places, but the mucus effectively preserves the burrow even where it is wider. Eventually the walls become hard enough for the burrow to remain preserved for some time after the fish has abandoned it. Sediment entering from both sides then fills the burrow, and the filling can be bedded.

Bony fishes inhale breathing water through the mouth and expel it through the opercular slits. In *Anguilla* the pressure of the discharged water is so high that it forces its way through the ceiling of the burrow to the surface. An occupied burrow of an eel can therefore be recognised by the two openings of different size lying close together; small sediment clouds or veil-like shreds of mucus with adhering mud particles are driven out of the smaller hole in rhythm with the breathing. When the eel is disturbed it retracts by folding its body within the tube. Frequently the animal overstretches the ceiling wall of the shallow burrow and arches it upward; this causes extension cracks at the surface of the sediment (Schäfer 1954e); and the cracks can be preserved if they are subsequently filled with sediment.

The eel can build open dwelling burrows because of its abundant secretion of mucus. In pure sand such mucus-treatment of the walls does not always suffice to keep the tube open; it collapses with movement of the eel, and the body becomes tightly surrounded by sand. In that case the fish must leave at least the tip of the head above the sand so that it can breathe water through the mouth. The interstices of the sand permit disposal of the used breathing water which leaves the opercular slits.

Anguilla can rise to the surface through thick layers of sediment by undulating movements. Even where 1 to 2 m thick sediment covers form within a short time, as happens in the shallow sea, eels can penetrate them (Schäfer 1956b). In so doing the eels make burrowing trails; where the trails perforate beds, the edges of the beds are bent asymmetrically.

The agglutinated burrow

A single mucus-application may be sufficient to harden a digging trail, but this technique does not permit a dwelling tube to be extended above the surface of the sediment, or one to be built either lying entirely on the surface or standing freely upright. In these cases the benthonic organism must build structures as bricklayers do, grain by grain. The course of the burrow is not simply prescribed by the digging trail of the builder. Some species follow a definite 'blueprint' for the structure without the slightest deviation in carrying out the work; others are able to alter the blueprint within reasonable limits according to local circumstances. This is also in contrast to mucous structures which are quite uniform in building material and shape, regardless of the species by which they are built; so they rarely reveal the identity of the builders.

Masonry structures often indicate activities that are vital for the animals. Thus certain modes of feeding are linked to certain architectural features. In such cases the animals never leave the dwelling without strong external cause. Frequently they have no effective means of locomotion outside their self-made structure. Organisms that are so highly dependent on dwelling structures and their functions thus become sessile epibionts. This is not so with builders of mucous burrows. They can all move about outside their burrows, and that is where most of them feed. What they require during their stay in the sediment is simply an open tube suitable for respiration.

Polydora. Polychaete worms of the genus *Polydora* belonging to the spionids have such varied building techniques that it is difficult to classify them. Members of this genus build mucous burrows, construct agglutinated tubes, and also drill deep cavities into hard rock. The most conspicuous and therefore the best known are the U-shaped boreholes of *Polydora ciliata* in lamellibranch and gastropod shells or in exposed limestone. McIntosh (1886), Söderström (1920, 1923), Richter (1924c), Hofker (1930), and Linke (1939) also reported occurrences of *Polydora* in sediment. Hempel (1957a, b and 1960) investigated the tube construction and drilling activity of *Polydora quadrilobata*, *P. ligni*, *P. redeki*, and *P. ciliata*. Occasionally on the tidal flats, and also below the low-water line to a depth of 60 m (in all parts of the North Sea), muddy-sandy areas and zones of pure, fine-grained sand are densely populated by these worms. *Polydora quadrilobata* and *P. ligni* settle mainly in fine sand or mud in the deep zones near tidal flats. *Polydora redeki* occurs frequently in hard clay.

Where *Polydora ciliata* has settled the sediments show numerous fine tubes in cross-section, barely 1 mm thick, which can be recognised distinctly as U-shaped burrows where they lie far enough apart. Often the U-shape is not developed if the tubes lie closely together. The worm builds a vertical U-shaped tube with two openings if this is possible. The U-bends are wide, the two branches rarely parallel. The same holds for U-shaped burrows of the worm in lamellibranch valves. According to Linke (1939) an inhabited U-tube can be 8 cm deep; most tubes, however, only reach a depth of 3 to 5 cm. After erosion of several millimetres the ends of tubes appear and project freely above the surface, showing that the walls are well consolidated.

In captivity the animals like to build at the wall of the aquarium. It is therefore easy to observe their behaviour (Fig. 187). Just above the surface of a dense settlement a forest of tentacles can be seen because each animal sits with its head at the level of the tube opening and stirs the water with its tentacles. The tentacles are always held forward in open water and also in the tube after it is completed; there they are active like flexible arms, arranging and transporting particles. However, if the animal crawls head-forward through sediment or if it builds a new tube in hard sediment, the muscular mouth lips take the lead, and the tentacles are folded back. In this way the animal pushes the particles aside and ahead. Once a small space has been opened ahead, the tentacles move forward again and start scraping away the falling sand. The burrow is widened when the first four body segments suddenly swell and two hand-like, bristled organs on the fifth segment press the sediment outward by tilting movements; thus they make room for the posterior part of the body which carries ciliated gills on the dorsal side. The bundles of bristles not only push and press soft sediment but can also scrape and chisel clay and hard calcareous substrates.

The walls of the dwelling structure are built in a very strange way. When the tentacles are stretched forward and are digging, they are capable of transporting sediment grains along a ciliated groove following the length of the tentacles. Fig. 187 shows a tentacle with its transporting groove (see also Hempel 1957a, b, and 1960). After the tentacle loosens a particle it glides from the tentacle tip to the mouth. The lips can be everted and grasp the grains unless they are too large. If a free-standing chimney is to be built at the mouth of the hole the lips press the mucus-coated sand grain on the rim of the chimney. Frequently sand grains pass through the digestive tract and end up in a row in the terminal section where they can be observed because that part of the body is transparent.

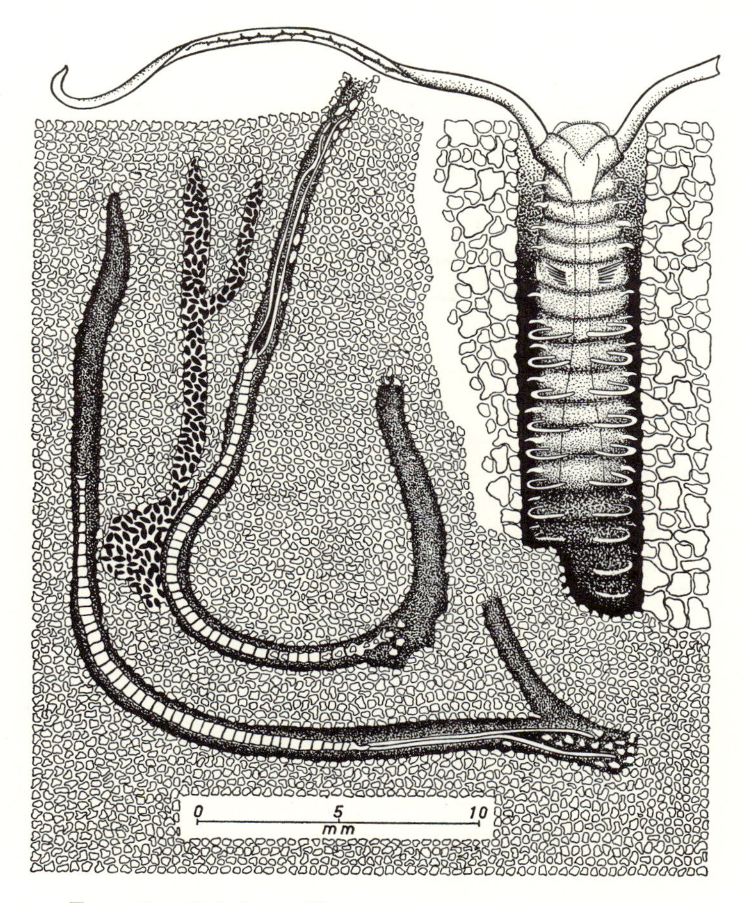

FIG. 187. *Polydora ciliata* burrowing in sediment. Left: mucus-covered burrows being built by the animals; one branch is filled with faecal pellets; right: enlarged view of anterior portion of the worm

The sand grains leave the intestine through a funnel-shaped, flexible anal process which pushes them sideways into the tube wall in such a way that their shapes and sizes fit together. To date it is unknown whether sediment grains undergoing this passage through the body are coated with mucus in the intestines. The process is continuous, the tentacles palpate the tube walls untiringly, collect loose particles or break off parts of the wall which are not firmly set, while the mouthward transport of collected sand grains goes on at the same time. Frequently, the direction of transport is inverted, and sand grains are moved toward the tentacle tip to be dumped into a dead section of the tube; particles can even be moved in both directions simultaneously on the same tentacle but along different paths.

Much of the waste material finally goes through the body again, except for grains that are too large, and it is coated with mucus and fitted into the wall. During this work the worm constantly moves up and down in its tube, it also turns on itself in the burrow so that it can work with the tentacles on the lowermost part of the structure and reach the upper section with the anal funnel. Where the tube is too narrow, the fifth segment widens the passage with its two bristle organs. Finally, glandular secretions from the eighth to twelfth segments form a mucous lining round the tube. This often becomes so hard that it holds the tube together as a whole even when it is washed out of the sediment.

A current of water created by the cilia of the gills flows through the tube as soon as the U-course of the dwelling tube is completed. The water is driven along the worm body from front to back. Both tube openings may serve as either inhalant or exhalant openings since the worm frequently turns in the tube. The oval faecal pellets are not carried to the surface but are pressed into dead-end sections of the burrow which are sealed off after being filled.

Since the worms constantly rearrange the sediment grains the original texture of the sediment in a dense settlement is rapidly destroyed and bedding is obliterated. If a thin layer of blue-dyed sand is placed on a *Polydora* settlement the coloured sand grains are, after a few days, evenly distributed over the entire thickness of the settlement, except for small remnants. The layers remain better preserved if new thin layers are added at short intervals and the animals have to move upward rapidly with sedimentation. Then the time spent in one level is insufficient for complete rearrangement. If 1 or 2 cm thick layers are added at once they are only gradually obliterated.

If there is large-scale planar erosion *Polydora* can shift the U-shaped tube downward, or it can move it upon the onset of new sedimentation. The animal closes the old tube and uses a new one from then on. Unlike Linke (1939) I have not noticed continuous up or down shifting of the U-bends. If the new sedimentation is too thick *Polydora* usually abandons the old burrow, moves up to the new surface without consolidating its digging trail and builds a new dwelling tube (Fig. 188). Where the rate of sedimentation is high, several horizons of U-shaped tubes lie above each other, each separated from the next by simple, more or less vertical digging trails. It has already been mentioned in another context that *Polydora* does produce U-shaped tubes with continuous bends in hard rock (Cretaceous chalk of Heligoland) or in calcareous lamellibranch valves, but using the building technique of the infilled flat pouch. The two types of construction, although used by the same species, should be kept strictly

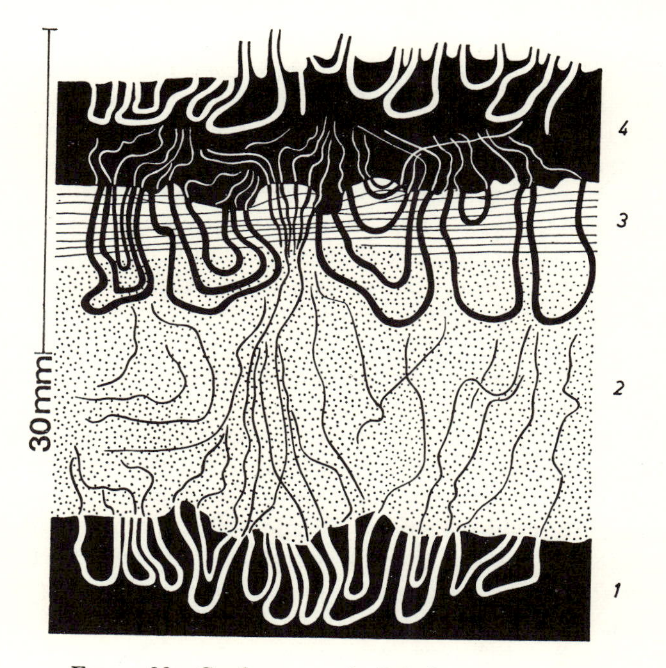

FIG. 188. Settlement of *Polydora* successively covered by several layers of new sediment. (1) originally settled layer; (2) rapid sedimentation forces the animals to leave the settlement, they build new burrows at the new surface; (3) slow sedimentation permits the worms to extend existing burrows gradually to the slowly rising surface; (4) renewed rapid sedimentation forces the animals to abandon these burrows and to build new ones at the present surface; burrow openings are extended above the present surface by chimney-shaped elevated structures

apart. Recent U-shaped pouch burrows made by *Polydora* with continuously shifted bends and fossil equivalents were described by Richter (1924, pp. 135 ff).

The dwelling burrows of *Polydora* in unconsolidated sediment can be built in pure, fine sands, muddy sands, consolidated mud and also in peat. In all these cases the building technique is probably the same, but observations have not yet been made in all sediment types. In mud and muddy sand the building work of the worm sorts the grain sizes; the burrow walls are almost exclusively made of whatever small sand grains are available whereas mud particles are not used. The rearranged building material of the walls thus differs distinctly from the surrounding material, and the tubes must be considered as agglutinated tubes.

Lanice. The polychaete worm *Lanice conchilega* (Terebellidae) is common in the southern North Sea. It settles individually or in large numbers on the tidal flats, in the permanently immersed channels of the littoral areas, and also frequently at the bottom of the open sea to a depth of 50 m. Its dwelling tubes have even been found at depths over 300 m (Wesenberg-Lund 1951). Their wide distribution and easy adaptability to laboratory conditions have made it possible for many investigators to study the animals in detail (Eales 1950, Ehlers 1875, König 1948, Linke 1939, Seilacher 1951, Watson 1890, Wilson 1950, Wohlenberg 1937). Most complete are Ziegelmeier's (1952) studies and his critical review of earlier results. His findings supplemented by some later investigations are the basis of the description that follows.

With some caution one can extract *Lanice* from its 30 to 40 cm deep dwelling tube. The 8 to 10 cm long body is distinctly segmented, the first eighteen to twenty segments bear laterally short, mobile parapodia, provided with bundles of strong bristles which can be protruded or retracted. Moving like legs they carry the worm up and down in its tube (chimney climbing). At the anterior end is a strong, spoon-shaped prostomium, or head lobe, that can be folded. A collar-shaped buccal segment surrounds the very extensible mouth. Numerous thread-like, contractile tentacles sit on a thick, muscular ring that surrounds the base of the prostomium. The tentacles near the mouth are somewhat shorter and thicker than the dorsal ones. The tree-like, branched, red gills sit in pairs on the dorsal side of the second, third, and fourth segments. On the opposite, ventral side, beginning at the third segment, lie eighteen to twenty flat, slightly bulging, bright red ventral plates; their size decreases from the most anterior plate toward the posterior end. Pairs of glands have ducts which lead to each of the plates; the secretion hardens the walls of the digging trail by cementing sand grains.

Digging a *Lanice* dwelling tube out of the sediment one usually recognises three distinct sections, the buried shaft, the protruding trunk, and on it the fringed crown. While the worm digs, making the buried shaft of the tube, it rotates about its long axis, secretes mucus which penetrates the grains of the sediment *in situ* and makes the wall self-supporting. In addition, a homogeneous mucous layer is placed on the inside of the walls. The result is a hard dwelling tube, hard enough to hold together if it is washed out of the sediment, and elastic enough to resist destruction for a long time when transported along the sea floor or over the sand of a beach. If the tube dries out, however, the cement becomes brittle and tears or splinters easily; the smallest movement causes the tube to disintegrate into a loose pile of sand and shell fragments. (For storage in collections the tubes must be kept moist.)

The free-standing trunk section of the tube that projects above the surface is constructed differently. The worm builds block by block, using small quartz grains and also small shell fragments. Fragments of bivalves and gastropod shells, often complete *Hydrobia* shells, sea urchin spines and Foraminifera tests are used as building material. The worm collects all these particles with its long tentacles, either while the objects drift by

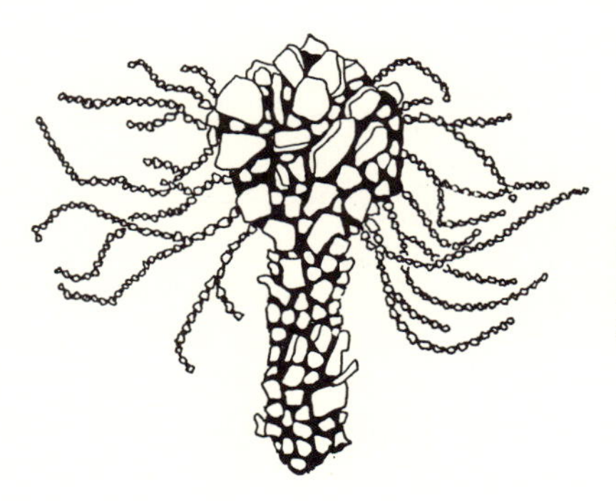

FIG. 189. Trunk and fringed crown of the elevated dwelling tube of *Lanice conchilega*. (Actual size × 0·5)

with a bottom current, or it picks them up from the bottom. The tentacles are provided with many mucous glands and grasp the building materials with two longitudinal ridges that face each other so that they form a groove between them. Once the particle is wedged into the groove, cilia in the groove transport the particle toward the mouth. Peristaltic movements and also rapid contractions of the tentacles supplement the work of the cilia. Now the smaller tentacles around the mouth grasp the building block and finally pass it to the lobe-shaped prostomium which can grasp and collect a number of building stones. Probably glands at the base of the prostomium and on the lower lip of the mouth coat the building material with mucus. Thus a package of mucus-coated building particles is made and then laid on the edge of the wall by the flexible prostomium and there distributed and smoothed as with a trowel. The free-standing trunk sections are usually 2 to 4 cm high. If the particles are large they are laid on one by one. Often they are fragments of barnacle armour plates or splinters of lamellibranch shells. Such flat objects are laid with their flat side horizontal so that the edges of the fragments project more or less outward. This gives the tube a scaly and bristly appearance (Fig. 189), conspicuously different from others built mainly of sand for lack of shell fragments in the environment.

The fringed crown is attached to the free edges of the trunk. The animal collects grains of building material in the same way as before. The prostomium arranges them in a single row, coats them with mucus, and thus forms a rigid little stick of sand which is attached at one end to the structure, projecting freely outward. One sediment stick after the other is formed and placed with those completed before, until an entire bouquet of thin, often branched bundles of fringes surrounds the top of the trunk.

The permanently elastic, buried mucous tube forms the base for the free-standing upper part of the dwelling tube, the trunk and the fringed crown, and supports them. This explains why the shaft must be so markedly different from the simple mucous burrows of the nereids or other mucous burrowers. It has a much stronger, reinforced and secretion-lined wall. Without this strong base in the sediment, the upper part of the dwelling tube with the fringed crown would collapse and become useless with the smallest shift or erosion of the sediment.

As mentioned above, worms that build mucous burrows are not completely tied to their burrows; many are predators or grazing animals and regularly leave their burrow in search of food, and the only function of the burrow is that of protection. *Lanice* is more closely tied to its home because the erect tube has a feeding function as well as a protective function. The fringed crown serves as a net in whose mesh phytoplankton and zooplankton are caught. The tentacles harvest the food off the fringes and bring it to the mouth. This explains why the branches of the fringed crown are always directed like a fan against the direction of a constant current (Seilacher 1951). In many cases the wall at the top of the tube is widened into solid plates, armoured with lamellibranch shell fragments, both on the up-current and the down-current side. The individual fringes sit on the edges of these plates. The tentacles can wave over the fringes in the space between the two plates without the head of the worm protruding from the tube.

Lanice, like several other tube dwellers, need not drive a water current through its tube. It has its gills at the head and can take advantage of the continuous exchange of water around its free-standing tube. For the discharge of faeces the animal bends the body into a hairpin curve, pushes the posterior end above the tube opening past the head, and squirts the faeces from the terminal intestine (Ziegelmeier 1952, p. 116, photograph). The sexual products are discharged from the tube opening by peristaltic movements. Therefore *Lanice* has no need for U-shaped tubes; a single vertical tube with a single opening at the top is enough. All the same, U- and even W-shaped tubes are frequently found according to Seilacher (1951). Seilacher, with König (1948), considered the

U-shaped tube the normal burrow form necessary for respiration. The W-shaped tubes, he thinks, are a consequence of the growth of the worm body; after the young worm has inhabited a first thin U-tube and later grown thicker and larger, it breaks through the wall of the burrow near the bend and makes a wider tube parallel to the old one to the surface, and provides it with a new trunk and fringed crown. Presumably, other factors (plugging, destruction, washing out) can cause the worm to break out from the old tube and to build a new branch. Ziegelmeier says that special living conditions cause the worm to construct U-shaped tubes (lack of food, competition for food, oxygen-poor water near the sea floor, or overabundant growth of byssal mesh of *Aloidis gibba* on the surface of the sediment). Whatever the number of constructed tubes, only one functions at any one time. Single tubes as well as the W- and U-shaped tubes with the subterranean connections are easily preserved in the sediment.

In one instance a tidal flat at the south coast of Mellum island, consisting of shell breccia mixed with sand and mud, was washed out deeply enough to expose dwelling tubes of *Lanice* forming a dense, multibranched network of vertical single tubes and W-shaped tubes closely knit together (Fig. 190). Apparently, many factors caused the great variety in the construction and frequent alteration and reconstruction of the tubes: numerous lamellibranch valves blocked some of the digging trails, new sedimentation and erosion, and probably also slumping, made certain tubes uninhabitable.

Ziegelmeier (1952) noted that *Lanice* prefers to settle in coarse sands and shell breccia. They are typical of areas with strong currents, and the worm must be able to cope with considerable shifts in the sediment and with new sedimentation. Rapid sedimentation at the place of settlement always inactivates the catching mechanism for plankton, the fringed crown. In such cases the animal must penetrate to the surface of the sediment, construct a new mucous tube section in the sediment, build a structure above the ground and provide it with the fringed crown.

Experimental sedimentation with dyed sand in an aquarium has shown that the new sediment layers abut at a right angle with the upper tube section or heap up against it. Above the fringed crown they form uninterrupted layers which were allowed to become several centimetres thick. The worm leaves its burrow and digs a cavity upward with its flexible, strong prostomium. Peristaltic movements push the body into this cavity. To supplement the mechanical digging, the worm directs strong water jets against the ceiling of sand by a so far unknown mechanism, in a fashion which Seilacher compares with the technique of flush-drilling. Possibly the peristaltic swellings of the body that run from the posterior end forward along the body in the rigid tube cause the jets; they could not be due to ciliated appendages, which many other polychaetes have, because *Lanice* does not have them. The worm tries to reach the new surface of the sediment as rapidly as possible. Beds perforated on the

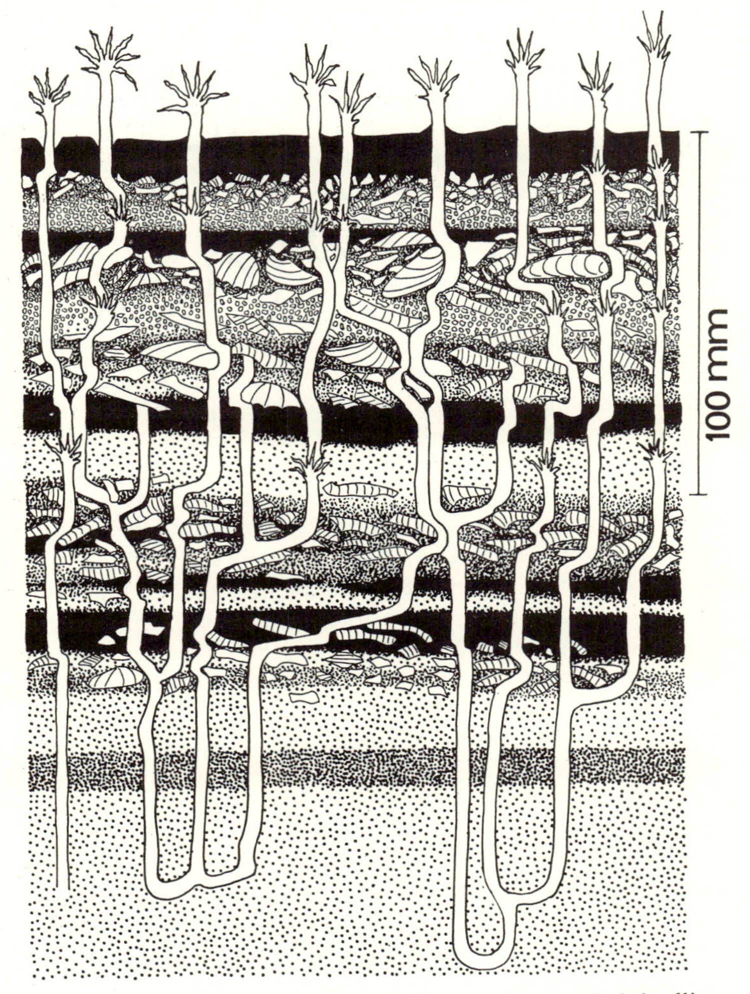

FIG. 190. Many-branched and frequently extended dwelling tubes of *Lanice conchilega*. The sediments are bedded muds, sands and shell deposits; periods of slow or no sedimentation are marked by fringed crowns

way sag slightly. They become hardened by the mucus in this position when the worm begins to secrete immediately above the fringed crown which thus forms a ring around an otherwise continuous tube. At the new surface the worm builds its free-standing tube-trunk block by block and constructs another fringed crown at the top. Whatever the number of episodes of new sedimentation, each time the layers below the fringed crown are at right angles with the tubes or slightly heaped against them, and above the crown they sag downward. Each time this alternation is distinctly marked by a fringed crown. *Lanice* tubes are frequently found in the swash marks of beaches; often they are multistoried with several fringed crowns indicating the number and thicknesses of sediment covers with which the builder of the tube had to cope.

With its unique body structure and behaviour, one should think that *Lanice* might adhere strictly to a single building technique. All the more remarkable, therefore, is a report by C. D. Müller (1956) which described an occurrence of *Lanice* on wooden posts which had been treated with a protective paint and were to be examined for resistance against *Teredo* growth. The posts were suspended for nine months (from December to September) in the harbour area of Norderney without touching the sea floor. After this period they were found to be covered by a dense, more or less vertical network of *Lanice* tubes, up to 10 cm thick. The number of the inhabited tubes ran into the thousands. Not a single U-bend could be discovered. The tubes were almost exclusively built from fine sand. The network was interwoven with *Sertularia* sticks and networks of byssal threads from *Mytilus*, and to it adhered various detritus and washed-in fragments of lamellibranch valves. Settling *Lanice* larvae apparently favoured wood surfaces that had become rough by being covered with basal plates of balanids. The shaft portion of the tube that normally sits in the sediment and is simply made by lining a digging trail with mucus, was in this case built in open water with the use of sand caught grain by grain. Another instance of the versatility of these animals is their behaviour if they touch the glass wall of an aquarium when they start building; in that case they build the tube leaning against the wall and construct the fringed crowns at a much greater height above the sediment than is usually the case. The author has seen especially tall bundles and meshes of *Lanice* tubes between the 'sand coral' structures of *Sabellaria spinulosa* in the channel of Varel in Jade Bay, described by Schuster (1952); here the *Lanice* tubes lean loosely upon the coral structures instead of rising freely above the surface as they generally do. On the flat sea floor there is usually no occasion to use this ability to lengthen the trunk in proportion to the buried shaft with the aid of a foreign support.

Sabellaria. This polychaete worm builds dwelling tubes in densely crowded, firmly connected colonies and creates extensive, resistant reefs on the sea floor. Occasionally these reefs cover several square kilometres and are inhabited by millions of individuals. Groups of 500 to 2000 larvae assemble and jointly begin to construct their free-standing structures which have a remote similarity with the arrangement of the calcareous skeletons of the tropical organ-pipe coral *Tubipora*. This similarity is the justification for the designation 'sand corals'. Interest in the structures and building techniques of *Sabellaria* arose when their tubes were compared by geologists with the Cambrian 'Pipe rock' quartzites of Scotland and Sweden and with the similar Lower Devonian quartzites of the Eifel mountains at Neroth. Thus the first reports from the North Sea about the life habits and the rock-forming 'sand coral' structures of this worm came from geologists and palaeontologists (Meyn 1859; Richter 1920a, b, 1921, 1927a) rather than from zoologists.

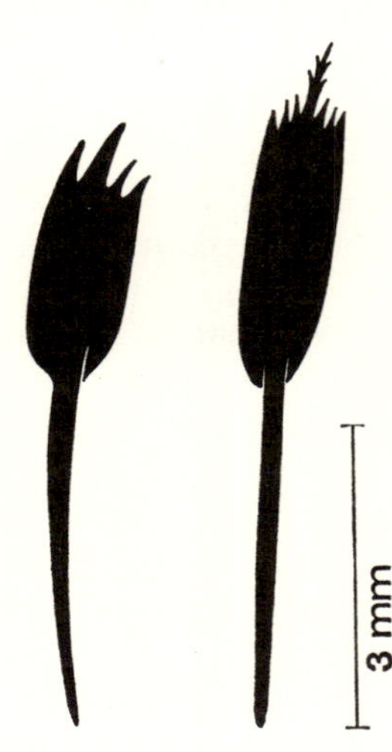

FIG. 191. Paleae from the heads of *Sabellaria alveolata* (left) and *S. spinulosa* (right).

Within a short time these animals can turn large quantities of sand into specific and preservable textures; according to Linke (1951) a single colony off Norderney island used about 800 to 1000 cubic metres of sand in six months. In so doing *Sabellaria* also changes the habitat for all other inhabitants of the area, transforming a sea floor of loose sand into a ragged reef, a new biotope inhabited by a specific fauna.

Two species of the genus *Sabellaria*, *spinulosa* and *alveolata*, live in the North Sea (according to Friedrich 1938). The two species differ mainly in the shape of the outer paleae (spoon-shaped, golden iridescent bristles in the two-piece head plate, Fig. 191). According to the older literature, only *S. alveolata* shows the organ-pipe growth of vertical, more or less elongated tubes cohering along their entire length (Saint-Joseph 1894, Richter 1920a, b, Abel 1935); *S. spinulosa*, on the other hand, is reported to produce disorderly arranged and often knotty tubes growing on lamellibranch valves ('Austerntod' or death of oysters in the

German vernacular). All recent observations from the southern North Sea relate to *S. spinulosa* (Schwarz 1932a, Schäfer 1949, Linke 1951, Caspers 1950a, and personal communications by E. Ziegelmeier). These investigators established that this one species can produce both organ-pipe growth and disorderly, knotty growth. The factors determining the differences in the construction are different living conditions, not different behaviour of separate species. The old descriptions, some quite detailed, about the kinds of structures of *S. alveolata* do in fact relate to *S. spinulosa*. According to Wilson (1950) *S. alveolata* builds honeycomb-like tubes from sand grains; in Britain, one therefore calls these worms, very

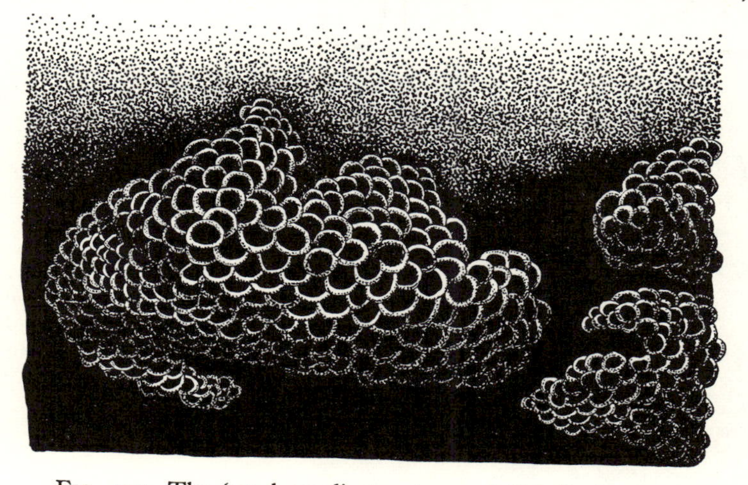

FIG. 192. The 'sand coral' structures of *Sabellaria alveolata* (after a photograph by E. Ziegelmeier). (Actual size)

appropriately, 'honeycomb worms'. When neighbouring worms are building their structures, they construct only one common wall between them, the outside of one tube serving as the inside of the neighbouring one. If a worm joins its structure to an earlier bundle of tubes it only constructs a wall in a half-cylinder, fitting its dwelling into the corner between two tubes and using their walls (Fig. 192). *S. alveolata* occurs on the English coasts on bedrock, often in areas just below the low-water line that are not much exposed to surf. Upon careful observation the two species can easily be distinguished. Since it builds mainly on rocks in zones with little sedimentation *S. alveolata* is less important in this context than *S. spinulosa*.

Fig. 193 is a longitudinal cross-section through *S. spinulosa* showing all the anatomical features one has to know to understand the animals' activities and behaviour. The worm has to perform several vital activities while it is sitting in its circular, cylindrical tube with a single opening.

(1) It must maintain a constant flow of water through the tube which it needs for breathing and for carrying faeces and sexual products out of the tube. The current is produced by numerous cilia that sit in spiral-shaped rows on the dorsal cirrae of the parapodia. The water flows

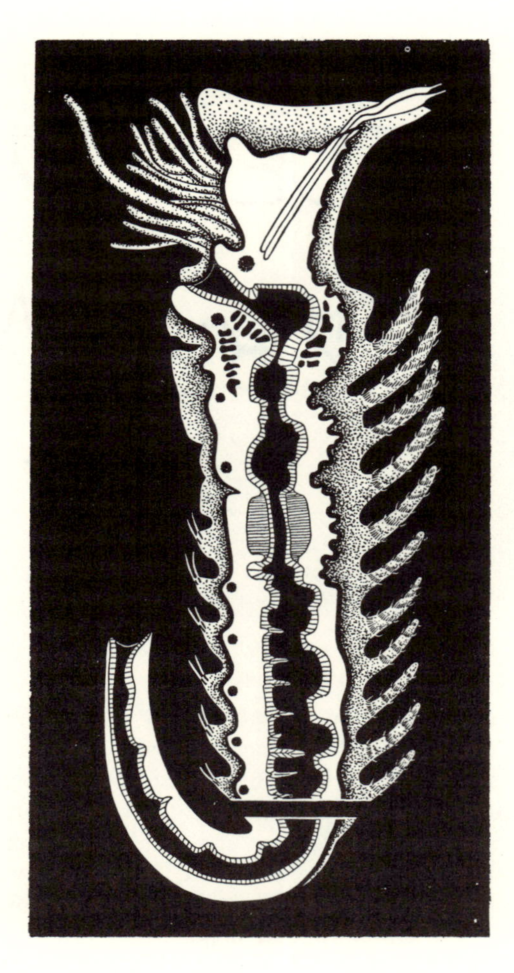

FIG. 193. Longitudinal section through *Sabellaria spinulosa*. Two paleae visible on the head; above the mouth tentacles; chain of ganglia along the ventral side; mucous glands near the oesophagus; dorsal appendages, spirally aligned, act as gills. (Actual size Schematic drawing)

downward along the dorsal side of the worm to the pygidium that is bent forward. There the water current turns, flows up and leaves the tube on the ventral side. The fresh, descending water at the dorsal side provides oxygen to the dorsal cirrae which function as breathing organs. On its ascending course the water carries the faeces from the anus in the pygidium to the outside. When the wall of the body cavity ruptures and liberates the sexual products they can freely flow out along the same course as the faeces (Fig. 194).

(2) The worm must move in the tube with suitable locomotory organs, the bristles of the three pairs of parapodia in the section below the head. The structure and the mechanism of these locomotory organs has been described in the section on chimney climbing (D I (*a*) (xi)).

(3) The worm must be able to procure plankton from the opening of the tube. It does this by means of the tentacles, arranged in groups on each of the two prostomium lobes above the mouth. The tentacles are provided with cilia that whirl the plankton toward the mouth (Johansson

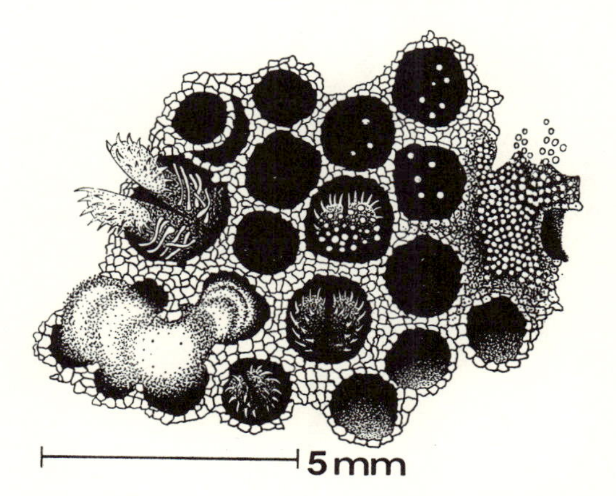

FIG. 194. Top view of colony of *Sabellaria*. In several tubes the inhabitants are more or less clearly visible, depending on how far up each sits in its tube; bottom left: a cloud of sperms being ejected; top right: isolated eggs float up from the tubes in the ascending current of spent breathing water; far right: accumulation of eggs

1927). While the worm is catching plankton it protrudes no further from the tube than will allow it still to hold on to its tube with the locomotory organs. *Sabellaria* can only be pulled from its tube if it is broken in such a way that the head and the portion with the bristled parapodia lie free. If the tube is broken off lower down and the posterior end of the worm protrudes from it, the worm lets itself be torn in two rather than let go, anchoring itself by bristles pressed against the tube wall.

When the worm stretches its anterior end out of the tube the mouth lies at the bottom of a funnel formed by the two parts of the prostomium. Only on the ventral side are the two prostomium lobes split down to the mouth. They can be bent far apart and slightly backward. As a result, the tentacles that sit on the edges of the lobes can be spread further apart and enlarge considerably the size of the catchment area of the funnel. Strong, lateral muscle strands that run from below the head region upward to below the rings of paleae, bend the prostomium lobes. When the worm retracts into its tube the prostomium halves fold together. The two half-circles of the paleae on the outer ends of the prostomium move toward each other until they form a single, closed disk, forming a lid for the tube as a kind of spiny operculum (Fig. 195).

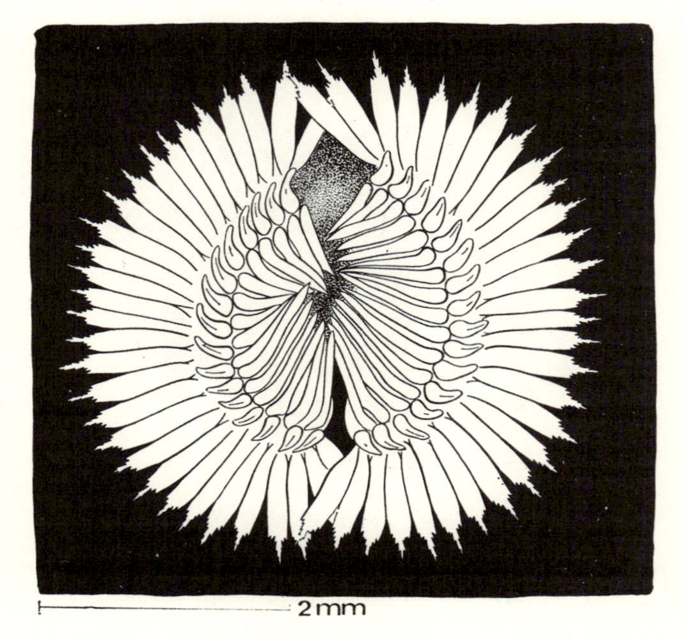

FIG. 195. The triple fans of paleae on the head of *Sabellaria spinulosa*.

(4) The worm must be capable of building and maintaining the tube. It must catch sand grains, coat them with mucus, and lay them as building blocks. Two rows of contractile palps, provided with tactile sensory organs and probably also with cement-secreting glands, are the main organs for collecting sand. They grasp a sand grain that dances in the current near the tube opening and bring it to the mouth. The lip ridges are very flexible and can take the grain from the palps. In a wide portion of the mouth cavity the sand grain is coated with mucus flowing from a large gland on the ventral side of the oesophagus (Fig. 193), while the musculature of the mouth turns and twists the grains. Then the lips grasp it again and transport it to where it is to be used. The grain is then pressed on to

its new substrate where the sticky mucus hardens in the sea water. The inside of the inhabited tube must constantly be lined with new mucus. The worm lets its lips glide along the walls of the tube each time it ascends or descends and smears new mucus on to the surface wherever there is need for it. The tubes project above the sea floor. They therefore need a stable base unless the whole structure is to be in danger; thus even deeper sections of the tube which no longer serve for dwelling are kept in good condition. Although the worm is less than 2 cm long, it maintains a 50 cm high tube in some cases.

The locomotory organs of *Sabellaria* are so perfectly adapted in shape and function to moving in the self-built tube that they are incapable of moving the worm outside. An adult *Sabellaria* is therefore nearly sessile. As with many sessile organisms, the sexual products of *Sabellaria* are simply discharged into the surrounding water, and there is no copulation. However, due to the large number of released sperms and eggs sufficient larvae come into existence to guarantee the preservation of the species; the probability of successful external fertilisation increases proportionally with the number of sexual products (Fig. 194). Settlement in dense colonies is another means of increasing the number of possible fertilisations. The larvae of one spawning period come to the end of the larval stage at the same time and simultaneously choose the place to build tubes for the adult period of their life.

This ability to choose has also been observed in larvae of many other animal groups as has been demonstrated in experiments (Kühl 1950b, Schäfer 1952a). In all cases the choice is double, one of a suitable substrate for settlement, and one of proximity to members of their own species. An incipient colony is therefore not simply a 'Platzgesellschaft' (literally, community of a place), or a homotype association or synchorium (Deegener 1918), in which each individual chooses the most favourable place for itself and its dwelling structure, and in which a multitude of individuals thus assembles at one place by accident. Rather, in addition to settling on a favourable substrate, the larva seeks intentionally the neighbourhood of members of its species. This serves a direct need, not of the larva but of the future colony which requires many individuals to build their structures so closely together that they touch each other. The physiology of the sensory organs necessary for this twofold choice is not yet clear. Anyway, this choice represents only one of many stages in a complete cycle of successive behaviours from larva to adult. As Richter (1921) said of the Devonian *Sabellarifex*, only one section of this cycle, the sand burrows of the adult worms, can be preserved, and observed by the palaeontologist.

Millions of larvae that have sunk to the sea floor encounter a bank of shell breccia at the bottom of a channel with a strong water current at a depth of 20 m (i.e. not far below the mean low-water line). There they settle in dense groups on individual lamellibranch valves and begin to build tubes. As building material they choose only small sand grains because their mouths are still small. They construct a narrow, barely thread-thick tube which they attach to the shell. The tubes spread in irregular, horizontal curves. When the young tubes touch they continue parallel or, occasionally, cross over one another, but in the end, they form a disorderly network. With the growth of the worms the diameters of the tubes become wider, and the walls consist of larger sand grains. There are extensive *Sabellaria* reefs that consist only of a network of tubes. In the Varel channel (Jade Bay) such a reef has formed within three summers (Schuster 1952). The tubes are built over an old reef of collapsed tubes with parallel growth.

According to Schwarz (1932a) network growth of *Sabellaria* is strictly tied to low density of the animals in a settlement. He submitted that, particularly in deep water, a high density of settlers induces organ-pipe growth with vertical tubes parallel and without intervening spaces. In his opinion, the incoming light causes the demonstrably phototropic animals to grow in this pattern. His interpretation is based on correct observations and on the correct idea. However, there are additional observations that demand modifications of Schwarz's work. The reef at the bottom of the Varel channel has been constantly observed over a period of ten years. The conditions were always the same; yet, the growth of the tubes and their interrelation have undergone such significant changes that the greater or smaller density of settlement combined with, or independently of, the effect of light cannot possibly have been the only cause for the change in structural form. At any rate, the effect of the incoming light is small in this instance because the reef lies at the bottom of a 12 to 15 m deep, sand-carrying tidal channel.

Experiments in the aquarium can be misleading if they employ overly strong stimuli and if the results are uncritically used to explain the behaviour of animals in their natural biotope. In experiments, a certain finite number of stimuli in a few given intensities, and the responses of the animals to these, are taken to be representative of the multiple, infinitely graded stimuli in nature and of the natural reactions to them. It seems more plausible to suspect that social behaviour or mass reaction plays a role in the building habits of *Sabellaria.* That lower animals do have such behavioural traits has been shown by Deegener (1918) and Schäfer (1951a, 1952a, 1956d). How else could it be explained that for many years some reefs do not get beyond the stage of prostrate net growth despite a very high density of settlers, yet that suddenly one spring the new structures join to form 40 cm high organ-pipe structures with the tubes running precisely parallel (Fig. 196)? Furthermore, what else could be the reason for organ-pipe growth continuing for many years in small individual colonies of about a hundred animals, and then, suddenly and on the same reef, more animals join to build a single, large organ-pipe structure, or that, finally, growth on the same reef at some stage becomes disorderly, individual tubes turning away from the parallel array? On

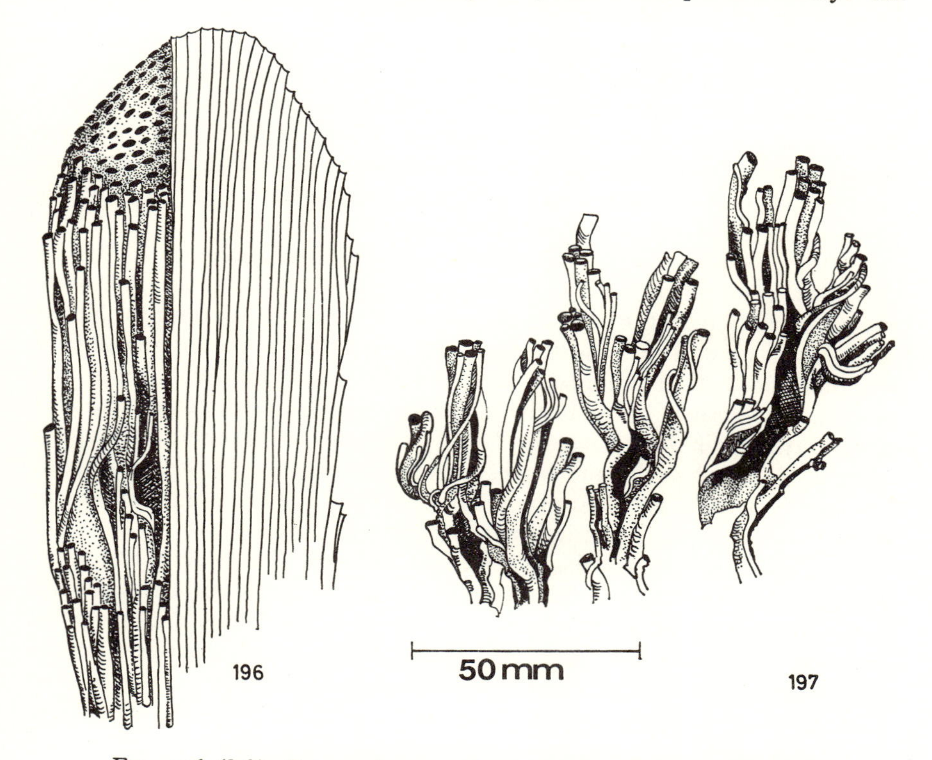

FIG. 196 (*left*). Organ-pipe growth of dwelling tubes of *Sabellaria spinulosa*. Left: external view; right: longitudinal section. The tubes are parallel and attached to each other

FIG. 197 (*right*). Organ-pipe growth of dwelling tubes of *Sabellaria spinulosa* splitting up into small stocks of mutually attached tubes.

other occasions closed organ-pipe structures separate and branch out as though the individual tubes no longer needed mutual support (Fig. 197). All these changes in the building style, affecting an entire colony of millions of animals at the same time, could not possibly be caused simply by illumination or differences in settling density alone; neither can they depend on the sexual development of the animals because sexually mature individuals are found both in network and in organ-pipe structures.

As the tubes separate, the inhabitant also lives an individual life. It does not have any contact with the members of the species. When the tubes join and run parallel, however, the individuals also live in contact with each other. Over the walls their palps and tentacles touch. It can be observed that the building work at the tube opening of each individual animal is adjusted to the building speed of the neighbours. In such a case an individual never pushes its structure far ahead of those of the neighbours although they are mechanically strong enough to stand up unsupported over 4 cm high, even in a strong current. The questions are: does a need for bodily contact exist only during certain stages of the development? Is it regularly succeeded by a diminishing need for contact which results in disorder in the building structure? Does the plankton content of the water affect the architecture of the structure, or the strength of currents, or the availability of building material?

After a colony numbering millions of individuals releases its sexual products, giant swarms of veliger larvae form and drift as an enormous plankton cloud over the tidal flats and coastal areas of the southern North Sea, driven by currents and the surf. However, the larvae do not all stay together, neither do all reach an area with favourable conditions for settlement together. Many of the larvae are separated from the cloud and driven away. They are mostly found building on exposed valves, on individual, large gastropod shells, stones, or on pieces of crustacean armour. However, such hermits are rare because they seldom survive the larval stage as individual living plankton organisms.

Small settlements consisting of 50 to 300 animals are more frequent. The worms often build on the carapaces of living *Cancer pagurus* or beach crabs. *Sabellaria* settle on crabs only in colonies, never individually. Probably they come from small larvae clouds that have sunk to the sandy bottom and have encountered no other solid place for settlement. Usually such settlements kill the host crustacean. Rapidly built, up to 15 cm high structures, often in regular organ-pipe growth, tower on the carapace and make the crustacean top-heavy so that it must fall on its side. Then the animal either starves, or one of its numerous enemies kills it (Fig. 198).

Sand coral structures can also grow in yet another form, the spiral or goblet growth. The specimens shown in Fig. 199 were collected by F. Trusheim in 1929 on groynes at Minsener Oldeoog and described by Schwarz (1932a). No more such structures have been found since then. In Schwarz's opinion the conspicuous forms are caused by oblique lighting on normal settlements of organ-pipe tubes immediately below the water surface. He says that the animals on the shady north side of an organ-pipe structure, being phototropic, are forced to turn sideways and around the tip of the organ-pipe structure; their spiral curvature through east (or west) toward south influences the neighbours in the same sense. The animals on the south side are crowded, evade crowding by twisting northward, thus arrive on the shady side and curve further toward the light, until, finally, all individuals are growing in a spiral. Once forced into this rotation, the animals continue building in this way. Logical as this interpretation of spiral growth appears, it implies a closer link to the

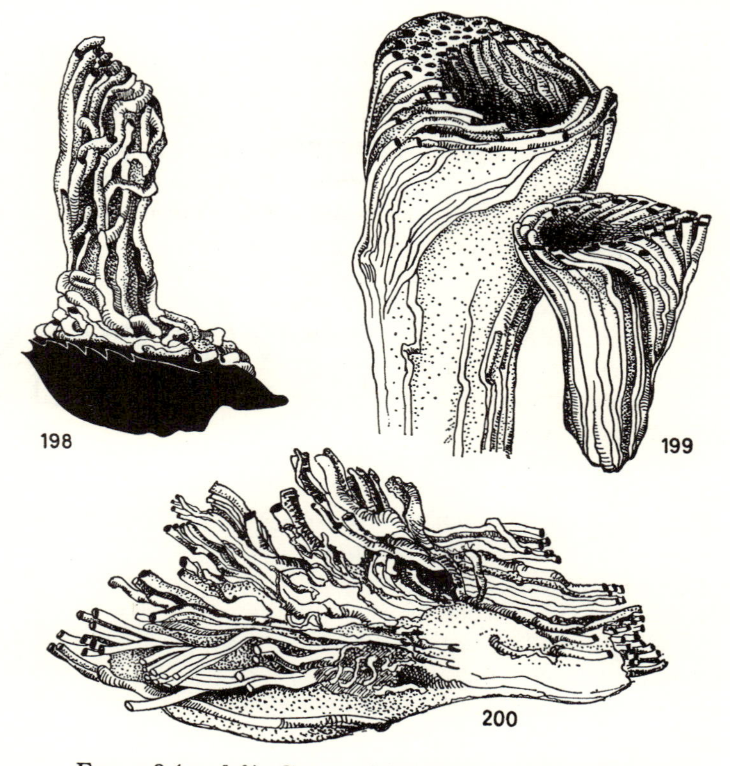

FIG. 198 (*top left*). Group of *Sabellaria* dwelling tubes on the carapace of a *Carcinides*. (Actual size)

FIG. 199 (*top right*). Group of spiralling dwelling tubes of *Sabellaria* in cup-shaped configuration. (Actual size)

FIG. 200 (*bottom*). Tumbled organ-pipe colony of *Sabellaria spinulosa* continues to be inhabited. Each tube curves into the new vertical direction, thus breaking up the mutual cohesion of the organ-pipe structure. (Actual size)

neighbour than seems justified; moreover, if phototropic influence is so strong, why should an inhabitant of the south side turn northward? Possibly, the inhabitants of eastern and western tubes approach the south side and overgrow the tubes there below their openings but without any damage to them, somewhat as they do in network growth. If oblique illumination were the cause of spiral growth, this would imply that the time of building activity is limited to morning and evening when the sun stands low. Also, illumination can affect *Sabellaria* only during low tide because a thick layer of plankton-rich water cannot be penetrated by enough light to affect the animals. Thus numerous questions concerning *Sabellaria* remain unanswered for the time being. However, several mainly descriptive statements can be made.

The single species *Sabellaria spinulosa* constructs unbranched dwelling tubes either as individual tubes, or as a prostrate network of tubes, or as a vertical organ-pipe structure, or as a twisted spiral growth, or, finally, as a structure of separating upright tubes. Whether building individually or in a colony, *Sabellaria* stays in the eulittoral area near the low-water line, from the lower margin of the periodically dry zone to depths of up to 50 m as on the oyster bank at Heligoland (Caspers 1950a). It always builds on solid but never drifting substrates; they can be lamellibranch valves, gastropod shells, whole or broken or as shell breccia, artificial, loose rock embankments or layers of branches, rarely marine peat, living brachyuran crustaceans, and also *Sabellaria* tubes. It requires strong, regular currents and water rich in plankton and avoids surf and the beach. It is phototropic, but this alone does not explain organ-pipe or spiral growth. Spiral growth may be restricted to periodically dry areas.

Sabellaria settlements occur in areas with active transportation of sand. Although living *Sabellaria* reefs are always in areas where there are currents, they must be free from sediment accumulation. Rapid sedimentation kills the animals but can preserve the tubes as fossils. Sudden bad weather is the normal cause of sedimentation in zones that were long enough free of it to allow settlements of *Sabellaria*. Yet storms can bring in new sediment in such quantities that even strong currents do not erode it or that the current pattern changes permanently. In other cases currents shift gradually and sedimentation may come to areas that had been free of it and inhabited by *Sabellaria*. In both cases the sand preserves the structures undamaged together with any accompanying epifauna.

Such events are not rare in the shallow part of the southern North Sea; nevertheless, the final preservation of *Sabellaria* structures is not ensured because so often sedimentation is followed by extensive erosion by shifts of channels; thus dead reefs that might have been fossilised reappear at the sea floor and are themselves eroded rapidly by currents which did no harm to them as long as they were alive. The organic cement of exposed, dead tubes is destroyed, and the tubes disintegrate to loose sand. This is probably the reason why fossil sand coral reefs are so rare. However, where they occur as fossils (in the Cambrian and Devonian) they have a wide, horizontal distribution. They are so rarely found because of rare preservation, not because they were rare. Shallow-water lamellibranch valves would be just as rare as *Sabellaria* structures (and all other structures from agglutinated tubes) in areas of the shallow sea, if they were equally fragile after re-exposure. Reineck found well-preserved fragments of *Sabellaria* structures in core samples 1 m below the surface of the sediment.

In a living reef many kinds of reef debris collect between the individual organ-pipes. Abandoned tubes collapse and disintegrate. Not all the sand grains carried by the passing current are caught by the worms; some are deposited between the reef structures. Lamellibranch valves get jammed in gaps between structures and form upright bodies of shell breccia. In a healthy sand coral reef these detritus layers do not interrupt the normal growth of the individual colony or of the reef as a whole. The animals can even survive deposition of a thin layer of sand. However, individual tubes often lose contact with each other in such cases. Like *Lanice*, *Sabellaria* builds from the inside by applying mucus to a digging trail, and not by placing grain on grain. Thus the bottom of a layer due to a relatively large influx of sand in a reef specimen is reflected in the sudden disorder of the organ-pipes. Too much sediment also frequently causes a change-over from organ-pipe growth to network growth. Instead of the neighbouring tube, the inflowing sediment supports the growing tube (Fig. 200).

Some *Sabellaria* reefs last for only one or two years; after the death of the initial builders the reefs disintegrate and soon disappear unless they are buried. Other reefs are resettled after each successive generation has died. Young larvae build either on the ruins or on still well-preserved or even partly inhabited older structures.

One reef can grow into an enormous, solidified sand body with a continuous succession of generations. Richter (1927a) described a giant sand coral reef built by *Sabellaria alveolata* in the narrows between Mont Saint Michel and the mainland of Brittany. It covers an area of 10 by 3 km; like a dam it almost connects the island with the mainland. According to Galeine & Houlbert (1922) the reef is 6 m thick. Several reefs of unknown thickness cover 3 to 9 km^2 at the bottom of the inner Jade Bay at a depth of 15 to 20 m.

Pectinaria. When an organism constructs a dwelling burrow in the sediment it generally becomes sessile. In turn, this link to one place creates new requirements. Organs and behaviour must be adapted to stationary feeding by angling, catching in nets, or filtering. The amphipectinid *Pectinaria coreni*, however, uses a dwelling burrow but gets food by collecting and picking up small organisms that live in the sediment. For this it needs liberal mobility in the sediment. In contrast to the scaphopod molluscs that have the same life habits, *Pectinaria* solves its special problems in the following way.

Since the gills lie on the outside of the body they need a protective cover. The worm builds a tube from cemented sand grains (Fauvel 1903). In the tube the worm sits in an open water bath in which it can spread the gills. The open upper end of the tube provides the necessary connection to the open water. Breathing water is renewed by movements of the body. The length of the tube therefore determines how deeply the worm can penetrate into the sediment without interrupting the supply of breathing water. Since the animal feeds by collecting it must be able to move together with its tube. A mobile tube cannot simply be built by applying mucus to sediment *in situ*, a technique that suffices for permanent but immobile structures in the sediment. Instead, the tube must be constructed so that it offers minimal resistance against sediment to be penetrated. The tube is therefore circular in cross-section and is smooth outside; it consists of a single layer of grains but none of them protrude beyond the regular, even outside of the wall. The tube is approximately 7 cm long, slightly longer than the worm that inhabits it; it resembles a quiver which can be dragged through the sediment. The tube has virtually become a part of the body in the same way as the gastropod shell has for the hermit crab, or the quiver for the larva of the caddis fly (Phryganeiclae; in German 'Köcherfliege', literally quiver-fly). The worm sits head-down obliquely or vertically in the sediment in its quiver, with a narrow opening of the quiver projecting a little above the surface of the sediment, and with another wide opening downward. In contrast to many other polychaetes, *Pectinaria* can build a quiver only once in a lifetime; if it is destroyed the worm dies. After a short, free-swimming stage, the young larvae, approximately 1 mm long, build a chitinous tube of assembled numerous scales and live in it for a short time. Soon this first tube becomes too small and is discarded as a whole after work on the final quiver has begun. This is built of sand and has a larger diameter (Scherf 1957). The animal moves by pushing and pulling movements with its setae, strong, golden, iridescent, awl-shaped bristles at the head, arranged so as to form a shovel-shape. The two groups of eight to ten setae are set in two continuous rows. They can be moved together or alternately (Fig. 201). This enables the animal to walk in the sediment on two bristle 'limbs', or to move by pushing or pulling movements if the two rows of bristles dig together.

In the head-downward, living position the feeding organs, the tentacles, and the mouth face the sediment where they are needed for searching food and ingesting. First the animal moves body and shell to the chosen feeding place. Then the setae-hands begin to scrape a small cavity to make room for the tentacles. They are club-shaped in rest but can be elongated. The tentacles palpate the walls of the small cavity in search of food; they loosen individual sand grains from the sediment and finally break entire clods from the ceiling of the cave, widening it sideways and upward. Every time sand falls down the tentacles examine it for food; when they find prey they grasp it and take it to the mouth opening. The tentacles are similar to those of *Lanice*; they are long muscular threads with a ciliated, longitudinal groove in which sand grains and food particles can be moved in either direction. Digging produces large quantities of waste that collect in the chamber until transported to the surface through the quiver. The waste accumulates behind the opening. Jerky contractions of muscles that run obliquely through the worm's body wall produce a strong current in the quiver capable of moving the waste sand. At the same time the head plate with its setae completely closes the lower aperture of the quiver and, if rapidly moved

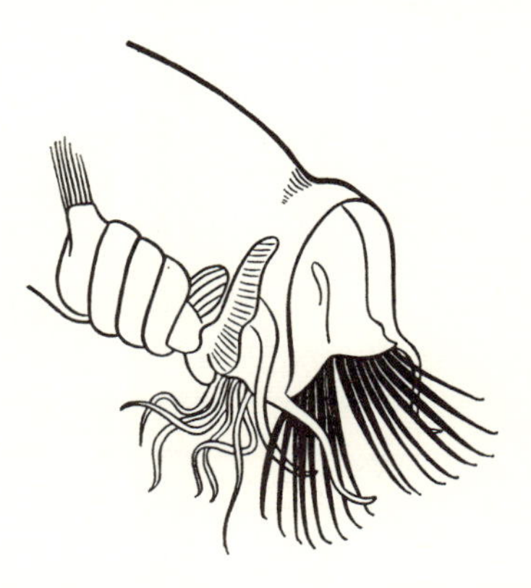

FIG. 201. Head of *Pectinaria coreni* in living position. (Black) two groups of setae. (Actual size × 3)

upward, acts as a piston. Sand suspended in water is pumped into the tube from below upward between the back of the animal and the quiver and expelled as a jet from the narrow, upper opening. Jet action is especially efficient due to the conical shape of this orifice. Wilcke (1952) first described the individual parts of the upper tube section of the quiver in some detail (see also Watson 1927). He calls this last section the chimney. It is approximately 1 cm long and is lined on the inside by several coats of fine detritus mixed with mucus, covered in turn by a layer of cemented grains. This section of the quiver like the rest is conical with the apex pointing upward. The cone is closed at the top by a mucous membrane (velum) with a circular hole in the centre. When the water is blown out the velum arches outward; when water enters it arches inward. The velum protects the inside of the quiver against intrusion of coarse sediment particles while the water is whirled in, and on the other hand, it increases the pressure inside the quiver when the water is expelled as a jet.

If the animal stays in the same place for some time it considerably widens the cavity in front and beside the mouth. The constant digging of the tentacles and the current of breathing water finally cause a break-through from the chamber to the surface. Wilcke (1952) calls this connection of the subterranean feeding cavity with the surface of the sea floor a 'pit'. The tentacles if stretched can reach from below through the chamber and the pit to the surface. Such a pit does not always exist; on the other hand, several pits can form. How long *Pectinaria* remains in one place depends on the food supply in the feeding cavity.

When the food gives out the animal moves and thereby distinctly marks any bed of sediment. Bedding usually disappears behind the moving quiver (Fig. 202). It is necessary to be familiar with the mode of locomotion of *Pectinaria* to analyse its traces in detail. It is impossible

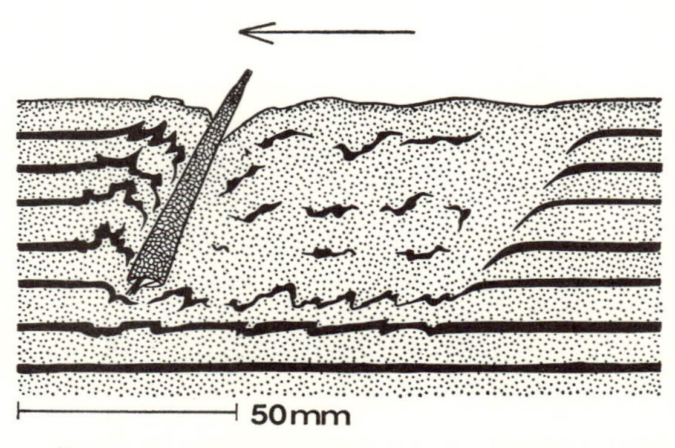

FIG. 202. *Pectinaria coreni* migrating through sediment and leaving a trail of perturbed burrowing texture. (Arrow) direction of movement

simply to drag a small, wooden stick as thick as *Pectinaria*'s quiver sideways through fine-grained sediment; its resistance is too high. We can, therefore, expect that *Pectinaria* does not simply move the quiver by pushing it sideways. Not only does the worm lack the strength to do so, but the stress on the quiver would break it.

In fact the animal sitting in its quiver pushes slightly backward and upward, i.e. lengthwise, with the setae. Thus the tube temporarily protrudes approximately 1 cm more than usual from the sediment. Then the setae dig obliquely downward and pull the lower, wider end of the quiver along, bringing the quiver to a more oblique orientation than before. Now the head of the worm anchors itself firmly in the sediment and, with a twist of the body, re-erects the upper end of the quiver. The movement is performed by two complicated muscle mechanisms on both sides of the body between the head and the base of the scapha. Each side has a strong longitudinal muscle which in contracting makes the body firm. Inside this long muscle, short bundles of muscles run upward in a zigzag pattern with each single bundle nearly at right angles to the dorsal side (Fig. 203). With successive contractions beginning at the head and ending at the scapha, these muscle bundles pull the body jerkily to the vertical. Since the head is anchored by the setae, and hooked bristles on the dorsal side of the worm are fastened to the inside wall, the zigzag muscles raise the quiver as if it were a boom.

FIG. 203. *Pectinaria coreni* in its mobile, agglutinated dwelling tube; the longitudinal and lateral muscles are also shown. (Actual size × 2)

The conical upward narrowing of the quiver facilitates the movement; as the diameter decreases from the point of attachment so does resistance by the sediment. Thus the conical shape of the quiver has a double advantage; it increases the velocity with which water and sand can be squirted and it helps locomotion in the sediment. The worm always advances ventral side forward because the muscles can erect the tube only toward that side. This also means that it advances toward the cavity it has made for feeding, the direction in which resistance is minimal.

Stationary dwelling tubes of animals living in sediment normally show wide variations of structure and orientation in response to external conditions, giving individual character to each tube. In *Pectinaria*, however, one tube resembles the next to the last detail in proportions, thickness of the walls, or diameter of apertures. This kind of dwelling tube is an instrument with a complicated use and is adapted to the organs of the inhabitant. It therefore cannot tolerate the least alteration from the blueprint without losing its effectiveness. For the same reason the quiver shapes of different species of the world-wide genus *Pectinaria* hardly differ from each other. What differences there are merely consist in variations of curvature of the tube. If the tube is curved the animal always moves convex side foward. Hessle (1917) listed the following species: *P. belgica*, with nearly straight tubes. It occurs on the Swedish west coast (Gullmar Fjord) and in the boreal parts of the Atlantic Ocean. *P. hyperborea* has slightly curved tubes and occurs on the Norwegian coast and in the Arctic, Atlantic and Pacific Oceans. *P. granulata* lives in the Arctic Ocean, and its tubes are weakly curved. *P. ehlersi* whose tubes are curved occurs near Tierra del Fuego. *P. auricoma* is found on the Swedish west coast, in the Mediterranean, Atlantic, Arctic and the northern part of the Pacific Ocean; its tubes are bent. *P. neapolitana* inhabits the Mediterranean and has slightly bent tubes. *P. bocki* whose tubes are unknown lives near Japan.

The animals can move vertically as well as sideways. This becomes necessary during rapid sedimentation at the place of settlement or during erosion. Upward or downward escape movements, i.e. movements parallel to the quiver, cause approximately vertical sag zones about the width of a pencil.

According to Linke (1939) mud-rich, fine sands are favoured by them. König (1948) calls them sandy sediments coloured grey from intermixed organic substances. Population density is never high; it can rise to eight or ten animals per square metre. The animals also settle individually. Areas that were free of these settlers for years may suddenly become populated.

Three groups of polychaetes can be distinguished with regard to *building material*:

(1) Dwelling tubes built from foreign substances that occur in the biotope. The building material consists of sand grains and various shell fragments which are obtained from the biotope. Mucus is used for gluing. It comes mostly from glands in the mouth cavity and the digestive tract, or from glandular organs at the head or immediately behind it.

Most species place individual sand grains into a mucous substance, and it is difficult to recognise the resulting tubes in a rock. However, in tubes with a wide diameter the course of the burrow is recognisable by bedded infilling; in narrow tubes this can be missing. In addition to *Magelona papillicornis*, *Diplocirrus glaucus*, *Sabellaria spinulosa*, *S. alveo-*

lata, *Pectinaria coreni*, *P. auricoma*, *Fabricia sabella* (see also Hagmeier 1930a), the species *Pygospio elegans* also belongs to this group. It settles in dense populations so that the barely 1 mm thick tubes form regular lawns; the individual tubes, however, do not touch each other. They go to 9 cm deep into the ground and are built of fine sand grains, cemented with mucus. In the absence of sand the tubes are made of lumps of mud, detritus, or plant remnants. Grains are often missing in the wall of the tube deep in the sediment, and it consists only of mucus and is less hard. The upper section of the tube has a brown stain from iron hydroxide which is lacking in the lower part. The tubes are frequently branched. According to Linke (1939) branches originate when partly washed-out tubes become kinked and then break. Then the worm builds a new tube from the break in continuation of the old tube. They may either go to the surface or end in the sediment. However, only one single tube is inhabited. Hempel (1957a and b) observed a three-piece filter net held at the tip of the tube. The net can be opened and closed; it probably serves to keep out foreign particles and for feeding. Tubes with such shutters never become filled with sediment but are compressed and preserved as flat features. This also holds for the tubes of *Lanice conchilega*.

In many cases a simple construction with sand grains is insufficient where isolated dwelling tubes project above the surface of the sea floor; then larger grains such as shell fragments, small gastropod shells, and sea urchin spines are used. Each species usually uses its own special material. Such tubes can easily be spotted in the rock because of the concentrated, orderly arrays of hard biogenic components. This kind of tube can also be preserved even if it is washed out and transported. To this group belong *Lanice conchilega*; *Owenia fusiformis* which arranges fragments of bivalves in roof-tile style in the wall of the tube, or sand grains if shell fragments are not available; *Onuphis conchilega* which builds with large fragments of bivalves laid flat on the top edge of the tube; *Onuphis quadricuspis*, *Hyalinoecia tubicola* and others.

(2) Dwelling tubes built of mud that is collected and kneaded with mucus; the mud is often shaped before it is used for construction. The tube can easily be spotted in the rock if the blocks are laid flat in the wall. *Neoamphitrite figulus* of the terebellids belongs to this group. It lives in the tidal flats of the North Sea coasts (König 1948) and also in mud areas to a depth of 60 m. The large worm builds its wide tube below and on the surface of the sea floor. Fig. 204 shows the arrangement of the shaped building components. The tube is multicurved and may have any orientation; it frequently branches out from breaks in the wall. The worm incorporates foreign tube fragments into its structure, fitting curved fragments neatly into the curvature of the wall. It also uses large mud

pebbles instead of shaped blocks (Schäfer 1956a). Members of the families Ampharetidae, Spionidae, and Disomidae also build tubes of mud. Those of *Manajunkia aestuarina* are horizontal.

The dwelling tubes of *Sabella pavonina* reach 50 cm deep into mud and protrude 8 to 10 cm or sometimes more above the surface. They are made of elastic and strong material consisting of stringy mucus encrusted with mud. The animals occur in the tidal zone but live mostly in deeper water. The tube with a diameter of up to 8 mm is open at both

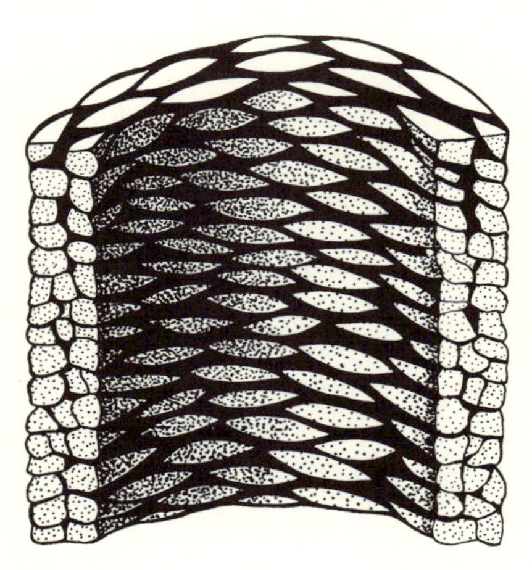

FIG. 204. Inside of the wall of a dwelling tube of *Neoamphitrite cirrosa*; shaped lumps of mud are cemented to each other (from Schäfer, 1956a). (Actual size × 3)

ends (Nicol 1931; Wells 1951, 1952b). The subterranean tube section curves sideways and narrows downward and is softer and more elastic than the vertical upper section in open water (Fig. 205). The worm pumps water into the tube from the top by rhythmic movements; the water leaves the lower, narrow opening and washes an irregular shaft to the surface through the sediment. In rare cases *Sabella* constructs its tube on rocky ground, in completely open water.

Sabella uses small detritus and mud particles as building blocks. They are caught by the tentacle crown, then taken to two pouches of the lower lip and collected there. The glandular fold of the collar provides mucous secretions which, with stored detritus, are pressed as a thick thread on to the edge of the tube. Thus each building component of the tube is a complete ring. From the outside the tube, therefore, is transversely grooved. It seems that a mucous lining secreted by the ventral, glandular plates smooths the inside of the tube.

The animal never leaves its tube during its lifetime and is thus almost sessile. Like all sabellids it has a faecal groove running ventrally from the posterior to the anterior but turning toward the dorsal side above the abdomen. The turn probably serves a hygienic function, keeping the faeces away from the respiratory and feeding organs. Regrettably, there are few and very incomplete reports about the dwelling tubes of other sabellids. Their tubes are probably somewhat similar to those of *Sabella pavonina*. Such sabellids are: *Laonome kroyeri*, *Euchone papillosa*, *Chone infundibuliformis*, *C. duneri*, *Myxicola steenstrupi*, and others (Hofsommer 1913).

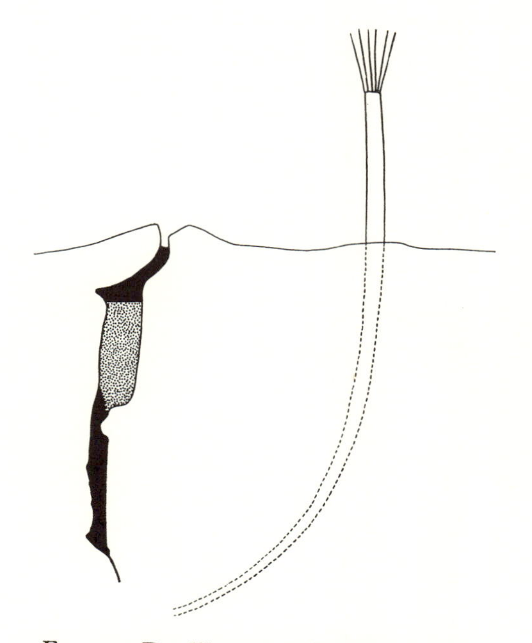

FIG. 205. Dwelling tube of *Sabella pavonina*. Cracks and cavities in the sediment were formed by water expelled by the worm at the lower end of the tube (after Wells, 1951). (Schematic drawing)

(3) Dwelling tubes built of secretions. The worm secretes a mostly calcareous tube in the way that a gastropod secretes its shell. The tube can be built only once in a lifetime and grows with the builder. The animal can close the tube with a calcareous or chitinous operculum. The walls of the burrow are so hard that the structure does not need to be built into the sediment for support and protection. Indeed it is invariably constructed above the surface, mostly leaning to or firmly attached to a substrate such as stone, cliff face, calcareous shell, or the thallus of a plant. Occasionally individual tubes are interwoven. The animals are restricted to areas with little sedimentation because as sessile animals they cannot cope with the danger of being buried.

The few important forms of this kind in the southern North Sea are all Serpulidae: *Pomatocerus triqueter* settles on solid substrates and rarely does so in isolation. Usually, many individuals join, and their three-edged calcareous tubes cover the substrate with a densely interwoven growth, frequently growing to the dimensions of a small calcareous reef. The larvae of the worm probably have an ability to choose, within limits, a place for settlement, as do the larvae of many sessile marine animals. The individual tubes combine into irregular, knotty networks without any order or mutual orientation; each fills a gap where it finds one. The openings also face in all directions; they are never overgrown by tubes of their own species. Other serpulids are: *Hydroides*

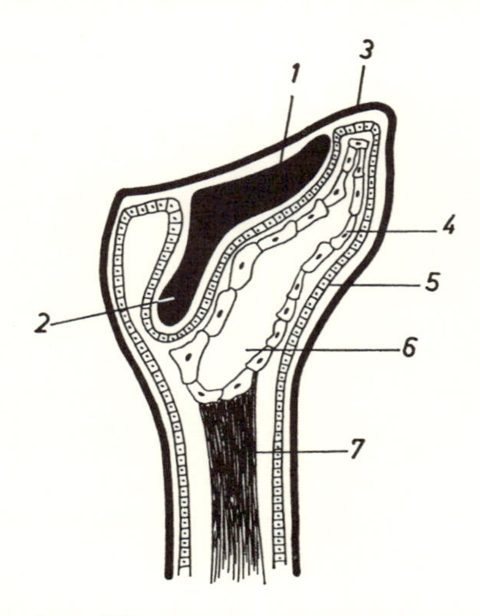

FIG. 206. Top portion of the dwelling tube of *Spirorbis*. (1) calcareous lid; (2) peg-shaped appendage on lid; (3) exterior cuticula; (4) epithelium of the coeloma; (5) epidermis; (6) ampulla; (7) stem muscle (after Eisler, 1907). (Schematic drawing)

norwegica, *Placostegus tridentatus*, *Filigrana implexa*, *Ditrupa arietina*, and *Salmacina dysteri*. All live on hard ground.

Spirorbis carinatus has an asymmetrical body which is coiled either to the right or left. The tube secreted by the worm is therefore similarly coiled. All whorls lie on the substrate. As a substrate the species favours the thalli of *Laminaria*, *Fucus*, and other seaweeds. Numerous tubular glands secrete the calcareous building material for the tube wall. Their ducts lead mainly to the collar of the head and to the thoracic membrane. The animal rotates about its axis and thus forms new parts of its tube. The tube can be closed with a calcareous lid (Fig. 206). The second ray of the gills of the head has lost its respiratory function; its end has become enlarged to a club shape, and it now serves as the lid. The thickened end fits exactly into the tube. It also serves as a breeding cavity (Elsler 1907).

According to Hessle (1917) the following members of the genus

Spirorbis also occur in the southern North Sea: *S. vitrens*—the shell is 2 mm in diameter and transparent, the animal lives on stones and lamellibranch shells at depths of 5 to 20 m and occurs in Greenland, England, and Newfoundland. *S. spirillum*—the shell is 1 to 1·5 mm in diameter, always coiled dextrally, and brilliant white with a reddish tinge; the opening is perfectly circular. The animal lives on *Laminaria*, red algae, or *Sertularia* at depths between 5 and 15 m and occurs off the Norwegian coast, in the Baltic, North Sea, Arctic Ocean, and off the coasts of Scotland, England, and Madeira. *S. borealis*—the shell is coiled sinistrally, has a diameter of 2·5 to 3·5 mm, is chalky white and opaque and the course of the tube varies. The animal lives on *Fucus* and lamellibranch valves and occurs in the Mediterranean, and near Madeira and Tenerife. *S. pagenstecheri*—the shell is sinistrally coiled, has a diameter of 1 to 2 mm, is chalky white, and narrows to a point at the opening. The animal lives on stones and lamellibranch valves, rarely on algae, crabs and lobsters, at depths between 10 and 30 m. *S. granulatus*—the shell is opaque, chalky white and has a diameter of 1·5 to 2·5 mm. The animal lives on stones and lamellibranch valves; it occurs near Sweden, Denmark, Norway, in the bay of Kiel, off England, Ireland, Greenland, Spitzbergen, and Newfoundland. The taxonomy of the genus *Spirorbis* has been discussed by Borg (1917).

Three groups of polychaetes can be distinguished with respect to *their surroundings*.

(1) Much of the dwelling tubes lies in the sediment. The builder leaves its tube only under special circumstances, such as sedimentation or lack of oxygen, and is thus almost completely sessile. Feeding is adapted to this sessile life and is either pulsing or whirling. Agglutinated tubes into which foreign particles are incorporated can be raised above the surface in contrast to mucous burrows which are restricted to completely subterranean structures. The hardened structure in the sediment serves as a base for the equally strong upper section which projects above the surface. It is frequently made of coarse components such as shell fragments. Many of the described polychaetes with agglutinated tubes belong to this group.

(2) The dwelling tube lies in the sediment but is mobile. Feeding by collecting small animals in the sediment is dependent on this mobility. The tube is smooth on the outside, and its shape is adapted to movement through the sediment. Members of this group are *Pectinaria*, *Owenia*, and *Onuphis*.

(3) The dwelling tube is constructed entirely on the surface. There are two distinct types. (*a*) If the dwelling tube is built of sand grains an individual tube cannot resist currents for any length of time. Therefore, the worms join to form colonies and extensive reefs of upright sand tubes,

solidly cemented together in bundles and thus resistant against erosion. Examples are *Sabellaria spinulosa*, *S. alveolata*, and *Fabricia sabella*. (*b*) If the tube is made of calcium carbonate secretions without foreign particles, the structure is strong and needs no protection by sediment. Members of this group build on rock, coarse pebbles and shell breccia, on living, vagile, shelled animals or on large thalli of algae. Examples are *Spirorbis* and *Pomatocerus*.

Phascolion. *Phascolion strombi* differs in its way of life from other sipunculids in one respect, it is almost completely sessile and lives either in self-made or foreign tubes. *Phascolion* is a relatively small sipunculid, being 1·5 to 2 cm long. It prefers empty *Dentalium* shells or large gastropod shells (*Buccinum* or *Lunatia*). The long, tube-like proboscis is provided with a ring of tentacles and serves to fetch detritus and dead organic material from the surroundings. The self-built tubes are open at both ends; their diameter is slightly larger than that of the animal body. The walls are built of small lumps of mud and calcareous splinters and are quite fragile. The animal attaches itself with a papilla at its posterior end. The tubes lie in the surface layers of the sediment, and their construction is unknown. The animals are widely distributed on the muddy sea floors of the North Sea. Fischer (1928) found them also in areas with sand and shell breccia together with *Amphioxus*.

The borehole

Whereas many endobionts must build mucous burrows or agglutinated tubes for respiration in loose sands or muds that are sometimes poor in oxygen or even reducing, the situation is different if the burrow is drilled into a hard substrate, such as wood, peat, rock, or biogenic calcium carbonate. No construction or hardening of the wall is necessary. The cavities, once drilled, enable the oxygen-rich sea water to enter freely and the inhabitants can live there permanently.

Whereas mucous and agglutinated tube-building endobionts can settle in extensive areas of the sea floor, boring endobionts are restricted. Abrasion and tidal erosion by a transgressing sea provide the habitat for the boring forms: bedrock, compacted peat, subfossil soils or wood, or else biogenic calcium carbonate (for a discussion of boring sponges, see Arndt 1943, 1952, and Schremmer 1954). Even if a borehole has been successfully drilled the settler in the shallow sea is in constant danger of renewed sedimentation. Only special types of adaptation succeed in the long run.

Lamellibranchiata. The most frequent species of boring lamellibranchs in the North Sea are *Barnea candida*, *Petricola pholadiformis*,

Zirfaea crispata, *Pholas dactylus*, *Saxicava rugosa*, and *S. arctica*. Of these, only *S. rugosa* bores exclusively in rock. It occurs in the southern North Sea only on the cliffs of Heligoland and bores there in the Lower Triassic ('Bunter') sandstones and in the Cretaceous chalk. It and many other boring lamellibranchs are also known along the rocky English coasts. The others can also live in those more entensive areas of the North Sea where sedimentation is not quite excluded, and where the limits of their biotope are not so narrowly delimited. This makes them particularly interesting for our actuopalaeontological studies.

Alluvial, submerged peats provide the principal substrate for boring in the southern North Sea where they extend to approximately 55° latitude. These peats have a short but eventful geological history. After the retreat of the last Pleistocene ice sheet ground and terminal moraines were subaerally exposed in the area of the present southern North Sea northward to beyond the Dogger and Jutland Banks. In early Postglacial times a series of peats formed, starting with the so-called basal peat. During the subsequent transgression from the north these peats remained partly preserved and were protected against further erosion by deposits of sand and mud, laid down by the primeval North Sea. According to Schütte (1935 and 1939) the mainland coast of the Early Postglacial (approximately along the present 50 m depth-contour) was transgressed by the ocean about 10 000 years ago; during the following 3000 years, the sea advanced to what is now the chain of the West and East Frisian Islands. With further transgression during the last 6000 to 7000 years the present tidal flats and marshes were deposited. Several regressions or times of standstill interrupted the advance of the ice. During these periods when the southern coastal areas became dry for some time, more (locally up to three) peat horizons were formed. They are domed bog, lake swamp, swamp forest, or *Phragmites* peats, separated by marine sediments. Newly deposited sediments caused considerable compaction, consolidation and dehydration of all these peats (see also Wildvang 1938, Haarnagel 1950, Grohne 1956).

These consolidated peats of different ages that are generally covered by young sands and muds, are occasionally re-exposed over wide areas by waves, or locally by currents or shifting tidal channels; and so these peats may lie exposed on the sea floor for a long time, and frequently also in the tidal flats (Fig. 207). These peat beds provide a substrate for the majority of the boring lamellibranchs. In the Jade area which is typical of the coastal zone, 61 out of approximately 600 dredging operations yielded fossil peat with lamellibranch holes, some with living animals in them (Schäfer 1939a). Fossil peat is probably alone responsible for the wide distribution of certain species of boring lamellibranchs. Only as long as new layers of Postglacial peat continue to be re-exposed can

Fig. 207. Four peat horizons in the intertidal North Sea, settled by boring bivalves where exposed

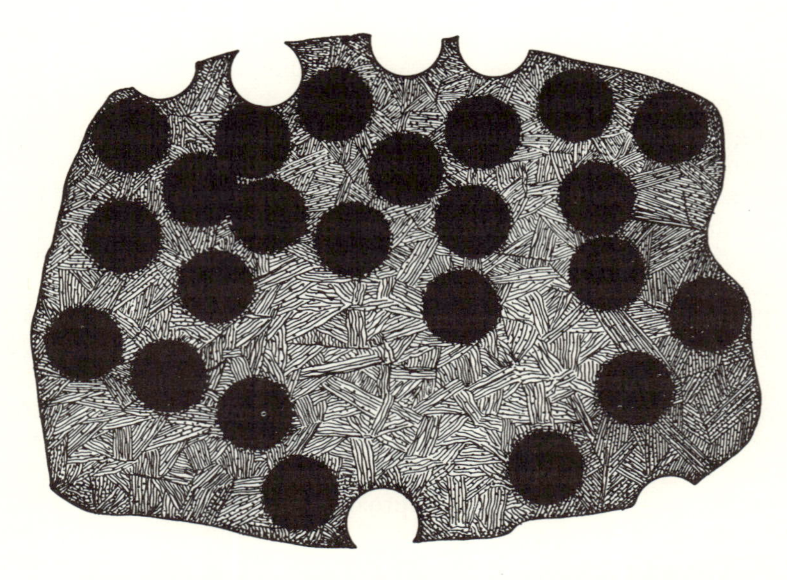

Fig. 208. Fragment of fossil peat perforated by boreholes produced by bivalves. (Actual size × 0·5)

boring lamellibranchs remain a substantial part of the North Sea fauna. If at some future time all peat were destroyed or none re-exposed the boring lamellibranchs would lose almost all of their biotope.

Two species, *Barnea candida* and *Petricola pholadiformis*, drill in fossil peats at the coasts and in the open sea. They sometimes settle very near each other (Fig. 208). The large number of their valves washed up in swash marks on tidal flats and on open beaches also indicate how dense the populations must be if one considers the limited extent of the peat outcrops. The dredging operations already mentioned have confirmed with ample material that *Barnea* and *Petricola* prefer peat as boring substrate, an observation already made by Boettger (1907), Roch (1926, 1927), Schwarz (1932b) and Wasmund (1926).

Barnea and *Petricola* bore mechanically, the only effective method in peat (see D I (*a*) (x)). Since the animals bore by scraping small particles from the walls of the hole with the teeth and file-like ridges on the outside of their valves, they cannot drill with equal ease through all kinds of peat, and not at all through some. If the peat particles are too tough and elastic, or if the constituent plant fibres are too loosely packed, then they yield under the pressure and the pull of the rasping teeth instead of tearing. Eventually such fibres get twisted around the teeth and clog the file. Only peats that were well compacted, dehydrated and compressed by a large overburden make a good substrate. Particularly suitable and plastic are impure peats that were formed in brackish water and mixed with clay. Recently formed peat is not a suitable substrate for boring animals. Therefore no boring lamellibranchs settled on the peat beds transgressed by the North Sea ca. 8000 years ago; they became a substrate only after being fossilised and compacted by a cover of sediment and then re-exposed. Boring lamellibranchs are therefore indicators of considerable erosion.

Some recent bogs are now exposed along the coasts of the North Sea. They lie outside the dykes, are up to 2 m thick, and clods of peat, often several square metres in size, drift away from them into the sea where they become embedded in sand or mud (for a description of such a bog near Sehestedt at Jade Bay see Künnemann 1936 and Haarnagel 1950). Although these recent peats may be exposed for decades they are free from boring lamellibranchs. One exposure of recent peat lies less than 6 km from a densely populated fossil peat area so that there should certainly be no lack of spat and larvae of boring bivalves. Whether a peat is affected by boring bivalves or not can therefore serve as a criterion to distinguish drifted recent peats from subfossil and fossil peats in the ocean without recourse to pollen analysis. No such maturing is needed for unweathered rock (e.g. clayey sandstone or limestone). Rock-boring bivalves can be the first marine settlers on terrestrial substrates as soon as the transgressing sea has covered them and removed the weathered soil.

Fossil peats can serve as substrate for boring bivalves for a limited time only. The peat surfaces are 'consumed' by the lamellibranchs. The more favourable the consistency of the peat is for boring, the denser the settlement. It frequently happens that finger-thick boreholes are adjacent. All boreholes over many square metres of peat surface have approximately the same diameter. The similarity of thickness, and therefore presumably of age, indicates that the boring lamellibranchs have settled on the peat at about the same time. Nowhere do two boreholes intersect. There is always some space between them, if only a millimetre or less thick. The larvae of the boring bivalves are apparently able to choose their position and to space themselves out during the hours of settlement. Such an ability has been demonstrated for balanid larvae. Once a peat has been densely populated by boring bivalves, a second settlement is impossible. The narrow, remaining ridges may possibly allow a young bivalve to bore for a short time, but with growth the hole would soon intersect a previous one. Then both the new and the old settler would lose support and either be washed out, or exposed in the too wide hole, become prey to their enemies, chiefly crustaceans. Once washed out, adult boring bivalves cannot start a new hole. Their foot can only push against the internal wall of the borehole and is not suitable for erecting the body, to bring it into drilling position, or to provide the firm anchoring needed for effective movements. However, if one places a boring bivalve into a prepared cavity, in proper boring position, the animal can and does continue to drill. If a first settlement on peat has been sparse, later generations of boring lamellibranchs can settle between existing holes. In that case boreholes of different sizes inhabited by animals of different age can be found on a single peat surface.

Boreholes considerably weaken the peat so that it is easily eroded. The lamellibranchs respond by drilling deeper. (Generally they drill so deep that their siphons just reach the surface; this means that the 4 cm long *Barnea candida* drills an 11 cm deep hole.) If erosion from the top continues the bivalves traverse the peat and end up in the subjacent clays. Boring lamellibranchs found in clays have invariably come from overlying peat that has been completely eroded. The animals continue to live there if the walls of their boreholes are solid enough to support themselves. Larvae, however, never initially settle on such substrates. Thus the layers of fossil peat rapidly disappear unless they are constantly re-exposed. The lamellibranchs which prefer to bore in peat accept few other substrates, amongst them subfossil wood, lignite-rich fresh-water clay, and the chalk and clayey sandstone of Heligoland; however, all these substrates provide but a small fraction of the presently exposed settling area in peat.

Traces of boring can be preserved permanently if settlements on peat are covered by new sediments. The boring animal at the bottom of its hole is able to remove small quantities of sediment by pushing it upward with its body. Larger quantities of sediment, however, stop movement of the bivalve in its hole, and a thick layer immediately kills the entire settlement; neither can the sediment in the boreholes be removed, nor can the siphons penetrate to the open water. The lamellibranch cannot actively leave its borehole either. Its foot, morphologically and physiologically suitable only for movements in the hole, is incapable of pushing the animal upward through sand and mud, unlike the foot of most other bivalves that live in the sediment.

If the sedimentation is substantial the drill holes of the boring bivalves are filled, usually in layers. Frequently the two valves of the dead lamellibranch at the bottom of the borehole remain within the plug of intruded sediment. If the infilling hardens it can be more resistant than the surrounding peat. The boreholes, by the way, are traps for small organisms and their skeletons. One frequently finds in them well-preserved (because not transported and rolled) valves of ostracods, complete or fragmentary armours of other small crustaceans, fish scales and vertebrae, Foraminifera, small gastropods, opercula of gastropods and spines of sea urchins.

Petricola pholadiformis is a newcomer to the North Sea. This boring relative of the Veneridae was accidentally introduced in 1896, initially to the English coasts, from the Atlantic coasts of North America, probably with imported oysters; ten years later it was reported from Juist, Amrum, and Sylt. From the beginning the lamellibranch has settled densely on peats *in situ*. It has spread with such speed that today it is the most common boring clam in the southern North Sea and its coasts. Today it can no longer be verified whether the advancing *Petricola pholadiformis* has crowded out *Barnea candida* by taking possession of its old settling places, or whether *Barnea* has always been as relatively rare as it is today. *Petricola* could spread so rapidly because its preferred boring substrate, peat, was in even better supply on this side of the Atlantic than on the North American east coast. Gould (1870) noted boggy layers at the Massachusetts coast with *Petricola* at about low-water line. According to Miner (1950) it occurs at the Atlantic coast from Prince Edward Island to the West Indies and Texas and bores in clay rich in plant material. In 1958 the author observed *Petricola* boring into clays up to 6 m thick and rich in *Spartina* at the Georgia coast.

Zirfaea crispata (Fig. 209) and *Pholas dactylus* can also occasionally be observed in submarine peats *in situ*. However, they do not form the extensive settlements that are known of *Petricola*. Their boring technique is the same as that of the species already discussed (Troschel 1913, Haas 1926). *Barnea* and *Petricola* are less common as rock borers than *Zirfaea* and *Pholas*. They are the main producers of the numerous boreholes in the bedrock and submarine screes of Heligoland. These species can drill obliquely and sideways as well as vertically downward into the boring substrate; they are therefore particularly well adapted to changes of position of their drilling substrates. Rotation of an inhabited beach boulder does not prevent the bivalves from continuing their drilling; occasionally one finds that boring animals have settled on all sides of a block (Seibold 1955, Adriatic Sea). Apparently *Barnea* and *Petricola* can only bore vertically downward into the sea floor and usually avoid

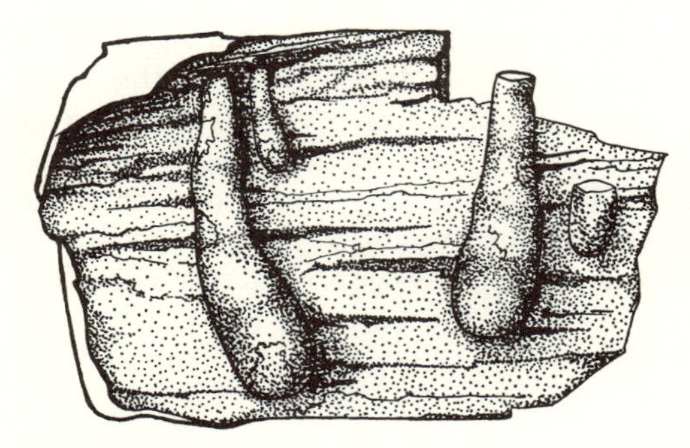

FIG. 209. Positive casts of *Zirfaea* boreholes in peat. (Actual size × 0·5)

even large boulders. Thus the loose pieces of peat and subfossil wood which are frequently washed ashore at the south shores of the North Sea show almost exclusively boreholes of *Pholas* and *Zirfaea*. *Zirfaea* in particular is regularly found in such wood pebbles, often in large numbers. But such wood must have become waterlogged and have remained in the water for many years. The difference in the size of several boreholes in a piece of wood indicates that the larvae always settle as individuals and not in swarms as those of *Barnea* and *Petricola*. Occasionally, especially near the coast, wood pebbles are embedded in sand and mud together with boring lamellibranchs. Upon consolidation the sediment which surrounds the pebble and fills the borehole can form a mould which shows every detail of the grain of the wood (Fig. 209).

Polydora. *Polydora ciliata* can bore into hard rock as just one of the several ways in which it constructs dwelling tubes, possibly the one demanding the most advanced behaviour (see section D I (*b*) (ii)). Boreholes are preferentially constructed in hard rock surrounded by loose sediment consisting of shell breccia of lamellibranch valves in fine sand.

If *Polydora* builds in loose sediments it tends to attach its tubes to solid walls. In the aquarium it likes to build at the glass walls where it needs to construct only a half-tube, the glass providing the other half. Stones, wood, and various plastic sheets are similarly used when they are placed into a settlement of *Polydora*. In nature, the worm builds in the same way and uses mostly shell fragments that are partly embedded in sediment and partly project above it (*Ostrea edulis*, *Cardium edule*, *Littorina littorea*, *Lunatia nitida*, *L. catena*, *Turritella communis*, or *Scala clathrus*). When attached to a shell the two U-branches for ingestion and egestion lie closer together than when built in loose sediment. The worm then widens the cavity at the expense of the shell, making it more nearly circular in cross-section. The part of the tube which is made of cemented sand grains appears as a high relief on the shell (Figs. 210 and 211). One frequently finds oyster shells that show this stage of drilling by *Polydora*; U-shaped grooves are sunk into the surface of the shell. The sand tubes which had originally been raised above the groove are destroyed during transportation before the dwelling burrow was finished in its final form. The grooves on the lamellibranch valves are the result of etching by mucous secretions (Söderström 1920, 1923). The position of the mucous glands and the chemical process involved are unknown. Söderström considers it probable that the organs providing the etching secretions are 'glandular complexes that lie in pairs in each of the seventh to tenth segments with bristles'. The author demonstrated that mucus from *Polydora* tissues, applied to valves in a moist chamber, quickly etches oyster shells. Richter (1924c) found traces of *Polydora* borings on Heligoland in clayey sandstone, in Middle Triassic limestone, Cretaceous chalk and lignitic chalk, and in Pleistocene clays rich in plant material. The boring traces in clayey sandstone prove that the animals must be able to bore mechanically as well as chemically by etching. Richter and Yonge (1949, p. 175) ascribe such boring capability to the bristles of the fifth segment which, they think, are used as scraping tools. The author has observed these bristles being used for pushing soft sediment aside to widen a hole, never for scraping.

The superficial etching so far described is the beginning of the construction of a dwelling tube in a calcareous substrate. Next the worm begins to etch deeper into the substrate at the U-bend deflecting it from the original plane. Since the etching advances over the width of the curved portion of the body a pouch-shaped cavity is progressively sunk into the calcareous shell (Figs. 210 and 211). Individual advances of the etching can be recognised as arch-shaped grooves on the inside wall of the finished pouch. At first the two separate sediment tubes on the surface simply join in a single pouch cavity constituting the U-bend portion of the structure. However, as the pouch advances deeper into

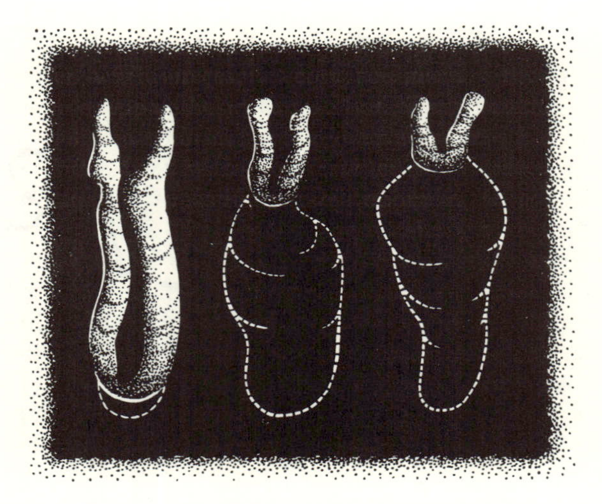

FIG. 210. Boreholes made by *Polydora* on an *Ostrea* valve. Left: superficial etching of the valve surface, at the U-bend formation of a pouch (broken line) begins below the surface. Centre and right: deep and wide pouches (broken lines). Infillings of cemented sand are lost from the specimen. (Actual size × 0·5)

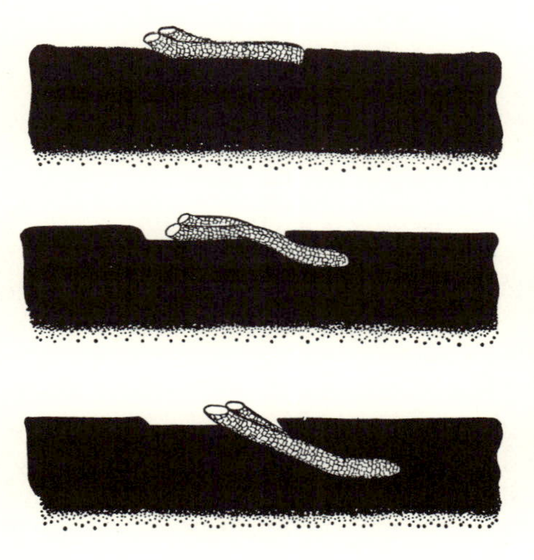

FIG. 211. Three progressive stages of etching of *Polydora* tubes into *Ostrea* valve in side view. Top: superficial etching at the surface of the valve, the tube is sunk into the etching pit; centre: a pouch is advanced into the interior of the valve and is lined with sediment; bottom: almost the whole structure now lies within the valve, superficially emerging parts of the U-shaped tube are short. (Schematic drawing)

the shell the worm fetches sand grains from the surface and builds a septum into the central portion of the pouch (Fig. 212). Thus parts of the straight branches of the U are also transferred from the surface into the pouch. According to Söderström (1920 and 1923) these septa can also be built of detritus other than sand and of particles scraped from the walls of the cavity. Eventually the worm can dwell exclusively in the pouch structure. The sand tubes above and outside the pouch are removed if the openings of the pouch lie flush with the surface of the

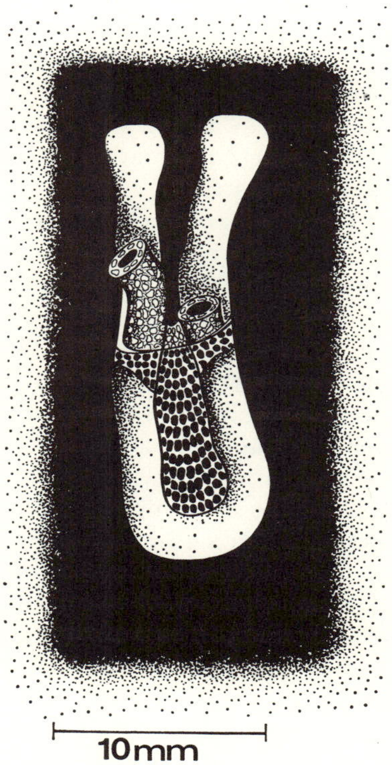

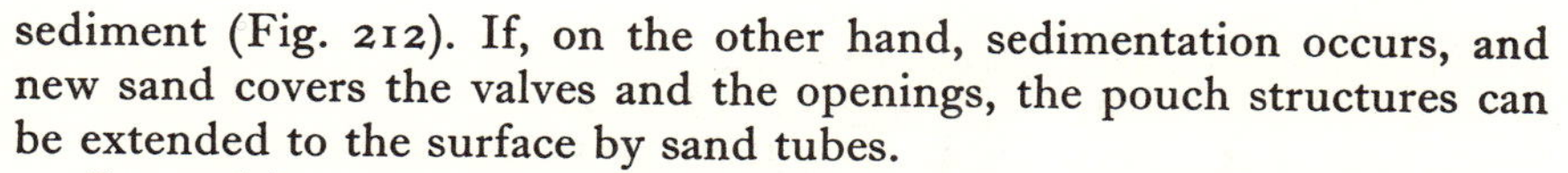

FIG. 212. U-shaped tube of *Polydora* in calcareous valve of a bivalve. The septum between the parallel branches of the tube and the emerging ends of the tube consists of cemented sand grains

sediment (Fig. 212). If, on the other hand, sedimentation occurs, and new sand covers the valves and the openings, the pouch structures can be extended to the surface by sand tubes.

In muddy areas such as Jade Bay inhabited by numerous *Polydora*, lamellibranch valves and limestones are unaffected by pouch holes. A *Polydora*, experimentally transplanted from the muddy sand of Jade Bay, immediately bores into a shell of a lamellibranch or gastropod in pure sand. On the other hand, *Polydora* does not seem to bore on exposed, submarine limestone cliffs or boulders unless the substrates are surrounded by sand.

The holes drilled into a single lamellibranch valve are often numerous. They are then similarly oriented, with all U-bends more or less downward. Parts of the valves that either project above the surface or lie more than 3 cm deep in the sediment do not bear any traces of etching.

Waste stowing

Seilacher (1956a and 1957) used the analogy with mining, and a miner's term, to describe a peculiar activity of sediment-dwelling animals; he says 'in a way similar to a miner stowing away waste in depleted tunnels (Bauversatz), many inhabitants of the sediment actively fill their own excavations'. Such activity leaves all formerly inhabited tunnels or sections of tunnels filled with sediment with the exception of a last, open, inhabited burrow. As the whole burrow, or a section of it, is constantly shifting while the immediately abandoned part is actively or coincidentally filled in, traces remain of some of the events affecting the builder and its burrow. Including cases in which an abandoned burrow gets more or less accidentally filled, however, Seilacher's definition of 'Bauversatz' has been somewhat extended. This seems justified because in fossils it is often impossible to determine whether an infilling has been the result of active stowing or of coincidental transportation.

Structures due to such infilling show a great variety of often quite characteristic forms; they were first found as fossils. In all cases, they consist of groups of tunnels or tunnel sections that show certain spatial relationships to each other (Fig. 213). Recent counterparts seemed to be missing until Seilacher pointed out that living organisms do indeed build such structures. What was missing according to Seilacher was a method of making these Recent structures visible. This is one of the cases in which a problem of Recent animal ethology was recognised and first investigated by a palaeontologist. Even now the finest details of infilling structures do not always show as clearly in Recent examples as in fossilised structures. Diagenesis and differential infiltration of colouring solutions can be helpful in emphasising small differences of texture. Reineck's methods (1957, 1958a, b, c, d) of sampling soft, marine sediments, and of hardening, grinding, and polishing them may help further palaeontologically significant investigations of these structures from the taxonomical point of view.

Genetically, two cases of infilling structures can be distinguished. (1) Structures that are due to external influences. Their shape is determined only by external conditions affecting the builder. They are usually caused by more sediment penetrating the dwelling burrow than could be removed, and the animal abandoning the burrow. The shape of the resulting structures is not specific to the species. (2) Structures that bear features which are characteristic for the species. Filled burrows are arranged in series, and their specific characteristics are independent of external events.

To group (1) belongs the filling structure of the burrow dug by *Nereis diversicolor* described by Seilacher. He found it in the clay cliffs of the banks of tidal channels in the tidal flats. The structures consist of 'vertical walls composed of overlapping, rain-gutter shaped lamellae of

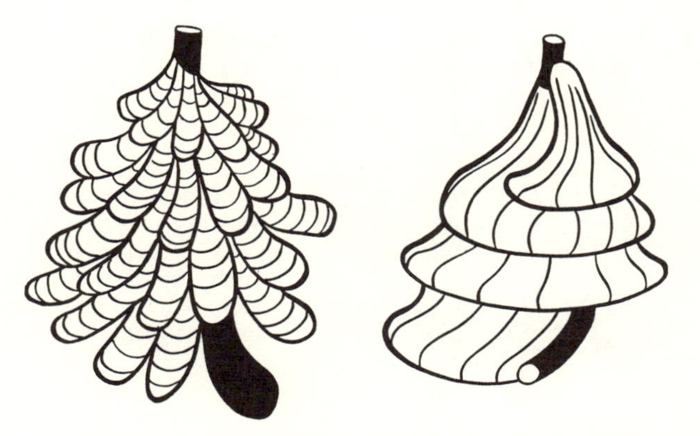

FIG. 213. Fossil structures due to waste-stowing techniques of burrowing. (Black) the last used dwelling tube; left: *Gryphillites*, right: *Daldalus* (after Seilacher, 1957). (Schematic drawing)

filling material (Versatzlamellen)'. They are horizontal, loop-shaped burrows into which sediment fell and collected at the sole; to preserve its habitation the worm began to dig at a level immediately above the filled burrow. As this process was repeated many times a whole pile of burrow loops built up as a result, one lying on top of the other and slightly shifted backward. From thin sections (Reineck 1958a) we know that *Nereis* always presses sediment which falls into the burrow sideways against the walls, so that a multilayered lining is formed (called 'Räumauskleidung' by Reineck; Fig. 168). The burrow walls become quite resistant because the worm mixes the material to be pressed against the wall with plenty of mucus. Even before any diagenesis the tubes are thus much more resistant against erosion than the surrounding sediment and can be washed out by currents.

On occasion the lining is applied to one side of the *Nereis* burrow only (section on mucous burrows, D I (*b*) (ii)). Then the burrow shifts sideways (Fig. 214). Reineck mentions this fact and the implication that one-sided application of successive linings represents a type of waste stowing. It has not yet been confirmed whether faecal material is incorporated into the linings.

Seilacher (1957) included the locomotory traces of lamellibranchs like those of *Mya arenaria* in the waste-stowing type of structures (Bauversatz). These traces are formed when sedimentation forces siphon-carrying lamellibranchs to move upward until they again reach the surface with their siphons. Seilacher described the resulting locomotory traces as 'cone-shaped lamellae, stacked like cups that sit vertically in clay' (see also Schäfer 1956a). In the author's opinion such locomotory traces should not be included in this group because this would lead to the inclusion of every rhythmically produced locomotory trace through

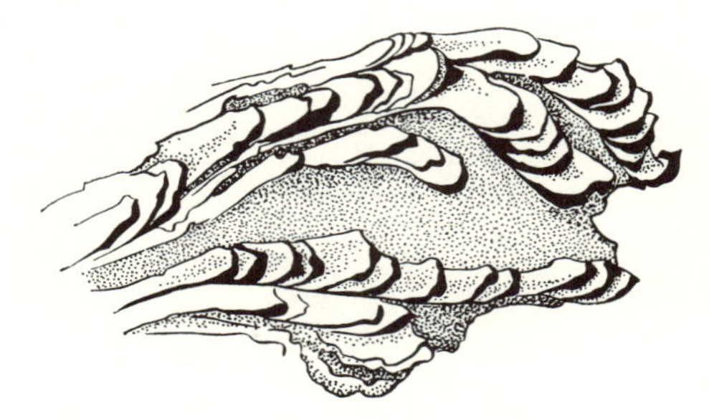

FIG. 214. Compressed mud lamellae, formed by lateral stowing of waste sediment by *Nereis*. Exposed by erosion, such lamellae withstand erosion longer than the surrounding sediment. (Actual size)

sediment. To avoid such all-inclusiveness, the group should be restricted to cases in which open burrows are gradually moved, not to those in which the moving animals simply displace the sediment on their way as in the case of the lamellibranch. Lamellibranch traces were therefore discussed in D I (*b*) (iii) on escape.

To group (2) belong the burrows of *Corophium volutator* that have been discussed by Seilacher, and all other burrows produced by shifting a U-bend downward while the old U-bend is filled up. The inhabitant stows into the previous cavity material specifically procured for this purpose. The way in which the tube is shifted and the material stowed or packed into the abandoned portion of the tube is distinctive for each species, and with one species the techniques are consistent and independent of external conditions. *Corophium* is a good example of the complex behaviour involved in producing a continuously shifting U-bend, in an invariable sequence of actions, as long as it builds in mud or clay. Yet, in clean sand the same animal builds a pit-burrow. The pouch burrows of *Polydora ciliata* in calcareous lamellibranch shells also belong to this group with continuously shifting U-bends.

It is to be expected that more Recent burrows of this type will be found in shallow seas. The numerous misinterpretations of burrow structures in the older literature and the late discovery of the first Recent examples prove how very difficult it is to see them in recent sediments.

Sediment-eating animals are more likely to stow waste material in disused tube sections but so far the author can prove only three cases, the polychaete worms *Heteromastus filiformis* and *Arenicola marina*, and the sea urchin *Echinocardium cordatum*. The burrow of *Heteromastus* has been described. It consists of a vertical faecal tube that leads from the surface down to several branched feeding tubes. In two instances the author could discern distinct stowing structures in connection with curved feeding tubes (Fig. 215). They were lying parallel and obliquely above a section of the latest feeding tube. This means that they were

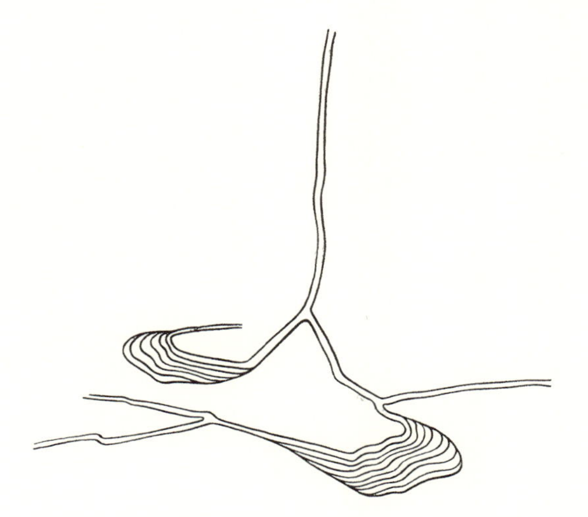

FIG. 215. Waste stowing in feeding trails of *Heteromastus filiformis*. (Schematic drawing)

formed when a loop of a feeding tube had been successively shifted outward and downward. It seems unlikely that the former loops were filled with faeces because *Heteromastus* usually takes the faeces to the surface. The filled tube sections were not hardened; in water they collapsed as rapidly as the walls of the burrow in use. They became visible only because the sediment happened to break along the plane of the filled burrows.

The stowing structures of *Arenicola marina* also belong to group (2); the author observed them in the muds of Jade Bay. In the section on mucous burrows (D I (*b*) (ii)) it has been explained how the shape of the *Arenicola* burrow changes from a U to a J, depending on how easily the sediment flows. The more sandy the sediment, the shorter the feeding tube; in mud the feeding tube continues nearly to the surface. If *Arenicola* settles in sand which is subsequently covered with mud, the animal is forced to ascend and ends up living in the mud, a less favourable substrate. It does not glide at all, and *Arenicola* needs a long feeding tube. Once the ingesting proboscis has penetrated from below to the surface of the sediment it must start another feeding tube from the bottom. The course of the new tube is slightly offset against the previous one (Fig. 216). Sediment that in the meantime has fallen into the old tube is

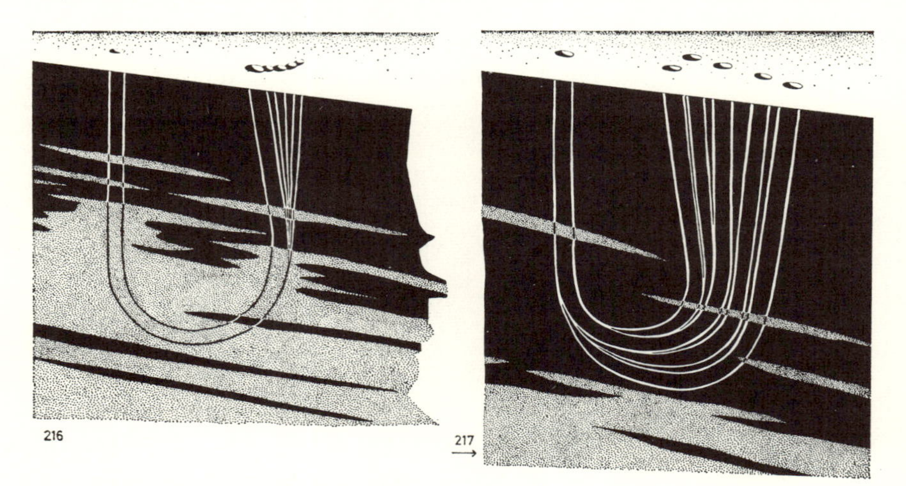

FIG. 216 (*left*). U-shaped burrow of *Arenicola*. Feeding funnels of subsiding sediment cannot form in a layer of mud (black) through which the top part of the burrow is built; therefore the worm must return to the surface repeatedly to obtain new sediment for feeding. The result is a stowing structure in the feeding tube

FIG. 217 (*right*). U-shaped burrow of *Arenicola* entirely in mud. Repetition of the digging of new feeding tubes, as in Fig. 216, here starts at the bottom of the straight portion of the faecal tube.

pressed against the wall of the older tube, producing what one could call a partly filled-in tube. To make room for the open space of the new, slightly offset tube, material is shifted so that it lines the old tube. Although *Arenicola* is a sediment eater the stowed material does not contain faecal matter.

If the entire burrow of *Arenicola* lies in firm mud the animal needs a whole new feeding tube from the lowest point of the burrow upward every time it wants to feed. Sitting head-down in the faecal tube, it deviates from the old feeding tube and moves up from there for feeding. A series of curved tubes is formed in this way, all of them starting from a single, vertical tube (Fig. 217). In very fluid mud the tubes can be observed three-dimensionally because the walls, having been compacted by pressure, turn into fairly solid shells (Fig. 218). It can then be seen that the waste-stowing structures consist of cylindrical tubes alternating with partly separate half-cylinders.

In the feeding tubes of *Echinocardium cordatum*, at least part of the material being stowed is feeding waste. However, the filled-in material,

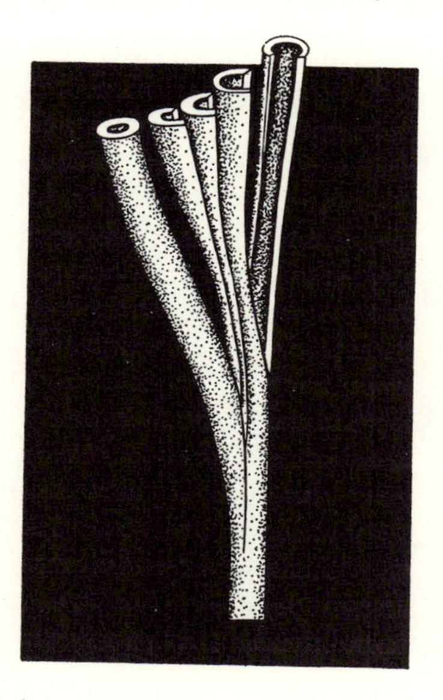

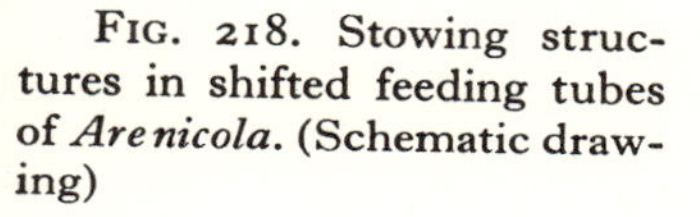
FIG. 218. Stowing structures in shifted feeding tubes of *Arenicola*. (Schematic drawing)

stowed in units that appear crescent-shaped in longitudinal section, contains sand-rich faeces and sediment transported rearward by the shovelling activity of the spines (Fig. 183).

(iii) *Escape*

Animals survive in their biotope, as individuals as well as species, because they are adjusted, as we usually call it, in their shape and function to the mechanical and chemical conditions of their physical habitat and of the living communities to which they belong. In many cases they can escape if things get too bad for them. Heidiger (1942, p. 26) says that escape is the alpha and omega of animal behaviour. An organism's functional relationship to space is of superior rank to all its other groups of functions, and this superiority has its expression in the appearance of the animal (Schäfer 1954a).

The term 'escape' shall be used here in a broader sense. While it is commonly restricted to movement away from an approaching enemy, it here includes any escape effort triggered by an external, adverse cause. Escaping behaviour is usually linked to locomotion. Movements of sessile animals in response to adverse stimuli have all the characteristics of escape but are omitted because no traces of such reactions could possibly be preserved.

The specific escape reactions from enemies of many lower marine animals have been described, e.g. how *Chlamys opercularis* flees from starfish, marine worms from *Lepidonotus squamatus*, *Crangon vulgaris* from higher crustaceans, the Pacific sand dollar, *Dendraster excentricus*,

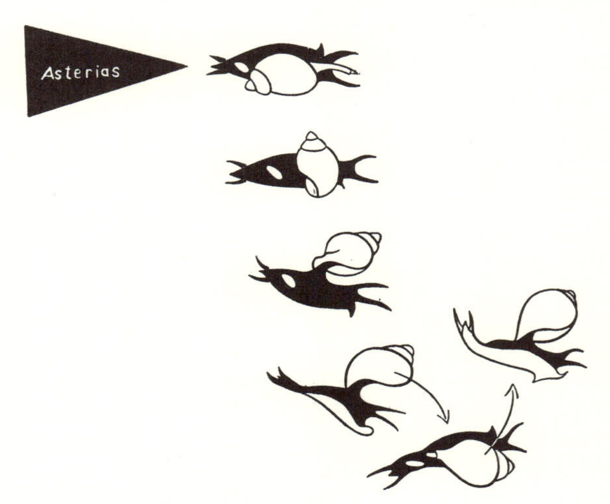

FIG. 219. Six stages of the escape reaction of *Nassa mutabilis* attacked by a starfish (after Hoffmann, 1930. (Schematic drawing)

from starfish (Tinbergen 1952), or how many animals flee from aggressive members of their own species.

Hoffmann (1930) described the escape reflexes in the prosobranch gastropods *Nassa mutabilis* and *N. reticulata* from starfish and sea urchins. The gastropod receives the stimulus with the central part of the back of its foot. Upon being touched at this place by an enemy, the animal flees in a series of up to ten 'jumps' achieved by flinging the shell from side to side (Fig. 219). Spectacular as they may be, cases of flight from living enemies occur much more rarely in nature than escape reactions from inanimate 'foes', such as changes in the conditions of the surroundings. In our context this aspect of escape behaviour is especially important because the locomotion during flight marks the sediment clearly. Fossil traces can occasionally be identified as expressions of such escape behaviour.

Sedimentation is the most frequent cause for escaping movements; this holds for all benthonic organisms, endobionts as well as epibionts. Each new layer of sediment removes the marine organism living on or in the ground from the oxygen-rich water above the sea floor. After the animal is covered by sediment it can only survive if it succeeds in escaping upward and reaching the new surface. *Many fossils are perfectly preserved for the very reason that they failed in the attempt.*

Because of their palaeontological importance this section will deal mainly with the reactions of lower marine animals confronted with sedimentation, and with the traces in the sediment that result. However, cases will also be described of planar and linear erosion inducing benthonic animals to move downward. When endobionts living in the top sediment layers are forced to try to reach a new, deeper level for settlement, their activities also produce preservable traces. Escaping locomotion from sedimentation usually causes perturbed burrowing textures.

Actiniaria. By far the majority of the actinians, and probably the most primitive amongst them, sit with the muscular sole on a solid substrate without being perfectly and permanently attached (Franz 1924). The solid substrate is usually bedrock, but calcareous shells lying on the sea floor also serve, and occasionally drifting or stationary pieces of wood. With movements of their soles, the sea anemones can achieve minor, slow locomotion on these substrates. They can thus choose the most favourable place for settlement or can correct their position if local conditions change. The animals crawl by lifting a small area of the sole from the substrate, advancing it and placing it down again. This movement continues in a wave over the entire width of the foot. Before a wave has completed the run over the sole a new wave begins to follow the first. This mode of locomotion thus resembles that of gastropods, called glide-crawling. Frequently, localities suitable for glide-crawling and free from sedimentation are densely populated with many different species of actinians. Many species are sensitive to a thin mud cover. Suitable biotopes for them are sediment-free rock surfaces with clear water and strong currents or surf.

Other actinians settle in the sediment. In spite of the different environment the method of feeding is the same. The animal catches swimming and suspended organisms, grasping them with its tentacle arms which are armed with stinging capsules. The body, however, may be more or less deeply embedded in sediment as long as the mouth disk with the tentacles is kept free to allow catching and ingestion of prey, and discharge of faeces. The animal breathes through the tissues of the mouth disk and through membranous septa in the gastro-vascular cavity. Wherever sediments cover the floor of the shallow sea they are constantly moved vertically and horizontally, and sedimentation and erosion occur. Sea anemones living there must therefore be highly mobile and always ready to escape from moving sediments to maintain the essential functions of feeding, respiration and discharge of faeces. Thus several groups of actinians turned into vagile endobionts when they conquered the open sea floor, while the majority still live as conditionally vagile epibionts.

Actinians adapted to life in the sediment use not only the muscular sole of the foot but the entire muscular tube of their body for locomotion. The body responds to horizontal movements of the sediment by changing its shape; it readjusts its position if laterally pushing sediment threatens to tilt it from its normal vertical position; it helps by pushing movements to penetrate the sediment. The animal digs deeper into the sediment by peristaltic movements to avoid being washed out; or if the animal has been washed out all the same it serves to dig the animal in again.

Some species, such as members of the genus *Peachia* and *Sagartia troglodytes*, can consolidate their base by cementing sand grains and shell fragments; warts at the aboral end of the body secrete a sticky substance for this purpose. Other species curve all the edges of the foot disk downward to produce a pouch in which they hold a lump of mud or sand; probably, this serves to weigh down the lower pole of the body, making it easier to maintain a vertical position in half-liquid mud; moreover, this widening of the lower pole of the body offers some resistance if the animal wants to push itself upward. (According to Pax 1928 and 1934, *Hormathia digitata* and *H. coronata* hold a pebble in the folded sole of the foot, probably for the same purpose.) Furthermore, there are taxa, such as *Peachia*, that have a 'physa' at the lower pole of the body; this is a hemispherical swelling of the foot which can bore and push by swelling and contracting, thus helping to keep the body in position. The physa swells and contracts by forcing water in or out through small pores in the muscular, external wall. In other species the entire muscular tube can perform peristaltic movements. Yet others, mostly worm-shaped and elongated, can move by slow, undulating movements, pushing the body upward, downward, or sideways, depending on the sediment.

Little is known about the locomotion of these animals apart from contributions by Osburn (1914), Parker (1915 and 1917), Portmann (1926), Siedentop (1927) and Willem (1927). Adaptations of individual species to life in sediment have not yet been investigated. The following species of actinians live in the North Sea in sediment (Pax 1928 and 1934).

Limnactinia laevis lives in the mud of the coastal waters below the tidal zone at a depth of 55 to 127 m. *Edwardsia andresi* inhabits soft ground, 150 to 600 m deep. *E. longicornis* favours soft grounds containing sand, fine gravel and fine shell breccia at depths between 15 and 28 m. *E. tuberculata* also lives in soft ground 11 to 326 m deep. *E. pallida* inhabits soft grounds mixed with fine sand, gravel and fine shell fragments at depths of 15 to 28 m. *E. danica* lives in 13 to 673 m deep soft ground. *Paraedwardsia arenia* inhabits muds at depths of 55 to 673 m. *P. sarsii* inhabits both sand and mud floors at depths of 11 to 200 m. *Nematostella vectensis* burrows in mud in brackish water. *Halcampoides purpura* lives in soft ground 21 to 110 m deep, *Peachia hastata* in 0 to 230 m deep soft ground, and *P. boeckii* in 146 to 684 m deep soft ground. *Eloactis mageli* also favours soft ground at depths of 3 to 647 m. *Bolocera tuediae* inhabits mostly muddy ground 10 to 2000 m deep. *Halcampa duodecimcirrata* lives in pure or sandy mud 10 to 85 m deep. *Actinauge richardi* lives on mud and sand but also on bivalves and stones at depths of 91 to 728 m; *Sagartia troglodytes* on stones, but also freely in sand and mud 0 to 50 m deep. *Actinothoë anguicoma* lives on sand and on mud, also on stones and lamellibranch shells 0 to 50 m deep. *Kodioides borleyi* is a mud inhabitant and lives at a depth of 36 m.

Without exception, sea anemones living in sediment can dig and thus counteract moderate movements of the sediment. If sediment covers the peristome with its tentacle crown the animal immediately reacts by pushing its body upward until the peristome has again reached the surface. This is an escape reaction. The successive, individual movements are as follows: the aboral end of the sea anemone swells with an intake of water and forms a broad base which offers a plane of resistance for all the movements in the body wall. Now a circular wave of contraction of the circular musculature runs through the wall from below to above; it stretches the whole body and thus pushes the oral body end upward. The tentacles are retracted and shortened during such movements and fold together into a narrow bundle. Next, the oral section broadens and thus anchors itself firmly in the sediment so that the whole body can be pulled upward. The pulling is done by the longitudinal muscles in the septa which normally serve as retractors for the tentacles and the mouth disk. In several pushes the sea anemone finally reaches the surface with its mouth and stops there.

The ability to dig upward varies. *Sagartia troglodytes*, for example, living in the beach areas of the tidal flats, is unable to penetrate 8 cm thick layers of sediment; it completely contracts its body and dies.

Since actinians do not secrete mucus while they are digging the digging trail does not become consolidated; the penetrated sediment flows round the body into the space left by the retreating animal. There the sediment comes to rest, still bedded according to the sequence of new layers which were penetrated. Thus distinct, vertical sag trails are produced in layered sediments. If waves in shallow water wash out sea anemones they rapidly burrow downward again wherever they land, using the same movements for upward burrowing in inverted sequence, until they have reached a suitable depth. During slow, planar erosion of the sea floor, they wander downward concurrently with the rate of erosion.

Polychaeta. As described in D I (*b*) (ii), many marine polychaete worms live in mucous burrows whose walls are consolidated so that fresh sea water can circulate inside the tube. The same animals, however, are also capable of moving in the sediment without secreting mucus and hardening the sediment. They never secrete mucus if their dwelling burrows are covered by sediment and have to be abandoned. The ascending trails to the new surface are strictly escape trails, and since they are free from sticky mucus, they are simple, often very thin sag trails or perturbed burrowing textures. They are always upward extensions of burrows with hardened walls and indicate sudden, rapid sedimentation in an area that is normally free from sedimentation. Once the worm has ascended to the surface and the escape trail has thus re-established connection with the open water, the animal begins to secrete mucus without leaving the sediment. It makes the walls self-supporting and thus turns the upper part of the escape trail into a new dwelling burrow (Fig. 169, see also Reineck 1958a).

When *Nereis* converts a flight trail into a dwelling burrow it simply secretes mucus and presses the body in peristaltic movements against the mucus-coated wall. Polychaetes whose dwelling burrows consist of agglutinated tubes, on the other hand, must rebuild a wall of individually laid grains. This is especially evident in *Lanice conchilega.* Here the upper part of the dwelling tube that projects above the surface is made of coarse sand grains and shell fragments and bears a crown-like cluster of fringes built of sediment particles. Whenever heavy sedimentation forces the worm to transfer its tube upward, a new fringed crown must be built because it is necessary for feeding. The old fringes, however, mark the levels at which additions were made and thus also the frequency and magnitude of episodes of sedimentation (see also Ziegelmeier 1952). If the sedimentation is particularly heavy and rapid and the old burrow can no longer be lengthened, *Lanice* also flees to the new surface; the path of escape is marked only by a narrow sag trail in the sediment.

Lanice and other inhabitants of self-made tubes are stationary because they can only leave their tube if they abandon it. However, in some cases, the self-built tube has almost become an organ of the animal and moves with it. The polychaete *Pectinaria coreni* is a good example; it lives in the sandy-muddy coastal areas of the southern North Sea and on the tidal flats. *Pectinaria* feeds on sediment and small animals (see

D I (*b*) (ii)). When feeding, the narrow, chimney-shaped posterior end of the quiver-like sand tube just protrudes above the sediment permitting respiration. The quiver is held vertically or slightly obliquely, and the grazing plane lies a quiver length below the surface of the sediment. All horizontal locomotion in this position serves for grazing and is so extensive that the beds within a *Pectinaria* settlement are partly destroyed by movements of the tubes. Locomotion is achieved by the scraping activity of the broad and hard setae of the head which can also push and pull. All other locomotion serves for escape and is always caused by new sediment covering the upper end of the quiver. The animal then pushes upward into the new layers until the top of the quiver again reaches the surface. Upon erosion the worm retracts into the sediment until the upper end of the quiver just protrudes a little above the surface.

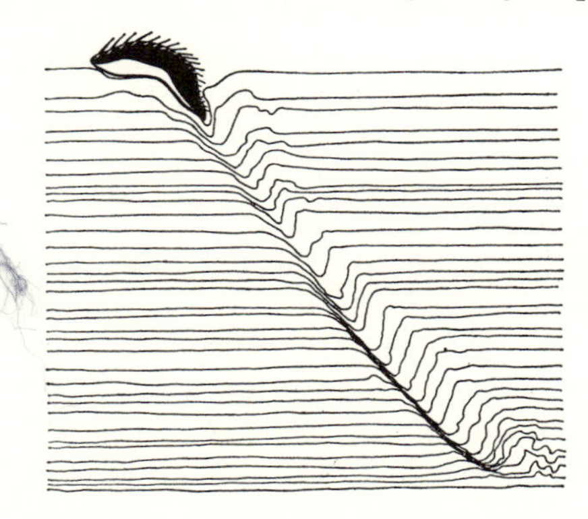

FIG. 220. Escape trail of *Aphrodite aculeata* in form of a sag trail. (Schematic drawing)

The large polychaete *Aphrodite aculeata* lives under the same conditions. This worm also needs a constant connection to the surface for breathing as it moves to feed on sediment and small animals. It brings the breathing water into its body in regular thrusts through the tail end which serves as a double siphon. Escape reaction sets in when this siphonal mechanism is covered by new sediment. If the covering layers are not thick *Aphrodite* pushes its body upward and backward by backward multiple, circular shovelling with its spiny bristles, until the siphon has regained connection with the surface and the animal can breathe freely. Thus constant moderate sedimentation results in a series of resting traces, one obliquely above the other (Fig. 4 in Schäfer 1956a). *Aphrodite* does not secrete mucus, leaving the sediment free to slide and flow. Each resting trace is filled by sediment that slides in from the sides. As a result, a broad, steeply oblique sag zone remains in the sediment. If a new sediment layer amassed above the resting place of *Aphrodite* is too thick, the animal crawls forward in a curve to the surface, leaves the sediment altogether, and buries itself at a new place, until only the tail which serves as the siphon protrudes to the surface. *Aphrodite* is very strong and can penetrate 25 to 30 cm thick layers of sediment by means of shovelling only. At times, the layers in the resulting digging trails are completely destroyed, turning them into perturbed burrowing textures. At other times, the sag zones resemble faults (Fig. 220).

Lamellibranchiata. Escape movements of lamellibranchs, triggered by sedimentation and erosion, are particularly pronounced. Whereas coelenterates and polychaetes possess locomotory organs capable of coping even with extreme movements of the sediment in their habitat, this does not hold for the lamellibranchs. The most frequent cause of death of lamellibranchs is erosion or sedimentation beyond their ability to escape. Proof of this is the frequent, sudden increase in lamellibranch shell breccia in the shallow sea during periods of bad weather; shifts of the local level of sediment are a constant danger to lamellibranchs.

The depth of burrow reflects the length of the siphon. In sediment that is neither eroded nor accumulated a sediment-living lamellibranch starts at the surface upon completion of its stage as a free-swimming larva. The animal digs down, foot ahead, according to the length of its siphon. The foot of the lamellibranch would have this one task, were the place of settlement free from erosion; and this task would be completed as soon as growth had ceased and the final depth had been reached (see also Reineck 1958a). Under these conditions, there is no need for escape locomotion. Such conditions do prevail in the deep sea. Reineck has described and illustrated the digging trail of a deep-sea lamellibranch. It is a conical digging core tapering toward the top, and the cut-off layers are dragged downward all around it. The originator always sits at the bottom of its burrowing trail.

Such digging cores are very rare in the shallow sea. Strong bottom currents rapidly erode the sea floor; to compensate, the lamellibranch must descend more rapidly than it normally would in response to the growth of its siphons; moreover, even an adult animal must keep its foot fit as a locomotory organ. If during periods of bad weather the rate of erosion is more rapid than the animal can descend it is eventually washed out. Not all endobiontic lamellibranchs are capable of burying themselves again; usually they die. If new layers are deposited on the sea floor the clam must give up its old resting place and move upward by pushing movements of the foot until its siphons again reach the open water. Because the youngest animals with the shortest siphons settle next to the surface they are in the greatest danger of being washed out and transported away by currents. Particularly during the early post-larval period they are in great danger, being barely capable of holding on to the sea floor in a bottom current.

Adult *Mya arenaria* sit 25 to 30 cm deep in the sediment. The young animals, to a shell length of approximately 10 mm, are provided with byssal glands whose long webs, to which sand grains adhere, hold the small animal in place like drag anchors. In very dense settlements of young animals, with 10 000 animals per square metre, the numerous byssal fibres cover the sea floor with a tightly woven network. This mesh considerably decreases the sediment mobility in the area of settlement (Kühl 1951, 1955). Nevertheless, if the young, growing animals

FIG. 221. A young *Mya arenaria* burrowing upward after being covered by sediment, in foot-forward position. (Schematic drawing)

are covered with mud they can dig upward to the surface by a special technique. Turning upside-down they dig upward foot-first by pulling movements of the foot (Fig. 221). In doing so they retract their short siphons. This ability to crawl upward foot-first is apparently lost when the siphons grow longer. According to Kühl (1951) these young lamellibranchs can penetrate 10 cm thick mud in two to ten hours. Kühl observed that the digging trail may either be straight or run in spirals (Fig. 222) when the lamellibranch repeats upward pulls with asymmetrical extensions of its foot.

Even after the byssal gland has atrophied and the ability to dig upward foot-first has been lost, the lamellibranch can still move up or down. The large and strong foot can still be used to push the body upward, pull it downward, or to dig in again after the animal has been washed out. *Mya* with 18 mm long valves has a foot of the same length. With increasing age, this proportion shifts, the valves becoming relatively bigger. Thus the older and larger the animal, the less effective is its foot for digging. But the necessity to dig downward becomes less as the lamellibranch settles at increasingly deeper levels where the probability of being washed out decreases.

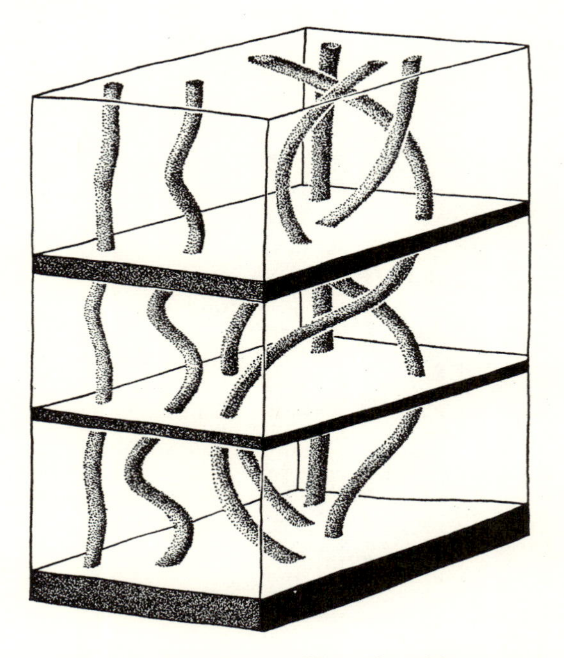

FIG. 222. Spiralling trails produced by young *Mya arenaria*, escaping upward from being covered by sediment. (Schematic drawing)

In the long run sediments frequently turned over by waves are unfavourable for the settlements of growing *Mya*. Therefore the conditions on the tidal flats of the eastern inner Jade Bay, the coast off Butjadingen, rarely permit a *Mya* generation to mature, even though the composition of the sediment and the food supply would certainly be suitable. Every year *Mya* larvae do indeed settle over the entire tidal flats, and the animals often reach a valve length of 2 to 3 cm; but every year, one or two strong, westerly gales dig up the settlements of young animals and wash them as shell breccia to the coast before the animals can burrow again into the sediment. The animals either die on their way to the coast and their clean valves end up in the swash marks, or great heaps of decomposing bodies accumulate along the coast if transportation is more rapid (Schäfer 1941b and 1950c). Only a few individuals who happened to have settled in sheltered places remain in the sediment and continue to grow; as they dig deeper, they withdraw from the zone of danger and finally reach, considerably reduced in numbers, the more permanent layers of the sediment. (For a history of settlements of *Mya arenaria* and estimates of the length of periods of their temporary absence from the coasts of the North Sea and the Baltic, see Hessland 1945; for *Macoma baltica* see Lammens 1967.)

Lamellibranchs without long siphons that must live near to the surface are always liable to be washed out. They therefore maintain the ability to burrow, in contrast to *Mya* which lives at a deeper level. The foot remains strong and large compared with the size and weight of body and valves. In many cases the valves are strongly curved and circular in outline, or flattened on one side; these features facilitate erection from a sideways lying position and thus burrowing. In addition, the outsides of the valves are frequently provided with scaly teeth or ridges which prevent slipping during erection. Where such supporting features of the shape of the valve are missing the foot is the only means of locomotion in the sediment. A comparative functional and morphological investigation of the lamellibranchs would furnish abundant relationships between the shape of the calcareous shell and the mode of locomotion and respiration; and many so-called ornamental features of the shells would be better understood. Until now such considerations have been applied almost exclusively to boring lamellibranchs in which the relation of valve shape and function are especially conspicuous. Seilacher (1954b) successfully applied such ecological and morphological considerations to the Triassic lamellibranch *Lima lineata*.

A Recent example is the genus *Cardium* all of whose species have very short siphons and therefore settle close to the surface: *Cardium edule*, *C. crassum*, *C. exiguum*, *C. echinatum*, *C. paucicostatum*, and *C. fasciatum* in the tidal flats as well as in the deeper areas of the southern North Sea. All are exposed to rapidly shifting sediments in the stormy, shallow sea. However, the more or less spherical shape of the closed shell and the strong foot enable them to burrow rapidly, if exposed, and thus to survive in their biotope (Kreger 1940). If they are covered by sediment they equally swiftly ascend to the surface by pushing movements. Great quantities of sediment, however, can immobilise them, and they die in the living position from lack of oxygen.

The Solenidae solve the same problem in a very different way. *Ensis ensis* and *E. siliqua* occur frequently in the North Sea. They have up to 16 cm long, very narrow, rectangular valves that are open at both ends and leave room for the passage of the foot at the anterior end and for the siphons of unequal length at the back. Everywhere else the mantle is closed. Generally, the lamellibranchs sit buried in sandy mud up to the upper edge of the valve so that the siphons can connect with the sea water. The animal can move swiftly, and in moments of danger, it digs so deep with its foot that its short siphons can no longer reach the surface. When it is withdrawn into the shell it takes up a third of the space between the valves. The lamellibranch digs with the pointed end of its foot fully extended downward in the direction of the long axis of the valves; then the animal broadens the tip of the foot into a thick disk. Subsequent contraction of the foot muscle pulls the soft body and shell downward. Locomotion by pulling (see D I (*a*) (vi)) is highly perfected in this case and is helped by the long narrow shape and smooth surface of the valves. Despite their great digging ability these lamellibranchs are quite frequently washed out; but they can rapidly bury themselves again by pulling with their foot, although the valves are not shaped to the best advantage for this purpose (Fig. 114). If the animal is covered by sediment it ascends by pushing with the foot. While the foot is being withdrawn and does not support the body, the lamellibranch opens its valves and jams them against the sediment, keeping the mantle closed. The Solenidae can also move sideways through the sediment. Being well adapted to life in the sediment and to quick escapes from unfavourable conditions they can survive in areas in which the sediments are frequently redeposited and shifted. Von Buddenbrock (1912 and 1913) pointed out that they have effective statocysts which play a role in burrowing.

Some lamellibranchs live constantly on the sediment surface. They have no siphons nor a strong foot for burrowing. They can therefore generally not penetrate through sediment that covers them and can only live where sedimentation is slight. The large, thick-shelled *Cyprina islandica* is an example. This animal avoids the southern coastal areas of the North Sea; probably the high summer water temperatures are not the only reason for this, but the animal also could not survive large-scale sedimentation. It lives in mud in the central and northern North Sea. *Ostrea edulis* is even more sensitive to sedimentation. *Mytilus edulis* can attach itself with byssal fibres to shells of other individuals of its own species, if sedimentation is very slight, and pull the body slightly upward to avoid burial. The individuals thus brace themselves against each other. *Chlamys opercularis*, on the other hand, never allows sediment to cover it; the animal can swim over wide distances by flapping with the valves. As soon as rapid sedimentation sets in *Chlamys* swims off (swimming of Pectinidae—see von Uexküll 1913, Bauer 1912, von Buddenbrock & Moller-Racke 1953).

The escape of lamellibranchs leaves very distinct digging trails in the sediment, the width of the trace corresponding to the size of the animal;

and where the density of settling of one species is high the digging trails must be numerous. However, the textures produced by wandering lamellibranchs can only be picked out by special treatment, such as Reineck's method (1957 and 1958c and d) of hardening with paraffin and then making thin sections. This explains why so little attention has been paid to them in Recent sediments.

Almost all sediment-burrowing lamellibranchs move in the same way; the valve-protected and therefore rigid body is moved vertically by pulling and pushing movements of the foot; the animal disrupts bedding layers in its path. According to Reineck's definition we can distinguish digging core and digging aureole. The digging core ('Wühlkern') comprises the sediment zone which roughly corresponds to the diameter of the closed valves. Occasionally coarse sediment particles and small fragments of lamellibranch and gastropod shells can be concentrated in the digging core; these coarser particles have been separated from the surrounding sediment by the beating movements of the lamellibranch foot and by the water discharged from the mantle cavity, while the finer grains are pushed or washed sideways into the interstices between sand grains. Both upward-pushing movements toward the surface and the downward-pulling movements loosen sandy or muddy sediment so much that penetrated layers sag downward, often quite far, regardless of whether the animal has moved up or down. Thus the direction of deflection of cut-off layers does not indicate the direction of movement of the animal. Top and bottom of a fossil specimen, however, are recognisable.

The curved portion of the cut-off layers is what Reineck calls the digging aureole ('Wühlhof') of the lamellibranch digging trail. If a digging trail is cut off-centre by a section, beds in the aureole appear as simple sag zones. The nearer a section comes to the centre of the burrowing trail, the deeper the beds. In a central section, both the digging aureole and core can be recognised (Fig. 223). The digging core is not always bedded but individual beds retain their identity if they are thicker than the height of the lamellibranch penetrating them; then only a sag trail indicates the passage of a lamellibranch through the layered sediment. Seilacher (1957) classified the textures produced along the trails of escaping bivalves together with those burrows whose disused sections have been stowed with waste.

Since the lamellibranch traces in sediment lack any features indicative of individual species, neither recent nor fossil traces can be used for comparative morphological studies. Only if the hard parts of a lamellibranch are found in living position at the end of a burrowing trail can such a trace be unambiguously identified. This case is unusual in fossils because most dead lamellibranchs are at one time or another washed out and transported away from their last digging trail, at least if they have lived in a shallow sea with the constant changes in erosion and sedimentation. The chances of death are unequal on an upward or downward escape path. In the course of upward flight the lamellibranch passes through successively younger sediments which are easier to penetrate the higher the lamellibranch digs. Downward escape, on the other hand, leads only too often into consolidated, subfossil layers through which the animal can no longer burrow. Erosion is therefore a more frequent cause of death than sedimentation. This also explains why in shallow seas, recent as well as fossil shell fragments and separated valves in secondary deposits are much more frequent than dead bivalves in living position.

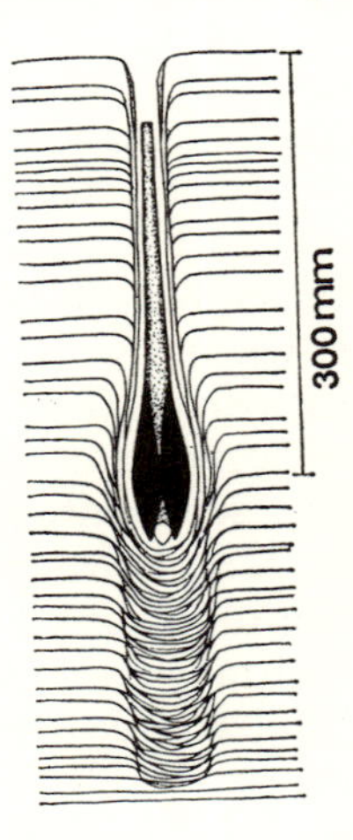

FIG. 223. Bivalve making an escape trail on its way up through bedded sediment. The digging core (completely perturbed) and digging aureole (deformed bedding) can be seen below the animal. Vertical layers around the valves and along the siphon are due to reaming and lining

Excessive sedimentation, however, can also be deadly. The weight of suddenly added sediment can arrest a lamellibranch and prevent it from reaching the surface; it then dies in living position in its burrowing trail. Elongate and oval lamellibranchs move upward only by the pushing with the foot, and their body remains vertical. However, *Cardium* which is circular in outline performs strong rolling movements with its valves in the sediment while it simultaneously pushes with the strong, bent foot. Even after the weight of new sediment has arrested upward progress the animal is still for a while able to roll from side to side, trying to find new leverage for another upward push. *Cardium* frequently die in this starting position. Then they are not found standing vertically at the end of a more or less long digging trail, but they lie on the right or left valve, as demonstrated in Reineck's experiments.

Where thick layers of sediment are laid down quickly, erosion on a similar scale normally follows a little later. Heavy sedimentation in the shallow sea is usually caused by storms producing bottom-touching waves. They take sediment from one place and deposit it nearby, perhaps above a lamellibranch which gets killed. But during the next gale, these same sediments are again transported away, organisms embedded in living position are washed out, and their last digging trails destroyed. Traces of living activities are rarely observed not because they are indistinct but because the textures themselves are rarely preserved.

Not only locomotion of the whole lamellibranch leaves an impression in the sediment, but the activity of the siphons also leaves traces and produces perturbation textures. The slender muscular tube of the siphons is strong enough to penetrate considerable thicknesses of sediment. Probably, the pushing movements of the siphons also help to loosen the sediment above a lamellibranch trying to reach the surface.

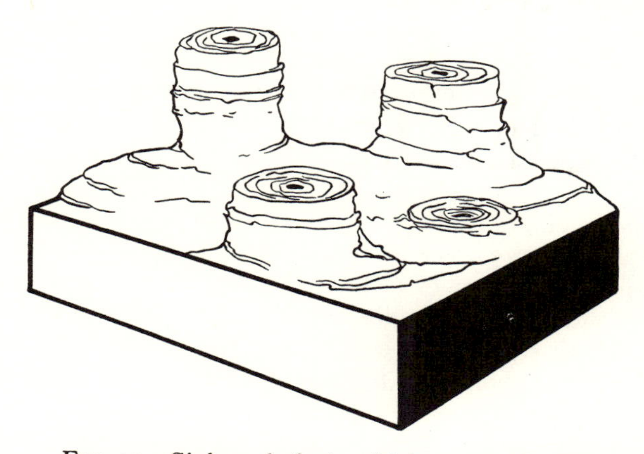

FIG. 224. Siphon shafts in which concentric linings with reamed material were compressed and consolidated. These more resistant bodies form chimney-like elevations upon erosion of the surrounding sediment. (Actual size)

This effect is comparable with that of the 'bolting' of worms. The epithelium of the siphons of many species secretes mucus which cements and consolidates the walls of the siphon pits. However, these traces are rarely preserved because they are usually destroyed before final sedimentation occurs, either by erosion or by the upward-moving lamellibranch itself.

Except for the detritus-eating forms, living lamellibranchs generally remain in one place as long as erosion or sedimentation permits. Weeks and months can pass before substantial changes occur in the level of the sediment surface. But even during such quiet periods the lamellibranch living in the sediment must cope with detritus and fine sediment particles that enter its siphonal pit. The siphon and the valves press these particles against the wall of the siphonal chimney and of the resting place of the animal; thus it keeps making room for itself. Each addition of sediment into the siphonal chimney results in a new lining of the walls with a new coat of sediment which is pressed on by opening the valves and moving the siphon. This lining is much more compacted than the surrounding sediment as a result of frequently applied pressure and of mucous coating (Fig. 223). Reineck calls this 'exogen-bedingte Mehrwandigkeit' (multiple coating dependent on foreign material). If such lamellibranch sites with several linings in mud are exposed by erosion their hard consistency becomes evident; the walls resist erosion and a circular rampart still remains long after its builder has been washed away (Fig. 224). If only minor erosion occurs the multilayered, consolidated walls of the siphonal chimneys project above the surrounding surface. Then new sedimentation may first fill the gaps between the chimneys of the lamellibranch site and the vertical, concentrically layered walls that were originally an internal trace, may be partly cut off by the flowing water and lie flush with the new surface. Finally, with more sedimentation, they lie flush with a bedding plane.

Scaphopoda. The scaphopodan families of the Dentaliidae and the Siphonodentaliidae are also endobiontic and capable of escaping vertically from addition or erosion of sediment that would otherwise kill them. Whereas the endobiontic lamellibranchs keep connection with the surface by means of their siphon, the scaphopods use the posterior end of their calcareous, tubular shell that in shape resembles somewhat an elephant tusk.

The organisation and locomotion of the scaphopods has been described above in the section on locomotion by pushing and pulling (D I (*a*) (vi); Fig. 115). They move horizontally through sediment to find fresh sand in which they collect small organisms with their tentacles and detritus from the interstitial spaces of the sediment. Movements of the foot loosen the sediment and greatly help the work of the fine tentacles. The animals move vertically as soon as planar erosion at the surface washes too much sediment away, or as new sediment covers the posterior opening of the shell which functions as a siphon; these movements are thus used to escape adverse conditions. Fig. 139 shows the trail produced by vertical movements of the animal, a narrow, unspecific sag zone in abutting sediment beds.

Gastropoda. Many marine gastropods (Prosobranchia and Opisthobranchia) live on the surface of the sediment or on plants. They are mostly grazing animals or predators. Prosobranchs are known to flee from enemies. They do so only in response to a specific, chemical stimulus. With mechanical stimuli the gastropod retreats into its shell.

However, the majority of marine gastropods, whether they live on the surface of the sediment or burrow through it, have to cope in some way with the consequences of sedimentation. Digging and burrowing forms such as *Lunatia*, *Natica*, and *Nassa* can easily dig through thick new sediments. Such vagile epibionts as *Buccinum*, *Neptunea*, *Hydrobia*, *Rissoa*, *Littorina*, *Nucella* and many others are also capable of doing the same.

An example of vagile epibiontic habits among the prosobranchs is the whelk *Buccinum undatum* (Schäfer 1956a). It lives from the lower littoral of the North Sea down to a depth of 1000 m. Heavy mucous secretion at the sole of the foot, mainly from glands at its frontal edge, cements the sediment wherever the animal crawls, and thus provides a safe crawling substrate on any surface. Moderate sedimentation does not disturb the locomotion of the animal.

The edge of the mantle forms a long, flexible siphon through which breathing water is whirled over the gills; it is discharged from the respiratory cavity at the right side of the mantle. In spite of its weight the shell with the siphonal groove and the siphon can be turned in any desired direction. These turns of the shell are executed by the columella muscle which connects the head and foot region with the shell and supports the shell while the animal crawls. The muscle is attached at a circular zone of the outermost whorl and runs forward from there over the smooth, cylindrical columella near the aperture and continues backward into the foot to the base of the operculum. Contraction or relaxation of the columella muscle not only raises or lowers the shell but also rotates it about its long axis because the columella over which the muscle is pulled acts obliquely both as a pivot and as a lever. As long as the gastropod rests on the sole the distal end of the muscle is anchored and the shell moves. However, if the shell rests on the sea floor the head and foot can be shifted in various directions by the action of the muscle which is now fixed at its attachment point in the shell.

Different combinations of actions of the columella muscle and of the ectodermal muscular tube (Schäfer 1943) produce the movements necessary for locomotion (Fig. 225). Upward digging after heavy sedimentation depends on the mutual displacements of the shell against the head and foot. When the shell rests, the columella muscle advances head and foot in a gliding, rotating movement, and when the foot is anchored in the sediment, contraction of the stretched columella muscle pulls the shell upward, turning and tipping it, toward the head while the outer edge of the aperture acts as a shovel. The foot secretes abundant mucus which forms a tough, sandy-mucous mattress which helps to anchor the foot in the sediment during the upward pull on the shell. In this way *Buccinum* can move through a 20 cm thick cover of sediment within a few minutes. Layers are completely destroyed where the shell has passed; and nearby they are warped upward. Short faults occur along the mucous mattresses that serve as shear planes. The mucus secreted during crawling does not form a cemented tube; as isolated shreds it remains in the sediment. The digging trail of *Buccinum* is almost vertical as a whole, but irregular in detail with different bends and kinks in different layers.

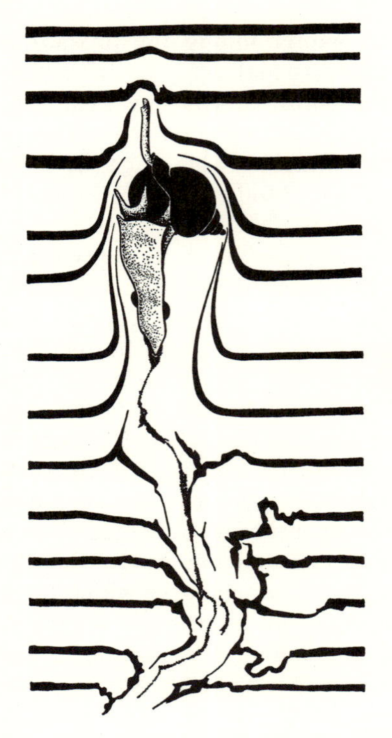

FIG. 225. *Buccinum undatum* piercing covering sediment layers (after Schäfer, 1956a) (Actual size × 0·3)

All other prosobranchs whose crawling soles also serve for locomotion on the surface and are not particularly adapted to digging, move in the same way as *Buccinum* when they are covered by sediment. The pulling movements of the soft body are especially effective in those species whose shell is tall and slender. In all cases, the digging trails are not specific and vary in the different sediments. However, they are characteristic of gastropod facies. Sediments containing *Hydrobia* are especially strongly disturbed because these gastropods live closely together.

(iv) *Resting*

Reduced activity or resting normally exposes animals to increased danger from predators. Therefore special resting places are sought either at the surface or inside the sediment. A resting trace can remain in the sediment or on its suface after an animal has left.

An animal resting on the sediment does not necessarily cause an impression on the surface because the body can be quite buoyant and hardly touch the sediment. Organisms that lie openly on the sediment either need no protection or are camouflaged or are almost transparent.

Lying down may not only serve for resting but also lying in wait for prey. Most animals combining resting and lurking can camouflage themselves. In addition, they often partly bury their body in the sediment. The resting traces produced in this way are especially important because they have the best chance of being preserved. Richter (1926) introduced the term 'resting traces' and Seilacher (1953b) first gave a detailed definition and the following classification for them (see also Seilacher 1953a).

(1) A resting trace is either (*a*) an isolated impression with a distinct outline which corresponds partly or entirely to the outline of the producer if the trace is stamped vertically into the sediment, or (*b*) a simple, local deformation of the original texture of the sediment.

(2) A resting trace occurs as a horizontal repetition of impressions, produced if the same animal shifts slightly sideways after first lying down, thus placing several resting traces beside one another, or produced if individual body parts move slightly after the body has come to rest, and the traces of the parts thus multiply.

(3) A resting trace occurs as a vertical repetition of impressions, produced if the same animal shifts upward or downward and leaves resting traces in several layers above one another. Resting traces produced in this way within the sediment can be particularly well preserved.

(4) Accumulations of resting traces that are oriented parallel to each other because the resting animals all face in the same direction in response to a current.

According to Seilacher, only the inhabitants of the sand surface (epipsammonts) are important as producers of resting traces because for ecological reasons they dig actively every time they come to rest on the surface. Certainly, the same holds for inhabitants of the surface of mud and soft detritus ground (epipelonts). Different digging tools and different organs for respiration are required at a resting place in mud. Resting traces are therefore produced in muds as well as in sands; but shape and preservability differ in the two cases.

The open water and the open surface of the sediment are the main habitat in which the producers of resting traces live and are active. In the author's opinion, it is inappropriate to include as producers of resting traces the entire group of animals normally living in sand (endopsammonts). For example, only in very rare cases can lamellibranchs produce resting traces in the true sense of the word (namely when they make an internal trace on muds below sand, Seilacher 1953b and 1957). If a lamellibranch actually remains in one place it does not produce a trace but it turns into a body fossil. If, on the other hand, a lamellibranch continues wandering through the sediment, grazing, sediment-eating, or filtering, it leaves a locomotory trace. Or it may escape from new sedimentation and leave an escape trail. Or erosion may wash the animal from the layer of sediment where it lived, and the texture of the resting place, although abandoned as required by definition, is destroyed. Animals that normally live lying in sand never leave the sediment of their own accord. There is thus no clear distinction between rest and locomotion, and it cannot be clearly determined where a trace must be defined as 'resting trace with horizontal or vertical repetition' (Seilacher) and where as locomotory trace. To avoid such ambiguities, it seems therefore more suitable to consider as 'rest' only the case of an animal which is normally freely mobile in the open water or on the surface of the sediment and occasionally stops moving, remaining temporarily in or on the sediment without producing locomotory traces. It should be considered as unimportant whether the animal rests to relax after locomotion, to hide from enemies, or to wait for prey.

Seilacher has mainly used the traces of echinoderms (Asterozoa and Ophiuroidea) to develop the definition of resting trace. This suggested itself because asterozoans have such a characteristic outline, and because they tend to shift their position by rotation about their axis of symmetry, and because their digging tools are small enough not to produce any wholesale destruction of the sediment surface. Their resting traces can therefore be easily recognised, both recent and fossil. The resting traces of all other animals, on the other hand, consist of more or less extensive, superficial destructions of the sediment surface and are quite unspecific. They deserve mention only because they are so frequent and are such a prominent cause of the destruction of bedding planes (Seilacher case 1b). Such a trace, seen on the bedding plane or in cross-section, is little more than a local destruction of layers. The reasons are several; either the digging tools of the originators are strong and large in comparison with the size of the body and treat the sediment and its original texture rather roughly, as in the case of the crustaceans; or the outline of the digging animal is characteristic, as in the case of the molluscs; or the body performs clumsy peristaltic, undulating, or pushing movements when it digs and thus destroys any impress of a specific body outline, as in the case of the fishes. Although fossil resting traces must exist in enormous numbers they therefore remain unnoticed and unrecognised as biogenic traces. Possibly, more sophisticated methods of specimen preparation such as thin sections and polished surfaces, or more intensive study of the fine stratigraphic details may make these traces perceptible as consequences of life activities, and may make them the subject of ecological interpretation. Even so, these structures may well be identified as resting traces but never classified taxonomically because they are morphologically unspecific (Seilacher 1953a, p. 440). Yet, the type of possible statement one could make about a bedding plane, say 'marked by resting traces, therefore an inhabited sea floor', would be useful for the student of palaeogeography.

If a burrowing animal impinges on a bed underlying the inhabited one and if this bed is a plastic mud the animal will leave impressions of its burrowing tools (Seilacher 1953a and b) as an internal trace. Such scraping or digging impressions are in some cases specific and also among the most likely to be preserved because they are produced deep in the sediment and therefore not so easily destroyed by currents. Indeed, sand from the overlying bed sifts into the depressions as soon as the producer has abandoned its resting place, filling them as if they were moulds (positive endogenic hyporelief, according to Seilacher). Seilacher thinks that although the formation of internal traces depends on somewhat special conditions, they are more frequently found as fossils than surface traces because of the better chances of preservation.

Crustacea. Fishes and crustaceans are the most liable to produce resting traces. Many crustaceans are highly mobile epibionts or nektonts which spend all periods of diminished activity resting on the sea floor or in the top sediment layers.

The clearest example is the crab *Carcinides maenas* which occurs frequently on beaches and tidal flats. The extensive wanderings of these brachyurans in search of food are described in D I (*b*) (i). Once their hunger is satisfied, the animals immediately dig into the ground and then usually remain motionless for a long time. Also after being seriously disturbed *Carcinides* tries to hide in a position analogous to that of resting; it burrows into the surface sediment of the open sea floor, or on stony ground it pushes its body between or under stones.

The desire to hide is proportional to the seriousness of the disturbance. When an enemy approaches to within about 5 m of *Carcinides*, the crab first increases its speed and tries to run away from the enemy. If the enemy catches up to approximately 2 m a warning or threatening behaviour is added to the rapid fleeing movements: the animal raises its clawed limbs, spreads them and opens the claws widely while it continues running sideways on its eight walking limbs. If the distance from the enemy is further reduced the crustacean finally halts but continues to warn and threaten; if touched it beats its claws against the enemy. *Carcinides* holds its own against an opponent of about equal strength if its chemoreceptors perceive that food is available nearby. In that case the visual stimuli which normally govern the fighting behaviour are reinforced by the chemical stimuli for search of food. Confronted by a superior enemy, on the other hand, who starts to press hard against it, to throw it over, or to grasp its limbs, the warning and threatening behaviour becomes less intensive, and soon all aggression ceases. Such an experience of defeat has a latent effect on the animal for several hours (Schäfer 1954a). This shows in an urgent drive to burrow from which nothing can divert the animal. After being forcibly extracted from its hiding place it immediately begins to burrow again. Even if lifted, the walking limbs continue with futile digging movements.

Only one circumstance can rout an anxious *Carcinides* from its resting and hiding place, covered by sediment. The very first sand grains rolling over its carapace make the animal restless. Before new sediment has completely closed in over the resting place the crustacean leaves but begins to dig at a new place. *Carcinides* is not adapted to life inside the sediment. The need for respiration permits only superficial burrowing. Two slits between the coxae of the last two pairs of walking limbs normally serve for aspiration. From there, the breathing water is pumped into the gill cavity by fast beating of the exopodites of the maxillas, and it is discharged through the mouth (Bohn 1897 and 1902). If the animal lies buried in the sediment, however, the opening that normally serves for respiration lies in the sediment, and only the mouth is sheltered from sediment by the claws. The flow of the breathing water is therefore reversed, and the mouth is the ingestive opening, and the water is then pumped through the coxal slits into the interstices of the sand. The frontal part of *Carcinides* with the eyes and the mouth must remain free. The same holds for many other decapod crustaceans; this includes all those which for resting burrow only superficially into the sediment. Species whose flow of breathing water cannot be reversed never burrow (*Hyas*, *Macropodia*, *Inachus*). If a wide dwelling burrow with self-supporting walls is constructed the inhabitant does not need to reverse the current of breathing water because the animal can breathe normally in the open space of the tube (*Eriocheir*).

Burrowing crustaceans always dig in the same way (*Carcinides*, *Cancer*, *Portunus*, *Geryon*, *Pilumnus*, *Thia*, *Atelecyclus*). The walking limbs, particularly the last pair, scrape the sediment from below toward the sides with the result that the carapace lies obliquely in the sediment. Now the anterior walking limbs supported by the palmae of the claws push the body backward and downward while the posterior edge of the carapace moves up and down in jerks. The displaced sediment is thrown over the descending carapace. Not all the species mentioned are equally efficient; the genus *Calappa* from the Mediterranean is functionally and anatomically particularly well adapted (Schäfer 1954a and 1956a). *Cancer*, on the other hand, is much more clumsy and less effective; *Portunus* generally leaves the swimming-paddles at the ends of its thoracic legs outside the sediment.

The brachyuran crustacean *Corystes cassivellaunus* digs in sandy sediment in the same way as the species described above. But a number of special morphological and functional adaptations to burrowing show that *Corystes* must burrow in the top layers of the sediment for more

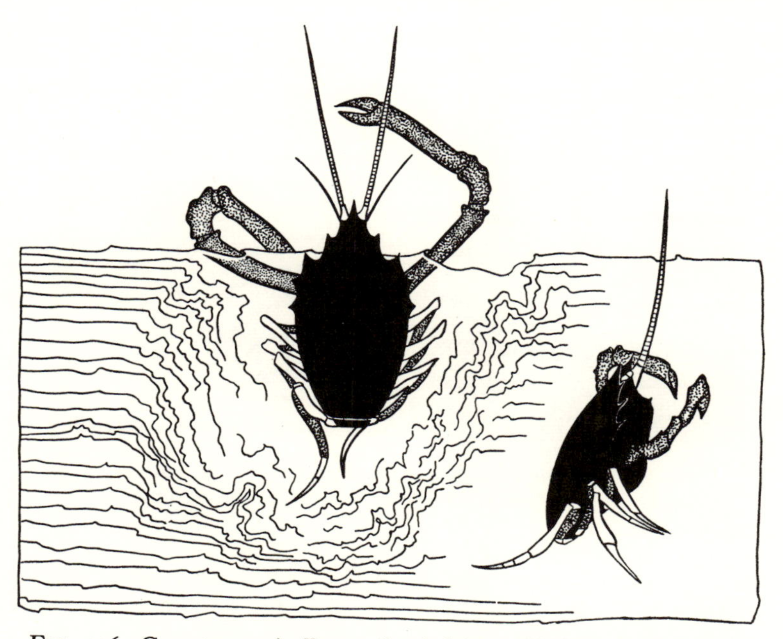

FIG. 226. *Corystes cassivellaunus* buried in sediment. (Actual size)

reasons than just for occasional shelter: at its resting place *Corystes* lies in wait for small organisms in and on the sand. It is advantageous for burrowing that the cephalothorax is elongated in the longitudinal axis, the direction of advance (Fig. 226); tooth-like, short spines on the cephalothorax point forward, away from the direction of possible enemy attacks; the walking limbs that serve for burrowing are short and strong; the fifth pair of legs, the leading one when burrowing, can turn far to the ventral and dorsal side; the clawed limbs are very long and serve to push the animal into the sediment; the antennae are greatly elongated and transformed into a siphonal mechanism (Garstang 1896a and b), and the animal can therefore dip into the sediment beyond the rostrum. *Corystes* spends most of its life in the sediment but does not move sideways in it. Rather, if it intends to move, it leaves the sediment for a short time. The walk is clumsy and yet strangely dance-like, with the body carried almost vertical and the long claws folded roughly at the level of the mouth. *Corystes* burrows very rapidly, it descends 5 to 6 cm in a few seconds, and only the red antenna tips show where the animal lies buried.

All these brachyuran crustaceans produce the same kind of resting traces in layered sediment: in sand, they are simple sagging and mixing zones surrounded by a zone in which the beds are either shortened and thickened or stretched and thinned; in mud, solid mud lumps lie frequently at the bottom of abandoned burrows that have been refilled with sediment, and in the surrounding zone the beds are shortened and

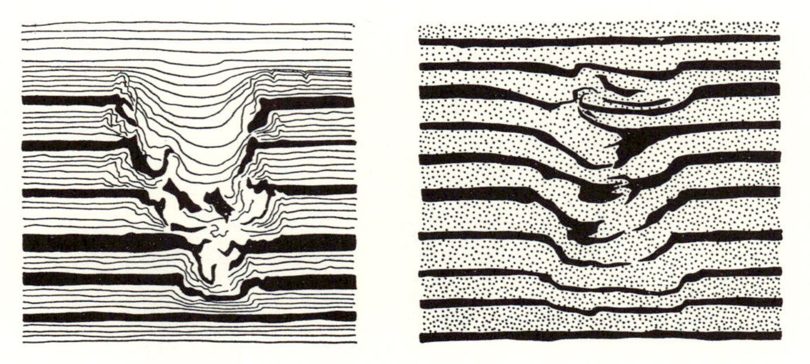

FIG. 227. Resting burrows of *Carcinides*. Left: in mud; right: in bedded sand; both covered by later sedimentation. (Actual size × 0·3)

stretched as in sand. In sand, the resting trace is immediately filled by sand flowing in from the sides, as soon as the originator of the trace leaves; in mud, a depression remains for a short while but it is also soon filled with mud which is bedded independently of the layering nearby. The resting trace has no characteristic outline either in the bedding plane or in cross-section (Fig. 227). Internal traces on sticky mud layers underlying the sand or soft mud in which the animals burrow are also similar from species to species; they consist of bundles of simple, wedge-shaped depressions which are frequently superimposed over one another. The bundles are arranged in array with a crude mirror symmetry (Fig. 141). *Corystes* avoids such mud layers altogether; if it comes into contact with mud while digging it leaves the sediment and searches for another place where the sand is thicker. Anyway, *Corystes* digs deeply for its resting place, often down to 8 cm.

Of the macruran crustaceans in the southern North Sea only *Nephrops norvegicus* retreats into a resting place in the sediment. In contrast to the crustaceans discussed in the preceding paragraphs, *Nephrops* is a territorial animal and does not roam over wide distances. Before choosing a resting place, it makes exploratory diggings into the top mud layers in an extended area. It loosens partly consolidated muds with the clawed front limbs and pushes the loose sediment aside into large piles with its long, narrow claws, holding them in front of the head. The animal works for many hours at such a cavity and always runs back and forth on the same path. The resulting pit is 10 to 20 cm deep and surrounded by gently sloping mud walls. The animal digs until it reaches consolidated mud and then pushes its body backward below overhanging mud clods. If stones are available it always builds its resting structure below a stone; it pushes so much sediment away from beside and underneath the stone that the stone eventually tilts and breaks into the cavity. *Nephrops* builds new burrows over and over again in a small area, partly levelling older structures so that finally the resting traces form an area of funnel-shaped depressions, each filled with independently layered sediment, with large and small mud clods between the depressions.

Crangon vulgaris (see Seilacher 1953a) burrows with its walking

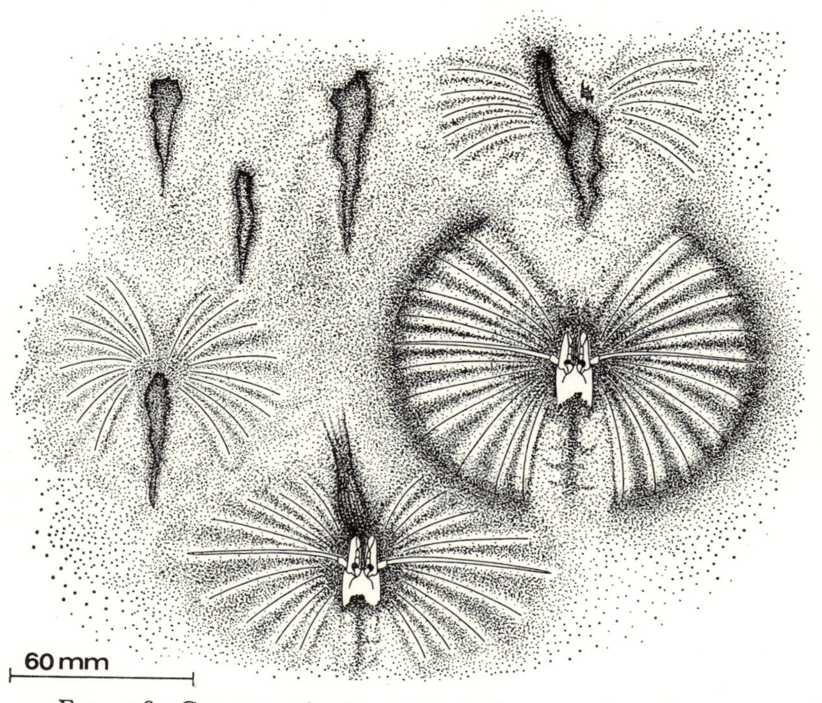

FIG. 228. *Crangon vulgaris* resting in burrows in sediment and producing circular scraping traces with their antennae

limbs until only the antennulae, antennae, eyes, and the top of the carapace emerge from the sediment that glides back over it from the sides (Fig. 228). In loose mud the animal pushes back and downward by whirling movements of the abdominal legs. It may remain in this position for many hours, motionless but for the long antennae which move constantly; they sweep closely over the surface and produce scraping circles and radial impressions. *Crangon* leaves its resting place always by moving forward or obliquely upward and forward; consequently, the abandoned resting place appears as a more or less flat, elongate depression with a

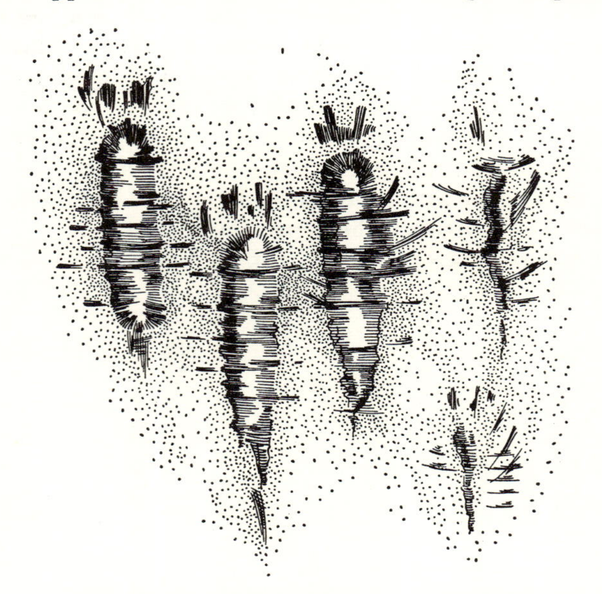

FIG. 229. Interior traces on a mud layer, covered by sand into which *Crangon vulgaris* has burrowed. (Actual size)

drag trail forward. In mud these resting traces have steep walls and can be up to 1 cm deep, particularly if the sea floor bears a continuous cover of diatoms. Frequently the resting traces of *Crangon* are well oriented with respect to a bottom current (rheotactic orientation). On muddy tidal flats near tidal channels hundreds of resting traces may have identical orientation.

Seilacher (1953a) has made a thorough experimental investigation of the phenomenon of the internal trace on *Crangon* and also *Trachinus*. He let the animal burrow in sand underlain by rather stiff clay. The burrowing limbs damage the mud layer with the terminal phalanges and produce characteristic, elongate depressions with cross-ridges (Fig. 229.)

There are also several species of Mysideae, Amphipodae and Cumaceae that leave resting traces produced by digging into the top layers of sand and mud. They use the articulated limbs of the cephalothorax for digging and therefore leave only simple, local deformations of the original texture of the sediment. The outline of the trace, seen either on the bedding plane or in cross-section, is ambiguous. Since these forms are small compared with decapod crustaceans their resting traces represent only minor alterations of the original bedding texture; but they are well seen in thinly bedded sediment. Moreover, wherever these crustaceans occur they are usually very numerous. Testing experimentally for internal traces in mud under sand, Seilacher (1953, 1959, 1960) found that their digging tools produce mostly small, paired depressions, the so-called bilobites. The shape of their mould resembles a coffee bean, and they are invariably produced by scraping with the symmetrically arranged walking limbs. If one of these animals moves forward in the sediment while it continues to scrape, the grooved double depression becomes a double trail which is usually delicately ribbed obliquely to the trail.

The Mysideae *Gastrosaccus spinifer* and *G. sanctus* burrow their prostrate, elongate bodies into the upper layers of the sediment. By shaking movements transverse to the body axis they manage to bring the sediment over them. The exopoditae of the walking limbs seem to perform these movements moving independently of the telson. The eyestalks, antennae, and antennulae always remain above the sediment. The abdomen is hinged at two points, between the cephalothorax and abdomen, and behind the fourth segment, and it is capable of throwing the body free of the sediment by a twitching movement. *G. spinifer* occurs quite frequently in the southern North Sea; according to Zimmer (1932, 1933a) it has been found on fine sand to a depth of 50 m. *G. sanctus* lives on muds and at greater depths.

The amphipods *Bathyporeia robertsoni*, *B. gracilis*, *B. guilliamsoniana*, *B. pelagica*, *Haustorius arenarius*, *Pontocrates arenarius* and *Monoculodes carinatus* all leave small, coffee-bean-shaped depressions, or bilobites as internal traces. The largest of these amphipods are barely 12 mm long, and most of them hardly reach a length of 6 to 8 mm. The animal can mark certain layers very intensely. 1750 *Bathyporeia* per square metre have been found in Lübeck Bay (cited from Dahl), 7500 per square metre (cited from Mortensen) near Cape Skagen (Stephenson 1929). These burrowing forms are noted for their strong thoracic limbs; the fifth to seventh are much broadened and have hard bristles for digging. The species mentioned are littoral inhabitants and many of them definitely favour the shallow water to a depth of 10 m. They search for food swimming in the water or crawling on the surface of the sediment and always hide in the sand after short outings.

The majority of the burrowing cumaceans prefer muddy substrates and live mainly at depths below 50 m, and almost all of them descend to depths of 1000 to 2000 m. Where these animals live they usually occur in great numbers. The cumaceans have life habits that let them produce many resting traces. Although they feed in the sediment, eating detritus and 'licking sand', they do not move from place to place as long as they sit in the sediment. When the animals intend to move they leave the sediment, swim about for some time and return to the sediment at a different place. During the night, and particularly during the mating season, they wander far and wide and can be found in large numbers in the plankton near the water surface. They are an important food for fishes.

FIG. 230. Cumacean crustaceans in the sediment. After rest they jump out of sediment like a released spring. A pit remains in the surface. (Schematic drawing)

Bodotria arenosa, *B. scorpioides*, *Cumopsis goodsiri*, *Lamprops fasciata*, *Pseudocoma similis*, and several *Diastylis* species are the most important that are known to burrow. The mode of life and burrowing techniques seem to be similar for all of them. All let the body sink to the sea floor with the abdomen bent to the dorsal side; they stretch the last four pairs of walking limbs downward and begin to scrape violently as soon as they touch ground (Fig. 230). In this way, the body sinks into the sediment, remaining curved at the centre as before. The animal continues to dig until head and telson with uropods can barely be seen. The more the body is bent, the deeper can it glide into the sediment. The gill apparatus, formed of the first maxillipeds with their rostral and caudal epipodial processes, pumps the water into the gill cavity. The gills are leaf-shaped and arranged in rows on a gill carrier; there is also an accessory gill element in the posterior section of the gill cavity. The used water is discharged through the siphonal tube, consisting of the pseudorostrum and the lancets of the gill apparatus.

While the animal rests in the sediment it remains bent so that the anterior part of the body stays almost vertical. This ensures easy breathing and also helps the animal to leave its resting place quickly. The cumaceans jump out of the sediment by suddenly straightening the curve of their body and then bending it ventrally. Enveloped by a cloud of scattered sediment the body shoots upward; then the abdomen stretches, and the walking limbs drive the body straight ahead with swift, alternate rowing beats. The thin, long abdomen occasionally helps by performing beating movements.

Despite their small size of 3 to 35 mm the animals can mark the superficial sediment layers distinctly because they occur in great numbers and move frequently from one resting place to another. If sedimentation sets in the animals leave their resting places and burrow again only after the water movement has ceased and the sediment has settled again. The internal traces of mysideans, amphipods and cumaceans are unknown. They should be visible in fine-grained sediments despite the small size of the originators.

Mollusca. Only a few molluscs produce resting traces; among those that do are the cephalopods *Sepia officinalis fillіouxi* and *Heterosepiola atlantica.*

Sepia officinalis fillіouxi is one of the most frequent shallow-water forms of the European shelf sea. In order to burrow it blows a shallow groove into the loose sand by a water jet from its siphon; while *Sepia* hovers closely above the ground it swings the flexible siphon forward, downward and backward until the depression in the sand is large enough to hold the elongate body. The broad edges of the fins steer and counteract the momentum due to the jet action, thus keeping the body in place. Once the body has been lowered into the depression, undulating movements of the fin edges throw some sand over the back as camouflage. Rapid and effective adjustment of the skin colour on head and body to the colour of the surroundings hides *Sepia* almost completely from both enemies and prey. Although disturbances of the top sediment layers of the sea floor due to *Sepia* are quite considerable they can probably never be perceptible as being biogenic resting traces; for the sand that is blown away laterally is deposited in the form of thin lenticular bodies in surroundings that already usually consist of lenticular beds. *Sepia* living in the North Sea or the Atlantic, by the way, seem to be much less inclined to rest on the sea floor than those of the Mediterranean. This may indicate that they are really two subspecies as was already suggested in view of the differences in the body proportions, patterns on the back and the outsides of the arms, and the morphology of the iridocytes in the skin (Schäfer 1937, 1938a).

Heterosepiola atlantica, a barely 3 cm long cephalopod, burrows differently. The animal pushes its body backward into the sand up to the head by peristaltic movements of the mantle while the arms supplement the movement by pushing. The arms are then held together so as to form a tube which prolongs the ventral siphon making it possible to breathe in the sediment. The animal stays in any one resting place only for a short time. After leaving the sand it swims about closely above the ground with fluttering movements of its round posterior fins, catching small crustaceans and fish fry. Even while the animal swims the mantle performs peristaltic movements, constantly altering the body shape. The animal interrupts its erratic swimming by short rests in the sediment so that the top sediment layers are soon marked with many scattered, thimble-sized, perturbed digging zones. If the bottom currents become strong enough to transport sand *Heterosepiola* rises to the open water and returns to the sea floor only after the water has calmed down.

The prosobranch *Hydrobia ulvae* also produces resting traces in the sediment. The original bedding on a sea floor settled by these snails is almost completely destroyed because in the zone of the tidal flats these animals usually occur in very large numbers (300 000 per square metre according to Linke 1939 and Smidt 1951). When the surface of the sediment begins to dry out the animal burrows more deeply to a maximum of 4 cm. It usually lies with the shell aperture at the bottom of its perishable burrow. It easily survives slight sedimentation, reaching the surface by pushing and pulling movements of its narrow, long foot. An opisthobranch gastropod, the tectibranch *Retusa trunculata*, lives in the same biotope as *Hydrobia*. The shell of this animal is less than 6 mm long. *Retusa* eats young *Hydrobia* which it swallows whole and cracks in the stomach between bar-shaped stomach-teeth. This animal produces up to 3 cm deep resting traces at low tide.

Echinodermata. Generally, little is known about the life habits of most echinoderms and least about their modes of locomotion and their relationship to the sediment and to sedimentation. Only a few asteroids produce resting traces, all belonging to the genera *Astropecten* and *Luidia*.

Astropecten dips into the sediment to search for food. It eats and digests at rest below the surface. This is certainly the only reason for burrowing vertically; the animal also hides during rest. Seilacher (1953b) described these activities in every detail for the Mediterranean starfish *Astropecten aurantiacus*. *A. irregularis* which is common in the North Sea can crawl swiftly over the surface of the sediment. For burrowing it stops and begins to dig by moving its suckerless tube feet transversely to the arms. Whereas during walking on the surface the tube feet on all arms work in a single direction, they now pendulate transversely to each arm. This change of movement thus requires a complete change of nervous co-ordination. The pendulating tube feet transport the sand out from under the arms, and the sand piles up on both sides of the arms. While the starfish subsides vertically the spines of the ambulacral plates are bent upward, distinctly narrowing the arms. In this way the starfish can bury itself completely, and undulating movements of the body rapidly remove the last traces on the surface. In most cases the animals sink into sediment above hidden food such as a small bivalve, gastropod, small brittle star or crustacean, or even a dead organism. The prey is ingested whole into the stomach which rapidly becomes filled. Occasionally, the body cover bulges so high that it re-emerges as a hump above the surface. The author has never observed an *Astropecten* move horizontally in the sediment. If it wants to move to a new place in the sediment it must return to the surface and dig downward somewhere else.

Whereas well-developed 'multiple, circular shovelling' can move a near-spherical or elongate body through the sediment in every direction (*Echinocardium*, *Aphrodite*), the reciprocating movements of the tube feet can move a starfish only upward or downward. Moreover, the shape

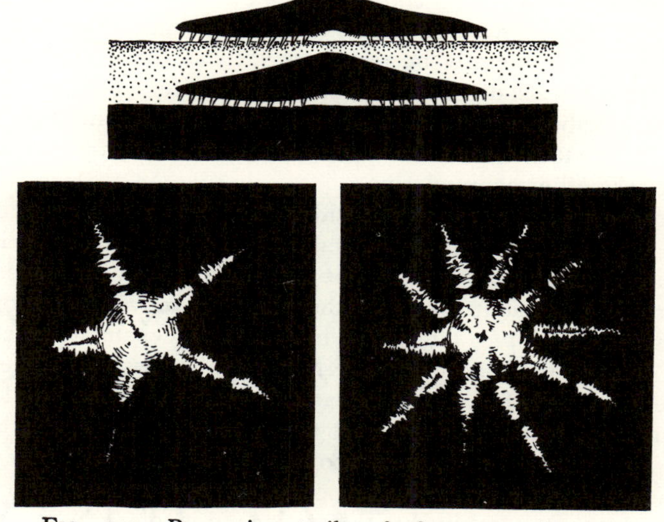

FIG. 231. Burrowing trails of *Astropecten irregularis*. Top: side view of the starfish on the sediment surface, and below buried in sand and touching a mud layer underneath; bottom left: the five-ray internal trace produced on the top of the mud layer; bottom right: ten-ray internal trace produced in the same way, but superimposing a second five-ray trace on the first after a rotation about the vertical axis. (Schematic drawing)

of the starfish with its five radially extended arms is extremely unsuitable for horizontal movement in the sediment, whatever the driving mechanism (shovelling movements or others). In experiments with *Astropecten aurantiacus* in sand underlain by stiff mud Seilacher (1953b) observed internal traces of the arms which showed distinct transverse grooves. They are evidently produced by the beats of the tube feet which move back and forth in the same direction. *Astropecten irregularis* leaves similar internal traces on mud layers (Fig. 231).

Astropecten is vulnerable to sedimentation because of the restricted mobility of the tube feet. They are unable to lift the body against an increasing weight of covering sediment. A load of 6 to 8 cm sand immobilises *Astropecten*; the short and stout arms can neither undulate, nor pull or push the body to the surface. If sedimentation sets in *Astropecten* therefore ascends without delay to the surface, there waits till deposition stops, and so survives. Although *irregularis* is the only species of *Astropecten* living in the North Sea it occurs as two distinct varieties, *pentacanthus* and *serratus*. According to Dahl (1928) these are not geographic varieties but are also found side by side. The species is very common on sand at depths between 10 and 1000 m.

The astropectinid *Luidia ciliaris* lives on mud in the northern North Sea and in the North Atlantic; it has seven long arms. The resting traces of *Luidia* could thus easily be misinterpreted as those of a five-armed starfish with the trace of two arms repeated. Its life habits are unknown. *Luidia sarsi* inhabits the muddy floors of the North Sea and the southern Kattegat. This starfish has five arms and is a voracious predator which mainly attacks other echinoderms.

The ophiuroid family, Amphiuridae, comprises several species that produce resting traces. Like others, these animals burrow into the superficial layers of the sediment not only to hide but also in search of food. For this purpose it is helpful that they can move horizontally over small distances in watery mud. The burrowing mechanism functions on the same principle as that of *Astropecten*. Double rows of tube feet without suckers sit on each arm; they beat and dig obliquely to the long axis of the arms and can move sediment out from under them. As a result, all arms subside. Distal tube feet also exist below the body disk and can remove sediment. In addition, the body disk can be pressed downward in jerky movements combined with rotations in its plane by means of the large, paired arm muscles on both sides of each arm. For a dextral movement the right muscles of all arms contract simultaneously, for a sinistral turn the left muscles. As Seilacher (1953b) has shown for *Ophiura texturata*, this contraction also causes a sideways shift of the bases of the arms. The traces of the bases become therefore considerably broader than those of the points, and the outline of the whole trace resembles more that of a starfish than of an ophiuroid particularly because the tube feet broaden the trace while they dig toward the sides and transport sand. If the points of the arms are shifted during the formation of the trace, the trace produces the misleading picture of a brittle star with branched arms, an euryalid. Thus the variable behaviour of one and the same animal can cause traces that suggest differences of family and genus (see also illustrations in Seilacher 1953b). The amphiurids that make this type of trace are *Acrocnida brachiata*, *Amphiura chiajei* and *A. filiformis*.

Although some echinoids produce resting traces, often (as with irregular sea urchins) involving major disturbances of the bedding, they have such unspecific body outlines that the organic origin of the traces can hardly be recognised. Locomotion in sand is always achieved by multiple, circular shovelling with the spines that are particularly adapted for the purpose. *Spatangus purpureus* is up to 10 cm long and lives in a burrow in sand; *Echinocardium flavescens* lives in coarse sand and shell breccia. *Brissopsis lyrifera* is 6 cm long and lives buried in mud. Whether the very small *Echinocyamus pusillus* can be ranked with this group is not clear; it occurs on sand and shell breccia. Little is known about its way of life; according to Smith (1932) 8·6 individuals are found on the average in 1000 cm^3 of shell breccia.

Several of the Holothurioidea (sea cucumbers) of the North Sea frequently stay in the sediment. Most of their traces can be considered as resting traces because these animals move very little.

The narrow, worm-shaped body of *Cucumaria elongata* tapers toward the back. The animal is bent to a U-shape when it is lying in mud and in very fine sand. The anterior end with the mouth and tentacles and the posterior end protrude above the surface. In this position the animal can feed on detritus and small organisms and it breathes with lungs that branch from the digestive tract near the anus which can be closed.

Strong circular and radial muscles are combined; together with five paired longitudinal muscle strands they perform the rare locomotory movements. Small calcareous bodies that lie closely packed in three layers below the thick and firm epidermis stiffen the body remarkably. All locomotory movements are slow and clumsy. The animals live at all depths in the central North Sea wherever the sediment is muddy enough, especially in the Kattegat. *Leptosynapta inhaerens*, a slender sea cucumber, up to 15 cm long, seems to prefer fine sand; it is found throughout the North Sea at depths from 2 to 50 m (Dahl 1928). *Leptosynapta bergensis*, 20 cm long, is frequently found in the fine sands near Heligoland. *Echinocucumis hispida* favours soft muds in which it lies bent to a U-shape.

Pisces. Resting traces produced by fish are in most cases only simple, local deformations of the original texture of the sediment because most fish penetrate into the top sediment layers by undulating or beating movements of their whole body. The few species that form specific internal traces because they are provided with special digging organs will be discussed separately.

Some of the most common fish of the North Sea that frequently rest on or in loose sediment are *Scyliorhinus caniculus*, *Squalus acanthias*, *Raja batis*, *R. clavata*, *R. radiata*, *R. fyllae*, *R. montagui*, *R. lineata*, *Squatina squatina*, *Anguilla anguilla*, *Conger conger*, *Zeus faber*, *Ammodytes lanceolatus*, *A. lancea*, *Trachinus draco*, *Gaidropsarus mustela*, *Pholis gunnellus*, *Zoarces viviparus*, *Myoxocephalus scorpius*, *Trigla corax*, *T. gurnardus*, *Gobius microps*, *Callionymus lyra*, *Anarhichas lupus*, *A. minor*, *Agonus cataphractus*, *Liparis liparis*, *Scophthalmus rhombus*, *S. maximus*, *Hippoglossus hippoglossus*, *Hippoglossoides platessoides*, *Pleuronectes platessa*, *P. microcephalus*, *Limanda limanda*, *Solea solea*, *Lophius piscatorius*, and *Pholis gunnellus*. The texture of the sediment must be considerably rearranged by the digging activities of fish.

Many fish show special adaptations to resting on or in sediment. The following types of adaptation can be recognised in common North Sea fish (Figs. 232 and 233): (1) Shape: the whole body is dorso-ventrally flattened (*Raja*, *Squatina*, *Lophius*). (2) Attitude, sometimes combined with shape; (*a*) serpentine attitude, the body is always long and either cylindrical or more or less flattened (*Conger*, *Anguilla*, *Zoarces*, *Pholis*); (*b*) sideways-lying attitude (*Zeus*, *Scophthalmus*, *Hippoglossus*, *Hippoglossoides*, *Pleuronectes*, *Limanda*, *Solea*). (3) Organs supporting the body; (*a*) paired fins (*Anarhichas*, *Scyliorhinus*, *Squalus*); (*b*) pectoral fins as limbs (*Trigla*). (4) Active digging; (*a*) with the anal fin (*Trachinus*, *Callionymus*); (*b*) by undulation of an elongated body (*Anguilla*, *Ammodytes*, *Branchiostoma*).

Some dorso-ventrally flattened fish can only lie loosely on the sea floor, particularly those that have a wide and flat underside or have a sucker mouth that needs a solid substrate for attachment. Thus *Raja* simply lies on the surface. However, *Squatina* and *Lophius* lurk, adjust their colour to that of the ground, and dig superficially into the sediment. *Lophius* even burrows quite deeply at times and erects a flag, the most anterior ray of the dorsal fin, to allure swimming fishes. Its mouth full of grasping teeth is gigantic, and the eyes lie near the top of the head at the surface. *Squatina* pours sand over its body by beating with the wing-like fins; on the sea floor it appears as an inconspicuous, smooth elevation because the pectoral and ventral fins taper gradually toward the sides. In addition, they adjust their skin colour to the environment. The animals grow to a length of 1 to 2·5 m. A 160 cm long *Squatina* maintained in the sea water aquarium at Wilhelmshaven lurks in its resting places and from there it can snatch young sharks. It changes its resting place about

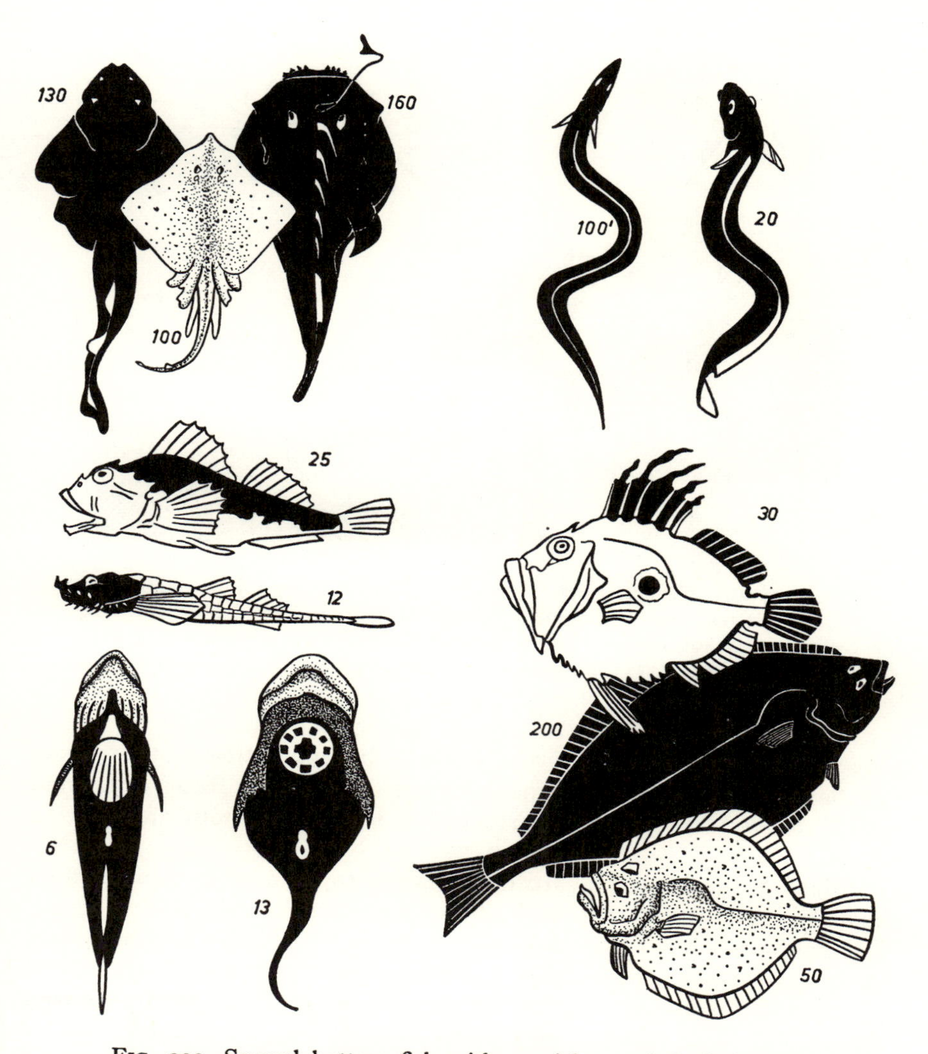

FIG. 232. Several bottom fish with special morphological and behavioural features. The figures indicate the length in centimetres and also serve for identification: (130) *Squatina squatina*; (100) *Raja clavata*; (160) *Lophius piscatorius*; (100[1]) *Conger conger*; (20) *Zoarces viviparus*; (25) *Myoxocephalus scorpius*; (12) *Agonus cataphractus*; (6) *Gobius microps*; (13) *Liparis liparis*; (30) *Zeus faber*; (200) *Hippoglossus hippoglossus*; (50) *Scophthalmus maximus*

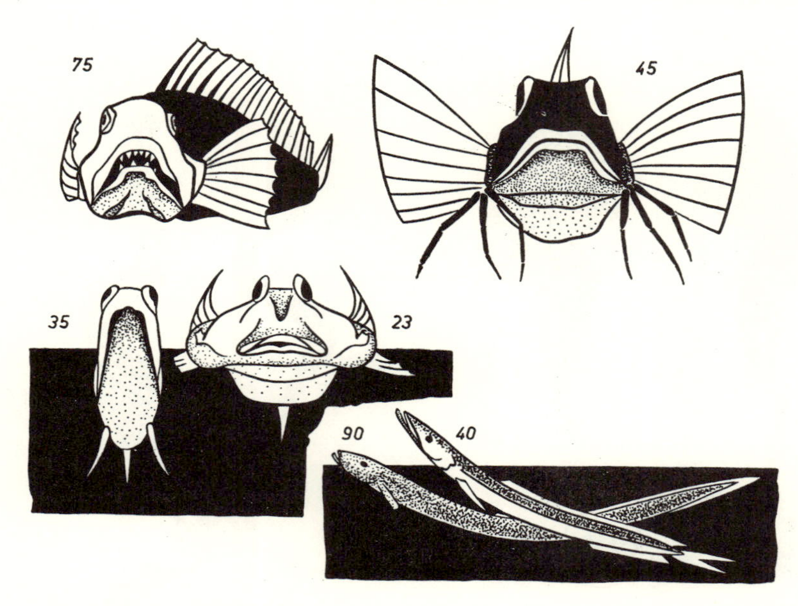

FIG. 233. Several bottom fish with special morphological and behavioural features. The figures indicate lengths in centimetres and also serve for identification: (75) *Anarhichas lupus*; (45) *Trigla corax*; (35) *Trachinus draco*; (23) *Callionymus lyra*; (90) *Anguilla anguilla*; (40) *Ammodytes lancea*

once a day and swims for several minutes by undulating movements of the tail, holding its pectoral fins immobile. The resting traces of *Lophius* and *Squatina* are extensive perturbation textures without any specific shape.

Many of the fishes with attitudes adapted to life on the bottom simply lie on the surface and never hide; they do not produce any resting traces. The flatfishes among them burrow by beating with the whole body and can change colour. Their resting traces are shallow, diffuse and unspecific.

None of the fish with supporting organs burrow; they hardly damage the surface of the sediment during rest.

Actively digging fish produce true resting traces. *Trachinus* and *Callionymus* burrow into the sediment by means of the anal fin whose strong rays beat alternately right and left and so shovel the sediment sideways from underneath the body. The fishes subside vertically over their whole length until only their dorsal edge projects above the surface. Both have their eyes on top of the head. The opercular gill slits are closed but for a narrow slit at the top. The mouth of *Trachinus* also opens obliquely upward; thus both intake and egestion of breathing water stay unblocked, even if the animal is deeply buried in sand (Fig. 234). *Trachinus* is up to 28 cm long and a predator that gets into lurking position in the sediment but catches its prey in the open water after a few, exceedingly quick swimming strokes. It thus is a bottom fish only conditionally. Seilacher (1953b) investigated the resting traces of

FIG. 234. *Trachinus draco* buried in sediment. (Arrows) directions of breathing current. (Schematic drawing)

this fish and illustrated the internal trace which the beating anal fin produces (Fig. 235); an elongate depression tapering toward the back has rhythmic, transverse grooves produced by scraping with the hard rays of the anal fins. The front end of the trace is occasionally marked by paired roundish depressions produced by the tips of the ventral fins.

For the trachiniform fish *Callionymus lyra* and *C. maculatus* the sediment plays a different role. These animals are bottom fish and hardly leave the sea floor except for their short mating dance. They burrow with their strong anal fin into the sand in the same way as *Trachinus*, and their eyes and gill slits are analogous. Their mode of feeding, however, differs radically from that of *Trachinus*. They eat small crustaceans and worms which they take from the sediment with their well-adapted mouth. A flexible clasp mechanism can fold the toothed jaws so that the open mouth points downward at a right angle to the body (Fig. 236). The head is broad and stays outside the sediment whereas the rest of the body is buried with the tail obliquely downward. The digging anal fin can produce an internal trace which consists of a depression with transverse grooves, similar to that of *Trachinus*, but it is deeper at the back than at the front (Fig. 235).

Fish belonging to groups (1) and (4*a*) lie more or less horizontally on or in the sediment and have their breathing apertures uncovered, either because they rest in a shallow position or because they have them on top of their head. This is not so with fishes that dig by undulation (4*b*). *Anguilla* either digs a burrow that is wide and strong enough to allow free breathing in the burrow or, if this is impossible, it must keep the head completely uncovered so that both mouth and gill slits are free for breathing. *Ammodytes* cannot strengthen its tubes with mucus and must invariably leave its head outside the sediment when resting. In coarse sediment with high permeability only the mouth (which can be stretched to form a tube) needs to lie at the surface because the used water can

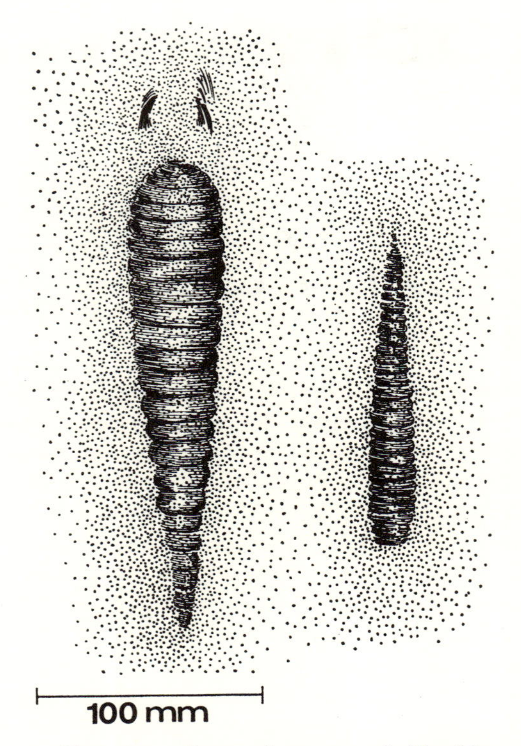

FIG. 235. Internal traces of *Trachinus draco* (left) and *Callionymus* (right)

be pumped through the gill slits into the sand. As Gripp observed, the animals bury themselves completely in the sand and breathe only interstitial water if they are caught by low tide and cannot escape into deeper water. The structure and the digging habits of *Branchiostoma lanceolatum* are discussed in the section on locomotion by undulation (D I (*a*) (iii)). *Anguilla* and *Ammodytes* dig in essentially the same way. The burrowing of all three animals causes narrow sag zones in the sediment. Only in rarest cases do they produce internal traces that could potentially be better preserved.

In summary only two, *Trachinus* and *Callionymus*, of all the North Sea fish that sometimes rest on the sea floor produce resting traces which can be identified under favourable conditions. This is because they are internal traces and have a specific shape. All the other fishes tend to destroy the original texture of the sediment. What remains of their

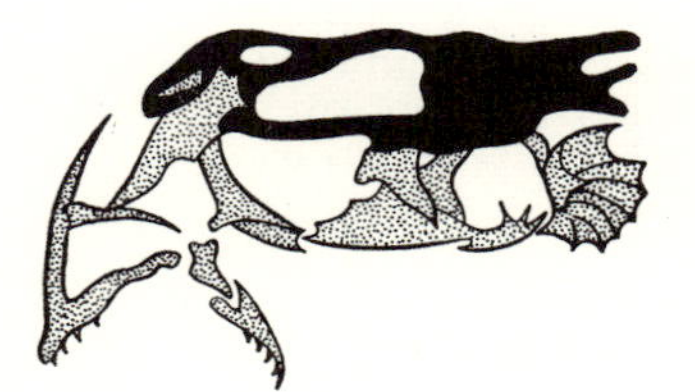

FIG. 236. Mouth of *Callionymus lyra*. Top: mouth closed; centre: mouth open; bottom: cranial (black) and visceral (dotted) portions of skull with open mouth. (Actual size)

activity in a bed or series of beds has no specific shape. All the same, it is important to know the cause of what must be numerous deformations and to take it into consideration when investigating fossil traces.

(*c*) LOCOMOTORY TRACES AS BIOSTRATINOMIC PHENOMENA

Several basic activities are important in the life of the animals, involve more or less extensive locomotory movements and can therefore produce traces in the sediment: these activities are searching for food, dwelling, escaping, and seeking a resting place. Sexual behaviour, important as it is for the life of the species, is omitted because it rarely involves locomotion and production of traces in lower marine animals.

Seilacher (1953a) chose the same ethological and ecological point of view to classify shapes of traces, considering them as expressions of animal behaviour, and therefore comes to almost the same results. He groups them into resting traces, dwelling and feeding burrows, grazing and crawling traces. To Seilacher the advantages of such an ecological interpretation are: (1) even poorly preserved and undifferentiated fossil specimens can be investigated in this way, (2) the ecological meaning of the features can usually be understood immediately because well-established, technical principles are sufficient to explain them, and (3) the ecological interrelations are generally valid for all kinds of animals at all geological times.

If, on the other hand, we consider a trace as a biostratinomic phenomenon, the picture is quite simple compared with the great variety of ethological-ecological factors; very few features produced by movements of organisms have to be distinguished. Richter (1952) called mechanical deformations of the sediment 'bioturbate textures', 'zooturbate', or 'phytoturbate', depending on whether animals or plants have caused them. These terms include surface and internal traces and the perturbations of the sediment due to burrowing (fossitextures). Combining elements of the classifications by Seilacher (1953a) and Schäfer (1956a) the following biostratinomical classification of Table 12 may be adopted.

TABLE 12

Bioturbate textures

1. Burrowing textures (fossitextura deformativa)
2. Complete structures (fossitextura figurativa)
    (a) sunk or internal structures
    (b) elevated structures
3. Half-structures
    (a) sculptured bedding planes
    (b) sculptured boundary planes
4. Mobile structures

(1) Burrowing textures. All locomotory movements within the sediment are called 'burrowing', because this term does not imply any specific technique. This agrees with Richter (1936, 1937b and 1952) who called all bioturbate textures caused by locomotion of benthonic animals in the sediment 'burrowing fabrics' (fossitextures). Burrowing in the sediment can have different consequences: if the burrowing is nothing but locomotion the original fabric is simply deformed (perturbed burrowing texture, or fossitextura deformativa); one or several laminae may be deformed. If, on the other hand, dwelling burrows are constructed or sediment-eating animals burrow repeatedly through the same area, then the layers may be completely destroyed in a limited area. In most cases, the laminae are deflected in a zone surrounding the one in which they have been completely destroyed.

(2) Complete structures. If the walls of passages burrowed through

sediment are hardened with mucus, cavities of a definite and permanent shape are formed (sunk structures). These hollow forms can be filled with sediment and thus turn into positive moulded forms (Seilacher: 'Vollform'. Schäfer: 'Gestaltungswühlgefüge'). The boundary against the surrounding sediment is distinct. The core consists of sediment which is not necessarily derived from the immediate environment. The infilling may or may not be bedded. Reineck (1958a) distinguished between the 'burrowing core', the part of the fossitexture that was originally occupied by the animal and directly shaped by it, and the 'burrowing aureole', the part in which the original texture was changed indirectly as a result of the burrowing activity.

In some cases mucus-hardened burrowing structures are raised above the surface of the sediment. Then the inhabitant must build it grain by grain in order to achieve a structure solid enough to withstand currents above the surface. Thus an elevated structure is erected as a continuation of a mucous burrow below (*Lanice*). Frequently, however, elevated structures are built on solid substrates, such as shells or stones (*Sabellaria*), or they lie openly on the surface of the sediment (*Cerapus crassicornis*, *Siphonocoetes* and others).

(3) Half structures. Seilacher paid particular attention to the trace in 'half form'. He distinguished between sculpted bedding planes, separating layers of equal material, and sculpted boundary planes, separating layers of different material, especially clay and sand. For this distinction of the half forms see Seilacher (1952a, p. 437). Such half structures are produced either as superficial or as internal traces.

(4) Mobile structures. Some organisms build tubes grain by grain in the sediment and carry them around as shells. The structure itself is a 'complete structure'; if moved, however, it also produces burrowing textures.

The different textures vary in abundance. Seilacher was the first to point out that half structures produced as superficial traces are most frequently observed on recent tidal flats; but water currents and waves often sweep over them or destroy them completely. Frequent as these traces are they are therefore rarely found as fossils. This is so although they have a better chance of being preserved than superficial traces in permanently immersed coastal sediments. On tidal flats at least the sediments become somewhat consolidated when they lie dry. Only below the level where bottom-touching waves are effective (in the North Sea below approximately 100 m depth) do the chances of preservation gradually increase ('critical plane', Rud. and E. Richter 1932, p. 370; Richter 1936, p. 235; Schäfer 1941a, p. 27).

The 'internal traces' (Seilacher) among the half forms have a good chance of being preserved, even in the shallow sea, if the series of beds that carry the traces remains preserved. An internal trace in half form can originate in two ways. Either an endobiontic animal moves in the sediment along the boundary between two layers and leaves its locomotory trace on the layer below; then the sediment of the overlying layer immediately fills the burrowed groove and thus preserves it permanently. Judging from the locomotory habits of the endobionts in the North Sea as we know them today, such a case must occur very rarely. Or an epibiontic animal burrows into the sediment and marks an underlying layer with its burrowing tools. As the trace is produced it is also immediately filled by sediment from the overlying layer and thus preserved from destruction. In his experiments with living animals in mud layers covered by sand Seilacher (1953a and b) produced numerous internal traces of this kind. Such internal traces are frequent and can be well preserved.

Burrowing textures are especially important because they can be far better preserved than half structures. Complete structures have long been found as fossils; in fact, until recently they were the only traces of life activities in sediment to which any attention was paid (Richter 1927b). By now, after intensive work on burrowing textures, it is well known that these textures are the most frequent of all bioturbate textures. If we take into account that almost all mobile benthos, endobiontic as well as epibiontic, are episodically forced to penetrate newly deposited sediment, such burrowing fabrics must evidently be very frequent. In each case the covered organism produces a local deformation of the new layers, either in form of a burrowing trail, gliding or sag zone, or as simple destruction or local compression of the layer. Such burrowing fabrics, however, cannot be identified. The only information conveyed by such a trace is that a bed was laid down on a sea floor inhabited by living organisms.

The lack of interest in burrowing textures may also be due to the difficulty of making them perceptible in a fossil series of beds, and to recognise them as biogenic in origin. Only thin sections perpendicular to the bedding of the specimens make it possible to study bioturbate textures in detail or to estimate their abundance. Before a method was developed of taking samples from Recent sediments without damaging their original fabric, and sectioning them, bioturbate textures were bound to be overlooked. Such methods developed by Reineck (1957 and 1958c) will become particularly important because they allow extensive comparative studies of the bioturbate textures.

In the foregoing sections 358 benthonic animal species of the southern North Sea were mentioned, arranged according to their mode of life. Of these, 358 species produce 579 different traces because some make more than one kind of bioturbate texture. Of the 358 species, 270 make burrowing textures, 95 make sunk and 6 elevated structures, 68 make internal traces and 132 superficial traces as half structures, and 8 construct mobile structures. Producers of superficial traces furnish internal traces; and all 68 species listed as producing internal traces also furnish surface traces. Animals that build sunk structures also produce burrowing textures whenever they are covered by sediment and ascend to the new surface. Producers of surface traces that are covered by sediments furnish burrowing textures in the same way. It is therefore clear that burrowing textures must be the most frequent bioturbate textures. The superficial traces, on the other hand, which 132 animal species produce, are almost all destroyed as soon as they are formed. Elevated structures and mobile tubes are highly interesting, specific adaptations ecologically, but they are of minor importance biostratinomically. What makes them notable is their solid construction and their occurrence in large numbers where they exist. Table 13 gives the figures for traces from 358 species

TABLE 13

Bioturbate textures and their palaeontological significance

	Percent of 579 traces	Preservation	Characteristic of constitutional type
Burrowing textures	46·7	good	very little
Superficial half-structures	22·8	very poor	very much
Sunk structures	16·4	very good	fairly
Internal half-structures	11·7	good	little
Elevated structures	1·0	very good	very much
Mobile structures	1·4	good	very much

recalculated as percentages with qualitative remarks about their chances of preservation and their reliability as indicators of the life habits of the animals that produced the trace. These percentages are based only upon species from the North Sea and hold only for the ecological and physical conditions that prevail there. It can be assumed, however, that similar or identical conditions exist in other shallow seas.

With increasing ocean depth the conditions change as a function of water agitation; if it decreases, proportionally more surface traces are preserved; if the rate of sedimentation decreases organisms must escape less frequently from under a cover of new sediment. Comparisons are impossible before the details of physical and biotic processes at various depths and in different oceans are better known.

II. Metabolism, regeneration and reproduction

Although trace fossils and body fossils seem to be entirely different phenomena both are causally related to metabolic processes in former living organisms. On the one hand, metabolic processes synthesise the living body; on the other, they provide the energy necessary for the work of the body by decomposition of energy-rich compounds. Whereas trace fossils are documents of locomotion and thus of former activity, body fossils are the material remains from constructive metabolic processes. The by-products and waste products of synthetic and decompositional metabolic processes are preserved as fossils; they can be food remnants, regurgitation pellets or faeces: or they may be the periodically or episodically discarded products of physiological regeneration or restitution. In contrast, there is also a pathological (traumatic, accidental or restorative) regeneration by which parts of the body that were damaged or lost by external or internal influences are repaired or reconstituted. Finally, sexual products and their enclosures (spawn, eggs or egg capsules) can on occasion be preserved. Whether spawn, eggs or egg clusters of vertebrates should be considered as independent individuals or as products of the life activities of their parents, is a matter of opinion; whichever way they are classified, they are body fossils like the waste products of metabolism or of regeneration.

(*a*) EATING AND REGURGITATING

Eating can produce traces visible only on parts of a prey too big to be swallowed whole and therefore divided into pieces before being swallowed or parts of a prey normally left uneaten. The technique of provision and ingestion of food and the nature of the feeding tools, teeth, claws, scrapers or boring or rasping tools, determine the shape of the discarded parts and the appearance of feeding traces on the prey. Knowing the tools and their mode of working often helps to identify the producer of feeding traces. Body fossils carrying feeding traces are doubly important; they may furnish information not only about the fossilised animals themselves but also about predators and their life habits. The following sections on food remnants are arranged systematically according to predators rather than to victims.

Regurgitation pellets are produced by some predators that remove indigestible food through their mouth because they are too large, hard, or awkward in shape to pass through the digestive tract.

(i) *Crustacea*

The decapod crustaceans are especially significant as producers of feeding traces because many of them are predators and because their prey is frequently large. This helps to recognise and on occasion identify the traces. Discussion will be limited to unambiguous and typical feeding traces.

In the Tortonian sands of the Vienna basin numerous gastropod shells are broken in the same, very characteristic fashion; Papp *et al.* (1947) called the pattern of breakage 'ribbon cuts'. The outer walls of the shells are removed along a strip beginning from the aperture and following the whorls so that several coils of the columella are visible (similar to Fig. 237). So many gastropod shells were broken in the same fashion, that the authors began to suspect that the breakages were feeding traces of gastropod-eating predators. They interpreted them as being made by hermit crabs, having seen *Pagurus striatus* break open gastropod shells in the aquarium of the Zoological Station at Naples. This animal produces similar marks to those on the Tertiary fossils near Vienna. The hermit crab grasps with one claw a gastropod, for example *Nassa mutabilis*, at the curved outer edge of the shell aperture and with the

FIG. 237. Two shells of *Buccinum undatum*, opened by *Palinurus vulgaris* with the ribbon-cut technique. (Actual size × 0·5)

other claw breaks a piece from the edge, somewhat as one tears a piece of paper. However, a brittle calcareous shell will break rather than tear. The crab breaks off one small piece after the other from the edge, rotating the shell at the same time. The snail retreats into its shell, but finally the crab grasps and eats it.

Eupagurus bernhardus, the hermit crab which lives in the North Sea, does not feed in this fashion but only eats dying or dead gastropods which it detects with its chemoreceptors. *Pagurus striatus*'s feeding by breaking gastropod shells is the more remarkable because all hermit crabs have a particularly close bond to gastropod shells which they use for housing. This has its expression in their behaviour; whenever a crab meets a shell it thoroughly examines it inside and out by palpating it with its claws, walking limbs and antennae. It does the same with any roundish and goblet-shaped object. Never does the animal damage or destroy a shell while it examines it. Then the shell is made part of the functional system of protection or search for shelter, and the activity of the animal has no connection with the search of food. As a consequence, a hermit crab never attacks and eats a gastropod whose unoccupied shell it subsequently occupies; one behavioural complex does not continuously merge into the other.

In hermit crabs of the North Sea the complex of behaviour relating to gastropod shells as potential dwellings seems to be more effective and general than in that of the Mediterranean *Pagurus* which eats gastropods after having broken their shell. Even so *Pagurus striatus* cracks only small gastropods. Probably this depends not only on the greater ease with which small gastropods can be overcome but also on the role of large gastropod shells as shelter. Absence of ribbon cuts on gastropod shells therefore does not prove the absence of hermit crabs.

Papp *et al.* (1947) mentioned only the Paguridae as producers of ribbon cuts in gastropod shells. The author has kept spiny lobsters (*Palinurus vulgaris*) in an aquarium for several years; the northern boundary of their area of distribution passes through the English Channel and along the Irish coast. These very strong crustaceans, up to 40 cm long, cut the shells of *Buccinum undatum* very neatly (Fig. 237). The crustaceans detect gastropods at a distance of 70 cm, approach them and immediately grasp them with their claws. They turn the shell so that the apex comes to lie between the mandibles and is effectively held by them. Then one of the claws grasps the gastropod shell at the next to last whorl while the other starts breaking the outside edge of the aperture by levering. The lobster breaks piece after piece from the arch of the shell along the whorls and keeps rotating the prey around its long axis. In securing the apex the mandibles frequently break off its tip (Fig. 237). The claws of the lobster are particularly suited for this breaking technique because the strong and pointed propodus process sits deeper at its palma than it does at the claws of other decapod crustaceans. Thus the propodus is an effective antagonist for the mobile dactylus (Fig. 238). Kessel (1938b) reports that the predatory beetle *Dytiscus marginalis* feeds on the freshwater gastropods *Limnaea stagnalis* and breaks their shells in the same way.

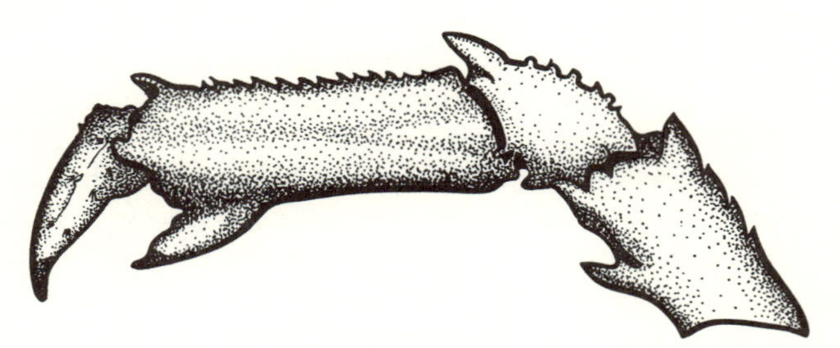

FIG. 238. Claw of *Palinurus vulgaris*. (Actual size)

Unfortunately gastropod shells broken in this fashion are not necessarily the result of predation. Shells of *Buccinum undatum* can be broken in the surf on stones and often produce the same end effect as the work of *Palinurus* (Fig. 239). This can even happen while the gastropods are still alive. Diving ducks have a special predilection for *Nassa*; they dive for them and swallow the shells whole; the musculature of the stomach wall crushes the shells with the help of stomach stones. In many cases the resulting breakages again resemble ribbon cuts. The gastropod shells are discharged without further damage in the faeces. Only one feature occasionally allows a distinction between the two similar types of damage caused either by predators or by the surf. The edge around the shell aperture can remain undamaged when the outside of the whorls is broken by the surf or under pressure in a muscular stomach (Fig. 239). This obviously cannot happen with the ribbon-cut technique of predation.

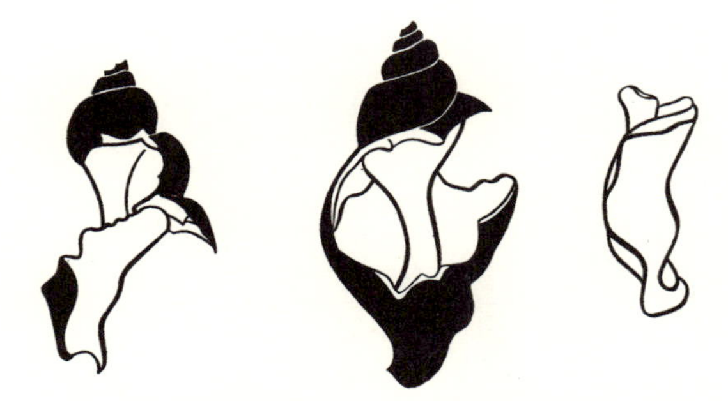

FIG. 239. Three shells of *Buccinum undatum* at various stages of being broken by the surf. (Actual size × 0·5)

On the intertidal cliffs of Heligoland and on the pebbly beach of the Heligoland dune, shells are rapidly destroyed. Nothing but the strong columella remains. Occasionally, *Homarus gammarus* breaks open a strong *Buccinum undatum*. The resulting fragments are completely irregular. Even the columella is broken. Hollmann (1969) described the specialised behaviour of *Homarus* while cracking shells of *Buccinum*. His description is precise and based on new observations.

(ii) *Mollusca*

Several species of predatory gastropods (*Lunatia*) live exclusively on bivalves. They bore from the outside through the shell, sink their proboscis into the borehole and eat the soft body.

Ziegelmeier (1954) discussed the population density of *Lunatia nitida* at several places in the southern North Sea. In a settlement of *Aloidis* he found one gastropod on $\frac{1}{10}$ square metre in spring, 1951, three in the autumn, five in spring, 1952, and two in the autumn of that year. This is a high density for a predator, and it can be concluded that many lamellibranchs must be killed by *Lunatia* and that many valves must be perforated by boreholes. Ziegelmeier lists the lamellibranch species attacked by *Lunatia nitida* in experiments in an aquarium: *Nucula nitida*, *N. nucleus*, *Spisula subtruncata*, *S. solida*, *Aloidis gibba* and *Mya arenaria* juv. Drilling holes were also found in the shells of *Lunatia nitida* itself. At the beach of Wangerooge the author observed boreholes caused by any of the several species of *Lunatia* in 15 per cent of 350 valves of *Donax vittatus* and 12 per cent of 350 valves of *Macoma baltica*.

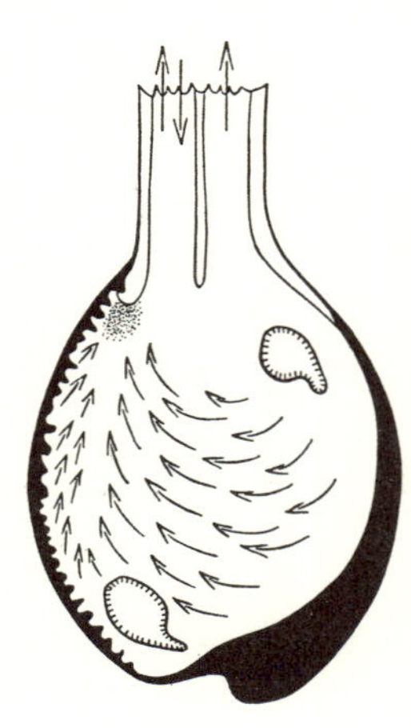

FIG. 240. Mantle cavity of a bivalve with siphons. Arrows indicate the beating directions of cilia. At the base of the ingestive siphon inedible detritus accumulates which is occasionally ejected through the ingestive siphon (indicated by double arrow at the mouth of this siphon) (after Kellog, 1915). (Schematic drawing)

The lamellibranchs which feed by whirling or grazing do not leave any identifiable food remnants. Yet many can separate edible from inedible and discard the useless material through the ingestive siphon (Kellogg 1915). Before the inedible material is discarded the cilia of the mantle collect it in a fold at the bottom of the ingestive siphon (Fig. 240). Others can regulate mucus production in the mantle cavity; the mucus production ceases when useless material is to be kept away from the mouth (McGinitie 1941, Jørgensen 1943).

The lamellibranch *Teredo navalis* bores into wood and feeds on it. It leaves very distinct feeding traces that are likely to be preserved (Miller 1924a and b, 1925, Miller & Boynton 1926, Dore & Miller 1923, Yonge 1925, Lasker & Lane 1953). What is preserved is generally a cast of the borehole consisting of hardened sediment which filled it, rather than the wood into which the holes were drilled.

The cephalopods *Loligo forbesi* and *Sepia officinalis* are predators in open water; two of their tentacles are very long, and with them they grasp free-swimming prey, mainly small fishes. They bring the prey near the mouth where it is surrounded by the arms, and there tear it to pieces small enough to be swallowed. Only a few fish scales fall on the sea floor. Jaws and radula shred the pieces mechanically, and then they pass through the digestive tract. (Bidder 1950, assumed that when *Loligo* eats crustaceans it digests them in its mouth rather than in the intestine and then spits out the hard parts.)

Extra-intestinal digestion has been proven to be the rule rather than an exception for *Octopus vulgaris*. The animal lives mostly on the sea floor and does not swim much; it collects stones and builds them into a high 'castle' in which it lives permanently (Fritsch 1938, Lane 1957). It feeds mainly on decapod crustaceans, small *Homarus gammarus*, *Carcinides maenas*, and *Cancer pagurus*, but also on many others. Sitting in its stone castle it detects the prey with its eyes, grasps it after a few, swift swimming strokes, surrounds it with arms, carries it back into the castle and, still holding it enclosed by the arms, covers it with mucous saliva from the two sets of saliva glands. The saliva contains a proteolytic enzyme. It is particularly effective on crustaceans, and its poisonous effect kills the prey rapidly (Bacq & Chiretti 1951 and 1953). Prolonged fighting occurs only with large prey animals. The author observed *Octopus* holding its prey in the described way for about twenty minutes; then it throws the empty carapace outside. The discarded carapace shows no trace of mechanical destruction and the parts are free from musculature and connective fibrous tissue. The individual armour parts are still partly connected by joint membranes. The armour is at first dyed bright yellow, mainly on the inside; this colouring fades after one or two days. The accumulation of crustacean armour around the castle of a single *Octopus* is so substantial that it forms a local accumulation. Eventually, armour parts discarded by *Octopus* cannot be distinguished from those left by crustaceans that have died naturally.

(iii) *Aves*

Many owls and other predatory birds regularly regurgitate indigestible material as pellets. The regurgitating behaviour of the herring gulls (*Larus argentatus*) is particularly important for this study because these large birds breed in extensive colonies on the coasts of the southern North Sea (Leege 1928, Goethe 1939, Schäfer 1954b). Sunkel (1925), Schwarz (1932c), Remane (1951) and Goethe (1958) also emphasise how geologically and palaeontologically significant herring gull pellets are because the birds are so numerous, strong, large and voracious. Teichert &

Serventy (1947) have noted regurgitated pellets of *Gabianus pacificus* on the coasts of western Australia.

The food of the herring gull consists mainly of invertebrates living in the intertidal area and on the coast. The birds collect their prey during low tide. Many of the prey animals possess strong, chitinous or calcareous parts. These are comminuted in the muscular stomach but do not pass through the intestinal tract. Instead, they are shaped into mucus-covered lumps and episodically regurgitated from the muscular stomach. The main prey animals from the tidal flats are small lamellibranchs (*Cardium*, *Macoma* and *Mytilus*), gastropods (*Littorina* and *Nassa*), crustaceans (*Carcinides* and *Crangon*), starfish, and flatfish (*Pleuronectes*). The hard parts of all these species are common in the regurgitated pellets. Goethe (1956a) listed 56 food animals, and more evidence for the extraordinary variety of the herring gull diet was given by Ehlert (1957), Leege (1917 and 1928) and Meijering (1954).

The food supply on the tidal flats of the North Sea coasts is so abundant that the animals have their choice. Consequently an egg-shaped pellet 4 cm long and 2½ cm wide, usually consists of remnants of a single species.

The birds comminute the food in their muscular stomach. Lamellibranch shells, starfish and gastropods are reduced to fragments. *Littorina* shells are broken if they are large. Gastropod shells broken in this way are similar to those destroyed by hermit crabs (Papp *et al.* 1947). The columella is still preserved and the arches of the whorls are broken. Some regurgitated pellets of the herring gull consist exclusively of *Littorina* shells with what look like ribbon cuts, and one pellet may contain up to 40 shells.

Remane (1951) calculated that if each herring gull ejects only one pellet a day, the approximately 100 000 herring gulls of Heligoland Bay produce about 1450 metric tons of shell breccia. All shell fragments from regurgitated pellets end up in the beach zone because the herring gull regurgitates only while it is sitting, never when swimming or flying.

Sometimes lamellibranchs and gastropods are ingested whole, and their shells are regurgitated undamaged from the gizzard. However, the whole shells are not indigestible remnants (Tinbergen 1936 and Goethe 1956a and 1958); rather, they are lamellibranchs, frequently *Cardium* or *Macoma* which are carried undamaged in the gizzard of the male herring gull and are intended as a gift for the female. At the future nesting site or at the favoured perch of the pair, the female induces the male by begging gestures to regurgitate the gifts. This ceremonial feeding takes place mainly during the period before breeding; actually, the female hardly eats the presents. Thus great quantities of unbroken and undigested lamellibranchs accumulate in the nesting sites on beach-grass-covered dunes, and they form continuous shell beds interbedded with the eolian dune sands. Year after year giant shell deposits, consisting mostly of paired valves, are thus deposited in the nesting colonies of up to 6000 pairs of herring gulls. Although these sediment bodies are formed terrestrially above the high-water line they are distinctly marked by these whole, marine shells and by the numerous pellets full of shell fragments. Together such marine fossil contents may conceal the eolian origin of such a sediment body if it were found in a geological section. Yet another source of marine shells at nesting sites is food for the young. After a few initial feedings with predigested food the parents feed the young, up to 15 000 a year in a large colony, by regurgitating from the gizzard. Much of this food is not eaten by the young and remains lying near the nest, eventually ending up in the dune sediments; these food particles consist mainly of crustacean parts.

(*b*) FAECES

The investigations of Moore (1931a, b and c, 1932a and b, 1933a and b, and 1939), Moore & Kruse (1956), Schwarz (1932b), Linke (1939), Manning & Kumpf (1959), Häntzschel, Farouk Le-Baz & Amstutz (1968) have provided much information on the form of faeces in many recent marine organisms. However, the more the faeces of different species became known, the more it became clear that faeces are not typical for species and not even for larger taxa (Schäfer 1953b).

Although this conclusion was certainly disappointing it could hardly be expected otherwise. For most multicellular animals the anatomical and functional conditions which determine the properties of faeces are the same. In spite of all the differences in general organisation and shape of the different invertebrates and vertebrates, they almost invariably have a membranous intestinal tube. Its walls are capable of peristaltic movements or they have cilia as in the Serpulidae and Sabellidae, for moving the faecal matter toward the anus. And so the number of possible modifications of the external shape of the faeces is small. They are the cylindrical pellet, the mucous sac, the segmented pellet, and the spherical or more or less elongate pellet. Occasionally, sculptured ribbons or grooves, transverse or in spirals, can occur on the outside of the faecal body; this is caused by the shape and sculpturing of the anus wall and by the mode in which the faeces are transported in the final section of the intestine.

The composition of the faeces is just as unspecific as their shape. The organogenic hard parts and the other food remnants that are contained in the faeces of various animals can hardly be distinguished from each other. Evidently, the structure of the animal, its life habits, its habitat and its way of feeding have little influence on the composition of the faeces of individual marine animal forms. Rauschenplat (1901) has investigated the stomach and intestinal contents of numerous lower marine animals in Kiel Bay. The compilation shows how uniform the constituents of faeces must be, whether they come from the intestines of crustaceans, gastropods, lamellibranchs or polychaetes:

Acera bullata (9 animals): diatoms (*Cocconeis*, naviculaceans, *Rhoicosphenia*, *Grammatophora*, *Synedra*), pieces of algae, and of florideans, individual small remnants of crustaceans, much detritus.

Amathilla sabini (11 animals): diatoms (*Melosira*, *Synedra*, *Cocconeis*, *Coscinodiscus*, *Grammatophora*, *Rhabdonema*, naviculaceans, *Rhoicosphenia*), quite large parts of algae, plant tissues, broken spicules of sponges, one halacarid, one empty egg cover, chaetopod bristles, small parts of crustaceans, shells of small gastropods and lamellibranchs.

Arenicola marina (10 animals): sand grains, detritus, diatom parts, pieces of plants, bristles of chaetopods, parts of crustaceans, shells of small gastropods and lamellibranchs.

Astarte borealis (12 animals): diatoms (*Coscinodiscus*, *Melosira*, *Rhabdonema*, *Cyclotella*, *Cocconeis*, *Synedra*), peridineans (*Ceratium*, *Prorocentrum*, *Distephanus*), remnants of seaweeds, spicules of sponges, unidentifiable material, much sand, egg cells.

Balanus crenatus: diatoms, *Ceratium*, pollen of conifers, balanid nauplius larvae, chaetopod bristles, detritus and much sand.

Balanus improvisus: diatoms (*Pleurosigma*, *Melosira*, *Coscinodiscus*, *Synedra*, naviculaceans), dinoflagellates, plant tissues, nauplius larvae, copepodan parts, balanid nauplius larvae, bristles of *Nereis*, spicules of sponges, detritus, much sand.

Cardium edule (12 animals): diatoms (*Cocconeis*), dinoflagellates, myxophyceans, pollen grains, spicules of sponges, chaetopod bristles, much sand and detritus.

Cerithium reticulatum (16 animals): diatoms (*Grammatophora*, *Synedra*, naviculaceans, *Rhoicosphenia*), plant remnants, small crustaceans, spicules of sponges, detritus, much sand.

Crangon vulgaris (24 animals): few diatoms, plant tissues, polychaete bristles, 2 ostracods, crustacean parts, fragments of small lamellibranchs, several gastropod shells, detritus and sand.

Cuma rathkei (30 animals): diatoms, crustaceans, spicules of sponges, pollen of conifers, *Ceratium tripos*, sand.

Cyprina islandica (22 animals): diatoms (*Sceletonema*, *Melosira*, *Cocconeis*, naviculaceans, *Synedra*, *Holsatia*, *Coscinodiscus*), dinoflagellates (*Prorocentrum*, *Distephanus*, *Ceratium tripos*), tintinnids, crustacean remnants, spicules of sponges, detritus and much sand.

Gammarus locusta (38 animals): diatoms, dinoflagellates, pieces of seaweed, *Ulva*, large pieces of algae, remnants of small and larger crustaceans, pollen of conifers, wood cells, a nereid bristle, spicules of sponges and sand.

Gobius ruthensparri (19 animals): copepods, nauplius larvae, ostracods, cladocerans, amphipod remnants, larvae of *Chironomus* and of lamellibranchs.

Halicryptus spinulosus (10 animals): broken diatoms, spicules of sponges, small pieces of seaweed, detritus and sand.

Harmothoe imbricata (13 animals): diatoms (*Cocconeis*, *Synedra*), seaweed, algae, thalli of algae, bristles of own species, bristles of *Nereis*, parts of crustacean armour, amphipod remnants, approximately 20 rows of an unidentifiable radula, little detritus, and sand.

Jaera marina (5 animals): diatoms (*Synedra*), algae and seaweed, small remnants of crustaceans.

Idothea tricuspidata (34 aninals): diatoms (mostly *Cocconeis* and *Synedra*), seaweed, filiform algae, *Ulva* floridean pieces, ostracods, remnants of large crustaceans, egg cells of metazoans.

Leander adspersus (26 animals): diatoms, pieces of seaweed and algae, florideans, bristles of polychaetes, parts of crustaceans and amphipods, fragments of lamellibranch and gastropod shells, and 2 mm long *Mytilus*.

Lepidonotus squamatus (12 animals): diatoms, dinoflagellates (*Ceratium*), remnants of seaweed, thalli of algae, bristles of polychaetes, remnants of crustaceans, much detritus, sand.

Littorina littorea (24 animals): diatoms (*Cocconeis*, naviculaceans, *Synedra*, *Grammatophora*), plant tissues, pieces of seaweed, 1 nauplius larva of balanids, *Amorphina panicea*, spicules of sponges, much detritus, sand.

Macoma baltica (25 animals): diatoms (*Coscinodiscus*, *Melosira*, *Grammatophora*, naviculaceans, *Synedra*, *Pleurosigma*), dinoflagellates (*Ceratium*), plant remnants, spicules of sponges, much detritus and sand.

Mya arenaria (16 animals): diatoms (*Cocconeis*, *Coscinodiscus*, *Synedra*, *Chaetoceros*, naviculaceans, *Melosira*), dinoflagellates (*Ceratium*, *Dinophysis*), myxocypheans (*Merismopedia*), tintinnids, algae, seaweed, pollen, remnants of crustaceans, bristles and jaws of chaetopods, nauplius larvae, spicules of sponges.

Mytilus edulis (30 animals): diatoms (*Coscinodiscus*, *Cocconeis*, *Melosira*, *Synedra*, *Rhizosolenia*), dinoflagellates (*Ceratium*, *Prorocentrum*, *Dinophysis*, *Distephanus*), tintinnids, pieces of seaweed and algae, spicules of sponges, remnants of crustaceans, bristles of chaetopods, detritus and much sand.

Nephthys (15 animals): diatoms, individual small plant tissues, bristles of *Harmothoe* and *Polynoe*, remnants of crustaceans and amphipods, detritus and sand.

Nereis diversicolor (20 animals): diatoms (*Synedra*), myxophyceans, pieces of seaweed, thalli of algae, bristles of chaetopods, small remnants of crustaceans, small lamellibranchs, much detritus and sand.

Nerophis ophidion (11 animals): copepods, ostracods, larvae of cirripeds, isopods (*Idothea*, *Jaera*), amphipods, remnants of crustaceans, larvae of lamellibranchs, radulae of hydrobians, small gastropods, seaweed.

Ophioglypha albida (42 animals): diatoms (*Coscinodiscus*, *Rhizosolenia*, *Melosira*, naviculaceans), dinoflagellates, bristles of *Terribellides stroemii*, remnants of crustaceans, spicules of sponges, copepods, bristles of polychaetes, filiformis algae, seaweed, much detritus and sand.

Orchestia litorea (10 animals): diatoms, pieces of seaweed, other plants, little sand.

Pleuronectes platessa (32 animals): copepods (harpactids), *Gammarus*, bristles of nereids, *Idothea*, larvae of lamellibranchs and small lamellibranchs (*Cardium*, *Mytilus*, *Mya*, *Lacuna*, *Doris*), little seaweed and sand.

Praunus flexuosus (17 animals): diatoms (*Chaetoceros*, *Coscinodiscus*, *Melosira*, *Cocconeis*, *Rhoicosphenia*), dinoflagellates (*Ceratium*, *Distephanus*), tintinnids, plant tissues, bristles of chaetopods, remnants of copepods and nereids.

Priapulus caudatus (16 animals): unidentifiable material and sand, small pieces of plants, bristles of chaetopods, spicules of sponges, pollen of conifers, bottom diatoms.

Scrobicularia plana (6 animals): diatoms (naviculaceans, *Coscinodiscus*), dinoflagellates, plant tissues, detritus and much sand.

Spinachia vulgaris (11 animals): copepods, larvae of balanids, isopods, *Gammarus*, mysideans, remnants of crustaceans, a piece of a hydroid polyp.

Kolosvary (1944) listed the plant and animal remnants in the faeces of *Cirripedia thoracica*.

Shape and consistency of the waste products of metabolism are important for the organism that discards them. These factors influence how and where the faeces are transported after they leave the anus and thus have hygienic consequences. The faeces must not re-enter the body because they are harmful to it, particularly to the respiratory organs whose effectiveness and proper functioning depend on clean water. Therefore, the faeces must either be transported as far away as possible, or, if they remain in the living area of the single producer or of a community of colony-forming producers, they must be well packed for transportation or storage. For aquatic organisms there is a danger that discarded waste may drift or spread by diffusion and re-enter the inhalant current that carries food and oxygen. Thus observable differences in shape and consistency of the faeces are ecologically significant.

(i) *Shapes of faeces*

The faeces of most aquatic animals are either shaped or packed in mucus, or they are carried to certain places under the control of special patterns of behaviour. Relatively few animals discharge unshaped faeces.

(1) Unshaped, water-rich and flaky faeces. This type is found in many animals which swim constantly. With a sudden contraction of the rectum the animal squirts loose faecal flakes into the water while it keeps swimming away from the contaminated water. Many free-swimming fishes behave in this way; so do several worms which rapidly migrate through sediment as mobile predators. All these animals belong to types (see A and E I (*a*)) which move quickly and rarely return to the same place until after some time has elapsed.

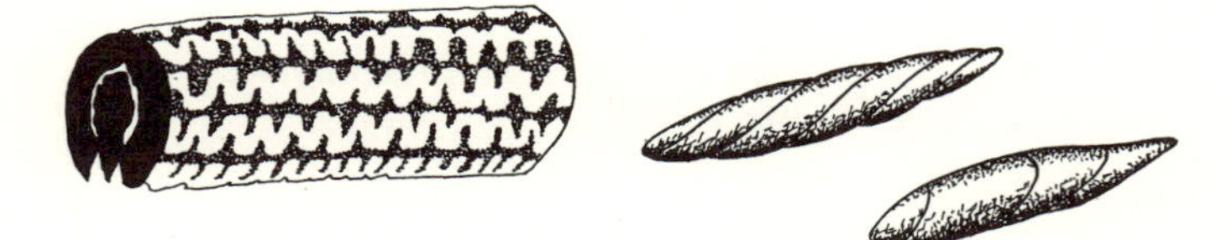

FIG. 241. Faeces of *Gibbula magus* (left) and *Acmaea testudinalis* (right) (after Ankel, 1938b). (Schematic drawing)

(2) Cylindrical faecal pellets of varying length (Fig. 241). They are frequently produced in prolific quantities by sediment eaters. The faeces are somewhat consolidated but are not covered with mucus. *Arenicola* produces up to 15 cm long cylinders of almost pure sand. These faeces must be piled up outside the burrow to prevent them from plugging the tube and from contaminating the supply of breathing water. They disintegrate quickly in flowing water because they contain little organic material and are not coated with mucus. However, if there is no current, large, gradually dissolving sand heaps often pile up around the aperture of the faecal tube of the burrow. The members of the genus *Lumbriconereis* that eat sediment like *Arenicola*, also produce long, cylindrical pellets as do also *Chaetopterus variopedatus* and the enteropneust worm *Balanoglossus clavigerus*.

(3) Mucous sacs containing unshaped faeces. This kind of faeces is found in cases in which it would be harmful to expel unshaped faeces, but in which no mechanism exists to produce water currents for washing away the faeces. It occurs frequently in gastropods; their intestine leads into the right part of the mantle cavity. The faeces are already enveloped in a mucous covering in the rectum and then flow together with mucus from the hypobranchial gland (Brock 1933 and 1936) from the mantle cavity over the sole of the foot toward the back and side of the animal

(Fig. 242). The mucus mixes with that from the glands at the front edge of the foot, and finally becomes part of the crawling trail (*Buccinum*, *Neptunea*, *Nassa*). Ankel (1936c) describes a glandular groove in several species of *Helcion*; it runs around the body above the sole of the foot. The unshaped faeces of this genus are also enclosed in mucous tubes. So are those of the opisthobranchs *Facellina drummondi* and *Dendronotus*

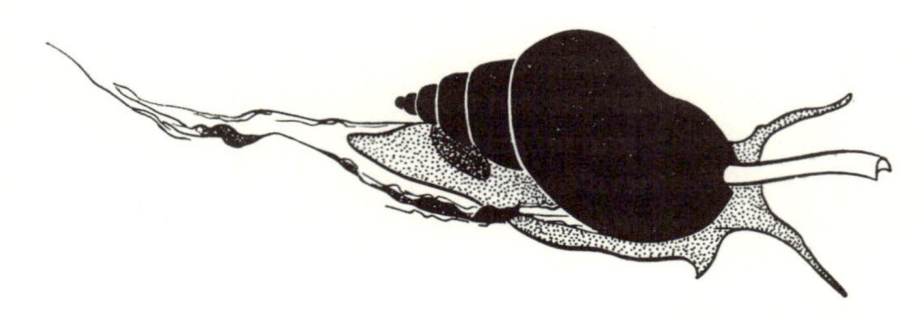

FIG. 242. *Buccinum undatum* with trailing faeces in mucous sacs. (Actual size × 0·3)

arborescens. The mucus-coated faeces sink to the sea floor without incommoding the animals that usually live on hydroid polyp colonies. *Aeolis papillosa* has the same form of faeces.

For the cephalopods, it is especially important that the faeces are coated with mucus because the anus lies in the completely closed mantle cavity which is also the gill cavity. The faeces-filled mucous sacs are discharged from the anus and then ejected through the funnel which is usually bent backward for this purpose.

The mucous sac also occurs frequently in decapod crustaceans. In *Cancer*, *Carcinides*, *Portunus*, *Hyas* and *Eupagurus* the faeces are contained in a continuous mucous tube; the powdery faecal material appears to be pressed in sections into this tube. If the mucous tube tears the powdery mass drops out (Fig. 243).

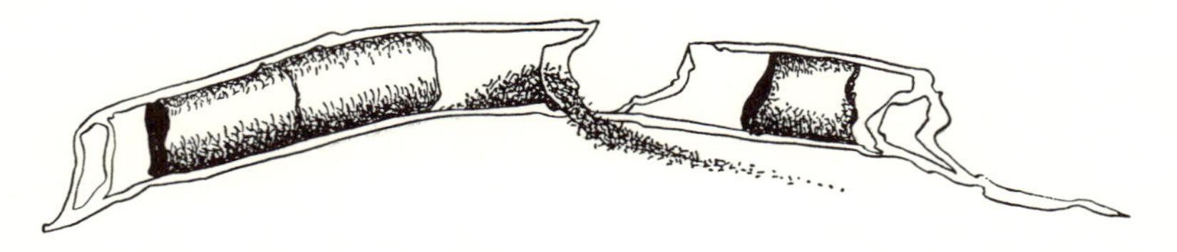

FIG. 243. Faeces of *Hyas araneus* fill a mucous tube. (Schematic drawing)

The mucous sacs of the actinians are buoyant and float to the surface of the water after being discarded from the upward-pointing mouth opening. Occasionally they contain calcareous skeletal remnants and then they slip outward between the tentacle arms over the body wall, and water currents take them away. Where currents are absent some mucous sacs may gather near the mouth, and the actinians do not thrive. This is also the reason why aquaria in which sea anemones are kept must be regularly cleaned (Fig. 244).

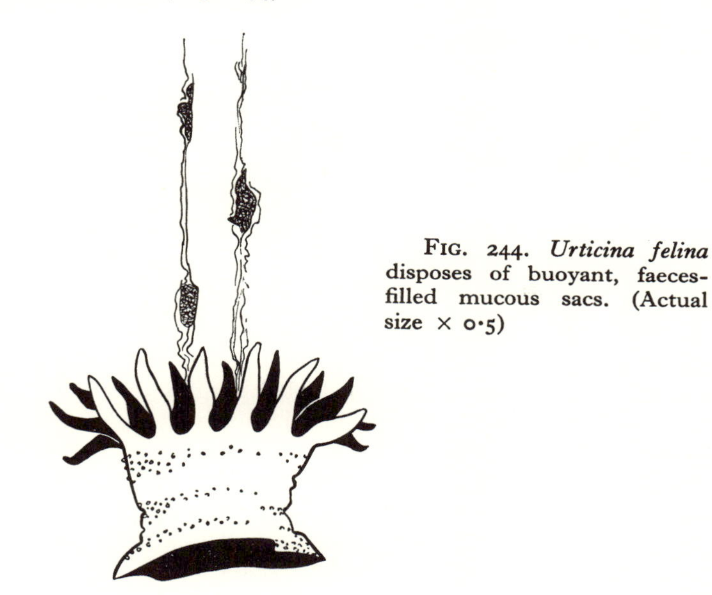

FIG. 244. *Urticina felina* disposes of buoyant, faeces-filled mucous sacs. (Actual size × 0·5)

The majority of fishes, especially if they are bottom-dwelling, produce mucous sacs. The consistency and shape of fish faeces vary with the food. Faeces of fish kept in an aquarium differ in shape and consistency from those of free-living animals because their food is different; captive fishes can therefore not be used for the study of faecal forms.

(4) Segmented faecal pellets terminating sharply and not tapering. Occasionally, the length to diameter ratio is 1 : 1; usually a section is longer than it is thick (Fig. 245). Segmented pellets are, as a rule, small, even if produced by large animals.

(5) Egg-shaped or elongate faecal pellets. They are always small (Fig. 246). Individual sections of segmented pellets differ from the egg-shaped pellets by their angular edges; the two can barely be distinguished without a hand lens. Whereas the egg-shaped pellets are already formed by peristaltic movements in the large intestine, the segments are produced only in the rectum by local contractions of the circular musculature which cuts through a continuous cylinder of faecal matter. Both types will be discussed together since the ecological implications are the same. The faecal bodies are highly dehydrated, pressed and formed. After being

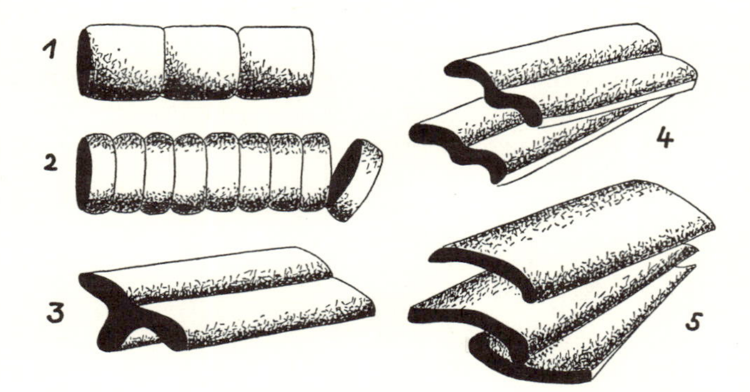

FIG. 245. Segmented faecal pellets of (1) *Cardium*; (2) *Venerupis*; (3) *Pecten*; (4) *Ostrea*; (5) *Mytilus*. (Schematic drawing)

thinly coated with mucus, each egg-shaped or segmented pellet leaves the anus either singly or with a few others; the pellets do not adhere to each other, even after some time has elapsed. Accumulations of these faecal sections and pellets can be rolled and whirled about by currents as if they were grains of sand. They are easily transported in water because they are small and not adhesive. It thus makes good sense that these forms of faeces occur chiefly with animals belonging to constitutional types without, or almost without locomotion; such animals have to rely on water currents for the removal of their faeces.

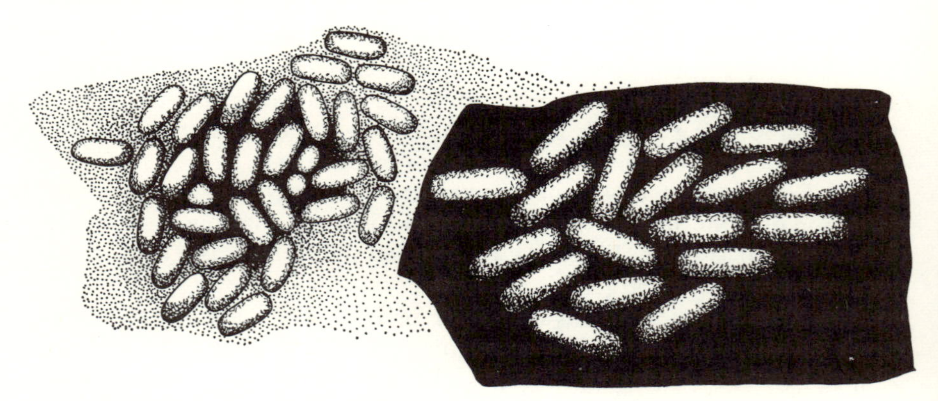

FIG. 246. Faecal pellets. Left: those of *Heteromastus* above the opening of a faecal tube; right: those of *Assiminea* (Actual size × 20)

Sessile or burrow-dwelling polychaete worms usually produce pellets. They are either discharged into the outflowing water current (*Sabellaria*, *Nereis*) or they are stowed in special deposits. Few polychaete worms produce a special water current for the transportation of faeces; this type of transportation seems to be restricted to certain of the tubicole forms. *Pomatocerus triqueter* has a ciliated groove on the ventral side which passes laterally to the back near the anterior end of the worm. The faecal pellets are transported along this groove to the dorsal side and from there between the back of the animal and the wall of the calcareous tube to the outside (Johansson 1927). Many polychaetes form a deposit of faecal pellets at the opening of a subterranean burrow. The most conspicuous example from the tidal flats are the faecal pellets of *Heteromastus filiformis* (see also Schäfer 1952b, 1953b, and Fig. 159).

Lamellibranchs with sessile or almost sessile life habits must necessarily produce a directed water current to supply them with breathing water and in many cases also with food; they do this by whirling, and they use the same current to wash their faeces away. The faecal pellets must be small so as to be easily transported by a gentle current. This holds for epibiontic lamellibranchs (*Mytilus*, *Pecten*, *Modiolus*) and is essential for animals that live buried in sediment. Here the force of the water current must be strong enough to lift the pellets from where they are discarded to the surface. For *Mya* this may mean a vertical distance of up to 40 cm.

Mann (1951) investigated the rate at which *Mytilus*, *Cardium*, *Mya*, *Macoma* and *Scrobicularia* ingest detritus suspended in water. He also studied the rates of breathing and of protein digestion in the digestive gland of the small intestine. He concludes that all three rates depend on the ecological situation of each species. Mann's investigations explain observations by the author that the rate of faeces production varies greatly from species to species, *Mytilus* and *Cardium* having the highest rates of the investigated sample (see also Verwey 1952 and Willemsen 1952). Segmented pellets are the most frequent faeces shape for lamellibranchs. Small, very short segments may occasionally adhere slightly to each other but they rapidly separate in a current (*Cardium*, *Venerupis*). More frequent are segments which are either flattened (*Barnea*), curved (*Mytilus*), cuspate (*Ostrea*), or T-shaped (*Pecten*) in cross-section (Fig. 245).

Many prosobranchs produce faecal pellets. They follow the same path as mucous sacs and move individually in a mucus stream starting from the right side of the respiratory cavity, and finally descending over the right, upper side of the foot. The faecal pellets are produced by peristaltic movements of the colon. This can be deduced from the fact that when a pellet-producing gastropod withdraws into its shell, tightly folding its body, it frequently expels a series of perfectly formed faecal pellets from its intestine just before it closes the operculum. Thus the pellets have been sitting in the rectum in their final shape. Pellets are produced mainly by prosobranchs that live on the tidal flats; mucous sacs would be disadvantageous for them because in air they would not detach themselves from the body as small, hard pellets do. A few prosobranchs, for example *Patella vulgata* and *Gibbula magus*, produce segmented pellets, and according to Moore so does *Gibbula cineraria*.

Echinoids produce spherical pellets. Those of *Echinus esculentus* and of *E. acutus* are usually green, probably because their diet consists almost exclusively of algae. Von Uexküll (1909) describes how the faecal pellets are finally removed after they leave the anus at the top of the domed test. If the faeces were left lying between the densely set spines they would eventually damage the skin. However, a chemical stimulus from the excrement causes the skin muscles which normally hold the spines erect to relax. The spines nearest the faeces slant downward, and the faecal spheres can roll off. The lower spines react in the same way but in response to the pressure exerted by the slanting upper spines, and not to a chemical stimulus.

A few faecal shapes do not fit the general classification. Ziegelmeier (1954) described *Lunatia* which bores into lamellibranchs; while the prosobranch drills it expels separate, white, spindle-shaped bodies from its anus. The faecal spindles consist of drilling mud which is scraped from the calcareous shell of the victim. These calcareous spindles are later followed by continuous faecal ribbons consisting of connected elements of the same shape and still containing much boring mud along with normal faecal matter. Individual segments still become detached from the ribbon when they are discarded. Eventually, the pointed faecal spindles are free of drilling mud and remain connected as a ribbon. Ziegelmeier observes that the number of the early, calcareous spindles depends on the thickness of the mollusc shell. He concludes from a comparison between the volume of the calcareous spindles and the quantity of the scraped-off shell material that most of the calcareous material swallowed during drilling is again expelled.

The fishes *Anarhichas lupus* and *A. minor* feed on gastropods, lamellibranchs and crustaceans and also ingest much calcium carbonate together with the soft bodies of their prey. The hard parts are eliminated as faeces. The coarse and sharp-edged shell fragments that result from the cracking of strong shells pass unchanged through the intestinal canal. There are no acidic gastric juices to cause even partial dissolution; this is apparently an adaptation to the extremely large quantities of ingested calcareous material which would neutralise gastric juices unless they were produced in excessive quantities. The faeces are shapeless, and the mollusc shells and crustacean armours that are eliminated in them do not differ from ordinary shell fragments. Eventually, considerable quantities of shell breccia collect at the localities to which the catfishes always return as is their habit.

The faeces of the herring gull (*Larus argentatus*) also frequently contain 75 per cent of small calcareous fragments of shells of ingested molluscs, although the large fragments are regurgitated. The faeces have a whitish and gritty appearance. The edges of almost all the shell fragments are rounded, and the surfaces are often polished. Regular, cylindrical, about 1 to 1·2 mm long calcareous bodies occur strikingly frequently; they are probably produced by prolonged, mutual grinding of shell fragments while they pass through the digestive tract. The largest calcareous splinters are about 3 to 5 mm long; occasionally, well-preserved, whole shells of *Hydrobia* and points of claws or whole dactyli of small brachyurans can be found. Fine quartz sand is another frequent constituent. A high percentage of organically produced calcareous material must thus be expected to accumulate every year on dune islands of eolian origin on which sometimes 10 000 herring gulls crowd together in less than half a square kilometre; they do so on Mellum island and other dune islands on the seaward side of the tidal flats of the North Sea.

Most sea birds feed like the herring gull almost exclusively on animals living on the tidal flats and the beach, others catch hard-shelled marine animals by diving; only the swans and geese are exceptions. Niethammer (1938 and 1942) described the food of sea birds. The diving ducks are particularly interesting; they fetch their mollusc food from the sea floor at depths of 3 to 7 m. Usually, they swallow the hard-shelled animals unbroken, and the muscular walls of the gizzard then open the bivalves and break the gastropod shells. Stomach stones and coarse gravel which help to comminute the food are frequently found in substantial quantities in the gizzards of these birds (personal communication, Dr F. Goethe, Heligoland bird station).

Foraminifera occasionally digest food in the pseudopodia strands before they are withdrawn into the test. Doflein & Reichenow (1953) report that ingested mud and detritus accumulate in the protoplasm in the shape of so-called stercomas. Movements of the protoplasm frequently shape these structures into spheres or ellipsoids; they are enclosed in iron-stained membranes similar to the cement of the tests and are expelled as faeces. Stercomas are supposedly very resistant against acids and bases. Where Foraminifera are numerous they can form deposits.

(ii) *Faeces as sediment constituents*

How well can the different types of faeces be preserved, can they be recognised as such in the sediment, and can they constitute beds?

(1) The components of unshaped faeces can neither be identified by

shape, nor chemically, because each individual originator of such faeces, usually a very mobile swimmer, inhabits a large area over which the faeces are spread sparsely. They cannot form perceptible beds.

(2) Cylindrical faeces are not usually coated with mucus and are often sandy so that moving water rapidly destroys them. They thus cannot be transported, individual faeces are rarely preserved, and formation of beds is impossible.

(3) Mucous sacs can be preserved but have shapes difficult to identify. Many of their producers are locomotory animals and expel them while moving. Thus mucous sacs cannot accumulate at the place of production. Neither can they be accumulated by being passively transported because dragging mucous threads become anchored. This type of faeces does not form deposits.

(4) Segmented pellets are dehydrated, very compact and suited for transportation. The segments can roll on a substrate or saltate in flowing water; this property is ecologically important for the producers. Many animal species produce segmented pellets and many of them occur in large numbers. The faeces can be preserved and can easily be recognised as faeces, but not specifically identified. They can form deposits because they can be transported.

(5) Egg-shaped pellets are also dehydrated, very compact and suited for transportation; they can therefore roll and saltate in flowing water. Animals of many species with large individual numbers produce them. They are preservable and recognisable as faeces but not specifically identified. Being easily transported, they can form deposits.

In the shallow sea where currents and waves act constantly, egg-shaped and segmented pellets are the only forms of faeces that can leave a mark on the sediment. They are particularly significant because of the quantities in which these faecal bodies are produced by all kinds of different animals and because of their preservability, ease of transportation and ability to form deposits. Furthermore, many of the animals that produce this kind of faeces feed on plankton and suspended detritus. Passage through the body and compaction into pellets rapidly transforms fine, suspended matter into particles that can be deposited on the sea floor. Schwarz (1932b) discussed this role of the benthonic plankton and detritus-eating filterers. It consists, he says, in turning suspended into more coarsely particulate matter.

As such particulate matter the pellets can become individual components of the sediment. Either individually or mixed with other grains or as homogeneous layers they can take part in the formation of a layered rock or one made up of lenticular sediment units. The coprogenic components in the sedimentation are not always traceable after the sediments are compacted. Although the faecal particles may be quite resistant during transportation in the water and on the sea floor, they do not always maintain their shape under the weight of an overburden.

Most of the faecal pellets are not embedded at the place of their discharge. It is therefore impossible to identify the producers from other traces or marks they may otherwise have left. Although the faeces consist essentially of clay they behave differently from disaggregated clay. Their grain size is comparable to that of coarse sand, and compared with mud they are differently distributed and deposited. Because such faeces are frequently produced by animals living in more or less pure, fine sand, they can decisively determine the eventual appearance of a sediment and can cause lenticular bedding where the sandy sea floor carries ripple marks.

On mud flats faecal pellets can be deposited as a homogeneous layer covering a smaller or larger area. Especially on fans at the mouth of tributary tidal channels and on the gently sloping convex banks of curved tidal channels, large areas can be covered with them. Thus dipping clay layers which appear dark in cross-section are formed as intercalations between the lighter-coloured layers of fine sand. Pellets can also be deposited in the troughs of ripples on a sandy sea floor and appear as mud lenses in the sandy sediment. Plath (1965) drew a map of the North Frisian tidal sea showing the production of faeces. He distinguished three main producers and places of production, banks of *Mytilus*, settlements of *Cardium*, and populations of tidal gastropods, consisting mainly of *Hydrobia ulvae*; he demonstrated that there is a relationship between these areas and areas of mud deposition. In the permanently immersed, deeper part of the shallow sea, the transportation and deposition history of faecal pellets have not yet been investigated. Most probably they are deposited in ripple troughs.

To the pellets produced by benthonic animals of the tidal flats and of the permanently immersed shallow sea, must be added the faeces produced in not inconsiderable quantities by planktonic animals. The larvae of many gastropods, worms and lamellibranchs produce faeces in the shape of small egg-shaped or segmented pellets, similar except for the size to those of the adult animals. Shaped faeces are not an ecological requirement for these planktonic larvae themselves; their digestive tract, however, is already organised for the requirements of the adult animals. Wetzel (1937a) investigated faeces of planktonic animals and distinguished regularly oval (0·07 to 0·28 mm), irregularly shaped, often spindle-like (0·21 to 0·61 mm) and rod-shaped faeces (0·12 to 0·23 mm).

Occasionally faecal particles remain where they were originally produced. On tidal flats near the high-water line, the water occasionally flows so quietly that the faecal pellets deposited on the surface of the sediment by polychaetes are not washed away; consequently, they accumulate quite rapidly and form a continuous layer. If such a faecal layer is then secured and consolidated by the growth of a continuous skin of diatoms, it can remain undisturbed until it is covered by fine sand (Schäfer 1952b).

More important are the cases in which endobiontic producers of faeces stow them in specific sections of their burrows. The genus *Polydora* is the only recent animal for which such behaviour has been demonstrated. With the funnel-shaped, very mobile last segment of the body, the worm stows faecal pellets into abandoned sections of its dwelling burrow, and then closes these sections with sand grains. These worms settle densely and frequently produce connections between neighbouring tubes. Such connecting burrow sections are often plugged with faecal pellets (Fig. 187). Presumably, *Nereis* also occasionally stows faeces into certain sections of its burrow. However, the faecal fillings of sections of dwelling burrows that can be occasionally observed on Reineck's thin sections through artificially hardened sediment, are mostly filled passively; faecal pellets are washed into the vertically descending sections of abandoned burrows and gradually fill them completely. It is not clear what the ecological significance of actively stowing faeces into parts of a burrow may be. For *Polydora* the reason may possibly be that it already uses the posterior end of its body for its building activity; it attaches sand grains to the burrow walls with directed movements, holding them in a funnel-shaped depression of its last segment. The animal may just as well handle faecal pellets that pass through the same funnel in an analogous way.

It is necessary for identification that the faeces remain where the producer has deposited them. Thus a burrow in which faeces are stowed indicates with high probability that they are faeces of a sediment-inhabiting worm and not of a lamellibranch or gastropod, although both produce similar pellets.

Gastropods can deposit their faecal pellets in such a way that it can tell us something about the animal. The animals can expel their faeces during locomotion. As a gastropod crawls along the surface of the sediment, one pellet after the other slips out of the shell at the right side of the body and drops to the ground. One faecal pellet lies behind the next if the gastropod crawls quickly. If it crawls slowly the line of pellets becomes denser. The bent and curved line of pellets shows the course taken by the animal. Such a line of faecal pellets definitely indicates gastropods because neither lamellibranchs nor polychaete worms can expel faeces while they move. The lamellibranchs which are more or less linked to one place eject the faecal pellets through the ex-current siphon, whereas the polychaete worms can only discharge faeces when at rest; they therefore produce heaps but never lines of pellets. The different behaviour has anatomical and functional reasons; the worm body is composed of two coaxial tubes; the muscular tube of the skin which serves for locomotion by undulating or peristaltic movements, and the intestinal tube whose own musculature affects the peristaltic movements of the intestine. Thus 'external' and 'internal' peristaltic movements are distinct from each other. The internal peristaltic movements of the intestinal tube move the faeces and shape the faecal pellets on the way to the rectum. The proximity of the outer and inner muscular mechanisms, and also the kind of innervation, exclude simultaneous activation of the two muscular layers. Consequently, a polychaete worm can either crawl or discharge faeces (Schäfer 1953b). Because only the gastropods are able to crawl and to expel faeces at the same time, a line of pellets is specific to gastropods. However, such lines of pellets are easily washed away. All the more remarkable are the finds of well-preserved lines of faecal pellets, known as *Tomaculum* Groon, in the Ordovician Herscheid Shales of the Rhenish Schiefergebirge (R. &. E. Richter 1939 and 1941).

(iii) *Contents of stomach and intestine*

As many fossil examples prove, the contents of the digestive tract can be preserved with the rest of the organism. Because material in the digestive tract disintegrates easily it is essential for the preservation of its shape and contents that the walls of stomach and intestine remain strong even after death and enclose the contents.

On the way through the digestive tract, food and hard parts are increasingly mechanically comminuted and chemically decomposed. Therefore the contents of the stomach, as the first link of the tract, can be most easily identified. The stomachs of predatory fishes, dolphins and seals contain identifiable parts of fishes, mainly head skeletons and vertebrae, cuttle-bones and jaws of cephalopods; the stomachs of fishes specialising in hard food contain hard parts of gastropods, lamellibranchs and crustaceans. Sea birds frequently carry in their gizzard single stomach stones and coarse sand. A. H. Müller (1951) cited Schmidt as describing the stomach content of a dolphin from the Spanish Mediterranean coast; bones, eyes and otoliths of approximately 7600 fish of five different species were found. Meyer (1931) reports on a large codfish of approximately 4 kg caught at the mouth of the river Jade. The stomach content weighed 0·5 kg and consisted of 16 plaice, up to 20 cm long. The fish all lay lengthwise parallel to each other and were closely packed; Meyer conjectures that the well-ordered arrangement in the stomach may be the result of active movements of the plaice while they were still alive in the predator.

Since chemical decomposition begins in the stomach, recognisable remnants of prey are rarely found in the intestine. In the small intestine, the contents are gradually dehydrated and shaped into distinctly separate masses. In the case of vertebrates these masses have little chance of being preserved in the water after they have left the intestine as faeces; often they are loose, and upon discharge look like a cloud of mud being blown from the anus and then drizzling to the ground. Thus this type of faeces can only be preserved as long as it is still in the intestine. The contents of the spiral-shaped intestines of sharks, rays and skates are occasionally preserved in shape, as long as they stay enclosed by the intestine, and therefore also within the dead fish. Well-shaped intestinal contents of polychaete worms have also been found preserved.

(*c*) PHYSIOLOGICAL AND PATHOLOGICAL REGENERATION

Lost indispensable parts of the body are regenerated by many animals. In physiological regeneration, old parts are replaced by new either, and most commonly, periodically, or episodically (moulting, change of teeth). If loss of an organ is not part of a normal function but caused essentially by external forces, and the organ is then replaced in its original size, shape and working ability, the regeneration is called pathological. The immediate palaeontological interest is not so much in the cell physiology and the mechanical processes or regeneration itself as with the fact that the replaced organs have been released into the habitat. Frequently they represent a significant and sometimes typical constituent of the assemblage of organic parts in marine sediments.

(i) *Moulting*

All growing arthropods moult. The higher crustaceans usually continue to grow through most of their long life and therefore also continue to moult. Sexual maturity which occurs long before the animals have reached maximum size does not stop the moulting. In the course of its life, a single animal thus furnishes many exuviae during all stages of growth; the size of each subsequent exuvia slightly exceeds that of the preceding one. It is often difficult or even impossible to distinguish between a fossil carcass and exuvia because the fossil properties are identical. Thus multiple exuviae from a single animal provide multiple body fossils of this individual.

The number of crustacean body fossils is no accurate clue to the former number of living animals. According to Ehrenbaum (1907) the lobster moults nine times during the first year of life, five times during the second, four times during the third, twice during each of the fourth and fifth years; and from then on the male lobster moults once yearly. If we assume a life of forty years before the lobster stops moulting it releases 57 exuviae into the sediment of its habitat in the course of its life. (For the moulting frequency in *Gammarus* and factors that influence it, see Kinne 1954a.)

Moulting of crustaceans and the histological and physiological processes involved in it have been frequently described (Tullberg 1822, Couch 1838, Jones 1844, Braun 1875, Vitzou 1882, Herrick 1895, Baumberger & Dill 1931, Drach 1935, Darby 1938, Schubert 1938, von Buddenbrock 1954, and many others). The old armour splits along definite lines so that the animal hardly damages the exuvia with the new armour.

The exuviae of malacostracan crustaceans are infrequently preserved and are discussed first. Moulting of Natantia has rarely been observed but is similar to that of the Reptantia. The carapace first tears along the median line, starting at the back and simultaneously through the joint membrane between the cephalothorax and the abdomen. Thus the animal can slip out of the exuvia top first. The cephalothorax with all its limbs, antennae, eyes and mouth appendages leaves the exuvia first; the abdominal section is freed last. Most Natantia moult sitting on the sediment, but some forms cling to hydroid polyps or algae, and the exuvia may be left hanging there or may drop to the sea floor. The various parts of an exuvia are not equally well preserved because the exoskeleton is not everywhere equally reinforced with calcium carbonate. The carapace and the abdomen of the Natantia have a tender skin and contain little calcareous material; thus even slight transportation destroys these parts of the exuviae. What remains are the most heavily calcified rostral areas and especially the hard claws. Glaessner (1929) published illustrations of several fossil exuviae of Natantia.

The moulting mechanism of the Reptantia is better known, especially for the lobster. The first slit on a moulting lobster appears as a transverse gap across the back between cephalothorax and abdomen where the thin membranous joint had been previously weakened by resorption; the new, still soft back immediately bulges forth through the slit. A little later the dorsal part of the cephalothorax tears along the median line, and the dorsal section of the body pushes further up. During the first movements of moulting the lobster lies down on one side and straightens its claws and limbs.

As is the case with the Natantia, the cephalothorax with its limbs and all appendages is freed first; later the lobster moves the abdomen in jerks and completely slips out of the exuvia. During moulting blood is withdrawn from the stronger limbs, particularly from the chelae of the claws; as a result, these parts shrink considerably and can slip through the narrowest passages of the exuvia at the coxa, basis and ischium. Furthermore, there is a thin and very extensible zone in the armour at the proximal sides of the skeleton of the limbs so that their diameter can be enlarged to let the voluminous claw musculature pass through (Figs. 247 and 248).

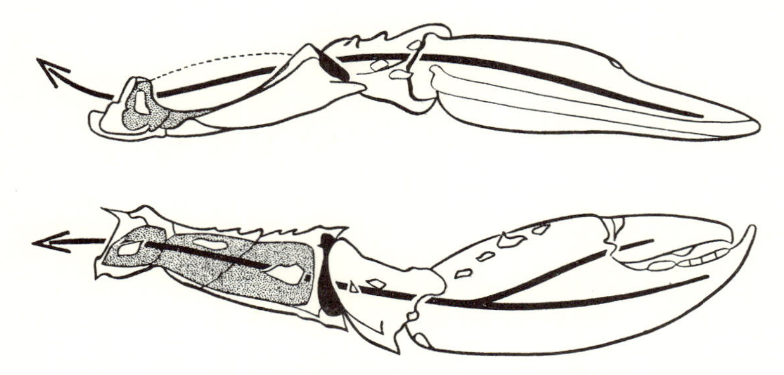

FIG. 247. Clawed limb of *Homarus* during moulting. Top: dorsal view; (black arrow) direction of movement of the limb through the exuvia; (dotted line) maximum swelling of soft-skinned zone (dotted area) during moulting. Bottom: lateral view. (Schematic drawing)

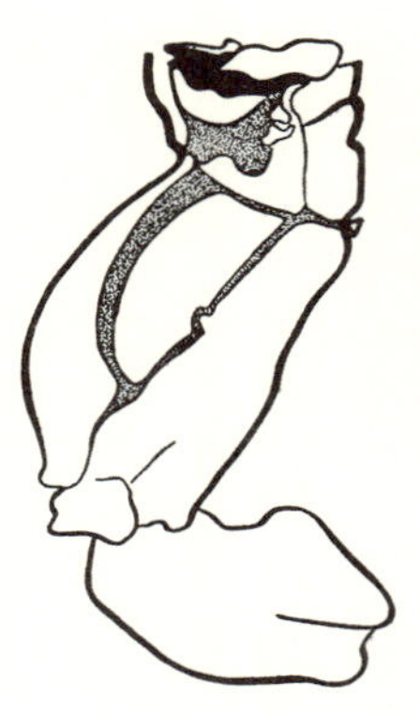

FIG. 248. Soft-skinned zone (dotted) in the exuvia of a clawed limb of a brachyuran crustacean. It is less extensive than in macrurans. (Schematic drawing)

The carapace of moulting brachyuran crustaceans does not split lengthwise; it is so broad and short that the crustacean can slip out of it backward and upward. The fourth segment of the carapace, however, which covers the gills from obliquely below remains stuck on the cephalothorax and separates along a suture from the upper parts of the carapace (Figs. 249 and 250). Thus the bottom opening of the dish-shaped carapace is wide enough for the crustacean to slip through. This separation along the suture between the third and the fourth segments of

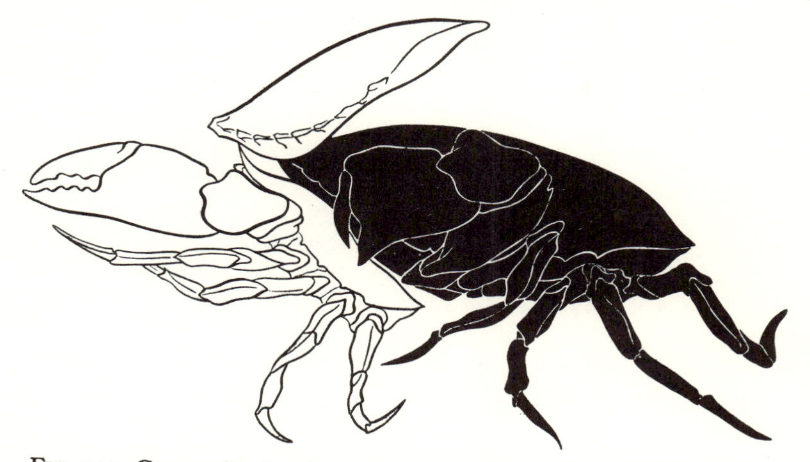

Fig. 249. *Cancer* slipping from its exuvia (white). (Schematic drawing)

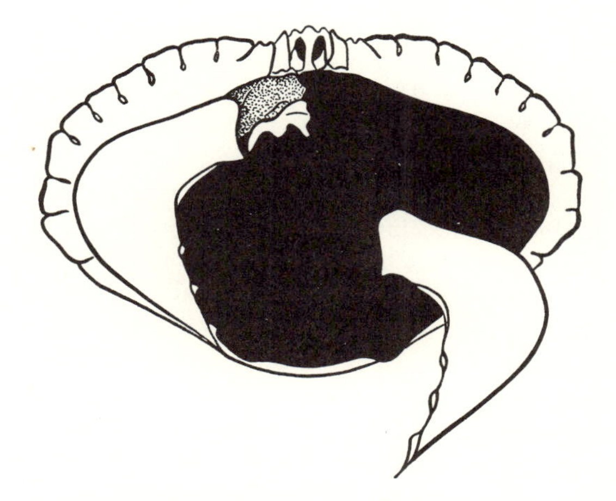

Fig. 250. Ventral view of carapace of *Cancer pagurus*. The fourth segment is in place on the left, removed on the right. A carapace without both these segments is part of an exuvia, a complete one is a remnant of a dead animal. (Schematic drawing)

the carapace occurs only in exuviae of the brachyurans (and not in carapaces of dead animals). Fossils with this feature are therefore certainly preserved exuviae (Schäfer 1951b).

Preliminary changes in preparation for growth begin while the old armour still functions perfectly. A new layer of hypodermis cells is formed beneath the old cuticle and is destined to become the new one. It consists of a larger number of cells than the old one, and because its area is also increased it is laid into numerous folds. To facilitate separation of the two cuticles the inner chitinous layer of the old one is then transformed into a mucous, lubricating exuvial fluid. After the animal has shed its exuvia the young cuticle can stretch, and the crustacean grows visibly larger. It does so by increasing the blood volume with water taken up into the body, and the resulting pressure stretches the cuticle (Paul & Sharpe 1916). Thus the new armour is essentially preshaped by the earlier growth of the new cuticle before the moulting. Only the final shaping of certain regions of the claws occurs after the moulting (Schäfer 1954a).

Although a calcified cuticle shed as exuvia has the same chance of being preserved as the functional armour of a dead animal, there are differences between the two. In preparation for the shedding the armour has undergone changes. The calcareous material between the chitinous lamellae (W. J. Schmidt 1924) is locally resorbed, making certain regions less hard and brittle (Mann & Pieplow 1939), mainly at the bases of the limbs and claws. Resorption is spread over certain areas in the lobster, it is missing altogether in *Palinurus vulgaris* and in *Nephrops norvegicus*, and also in those hermit crabs that do not have especially large claws; in the brachyurans, it is concentrated on certain lines on the inside of the meropodite (Fig. 248). Elsewhere, especially at the points of teeth of the claws, at the bulges of the claw joints and at the rostral spines, no resorption occurs, and these parts of the exuvia retain their original resistance and can well be preserved as fossils. The local character of the resorption seems to suggest that it is useful for the moulting technique but unimportant as a means of conserving calcium carbonate.

In both the brachyurans and macrurans the carapace tends to separate from the cephalothorax after the crustacean has slipped out of the exuvia. The dorsal plate either separates totally from the rest of the armour or remains loosely connected to it. In the macrurans the connection between cephalothorax and abdomen is also loosened, and even minor transportation either separates them, or the abdomen is folded toward the ventral side (Fig. 251). The exuvia of many macrurans loses its resistance because it partially rips during shedding along the dorsal, median line (but not in *Nephrops*). Such an exuvia is soon compressed, or parts break off, or it completely splits lengthwise.

Conditions are different for the exuviae of brachyuran crustaceans. Because of the different moulting technique the carapace splits along the curved suture between the third and fourth segments, and a loose membranous connection of the top of the carapace with the ventral, facial

Pl. 1*a*. *Delphinus delphis*. Note the bare bones in the flipper and jaw regions (photograph by F. Wunderlich). *p. 15*

Pl. 1*b*. *Phocoena phocoena*. *p. 16*

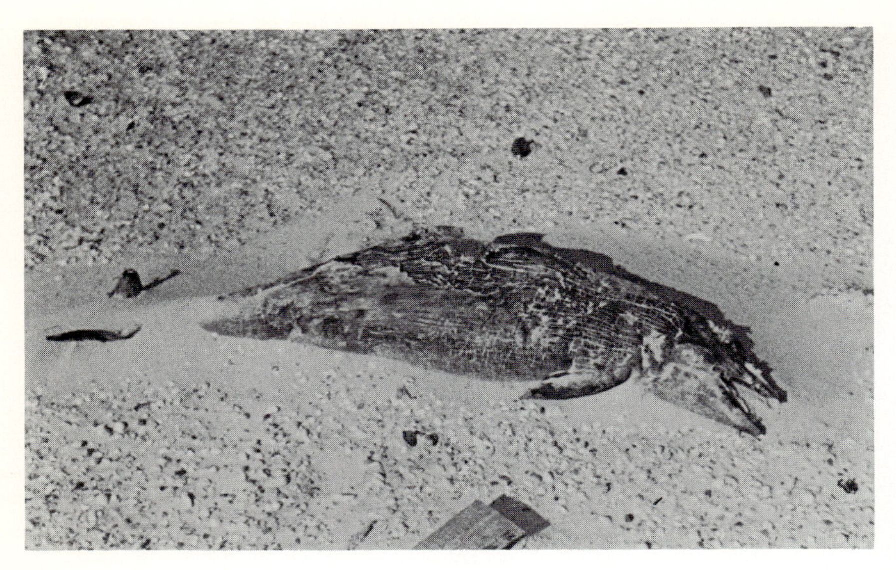

Pl. 2*a*. *Phocoena phocoena* being mummified. Note patterns formed by tears of the skin. *p. 22*

Pl. 2*b*. *Phocoena phocoena*. Tail fin and tail stem twisted by mummification. *p. 22*

PLATE 2

Pl. 3*a*. Drifting carcass of *Phocoena phocoena*. *p. 20*

Pl. 3*b*. Mummified carcass of *Phocoena phocoena*. *p. 23*

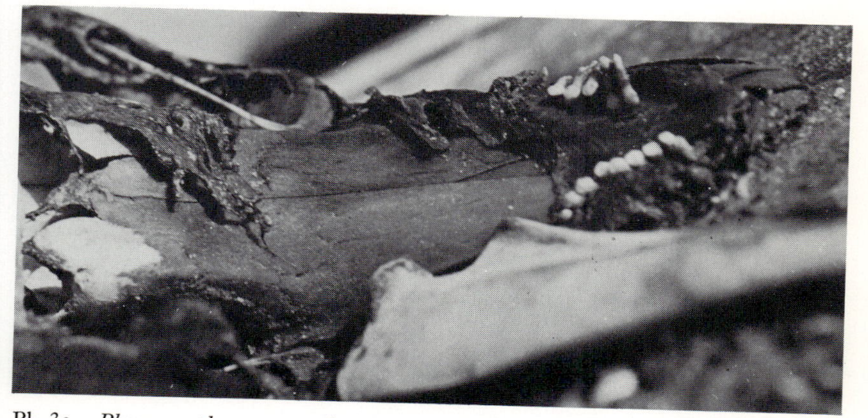

Pl. 3*c*. *Phocoena phocoena*, teeth coming loose from the jaws but held together in rows by gum tissues. *p. 23*

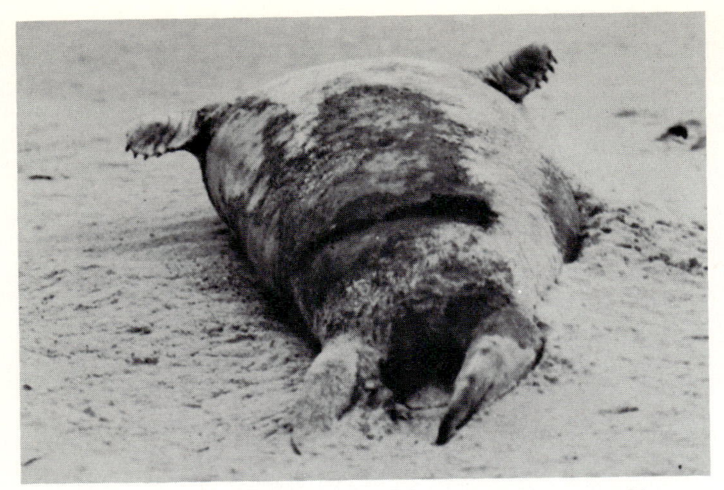

Pl. 4*a*. Carcass of *Phoca vitulina*, bloated by gas. *p. 34*

Pl. 4*b*. Carcass of *Phoca vitulina*, deflated. *p. 34*

Pl. 4*c*. Maggots on *Phoca vitulina* skin. Hair of fur has fallen off almost completely. *p. 36*

PLATE 4

Pl. 5*a*. Air-photograph of flock of *Tadorna tadorna* on the tidal flats of Knechtsand (north of the mouth of the Weser River). The birds are moulting and cannot fly (photograph by F. Goethe). *p. 40*

Pl. 5*b*. Traces made by *Larus argentatus* by standing in shallow, wind-agitated water (photograph by H. E. Reineck). *p. 39*

Pl. 6*a*. Picking traces made by broad-billed sandpipers on tidal flat. *p. 38*

Pl. 6*b*. Flock of broad-billed sandpipers on Mellum island (Rivers Jade and Weser) (photograph by H. Böcker). *p. 38*.

Pl. 7*a*. Wing of broad-billed sandpiper, mummified with quills still intact. *p. 48.*

Pl. 7*b*. Carcass of *Larus argentatus* that has been frequently moved at high water. *p. 48.*

Pl. 7*c*. *Larus argentatus*, in wind-blown sand. *p. 47.*

PLATE 7

Pl. 8*a*. *Larus argentatus*, whole wings lifted from the carcass. *p. 46*

Pl. 8*b*. Pellets regurgitated by *Larus argentatus*. *p. 413*

Pl. 8*c*. Faeces of diving ducks, consisting of broken lamelli-branch shells. *p. 425*.

PLATE 8

Pl. 9*a*. Skeleton of plaice (*Pleuronectes platessa*). Almost undamaged because the carcass was never buoyed to the surface during decomposition. *p. 55*

Pl. 9*b*. Statoliths concentrated in pellets regurgitated by *Larus argentatus*. *p. 91*

Pl. 10*a*. Star-shaped trace produced by a fish (*Gobius microps*). *p. 398*

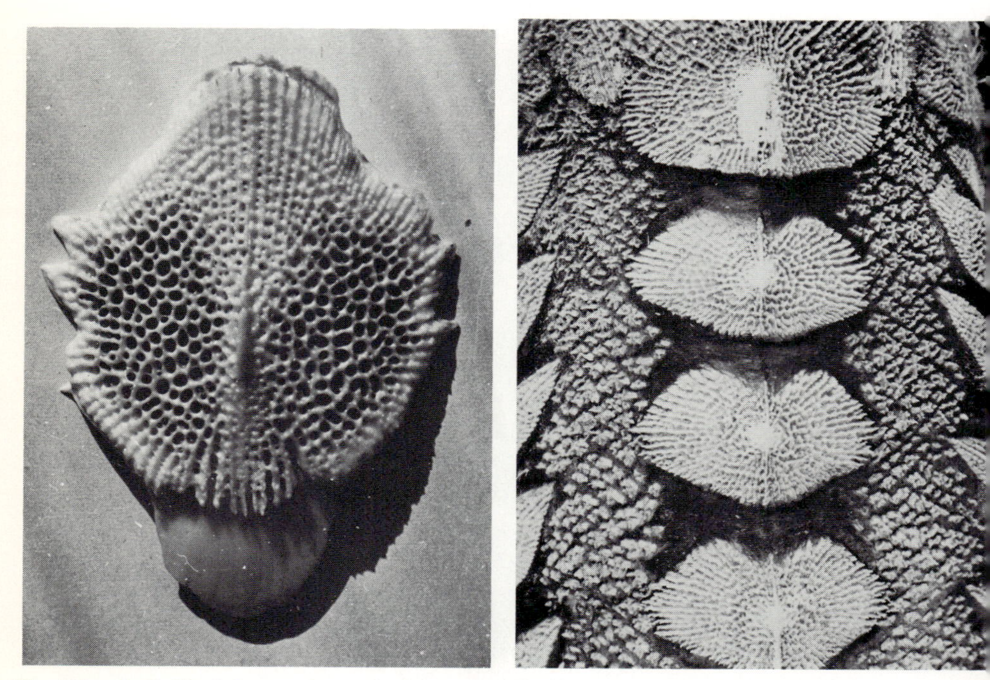

Pl. 10*b*. Individual bony plate of *Acipenser sturio*. *p. 79*

Pl. 10*c*. Bony plates of *Acipenser sturio*. *p. 78*

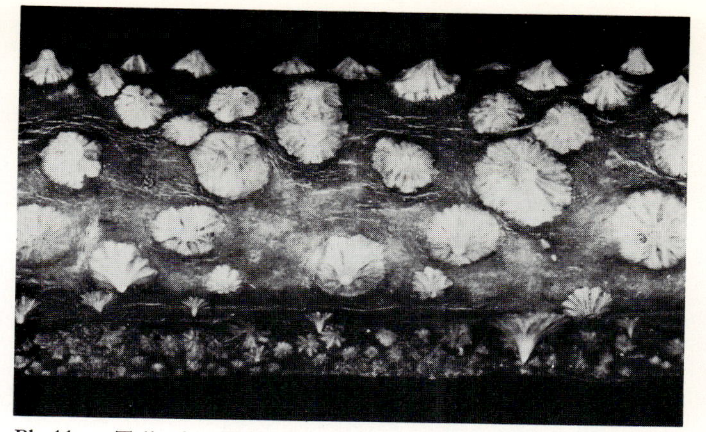

Pl. 11*a*. Tail of a *Raja* with dermal teeth of various shapes and sizes. *p. 75*

Pl. 11*b*. Skin with dermal bones of *Cyclopterus lumpus*. *p. 84*

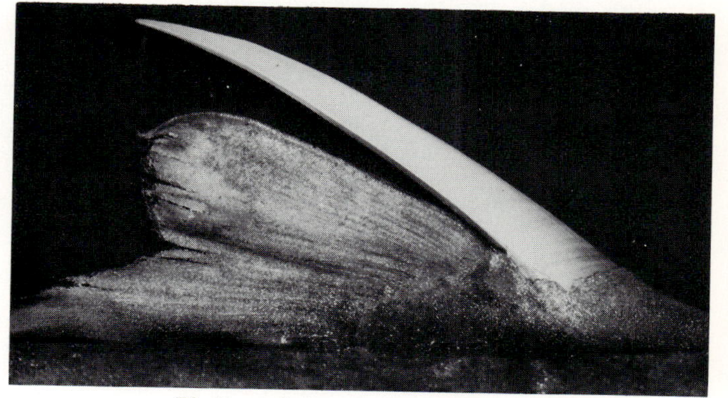

Pl. 11*c*. Spine of a shark's fin. *p. 77*

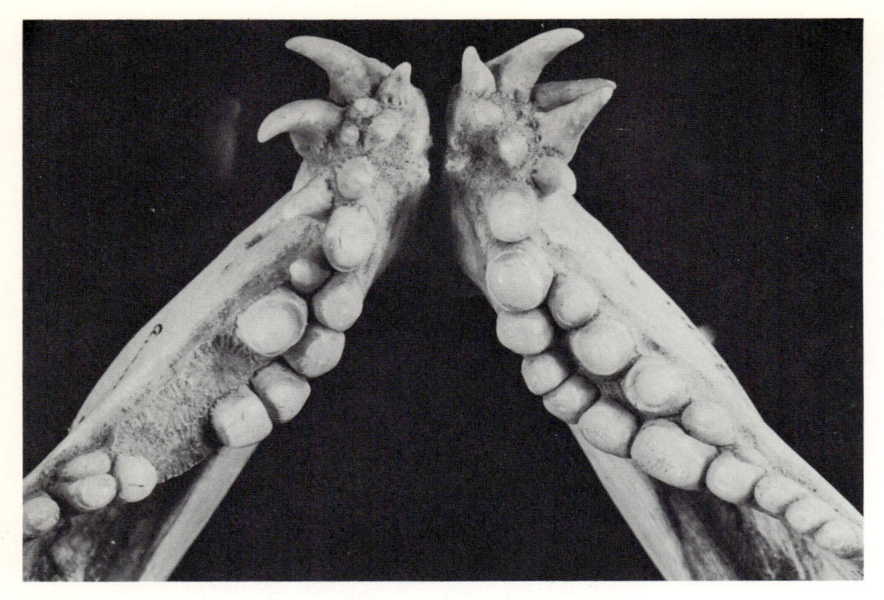

Pl. 12*a*. Lower jaw of *Anarhichas lupus* with grasping teeth and grinding teeth. *p. 72*

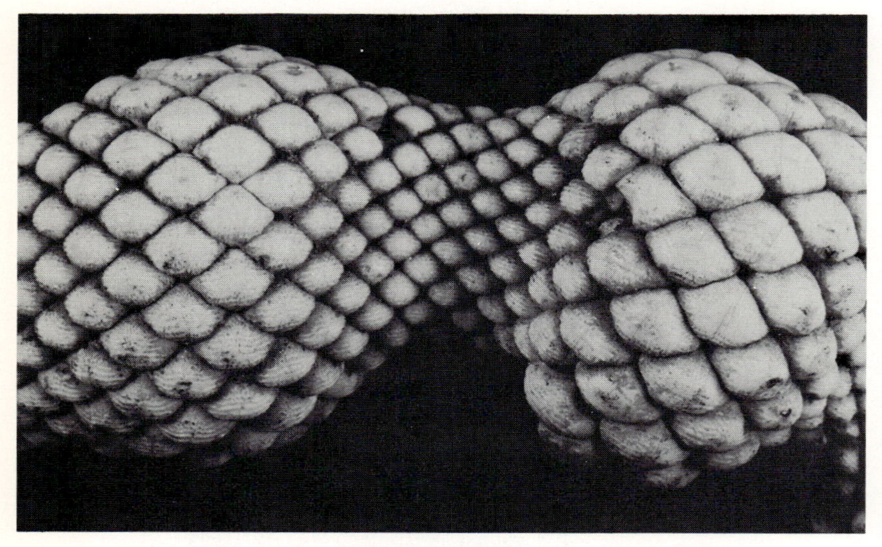

Pl. 12*b*. Grinding teeth of a *Raja*. *p. 68*

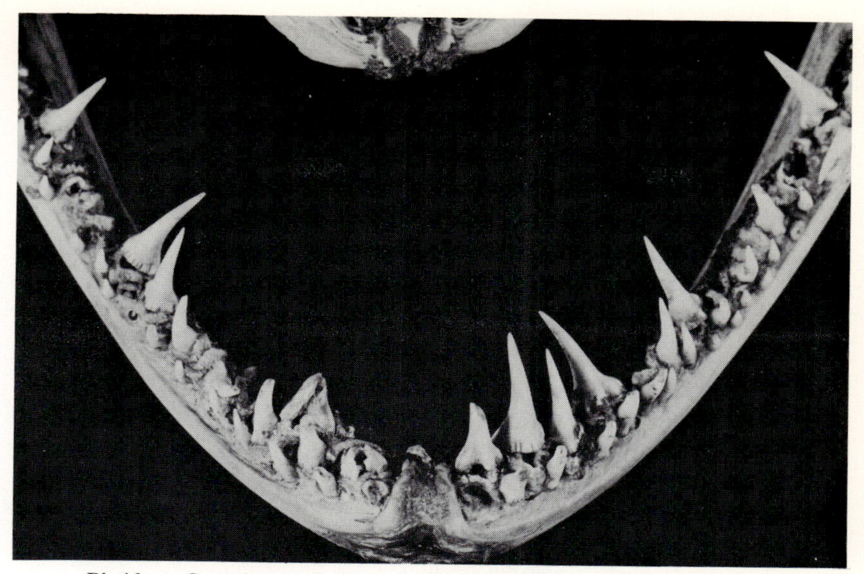

Pl. 13*a*. Grasping teeth from the lower jaw of *Squatina squatina*. *p. 67*

Pl. 13*b*. Movement and replacement of teeth in the lower jaw of a shark. *p. 64*

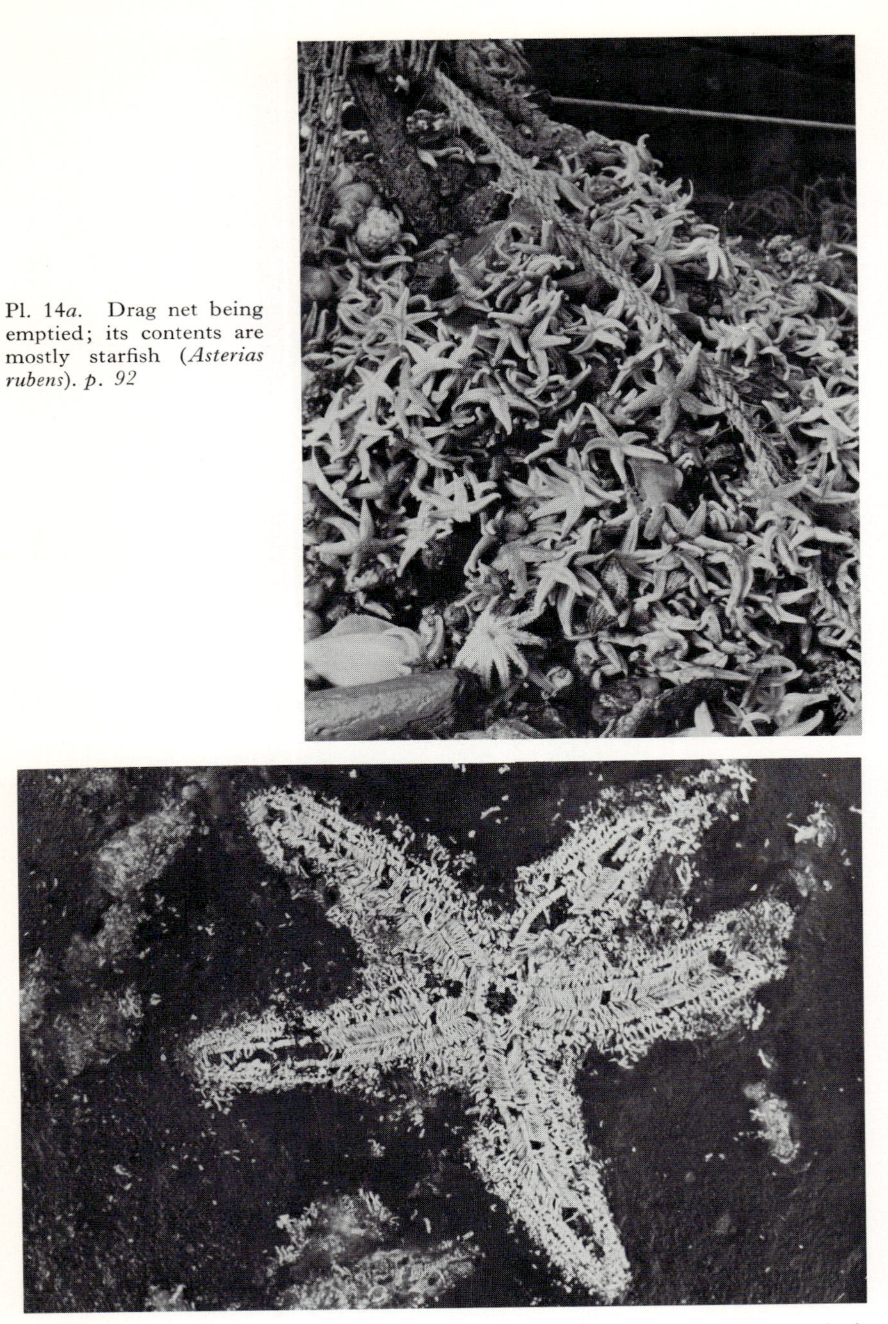

Pl. 14*a*. Drag net being emptied; its contents are mostly starfish (*Asterias rubens*). *p. 92*

Pl. 14*b*. Ventral portion of the skeleton of an *Asterias rubens*. The dorsal portion had earlier, upon decomposition, become buoyant and drifted away. *p. 96*

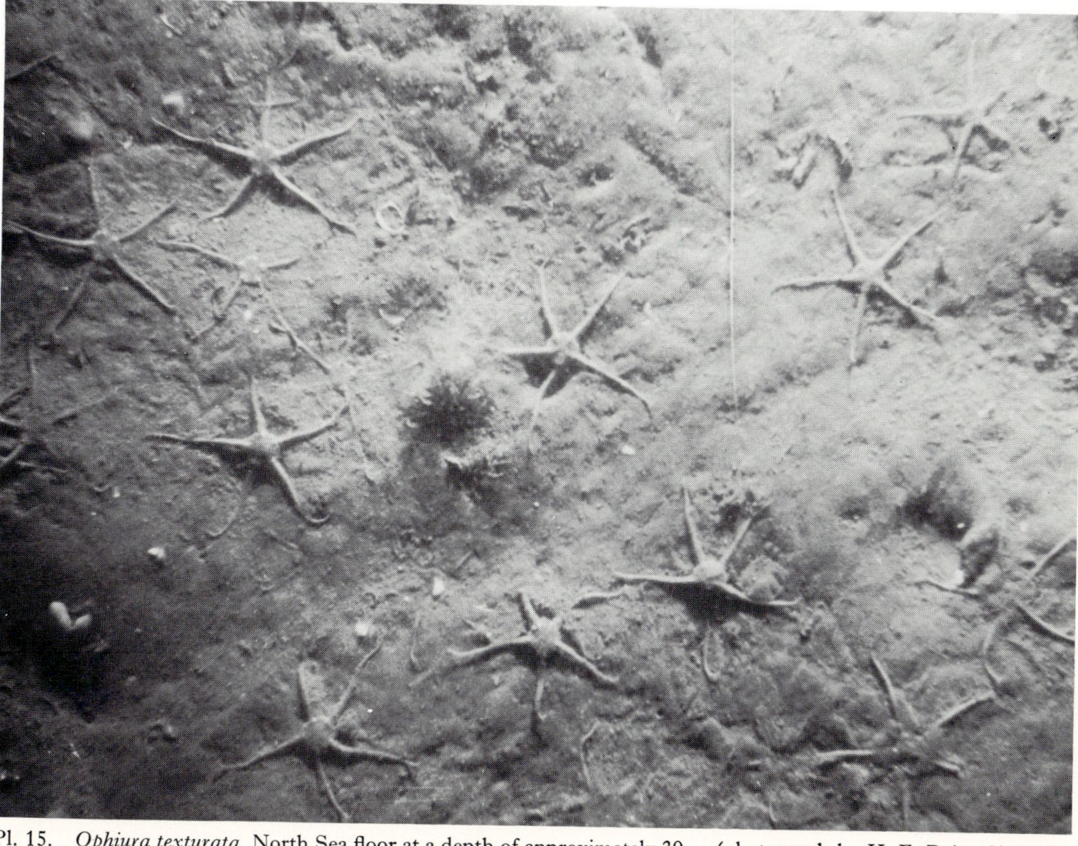

Pl. 15. *Ophiura texturata*. North Sea floor at a depth of approximately 30 m (photograph by H. E. Reineck). *p. 97*

PLATE 15

Pl. 16. Layered, sandy sediment, the upper portion disturbed by burrowing *Echinocardium cordatum*. *p. 310*

PLATE 16

Pl. 17*a*. Accumulations of platelets from brittle star skeletons. *p. 99*

Pl. 17*b*. Crawling trail of *Macoma baltica* (photograph by F. Wunderlich). *p. 267*

Pl. 18*a* (*above*). Crawling trails and boreholes of *Hydrobia ulvae* (photograph by H. E. Reineck). *p. 152*

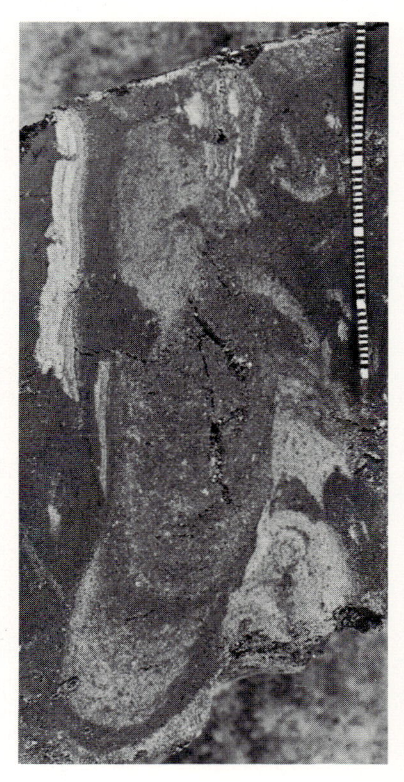

Pl. 18*b*. Trails in the sediment produced by *Mya arenaria* (photograph by H. E. Reineck). *p. 373*

PLATE 18

Pl. 19*a*. *Mya* in living position in sediment (photograph by H. E. Reineck). *p. 207*

Pl. 19*b* (*below*). Bedding bent by burrowing lamellibranchs (photograph by H. E. Reineck). *p. 378*

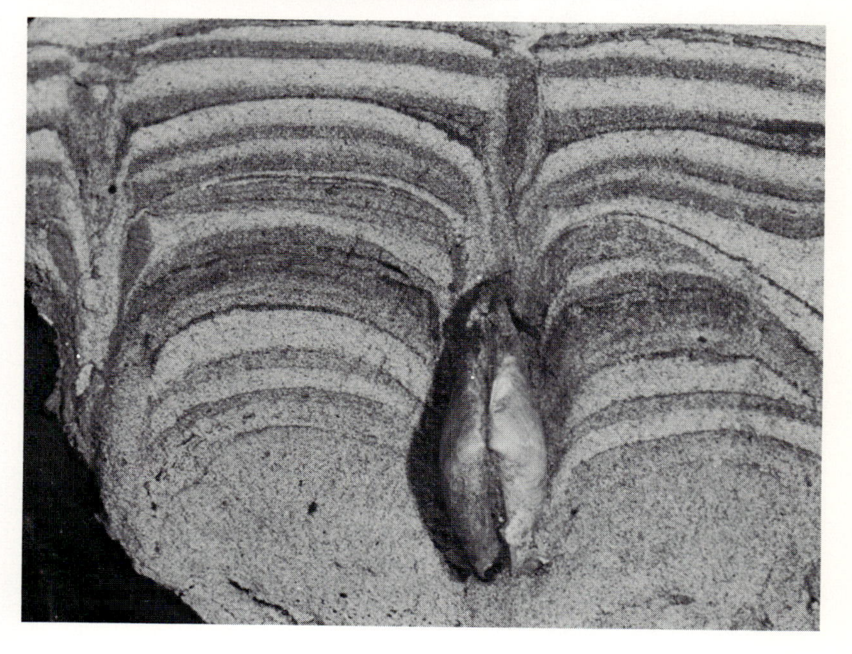

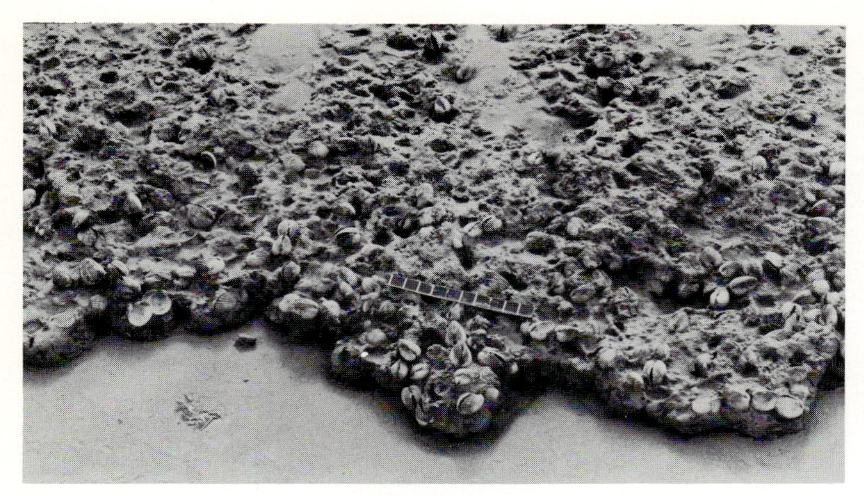

Pl. 20*a*. Colony of *Cardium edule* in living position, having died in a mud bank and then having been washed out. *p. 158*

Pl. 20*b*. Pavement of *Mytilus* valves. *p. 165*

PLATE 20

Pl. 21. Breccia of lamellibranch shells; note two *Mya arenaria* that have died in living position (photograph by F. Wunderlich).
p. 159

Pl. 22. Valves of *Mya arenaria*, many raised to an upright position by lateral compression (photograph by F. Wunderlich).
p. 165

Pl. 23*a*. Boreholes of *Pholas dactylus* in Heligoland chalk. *p. 356*

Pl. 23*b*. Moulds of tubes drilled by *Zirphaea crispata* into wood. *p 356.*

Pl. 23*c*. Lamellibranch valves perforated by the boring gastropod *Lunatia nitida. p. 242*

Pl. 24*a*. Cuttle bone of *Sepia officinalis* in a swash mark. *p. 168*

Pl. 24*b*. Exuviae of *Corophium volutator* in a swash mark. *p. 436*

Pl. 25*a*. Star-shaped traces made in sediment by *Corophium volutator*. *p. 307*

Pl. 25*b*. U-shaped burrow made in sediment by *Corophium volutator*. Note scraping traces made during work on infilling (photograph by H. E. Reineck). *p. 307*

Pl. 26*a*. Burrowing traces made by a *Carcinides maenas* on the way out of a new cover of sediments. *p. 388*

Pl. 26*b* (*below*). Clawed limb of *Cancer pagurus* in a swash mark. *p. 138*

Pl. 27*a*. Pieces of the carapace of a *Cancer pagurus* in a swash mark. *p. 138*

Pl. 27*b*. Ventral portion of the skeleton of the cephalothorax and limbs of a *Carcinides maenas*. *p. 138*

Pl. 28*a*. Sandy material disposed from burrows of the beetle *Bledius spectabilis*. *p. 141*

Pl. 28*b* (*below*). Bedding disturbed by burrowing *Nereis* (photograph by H. E. Reineck). *p. 290*

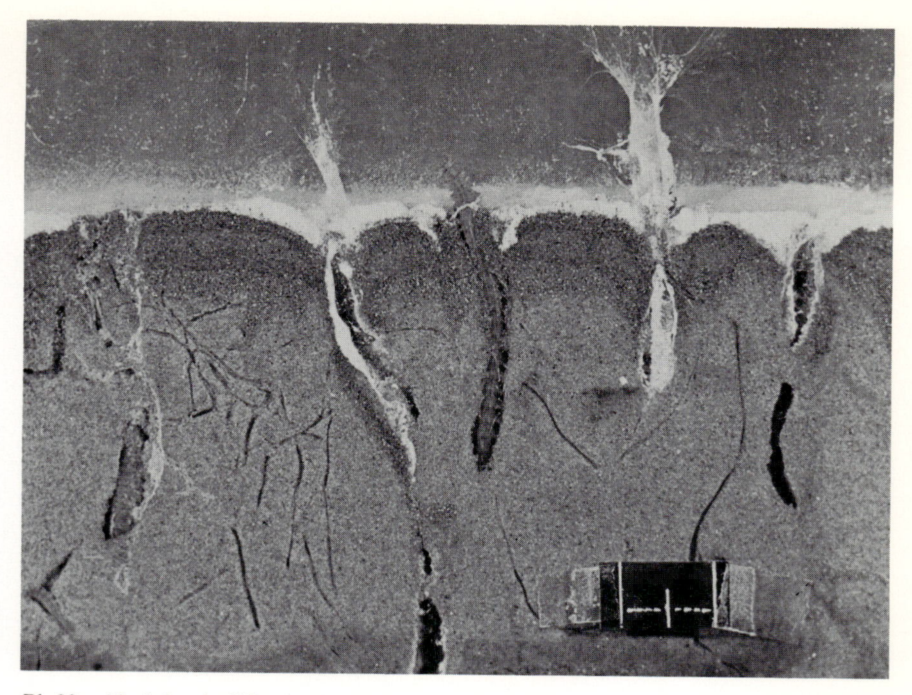

Pl. 29*a*. *Nephthys*, building mucous burrows. Note how white, washed-in sediment is used as lining for the dwelling structures (photograph by H. E. Reineck). *p. 298*

Pl. 29*b*. Sediments of tidal flat perturbed by burrows. *p. 283*

Pl. 30*a*. Plastic moulds of *Corophium* burrows (photograph by H. E. Reineck). *p. 293*

Pl. 30*b*. Spiral-shaped mucous tube of *Paraonis fulgens*. *p. 296*

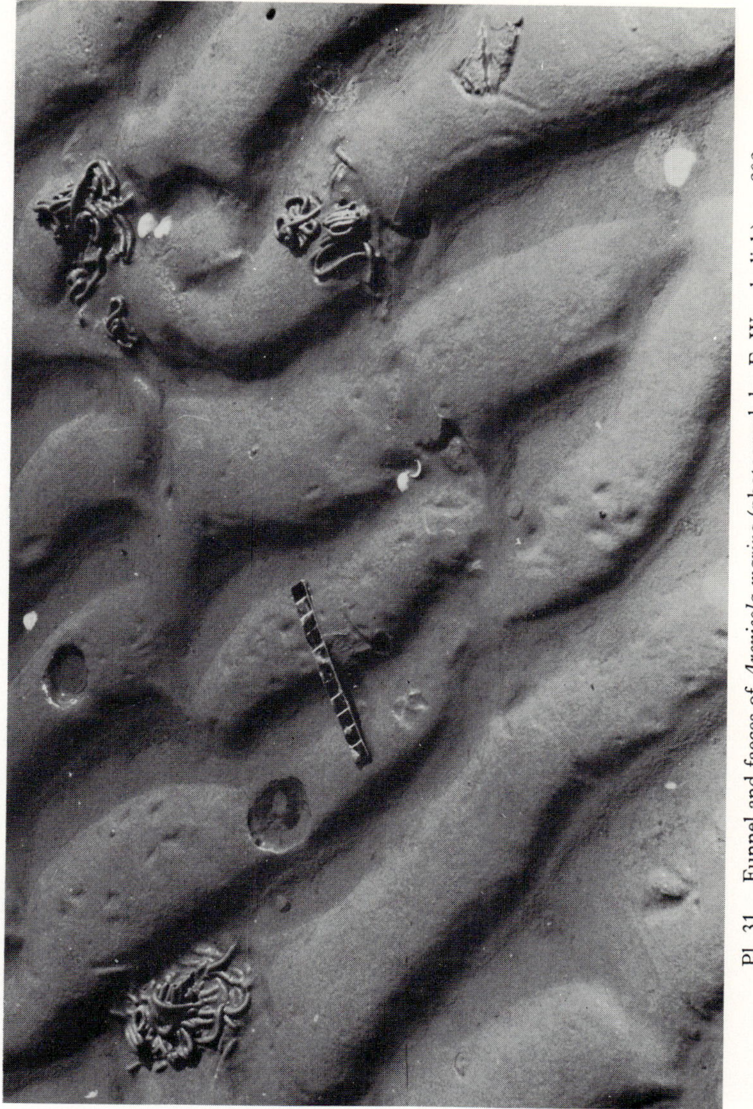

Pl. 31. Funnel and faeces of *Arenicola marina* (photograph by F. Wunderlich). *p. 302*

PLATE 31

Pl. 32*a*. *Arenicola*, progressing toward mummification on dry tidal flat. Note results of repeated autotomies. *p. 179*

Pl. 32*b* (*below*). Piles of faeces of *Heteromastus filiformis* on muddy tidal flat. *p. 423*

PLATE 32

Pl. 33*a*. *Nereis*, mummified on pane of glass. Note formation of faecal pellets in intestine. *p. 179*

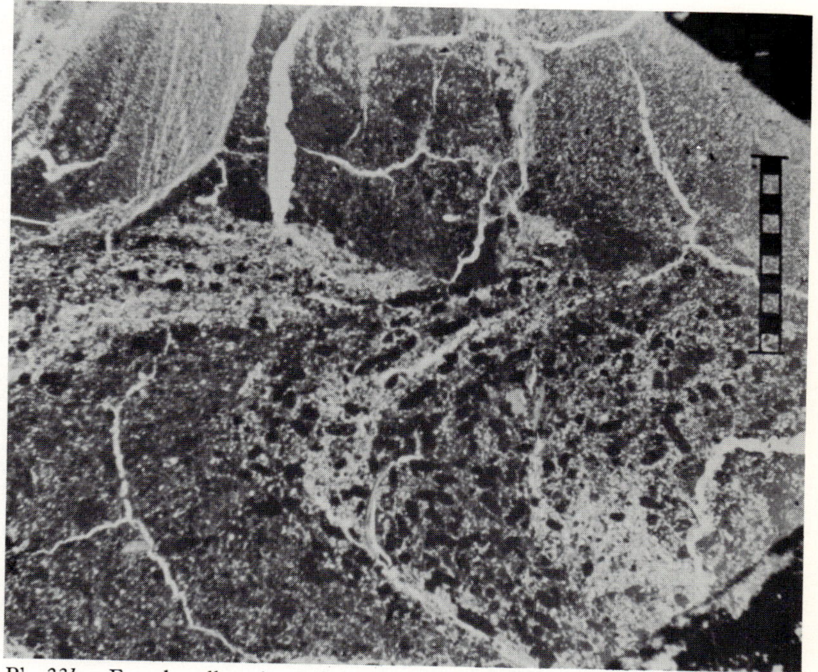

Pl. 33*b*. Faecal pellets in sediment (photograph by H. E. Reineck). *p. 419*

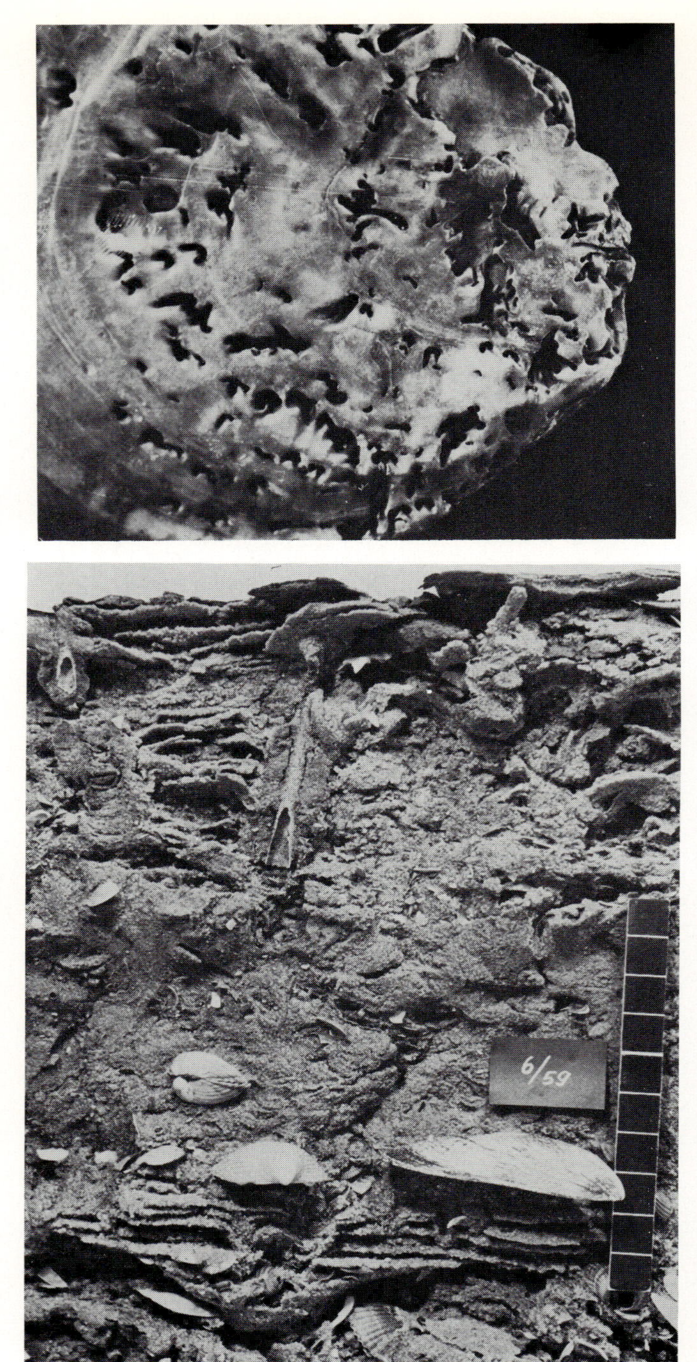

Pl. 34*a*. *Ostrea* perforated by *Polydora ciliata*. *p. 228*

Pl. 34*b*. *Pectinaria coreni* in living position in the sediment. *p. 340*

PLATE 34

Pl. 35. Agglutinated dwelling tubes of *Pygospio*. *p.* 322

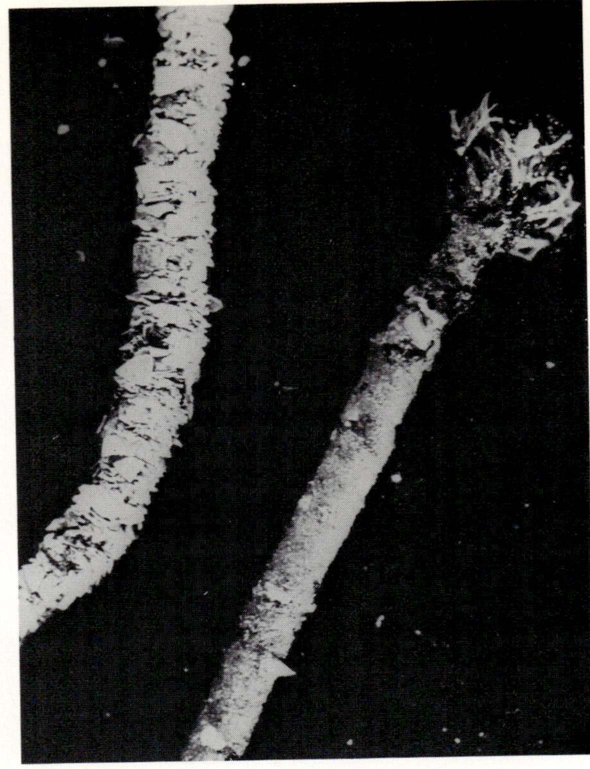

Pl. 36*a*. Washed-out agglutinated tubes of *Lanice conchilega* (photograph by H. E. Reineck) *p. 345*

Pl. 36*b*. *Pomatocerus* on valve of *Pecten*. *p. 350*

PLATE 36

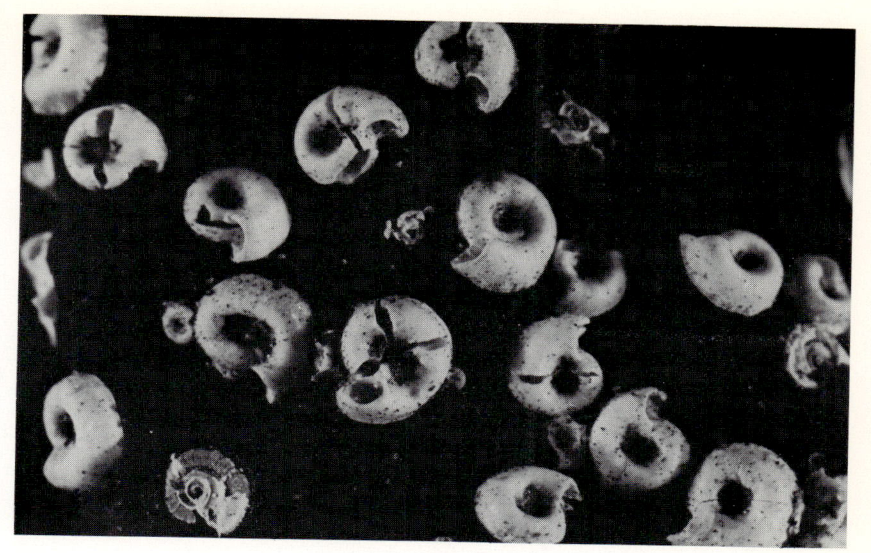

Pl. 37*a*. *Spirorbis. p. 348*

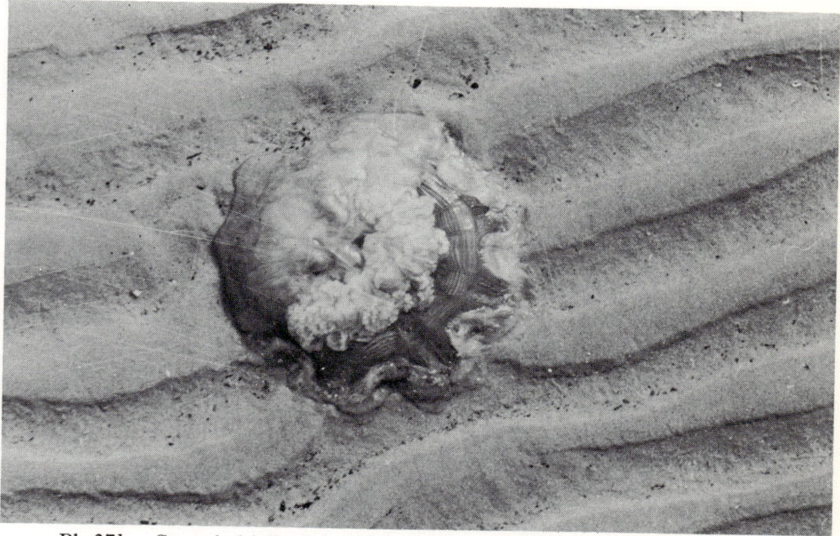

Pl. 37*b*. Stranded jelly-fish (*Cyanea lamarcki*) in beach sediment. *p. 187*

Pl. 38*a*. Stranded jelly-fish (*Cyanea*), beginning to dry and to shrink. *p. 187*

Pl. 38*b*. Decomposing jelly-fish, gas bubbles form under the umbrella. *p. 188*

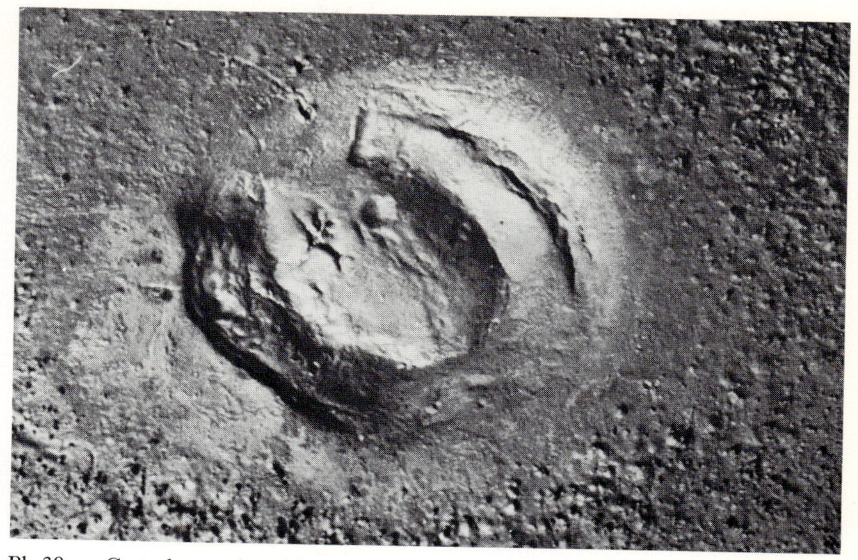

Pl. 39*a*. Craterlet produced by subsidence of mud layers above a decaying jelly-fish. *p. 188*

Pl. 39*b*. *Ostrea* and foot plates of *Balanus balanus* perforated by the sponge *Cliona cellata*. *p. 117*

part of the cephalothorax below the frontal edge serves as the hinge about which the carapace is folded by the departing crustacean. In most cases it simply flaps back down like the lid of a box on the empty cephalothorax cavity and closes it so tightly that the exuvia can hardly be distinguished from an undamaged carcass. If, however, moulting takes place in flowing water the carapace lid occasionally flaps over forward and then lies upside-down in front of the cephalothorax, held only by the membranous connection with the frontal edge. This position is the same as

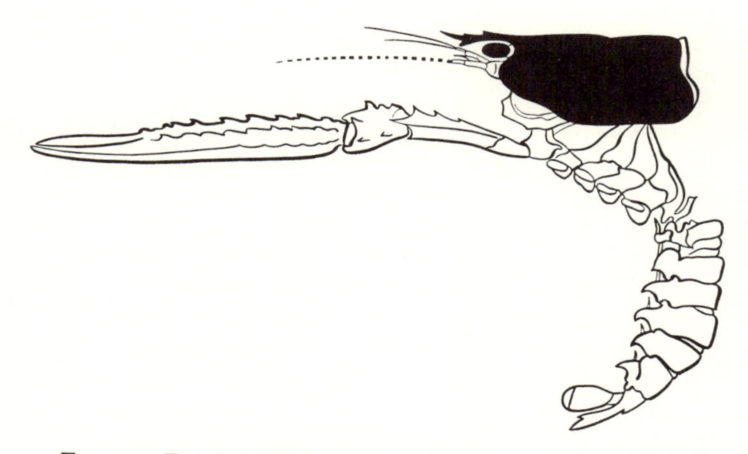

FIG. 251. Exuvia of *Nephrops*. (Black) rotated carapace. (Actual size × 0·3)

that occasionally found in armours of dead brachyurans, the only difference being that the fourth segment is not folded over (Fig. 83). The same term 'Salter's' position is now used for this attitude of exuviae and armours of dead animals. It can no longer be considered as diagnostic for exuviae as Richter (1937c) still assumed (for a discussion of the fossil occurrence of trilobites and brachyurans in Salter's position compared with recent conditions of embedding, see Schäfer 1951b). Obviously only those brachyuran crustaceans with a wide frontal edge can end up in Salter's position. If the cephalothorax is elongated rostrally the carapace cannot flap back and forth about a broad hinge, and the carapace can merely be shifted sideways against the ventral portion of the exuvia (e.g. *Hyas* or *Macropodia*).

Exuviae washed up on a beach can shrink upon being dried by wind and sun. The thin chitinous walls shrink, curl and eventually break up. Only the more heavily calcified supporting ridges, bulges and portions shaped like T-beams remain preserved. Thus the original shape of the exuvia changes rapidly and would be difficult to identify if it were preserved.

Exuviae immersed in water, on the other hand, maintain their original shape. During long-distance transportation, however, they separate into parts which are sorted out. In swash marks and shell deposits derived from deep water one can frequently find only dorsal plates of brachyurans. The limbs of exuviae hang as empty tubes loosely from the cephalothorax, attached only by the joint membranes; they yield readily to any push or pull. They thus may end up in positions that are unlike any that could occur in a living animal or in a carcass.

The exuviae of entomostracan crustaceans are usually only preserved where the animals are numerous. The sessile balanids and lepadids expel exuviae of limbs and cephalothorax into the water; they drift around for long periods as skinny, feathery bundles. Occasionally they can accumulate and be fossilised. The isopod *Ligia oceanica* moults in two steps; the anterior body section moults first and the posterior some time later (for data concerning the ecology of *Ligia* see Jöns 1965). The moulting mechanism of the amphipod crustaceans *Corophium volutator* is unknown; they inhabit the muddy tidal flats in extraordinarily populous settlements. Their exuviae accumulate frequently at coasts bordered by tidal flats and form extensive swash marks; these swash marks often contain millions of exuviae matted together into a dense, straw-like mattress (Schäfer 1941a).

Accumulations of exuviae, whether from entomostracan or malacostracan crustaceans, should not be interpreted as representing localities where the crustaceans go in order to moult. The majority of decapodan crustaceans live a solitary life; they do not form groups, and unlike many community-forming arthropods they are not inhibited from killing members of their own species. During moulting all crustaceans are incapable of defending themselves against attackers, and for a period after moulting they remain vulnerable. Members of the same species are particularly dangerous because they are good fighters, live in the same biotope and can and do attack and kill defenceless moulting animals of their species (Schäfer 1951a and 1956d). Decapodan crustaceans intending to moult have a specific behaviour; they attack members of their species especially violently and chase them away. The moulting animal, while it still can, wants to establish an area free from members of its species. This fight against conspecific animals can thus be considered a 'preventive war'. For the same protective reason cave-dwelling forms usually moult in their dwelling burrow. This explains why claw and other armour remnants are so frequently found in the burrows of fossil *Callianassa*. These parts are the most resistant and heavily calcified parts of the exuviae. They are not remnants of dead animals because, like many other burrow-dwellers, *Callianassa* leaves its burrow with approaching death. Thus moulting necessarily takes place in solitude, whether at the surface of the sediment or in a burrow. Marine arthropods do not use special localities for moulting as Deecke (1915), Mertin (1941) and Richter (1916 and 1942) assumed.

Accumulations of exuviae are thus always caused by transportation. The exuviae are washed together over a shorter or longer time. Frequently large numbers of animals of one species are born at the same time, often on the same day; they complete their larval life at the same time and during their first year of life they also moult simultaneously. Later in life this simultaneity is broken because the periods of moulting get out of step among individuals of the same age. Simultaneous release of numerous exuviae favours accumulation. Thus accumulations of brachyuran exuviae are exclusively from young animals.

(ii) *Change of teeth*

The regeneration of fish teeth has been discussed in C I (*d*), especially for the selachians. The periodical or aperiodical replacement of mouth teeth greatly increases the number of extremely hard, resistant and specific hard parts that are released into the marine sediments. It is safe to say that nine-tenths of all fossil fish teeth in marine sediments are teeth that were discarded to be replaced by new ones.

(iii) *Moulting of birds*

Numerous birds permanently inhabit the southern North Sea, both in terms of species and number of individuals. Larger numbers migrate through the area in spring and autumn, and often linger for several weeks. Thus feathers discarded by moulting represent a noteworthy component of the contents of the sediments of the area.

Feather-bearing vertebrates can be retraced to Jurassic time but they began to diversify during the Cretaceous and Tertiary and have by now reached a great variety of morphology and function. Fossil feathers, either together with the carcass and skeleton or alone, can be expected to be found in deposits from all these periods. Lambrecht (1933) reported several finds of fossil feathers, such as a feather of the group of the Archaeoopteryges in the Solnhofen limestone, impressions of feathers in the brackish to marine limestones of the Oligocene *Cyrena* marls at Bellingen near Lörrach, in the *Indusia* limestone of the Auvergne and in the Miocene *Cypris* shales at Krottensee in the Tertiary basin of Eger.

Moulting is a periodical replacement of the plumage of birds. The quill at the base of the feather is embedded in a tube-shaped feather follicle. Before a feather, still in perfect functioning condition, is discarded an epidermal papilla that is sunk into the cutis at the base of the follicle begins to form a new feather after a resting period corresponding to the interval between moultings. The new feather pushes the old feather shaft out of the follicle. This mechanism is quite different from that of the replacement of the scales of fish. The moulting of birds is governed by a hormone-controlled time schedule, and changes between young and adult birds where it is seasonal. The feather-forming epidermal papillae are differentiated from place to place on the body and produce feathers of locally different shapes or sequences of such different feathers during the development of an individual.

In many birds the periods of moulting are sharply defined; in others, the change, mainly of the pinion and steering feathers, is spread over the whole year. Many species moult completely after breeding. Complete moulting is also particularly frequent in young animals when the juvenile plumage is replaced by the adult plumage. Many migratory birds moult twice a year; the first moult starts at the time of incipient gonad activity, the second moult at the close of the reproductory period so that there are distinct breeding and resting plumages. Frequently, the second is a complete moult. (For details about moulting of birds see Stresemann 1934; for morphological details and changes in the shape of the feathers, see Sick 1937.)

Some species cannot fly at all during a complete moult. Specific adaptations in their behaviour help to protect the animals against hostile influences during this period. Thus the shelduck (*Tadorna tadorna*) which breeds along all coasts of the North Sea, usually in caves, completely moults after the breeding period. Then these animals move to specific moulting areas where they remain throughout the time in which they are unable to fly. The largest moulting area is the isolated Grosse Knechtsand shoal which covers an extensive area of tidal flats between the mouths of the rivers Elbe and Weser; the tidal flats there cover an area of about 50 by 30 km (see also Schäfer 1941a). Here the birds are protected from enemies and find plenty of food while they cannot fly. Every year, tens of thousands and occasionally up to 70 000 of them flock together there. The Heligoland bird station at Wilhelmshaven has undertaken to study the birds during the moulting period from the air, sea and on the ground; the birds were counted and arrivals and departures observed (Goethe 1957). The first birds ready to moult arrive at the end of July, the last animals in their new plumage leave in October. Banding of birds on the Knechtsand has helped to determine the origin of the birds, but because the animals are protected all year round, relatively few return reports have come in. Of 793 bandings on the Knechtsand (Freemann 1954) twenty have been reported from Great Britain and

Ireland, ten from the Netherlands, eight from France, five from Germany and two from Denmark. Since the shelducks from other parts of Scandinavia also moult in Heligoland Bay it becomes understandable why so many of these large birds concentrate every year in Heligoland Bay for moulting (Fig. 252). Goethe discusses the hypothesis that the breeding and the moulting areas were closer together at the end of the glacial period, when the coastline of the North Sea was at the present 50 m depth contour (Fig. 277). As the North Sea gradually transgressed southward the distance between breeding and moulting areas gradually increased for the birds living in Scotland, Denmark, Sweden and

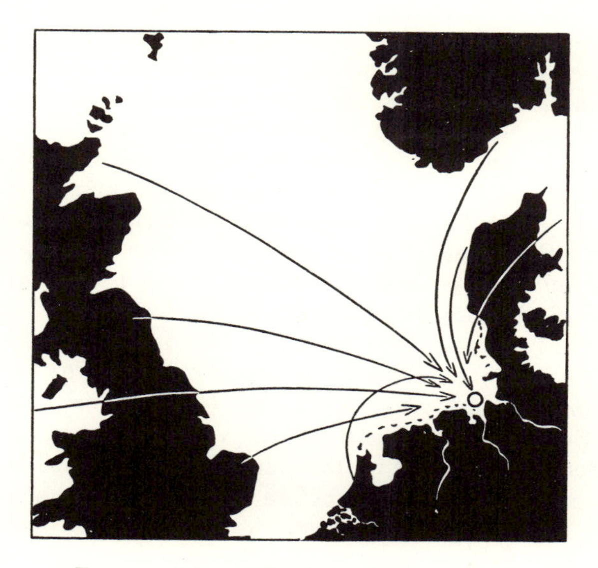

FIG. 252. Map of North Sea showing the directions from which *Tadorna* migrate toward their moulting refuge in Heligoland Bay

Norway. The assumption is that the birds remained true to their original breeding regions but had to migrate increasingly further south to find a biotope that corresponded to the original one they needed for moulting.

It is obvious that such a concentration of large moulting birds furnishes each year great amounts of feathers; great enough to become palaeontologically significant. The congregation of birds is also effective in another way; in their search for clams the birds make holes by repeatedly stepping on the wet sand surface. The top sand layers are thus turned over (Freemann 1960).

Herring gulls (*Larus argentatus*) moult twice a year (Goethe 1956a); in winter they renew the smaller feathers and in summer after the breeding period they change the entire plumage. Great quantities of large and very strong feathers are released into coastal sediments. Along the German sectors of the North Sea coast alone, 19 810 pairs breed every year (Goethe 1956a); if individuals that do not breed are included this would mean about 50 000 individuals. Goethe (1956a, p. 12) photographed an accumulation of moulting feathers near where herring gulls usually rest on the island of Memmert.

During the time of the summer moult the Scolopacidae, the curlew (*Numenius arquata*) and related species tend to form swarms which stay on coastal tidal flats. At that time their hard flight-feathers can be found in great quantities where these swarms have their resting places.

Birds living on the open sea or spending much of their life far from the coasts are important because their discarded feathers come to lie more frequently in the sediments of the open sea than those of the coastal birds. Finds of feathers do not indicate proximity of the coast. Remmert (1957) has listed the pelagic species of the North Sea and mentions: *Fulmarus glacialis* (Fulmar), *Hydrobates pelagicus* (Storm Petrel), *Oceanodroma leucorrhoa* (Leach's Petrel), *Puffinus puffinus* (Manx Shearwater), *P. gravis* (Great Shearwater), *P. griseus* (Sooty Shearwater), *Sula bassana* (Gannet), *Phalacrocorax aristotelis* (Shag), *Rissa tridactyla* (Kittiwake), *Sterna dougalli* (Roseate Tern), *Fratercula arctica grabae* (Lund), *Plautus alle* (Little Auk), *Uria aalge albionis* (Guillemot), *Alca torda* (Razorbill), *Cephus grylle* (Black Guillemot), *Branta bernicla* (Brent Goose), *Somateria mollissima* (Eider), *Stercorarius skua* (Great Skua), *Melanitta nigra* (Common Scoter) and *Melanitta fusca* (Velvet Scoter).

Discarded feathers drift on the water for a long time and can therefore travel over wide distances before they sink to the sea floor. In an experiment by the author a contour feather from the hand wing of a common gull drifted on the sea water for more than four months. The lower parts of the shaft fill earlier with water than the upper parts; therefore the feather eventually sinks to the sea floor shaft-first and stays in this position even on the sea floor. This is probably the reason why feathers are so rarely embedded in sediments of the deep water. They probably disintegrate into parts which may then come to lie horizontal and to be covered by sediments.

When the feather disintegrates, the barbs (rami) break first because they have become brittle; the broken barbs are still held together by the distal barbules (radii). As a result, whole pieces of feather web may separate from the shaft (rhachis). While examining 11 cm diameter core samples from the basin of the inner Jade Bay open to the tides, Reineck (personal communication) found a whole feather with a strong shaft, probably from a herring gull, at a depth of approximately 1 m below the surface in sandy mud. This proves that feathers can exceptionally be embedded undamaged on a permanently covered sea floor and not only on periodically dry beaches and on tidal flats. Moulds of fossil feathers are more frequently found than bodily preserved feathers.

(iv) *Autotomy*

Members of several phyla can commit autotomy of organs in perfect working condition (worms, molluscs, echinoderms, arthropods and several vertebrates). Discarding entire limbs and regenerating lost parts later on is in many cases biologically advantageous because an animal can abandon a limb grasped by an enemy and so escape. In some cases, animals commit autotomy to rid themselves of tissue under poor living conditions. Thus starfish frequently dispose of one or several arms if the oxygen content of the water is low in order to reduce the number of cells needing to be supplied. (In an aquarium, it is easier to keep small or moderately large *Asterias rubens* which require less oxygen than large animals with a diameter of 30 cm.) Accumulations of mutilated starfish in nature indicate bad living conditions, generally a lack of oxygen. The polychaete *Arenicola marina* also reacts to deficiency of oxygen (but also to rough handling) by sometimes committing repeated autotomy of the tail section. Sudden chemical stimuli can also induce autotomy (decapod crustaceans should therefore not be fixed in formalin. Usually, however, autotomy is triggered by some strong mechanical stimulus on the discarded organ.

Decapodan crustaceans are capable of autotomy of their clawed and walking limbs. The limbs that are used for feeding and the abdominal limbs cannot be severed. Autotomy always affects a whole limb, never its individual sections. The anatomically predetermined line of separation is the suture between basis and ischium; thus the coxa always remains on the thorax. If, on the other hand, a limb separates from the body of a carcass, the coxa always separates with the limb.

As one would expect, a complex of cells capable of regenerating a new limb lies at the basis and at no other sutures. Therefore a limb is never partly destroyed by autotomy. The functional unit 'walking limb' or 'claw-bearing limb' either exists as a whole, or it is altogether missing and then renewed as a whole. Thus if one finds individual parts of limbs or individual claws in the sediment, one can be certain that they have been destroyed or have disintegrated at some later stage and have not been discarded by autotomy. Also, living crustaceans are never found with legs mutilated so that one or several distal parts are missing; and dead bodies so mutilated must have lost their missing parts after death. While a limb is being regenerated, although first small and incapable of functioning, it is complete in all details, and it has the right proportions.

It takes two to four moults until a new limb has reached its final size. The frequency of autotomy can be determined from the great number of living crustaceans with a missing walking limb or clawed limb. Evidently, each autotomy results in a temporarily somewhat reduced fitness for life. Decapodan crustaceans missing walking limbs are hardly restricted because a change in co-ordination of the remaining limbs largely makes up for the damage. Noticeable restrictions occur only if many limbs are missing and if their distribution is unfavourable, say, all on one side of the body. This is not so with the claw-bearing limbs; the loss of one of them significantly influences the activity. If both clawed limbs are lost the animal generally dies. In captivity, however, with some help with feeding, both claw-bearing limbs can regenerate. Autotomies of clawed limbs are particularly bad for animals that use them not only for grasping but also for support, locomotion or for burrowing. Swimming crabs, on the other hand, can even swim with only a single swimming limb.

(*d*) EGG LAYING

The majority of marine animals release their eggs one by one into the sea. The eggs are invariably provided with a yolk and capsule. The egg and the larva live as plankton or benthos; after the larval stage most animals can choose the place for their future life. The egg cells themselves are rarely preserved.

(i) *Gastropoda*

Ehlers (1868), Fuchs (1895) and Richter (1937a) pointed out that gastropod egg clusters and capsule cocoons could possibly become fossils. Richter illustrated a string of box-shaped egg capsules, interconnected by a strong band, of *Busycon canaliculatum* from the east coast of North America. He also noted the large egg lumps of *Buccinum undatum* from the North Sea.

The young animals pass their entire larval period inside these capsules until they grow their own shell. The capsules outlast their actual use. The egg clusters of *Buccinum undatum*, *Neptunea antiqua*, *N. turtoni*, *N. norvegica* and *N. islandica* can be found in the sediments of the North Sea floor more than a year after spawning and thus long abandoned by the young animals; or they are found dried up in swash marks.

Capsules and cocoons have the same basic structure (according to Ankel 1935, 1936a, 1937a). They consist of two halves joined by a suture. Each half capsule is formed in the tubular uterus by one of two separate glandular zones; two additional glandular zones that lie between the others produce the cement that makes the suture (Fig. 253). In many cases the cocoons have a prepared opening for hatching. This opening is closed by a plug, again formed in halves, which can later be dissolved by enzymes produced in the embryo when it is ready to hatch. During its passage through the uterus the cocoon is still soft and gelatinous. After leaving the ovipore it slips along a groove on the foot and then below the sole. There the shapeless cocoon is pressed against a glandular zone whose secretion hardens its wall and gives it the final shape. In *Buccinum* the glandular zone can only be observed during the time of oviposition. A kneading process at the time of hardening produces ridges, folds and depressions in the wall that are typical of cocoons. Cocoons of prosobranchs are released in batches, in many cases from several females, and are attached to some solid substrate, a lamellibranch valve, gastropod shell, stone or submerged branch. *Neptunea* attaches them preferably to

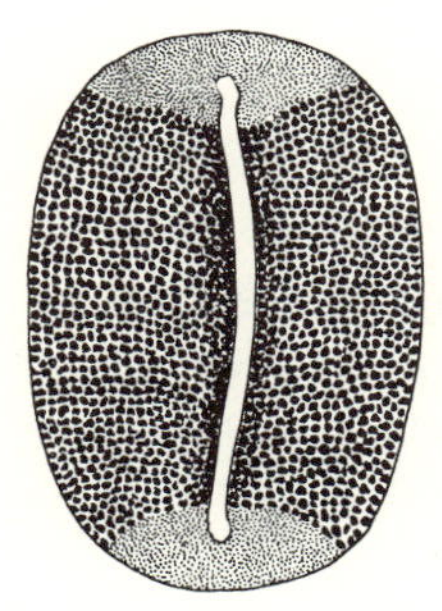

FIG. 253. Mud-filled cocoon capsules of *Buccinum* in bedded sediments.

the shells of its own species. While the gastropod spawns it crawls over the substrate along a definite spiral course and piles the individual cocoons in spiral rows on top of each other (for *Buccinum* see Schäfer 1955a). The cocoons of *Buccinum undatum*, each of which has a diameter of about 1 cm, are agglutinated into clusters the size of a large coconut.

Frequently, strong currents detach the cocoon clusters from their substrate. Then they are rolled about on the sea floor until they finally become covered by sediment. If the young gastropods have already hatched and the openings of the capsules are unplugged, fine mud rapidly enters the cocoons and fills them. This filling prevents collapse of the cocoons under pressure. In bedded sediment they appear as piles of flat, lentil-size cakes, and they can be distinctly recognised, especially in sand, even after the wall of the cocoon is eventually dissolved; this takes two to three years in the reduction zone (Fig. 254).

Loose and porous clusters of cocoons frequently serve as hiding-places for brittle stars (mainly *Ophiothrix fragilis*). Some prosobranchs also withdraw into them. If such a near-spherical cluster rolls back and forth on a beach numerous bird feathers often become inserted into its openings. When the clusters are eventually embedded, so are all organisms or parts of them that have settled on it or have become accidentally attached to it.

The spawn of most prosobranchs and opisthobranchs is gelatinous and often very small and is frequently attached to stones or shell fragments. Such eggs do not leave any traces, with the possible exception of the collar-shaped spawn ribbons of *Lunatia*. *L. catena* (Ziegelmeier 1961)

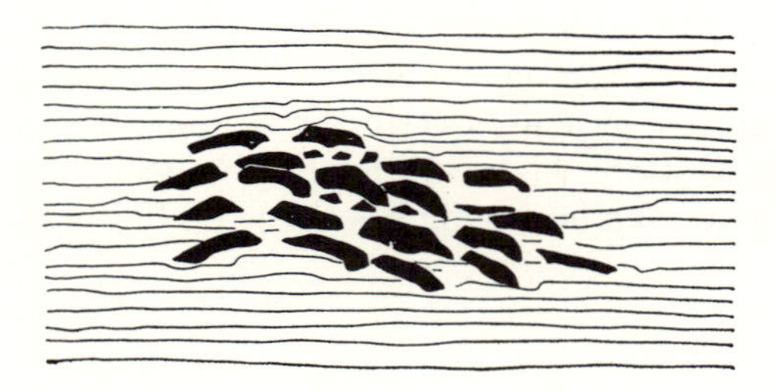

FIG. 254. Egg capsules of *Buccinum* full of fine mud in bedded sediment. (Actual size)

incorporates sand grains into the gelatinous matrix of its spawning ribbon giving it a certain stability. Small spaces are left to serve as chambers for eggs. The ribbon is bent into a circular loop open on one side, is approximately 1·8 cm high and has a lower diameter of 6 to 7 cm. *Lunatia nitida* produces similar spawning rings 2·5 cm in diameter and with a height of approximately 0·8 cm. These spawning rings are deposited on the sea floor.

(ii) *Cephalopoda*

Only a few cephalopods spawn regularly in the North Sea; amongst them are *Alloteuthis subulata* and *Loligo forbesi*. Their spawn, like that of many other cephalopods, consists of bundles of gelatinous sacs. One bundle of *Alloteuthis* contains about 20 to 30 eggs, and later embryos, those of *Loligo forbesi* are strung into a long, ribbon-shaped cluster and can contain up to 80 eggs.

These bundles of spawn sacs float or drift over the sea floor if they are torn from the seaweeds or colonies of hydroid polyps to which they were originally attached. They never leave a trace on the sea floor but furnish characteristic impressions if they are thrown on a beach or carried over a tidal flat (from June to September). In clays formed from faecal pellets the spawn and the embryos, still carrying large and rather hard yolk sacs, can be moulded into the substrate and possibly preserved (Schäfer 1941a and b).

(iii) *Pisces*

In the North Sea sharks and, particularly, rays and skates furnish egg capsules that can be preserved. Almost on every walk along the swash marks of the beach, and at any season, one can collect one or more horny egg capsules from one of the numerous rays or skates. Because these egg capsules are not released at all seasons (loaded females of *Raja clavata* being observed from May to July, and *Raja batis* from February to July) this shows how well the capsules are preserved. They retain their shape and consistency long after the embryo has hatched.

After internal fertilisation, often accompanied by a mating ceremony, the sperm of several of the rays and sharks is stored in the upper parts of the oviduct. Therefore it is not necessary for every oviposition to be preceded by a new act of fertilisation. A number of egg capsules are periodically released together; *Raja clavata* lays 15 to 20 egg capsules at intervals of several weeks.

The females of several species lay their eggs in special localities or attach them to seaweeds or animal colonies on the sea floor. When the sharks *Scyliorhinus caniculus* and *S. catulus* lay an egg capsule two flexible soft threads first egress from the cloaca. These threads are part of the egg capsule, the bulk of which remains in the cloaca. The spawning female now swims several times around some solid object that stands up on the sea floor and so attaches the threads. While the animal continues to swim around the point of attachment the egg capsule is pulled from the cloaca (Kopsch 1897). The capsule is narrow and elongated and bears double horns. Later the threads become rigid and contract slightly and thus attach the egg capsule securely to the substrate. The egg capsules of rays are also double-horned. Several egg capsules of *Raja batis* are occasionally found as clusters on the sea floor with matted thread-like appendages. The spawning mechanism of this species is not known; apparently, the egg capsules become entangled during spawning.

Raja clavata lays the individual capsules at longer intervals. The female ray is not selective as to where the capsules are deposited. The animal swims about in open water, and at first two horns at the front of the capsule slightly bent toward each other begin to protrude from the cloaca. Over the next few hours the entire capsule is expelled and drops to the sandy sea floor wherever the fish happens to be (Schäfer 1953a). Thus a spawning female spreads about 20 capsules in the course of several weeks over a wide area. The black egg capsules at both sides bear broad appendages which move like loose sails in the water (Fig. 255). These web-like, ragged appendages feel sticky; they cling to anything with which they come into contact unless it is very smooth. Like the egg capsules themselves, the appendages consist of numerous, fine, horny threads that are not cemented together but can move in connection with each other like the threads of a thin veil. The individual threads anchor themselves on any rough object over which they glide, thus giving the impression of stickiness. On the sea floor the threads are usually anchored to loose pebbles, gastropod or lamellibranch shells while a light current moves the capsule along the sea floor. If thick layers of sediment cover these capsules they can become fossilised. If the capsule remains lying on the open sea floor the young ray hatches after about

FIG. 255. Egg capsule of *Raja clavata*. (Actual size × 0·5)

four months. The young fish slips through a slit between the two longer horns of the capsule. In water the capsule disintegrates after about a year, long after the fish has abandoned it. It breaks down into bundles of horny threads that eventually become brittle and thus can no longer maintain the shape of the capsule because they are free to slip over each other. The capsule disintegrates more rapidly in mud, rich in iron sulphide. According to Jensen (1951a and b) naticids occasionally bore into the egg capsules of the rays.

The following rays usually spawn in the southern North Sea: *Raja clavata*; *R. radiata* rarely in the southern North Sea; *R. montagui* more frequently in the southwestern than in the eastern North Sea; *R. naevus* infrequently; *R. blanda* spawns only in the southwestern North Sea; *R. batis*; *R. oxyrhynchus* rarely in the southern North Sea.

(iv) *Aves*

The majority of fossil birds' eggs come from lacustrine sediments or they are embedded in tufa as in the case of entire layers of birds' eggs in the Miocene fresh-water tufa at Steinheim (Lambrecht 1933). Lambrecht also mentions occurrences of birds' eggs in the Oligocene *Indusia* limestone in the Auvergne, in the Mainz Basin (at Weisenau) and in obvious marine beds on the island of Seymour. According to observations by the author on Recent beaches the conditions for the fossilisation of birds' eggs are particularly favourable at the very edge of the sea; however, beach sediments including the fossils embedded in them are frequently reworked. On the other hand, if the formation of the sediments is followed by a phase of regression they are totally destroyed by subaerial erosion. The fact that birds' eggs do not frequently occur as fossils (at least since the early Tertiary) is therefore due only to the rare preservation of both subaerial and subaquatic beach sediments; it is not due to eggs being hard to fossilise or to especially unfavourable conditions at the time of burial.

Thus the geological significance of the reproductive activity of birds is not so much fossil preservation as the contribution of large amounts of biogenic calcium carbonate to the sediments of the littoral zone. The yearly output of eggs or egg-shells is in fact large. F. Goethe of the Heligoland bird station (personal communication) has computed the number of eggs laid annually by sea birds at the German coasts on the basis of data collected by his institution for 1958. The list comprises all species which breed regularly in the coastal regions, i.e. on the open beaches, salt marshes and dunes of the German part of the North Sea and western Baltic (Table 14).

TABLE 14

Birds' eggs laid along the German coast in 1958

	Breeding pairs	Number of eggs including later clutches
Ringed plover (*Charadrius hiaticula*)	600	2 340
Kentish plover (*C. alexandrinus*)	500	1 950
Redshank (*Tringa totanus*)	2 000	6 600
Avocet (*Recurvirostra avosetta*)	900	5 040
Oystercatcher (*Haematopus ostralegus*)	2 500	10 500
Sandwich tern (*Sterna sandvicensis*)	3 000	5 400
Common tern (*S. hirundo*) Arctic tern (*S. macrura*)	12 000	39 000
Little tern (*S. albifrons*)	9 000	3 510
Herring gull (*Larus argentatus*)	15 000	63 000
Lesser black-backed gull (*L. fuscus*)	20	66
Common gull (*L. canus*)	7 100	25 560
Kittiwake (*Rissa tridactyla*)	65	130
Black-headed gull (*L. ridibundus*)	780	2 760
Guillemot (*Uria aalge*)	1 000	1 100
Shelduck (*Tadorna tadorna*)	900	11 700
Eider (*Somateria mollissima*)	100	800
TOTAL	55 465	179 456

According to these figures, the shells of at least 179 456 eggs have been produced in 1958 as potential fossils or as possible calcareous constituents of marine, coastal dune sediments. Potential candidates for inclusion in the littoral sediment are chiefly those eggs that are laid in nests in the immediate vicinity of the mean high-water line. There exceptionally high water can flood the nests and wash hundreds of eggs away. According to Table 14, 56 450 eggs a year are laid in this environment. Occasionally, floods reach the breeding colonies of the little tern (*Sterna albifrons*), common tern (*S. hirundo*), arctic tern (*S. macrura*) and the sandwich tern (*S. sandvicensis*); the eggs of the Kentish plover (*Charadrius alexandrinus*), the ringed plover (*C. hiaticula*), and individual nests of the herring gull (*Larus argentatus*) which are frequently built on the exposed sand flats a little above the mean high-water line (Jungfer 1954); the nests of the oyster-catcher (*Haematopus ostralegus*) are also quite frequently flooded. Weckmann-Wittenburg (1931) illustrated sandwich tern eggs washed from the beach of the island of Norderoog. Kuhlemann (1942) described a spring tide disastrous for eggs on Scharhörn island. Goethe (1939) cited the same dangers for the eggs on the island of Mellum (on 20th and 22nd June 1927, 500 clutches of the common tern, three of the oyster-catcher; and on 16th and 17th June 1928, many clutches of the common gull, 300 of the common tern, 23 of the little tern, and four of the redshank were destroyed).

Partly incubated birds' eggs float very well. Ringleben (personal communication) described the fate of a guillemot egg (*Uria aalge albionis*) which must have dropped from a cliff on Heligoland into the water and floated to the island of Scharhörn at the mouth of the river Elbe. The egg was still alive, was placed into an incubator by the bird warden on Scharhörn, and hatched. The proof of long-distance drift is especially conclusive in this case because in the whole southern North Sea area guillemots breed only on Heligoland. Fractured eggs sink to the sea floor and can be buried there.

Above the reach of spring tides on sandbanks and dunes, nests of beach-breeding birds can be rapidly covered up by wind-blown sand. In the summer of 1959 Rittinghaus (Heligoland bird station at Wilhelmshaven) experimented on the small sand island Minsener Oldeoog to see how the Kentish plover (*Charadrius alexandrinus*) finds its nest if it is covered by drifting sand. According to him the breeding pair frequently finds the nesting place despite considerable changes in the environment. The Rittinghaus filmstrip (Film-Encyclopaedia of the Institute for Scientific Film, Göttingen) shows the searching movements of the bird; it examines the sand with its beak and scrapes in an attempt to free the eggs. Rittinghaus demonstrated that the bird is able to lift the eggs from a depth of 6 to 7 cm of sand. Not all the eggs, however, were found by the birds after a high wind. Lost eggs and all the shell fragments from the hatched birds end up in the dune and marsh sediments near the beach. According to Table 14, this adds up to 123 006 eggs annually, counting the eggs of redshank, avocet, herring gull, lesser black-backed gull, common gull, kittiwake, black-headed gull, shelduck and eider.

The calcareous shell of an egg can stay intact for many years if it is embedded in dune sand. The author found eggs of the herring gull in moist dune sand on Mellum island which were probably covered for more than five years. The shells were fragile but reinforced by the moist inner membrane of the shell which still clung tightly to its inside. The

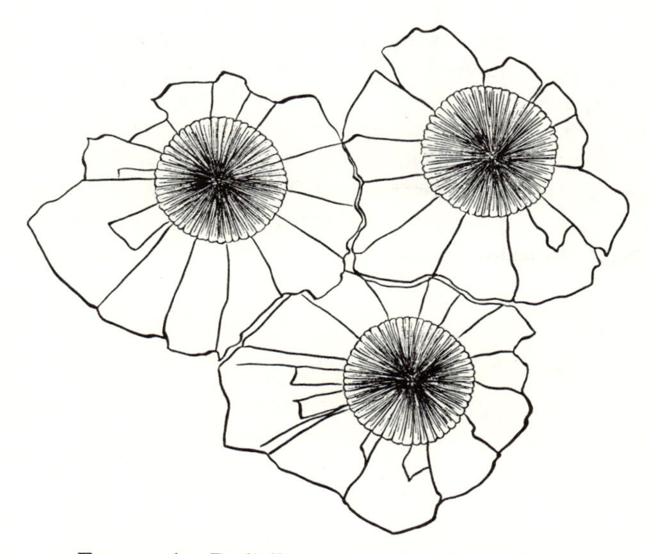

FIG. 256. Radially arranged spherulitic crystals as constituents of the calcareous outer layer of a bird's egg (after W. J. Schmidt, 1929). (Schematic drawing)

eggs contained feathered embryos and moist, yellowish-reddish yolks that were somewhat consolidated but not very odorous. As embedding experiments have shown, the calcite shell becomes soft after about ten months and can be cut cleanly with a knife without splintering as fresh eggs do. The change in the consistency of the shell is probably due to the decomposition of an organic cement originally holding together the polygonal calcite components of the egg-shell (Fig. 256). Each of the components that are usually four- to five-sided consists of a single layer of radially arranged spherulitic crystals of different sizes; a peg in the centre of the spherulite attaches it to the shell membrane below (Schmidt 1929, in a microscopic study of thin-sections). The individual structural elements of the shell are formed by crystallisation from a secretion of the oviduct.

If birds' eggs lie in the bluish-black reducing zone of muddy sediments the calcareous shell dissolves after approximately half a year. However, the egg itself, possibly with a partly developed embryo, is held together by the taut, inner shell membrane which is usually a dark yellow colour and is quite tough and resistant. It can be preserved as a mummy for years.

In contrast to the only other non-aquatic eggs, those of reptiles, birds' eggs contain the entire water supply necessary for development. Because there is also no surplus water, and all the supplied water is used by the growing body, no water exchange takes place through membrane and shell. This complete separation of the egg contents from the environment is probably responsible for the slow decomposition observed and frequent preservation. Another consequence of the perfect enclosure is that decomposition gases, whose pressure would otherwise burst the egg, do not develop. If a fresh egg breaks in the surf or if a broken egg comes to rest in littoral sediments it is rapidly and completely filled by sediment, making a perfect mould.

(*e*) THE INFLUENCE OF METABOLISM, REGENERATION AND REPRODUCTION ON THE FOSSIL CONTENTS OF A MARINE BED

The preceding paragraphs were mostly concerned with the physiology of the processes of feeding, regurgitating, forming of faeces, autotomy and spawning. Almost as an aside, the appearance of body fossils which result from these life processes, mostly as by-products, has also been described. For the geologist and the palaeontologist, such a survey is of value only if it also shows how common the various groups of body fossils are, and what proportion of them may become fossilised. Very generally one can say that this depends on the number of available remains and on how well the individual objects can be preserved and identified by their shape. In many cases, abundance and likelihood of preservation do not go hand-in-hand.

In quantity, faeces doubtlessly take first place. Wherever marine animals live food is ingested, metabolic processes occur and faeces are discarded. However, the majority of faeces, with the exception of those of a few higher animals, are not released as coprolites that can be identified by their shape and contents. Rather, the oval segmented pellets of gastropods, lamellibranchs, worms and planktonic animals make up by far the greatest proportion, corresponding to the enormous numbers of these animals. These faeces have the greatest lithological significance in the formation of beds, but little can be deduced from their shape even if preserved—and this is rarely the case.

Hard parts are most easily preserved and identified. Hard parts include the products of metabolic, regenerative and reproductive processes, feeding remnants of crustaceans and molluscs, exuviae of crustaceans, periodically discarded teeth, limbs discarded by autotomy, spawns of molluscs and fishes, and the eggs of birds. Despite their opportunities for preservation in marine sediments, these parts hardly influence the 'normal fossil contents' of a bed which are important for the geologist. They are too rare in rocks to mark a facies. It is a lucky accident to find one of them in a marine rock; and then often its origin and significance are hard to understand. Thus the informative value of the products of life processes seems to be negligible, and the necessity to study them in the present ocean equally so. Nevertheless, if one scans through a modern textbook of palaeozoology and looks for the accounts and results of such lucky accidents, one begins to realise how much of palaeontological knowledge is based on their interpretation. Many an accidental find and its correct interpretation has clarified a complicated ecological relationship, and provided the basis for a classification or for the recognition of an important lithological process. Quite recently the wait for lucky finds has been replaced by systematic search, and young palaeontologists are trained to rely less on luck and more on diligence; for in palaeontology as elsewhere fortune can often be forced by diligence and knowledge. The author hopes that the contents of this section may help in finding and interpreting fossil documents of life processes by pointing out the special conditions under which they may have been preserved.

PART E

Animals and animal communities as elements of the facies

IN the preceding sections habitat and behaviour of individual marine animals were frequently mentioned along with the details about their growth and death. However, although ecological problems were touched upon, they were always limited to the ecology of a single taxon. No doubt it is useful for the palaeontologist to know about the activities of individual organisms and about their relationship with their surroundings, but he must also be aware, when working on the fossils of an outcrop, that he is dealing not only with an accumulation of individual animals or their living traces but also frequently with the fossil remains of an entire community of organisms. In other words, he is confronted with the remnants of a biocoenosis. If he removes an isolated specimen in order to investigate it in more detail in the laboratory he necessarily destroys a fossil document of a biocoenosis. The study of fossil ecosystems and the description of biotic interdependence has so far played only a minor role in palaeontology. Only the ecosystems of communities of sessile organisms that obviously lived together (for example coral reefs) and thus must have formed a biocoenotic unit have been investigated. This is not necessarily the case when animals are simply embedded together as the result of passive, physical accumulation of their remnants; in that case one may speak of a community of graves, or taphocoenosis. In fact, true living communities preserved together as a whole are not frequent in the sediments of a shallow sea; there the ecosystem is in most cases destroyed before it can be preserved, or it is contaminated by addition of foreign elements which may be of the same geological age but transported from a different habitat. These difficulties are the reason why the study of ecosystems still plays such a minor role in palaeontology.

What the palaeontologist can really recognise on an outcrop is a lithologically definite rock unit formed by deposition in a period with constant conditions. Such a definite unit he calls a facies. If it is also, and prominently, marked by characteristic remnants of organisms he speaks of a biofacies.

I. Biocoenoses

The concept of biocoenosis, meaning living community, was first formulated by Karl Möbius, a zoologist at Kiel. Using an oyster-bank of the coastal North Sea as an example for an integral animal community, he developed this now indispensable concept in 1877, terming it 'living congregation'. Möbius (see also 1871) included in his concept a biotic and an abiotic element. The present term 'ecosystem' emphasises the systematic, whole unit which necessarily includes the interrelation between the biotic elements and the abiotic ones, the latter constituting the biotope (see also Tansley 1935, Tischler 1951 and 1955). Thus the concept covered by the two almost synonymous terms biocoenosis and ecosystem comprises ideas beyond the purely biological, including some that are important for the palaeontological problem of the interrelation between biocoenosis and biofacies. So understood, the concept belongs to Thienemann's (1942, 1954) general 'theory of the economical balance of nature' (see also Thienemann & Kiefer 1916).

Several postulations have resulted from the theory of biocoenosis (see Tischler 1951, 1955; Thienemann 1939, 1941; Remane 1943; Schwenke 1953 and Balogh 1958). (1) Because the requirements of each organism are definite and unalterable each species can only exist in places that fulfil these requirements. The range of conditions or 'ecological valence' can be wider or narrower, species with a wide range being called 'euryoecous' and those with a narrow one 'stenoecous'. (2) Identical associations of species recur in places that are equal ecologically. Species occurring in the same biotope are usually also interdependent. A lasting association of species is based on considerable self-regulation. The active factor in this regulation, however, is the individual organism. This ensures a 'synecological' balance in any given biotope. (3) Living and lifeless components in the whole complex are interrelated and therefore form an organised unit. (4) A biocoenosis is not a biological form in the sense in which the cell, the multicellular organism or social colonies are biological forms.

Since a biocoenosis (or ecosystem) encompasses a biotope and a community of all organisms living in it, biofacies is a closely related concept. It is therefore necessary to study biocoenosis to understand biofacies.

(*a*) BIOCOENOSIS, A PROCESS IN TIME

To describe a living community, or biocoenosis, one must begin by listing and counting its members and noting their features. It is also essential to observe which of the properties of each of the members affect other members, and in which subarea of the biotope each is active. Finally, this leads to a description of the interrelation between members and the inanimate surroundings; these relationships may be divided into separate influences of surroundings on organism, organism on surroundings, individual on individual, species on species, and of one part of the surroundings on the other.

A biocoenosis is maintained by a constant flow of energy to all its living members. Energy is transmitted from member to member in the form of food. Each member of the community as it ingests food is linked to the others according to its status as a consumer or object of consumption, or both. Thus definite food chains arise in the community, and every member becomes a link in one of the chains. Green plants are always at the beginning of the chain because only they can, by means of photosynthesis, transform radiated energy of the sun into chemical potential energy and store it. Green plants are the principal objects of consumption in the community (with the exception of a few semi-parasites and parasites which must be considered as partial consumers). Next in the chain are herbivores followed by the carnivores and finally by detritus-eaters. The chemical potential energy produced by green plants is put to use successively by all members of the community in sequence according to their position in the food chain. Part of the energy is used up by each of the individual members of the biocoenosis to sustain its living functions as metabolic fuel. This part of the energy is dissipated into heat, passes into the atmosphere or the surrounding water and is lost for the biocoenosis as a whole; therefore a constant supply of newly synthesised food is required and supplied by sustained photosynthesis. Another portion of the energy is built into the bodies of living organisms as tissue or as stored supply and is thus fixed temporarily. Yet another portion goes into synthesis of organic material of low food value, for example secretions or excrements; however, at least in part they are still consumed by other organisms.

The number of plants and animals that make up a biocoenosis determines the quantity of energy that flows through a biocoenosis at a given time. In economic biology one therefore speaks of the 'biomass' which is invested with living beings at a given moment in a certain area, say one square metre, of the biocoenosis; similarly, 'bioenergy' is the amount of potential chemical energy represented by the living beings. The biomass can be expressed as total weight per unit area or volume; a separation of biomass by species normally reveals that certain species of the biocoenosis make up most of the mass.

In a marine biocoenosis, the phytoplankton living in the topmost, well-lit water layers does almost all the photosynthesis. In the littoral areas the marine phytoplankton is supplemented by the fresh-water phytoplankton of rivers and sometimes of sewage from the mainland. All marine organic life, at whatever depth, and whether living as nekton or benthon, always depends on the phytoplankton although this dependence may be quite indirect and through a complex food net rather than a simple food chain. The transport of food in the form of plants or animals can either be vertical, downward from the scene of photosynthesis, or horizontal from areas with abundant photosynthesis to others where the production of phytoplankton is poor or missing altogether. New settlement or maintenance of the numerous animal communities in such areas far from the primary food production depends on the transport of secondary food organisms through the sea with its world-wide system of currents and passages. Actually, the majority of the marine biocoenoses are 'open', they obtain energy from outside their boundaries. Such biocoenoses do not themselves contain 'constructive' organisms but only such organisms that consume, temporarily store, and ultimately spend chemical energy.

By its nature a biocoenosis is theoretically immortal as long as there are members with the necessary attributes; the balance that ensures its continuity is independent of factors determining the birth or death of any of its individual constituent organisms. However, even the relatively more stable, balanced interrelationships that are the essence of a biocoenosis are in turn subject to change in time. Conditions, both the local ones within the biotope and the physical and biotic conditions at a distance, change with time, so that any given biocoenosis has a beginning in time, and after a certain period an end. A historical record of the temporal existence of such a biocoenosis may be preserved in material form; the palaeontologist's problem then is to recognise what may be intelligible in such a record remaining after a biocoenosis has long ceased to exist.

In a self-regulating ideal biocoenosis, energy and organic materials should be used up completely, and thus no trace of the existence of such a biocoenosis should be left. However, usually a biocoenosis affects its physical environment in some way, sometimes permanently, by the manifold life activities in it. A biocoenosis on the sea floor produces a permanent record in several ways. (1) It furnishes organogenic hard parts; (2) its living manifestations can be recorded as traces and bioturbate textures in bedded sediments; (3) it furnishes partly decomposed organic products, in some cases with chemical properties which are specific for the biotope. Gerlach (1968) showed the interrelation in the production of organic materials in the ocean ('food pyramid') including the role of detritus, dead organic material and their ultimate disposal.

(1) The biomass of a biocoenosis may consist to a considerable extent

of skeletons or other organogenic hard parts. The higher the productivity of a marine biocoenosis the higher usually is the proportion of structural hard parts. They are built by life processes at the expense of the total energy supply of the biocoenosis. This energy thus can no longer serve as food for other organisms. The fate of such hard structural parts after the death of the organisms has been discussed. Their preservation depends on sedimentation; under certain and frequently occurring conditions a steady supply of sediment continually buries and preserves the hard parts, with the exception of some that are dissolved before and during sedimentation and during diagenesis, thus reducing the total organogenic contents of the sediments. Beds formed during the lifetime of a marine biocoenosis are like continuously added pages of a chronicle of the biocoenosis, a chronicle that is hardly ever written of terrestrial biocoenoses. This kind of record is destroyed only by the loss of the pages themselves, by erosion or reworking of sedimentary beds.

(2) Sediments not only cover and enclose the hard parts produced in a biocoenosis, they can also preserve on their bedding planes traces of certain life activities in the biocoenosis, such as locomotory traces, burrows or food remnants. The activities of the animals of a former biocoenosis can modify the texture of contemporaneous sediment layers. In terms of economic biology one can say that a part of the total energy expended in a biocoenosis has been converted into mechanical energy, used to create a trace or a burrowing texture.

(3) Most of the continuous supply of energy originally introduced into a biocoenosis by photosynthesis is usually completely expended. In many cases, however, not all the organic material is completely decomposed, and it may be preserved in various stages of decomposition, as partially decayed carcasses, faeces, or in other forms; these organic products can form whole layers or they can be added to contemporary sediments and modify them physically and chemically. Increases in the rate of sedimentation after deposition of such materials help to preserve them because an impermeable cover reduces the oxygen supply and slows or modifies the decomposition of the organic material. Another factor favourable to preservation of organic material is an oxygen-free or oxygen-poor environment at the place of the primary deposit such as exists at the bottom of an anaerobic water body in a closed basin. Specific chemical conditions of the biotope may enable lacustrine, brackish or marine sediments to be recognised in the fossil state.

Any change in the physical and chemical conditions in a given region of the sea floor, as long as it does not exclude marine life altogether, permits certain, selected species to settle there. Individuals of these species are always available for such a first settlement because most have a mobile youth stage as eggs or larvae, or the adults themselves are locomotory and can move widely enough to find new habitats. Such newcomers thus form a new biocoenosis shortly after a previous system has been destroyed.

This initial phase of a biocoenosis may be called its foundation stage. The more changeable the conditions of a habitat the shorter is the life span of any biocoenosis in it, and the smaller is its effect on the forming layers of sediment. Thus the opportunity to form a biocoenosis is less important for the stratigrapher than the problem of how long a biocoenosis can exist, and when and why it collapses. The causes for collapse can be physical or biological. In the North Sea several distinct physical events can cause the collapse of a biocoenosis:

(1) Destruction by a sudden high rate of sedimentation, usually caused by a strong storm. In this case the biocoenosis is destroyed, but a similar one develops rapidly by immigration from neighbouring areas. (2) Destruction by removal of the biotope through deep erosion, also usually caused by a strong storm. In this case a similar or dissimilar new biocoenosis rapidly develops. (3) Destruction by a sudden change of grain size or composition of the sediment (or both) without a change in the rate of sedimentation; such changes are generally caused by changes in the direction of currents. In this case a dissimilar new biocoenosis develops. (4) Destruction by a gradual change in grain size or composition of continuously deposited sediments, generally caused by changes in the direction of currents. In this case a new biocoenosis develops gradually while the members of the old one gradually perish. (5) Destruction by the onset of sedimentation in areas in which no sediments were previously deposited, usually caused by a change in the direction of currents. In an extreme case the sea floor consists of bedrock before the sudden onset of sedimentation. In this case a completely different new biocoenosis develops. (6) Destruction by an increase of excretory substances, especially faeces, caused by a decrease of the velocity of bottom currents. A dissimilar new biocoenosis develops.

(7) Destruction by decrease or increase of salinity, depth of the sea, oxygen content of the water, temperature or climate. Unlike (1) to (6), these causes are long-term and long-range and fundamental geological or climatic changes, affecting more than a single biotope.

Several causes of collapse of biocoenoses in the North Sea are biological: (1) Destruction by an increase of the amount of hard structural parts from dead members of the biocoenosis, caused either by insufficient transportation of shell fragments, or by too low a rate of sedimentation to raise the biotope gradually. A dissimilar biocoenosis develops gradually. (2) Destruction by intrusion of one or several new species, caused by a high production of larvae or by the change of physical conditions in distant or neighbouring biotopes which force their inhabitants to emigrate. A partly or completely dissimilar biocoenosis develops.

Slight changes in time of the structure of a biocoenosis and of the interrelation between its members, involving small departures from a median, constitute a so-called sequence of aspects. The term indicates that the biocoenosis as a whole continues to exist through these minor changes. In the shallow sea such changes in structure are usually due to seasonal variations in the interplay of forces within the biotope. A sequence in time of fundamentally different biocoenoses is called a succession; one biocoenosis at a locality is replaced by another. The physical changes (1) to (7) and the biological changes (1) and (2) cause true successions, and those caused by the changes of case (7) may be called secular successions. According to the nature of the cause one may classify successions as physical or biological.

A biocoenosis invariably comprises a certain number of distinct functional groups of organisms. Each group performs a particular and different activity, and together these groups constitute the animate component of the biocoenosis, guaranteeing its continuation and establishing a mutual balance. What characterises a group is its life activity. The number and the life habits of collaborating or antagonistic groups characterise a biocoenosis. Each biocoenosis has what may be called a 'Stellenplan' (roster of organisation, Kühnelt 1954). Each group fills a definite 'niche' in the structure of the biocoenosis. Remane (1943, 1952) called species with similar life habits, distinct from those of other species, 'Lebensweisetypen', behavioural types; if, furthermore, such species with similar habits and functions also resemble each other morphologically and structurally he calls them 'Lebensformtypen', constitutional types. The morphological structure of animals of a constitutional type tells us something about their activities and life habits. However, before a species can be assigned to a behavioural or constitutional type its functions and morphology must be thoroughly studied from the ecological point of view, both with respect to the ecology of the species itself and with respect to its ecological interactions with other species. In the preceding sections on individual marine animals of the southern North Sea such an assignment was made, sometimes only tentatively, whenever possible. This was done with the double purpose of elucidating both the ecology of the individual species and of providing material for an eventual classification of the biocoenoses and biofacies of the southern North Sea area.

Useful as the concept of constitutional types may be, it hardly seems possible to characterise any biocoenosis by the constitutional types of its member organisms. In every single case, only a very small number of all the species that participate in the structure of a biocoenosis can be assigned to such a type; these few species are insufficient to characterise such a large entity as an entire biocoenosis or distinguish it from a related one. Thus in addition to the concept of the constitutional types it will always be necessary to consider also the geographic boundaries of biotopes with distinct physical features whenever a biocoenosis must be demarcated.

If two or more geographically separate biocoenoses have structures that can be represented by identical or similar 'rosters of organisation' such biocoenoses are considered 'equivalent in organisation'. One might expect that the same behavioural and constitutional types should be found in biocoenoses with equivalent organisation, even where the identical types are actually represented by different species; however, by definition the similarities of a 'type' ought to be due to similarity of function rather than of phylogenetic relationship. Biocoenoses with analogous structures and functions are also frequently called isocoenoses. They can be widely apart in space and different in their edaphic and climatic conditions. The isocoenoses of the floors of the world oceans have not yet been classified; only a beginning of such a classification has been made for a few particularly conspicuous biocoenoses, such as the coral reefs.

(*b*) DOMAINS OF BIOCOENOSES

In the preceding discussion of the principles governing biocoenoses the problem of the geographical extent of individual biocoenoses has not yet been touched, and neither has there been an attempt to classify biocoenoses. The first classifications of biocoenoses by geographical domains and subdomains in them were made on land where it is undoubtedly easier to mark boundaries than in the sea. Thus a 'bioregion' has been taken to be the largest geographic domain. It comprises a whole landscape, populated by the most general phytocoenoses and zoocoenoses, the so-called 'biomes'. On land these are mainly determined by the wide climatic zones on the earth. Tischler (1955) distinguishes eight living zones, differing in temperature and in moisture, with equally different communities of living organisms. The phytocoenoses of the regions show particularly striking contrasts, but in turn they considerably influence the animal populations they support.

No similar bioregions have been established for the oceans, if only because the physical and biological properties of so few sea floors are yet known. Instead, marine biologists have tried to establish smaller geographic domains. Caspers (1950b) discussed the concepts of biocoenosis and biotope in the sea; today his terms are generally used in marine and fresh-water biology. Möbius himself (1877) developed the concept of biocoenosis for an oyster-bank. But the practical work of the following decades proved that it is not always possible to demarcate valid domains of biocoenoses on the inaccessible sea floor. Instead investigators began simply to list and describe assemblages of organisms and the substrates on which they were found.

For many years the most important instrument for fetching organisms and sediment from the sea floor was the Petersen sampler (for present estimates of productivity of Heligoland Bay see Ziegelmeier 1963). Organisms found in a sample were identified and grouped by quantitative methods, and the groups were called 'associations'. These associations referred only to bottom-dwelling organisms because it was only possible to investigate the grab samples completely. It was found that fauna living in the sediment is much more regularly distributed over an entire area than animals living above the ground which are mostly locomotory and travel erratically over wider distances.

As so often in the research on biocoenoses, practical considerations provided the first motive for intensive studies, and it was mostly in connection with fisheries that work on the benthos of the ocean was done. Thus distinctly recognisable, closed units of bottom-dwelling animals were demarcated; they are the most important source of food for all bottom-dwelling fishes and particularly for the commercially significant flatfishes. At the beginning of this century, Petersen (1915, 1918) estimated the productivity of the sea floor in the Danish and Norwegian waters. In the years following the first world war the Deutsche Wissenschaftliche Kommission für Meeresforschung and the Biologische Anstalt Helgoland (Hagmeier 1925) undertook similar estimates of productivity in the southern North Sea area and especially in Heligoland Bay. Even today productivity research has remained the most important and most reliable basis for the classification of bottom faunas; and where no such research has been done no classification of associations has ever been established, and even less can be said on the biocoenoses of the sea floor. Although the North Sea is a relatively well-investigated ocean little has been achieved beyond this first stage of delimiting associations as statistically defined units. However, these studies show that the associations on the sea floor are quite distinct units, each composed of typical groups of organisms. Wherever it was possible to understand the ecological and functional conditions that govern such associations, the step was made from the definition of a simple association to its comprehension as a biocoenosis. Only recently have investigators begun to compare related communities and to study the conditions that influence their composition (Thorson 1966).

Caspers (1950b, following Möbius) thought that functional considerations are useful for characterising a biocoenosis but less so for defining its geographic boundaries. For example, a benthonic community requires phytoplankton as a source of food, but the phytoplankton is not a component of a benthonic biocoenosis. A biocoenosis on the sea floor includes only the two other large ecological groups of organisms, the consumers and detritus-eaters. Most biocoenoses of the sea floor are therefore 'open' or 'dependent'; especially if they are pure animal communities. Thus it seems all the more appropriate to define bottom communities simply according to their geographic domains. Subdivisions of communities with more narrowly defined common traits in species composition are called 'variations' or 'variants' (or 'sub-communities' according to Hagmeier). Important individual, characteristic species may demarcate groups, but their presence or absence alone is insufficient to establish an ecological unit. The ecological structure characteristic of a biocoenosis requires a typical combination of species in a well-delimited biotope.

(*c*) BIOCOENOSIS AND SOCIETY

The preceding sections have shown that a biocoenosis is an ordered unit in nature. Living organisms, the most important components of a biocoenosis, achieve synecological balance each by its own activities which it performs for the sake of the preservation of its own life and for the propagation of its species; the members of a biocoenosis do not collaborate actively. Thus a biocoenosis is not a biological system. This view is not unanimously accepted by all synecologists (see also Peus 1954).

Animal societies are different from biocoenoses. They are congregations of individuals of the same species, representing a biological system of a higher order than the multicellular individual; the existence of each individual in space and time is completely subordinated to the superior unit. Since palaeontologists are frequently confronted with remnants of animal societies, and since a society as a whole often produces characteristically shaped fossils, the principles governing their functions are geologically important. The following discussion of animal societies is based on investigations and observations by the author (Schäfer 1956d).

Not every assemblage of animals of a single species can be considered as a society. In many cases such assemblages are due simply to local favourable conditions; the place may be climatically favourable, it may furnish good dwelling space or abundant food, or it may possess some other qualities which distinguish it from the surroundings. As soon as these conditions become less than optimal at such a locality the assemblage disperses, or remaining individuals eventually die without progeny. If individuals of a species do not maintain lasting relationships with conspecific animals they must be considered solitary. The behaviour of each individual is determined exclusively by its own requirements, in contrast to social animals.

The difference between social and solitary animals is frequently not externally apparent but it shows clearly in the behaviour of individuals under conditions of crowding. Crowding may be due to a reduction of a formerly larger living space, or, more frequently, to an increase in the number of individuals. Crowding may encroach on the 'critical space', the minimum area which an individual requires for its existence. Encroachment into the critical space causes a 'critical situation' which if it

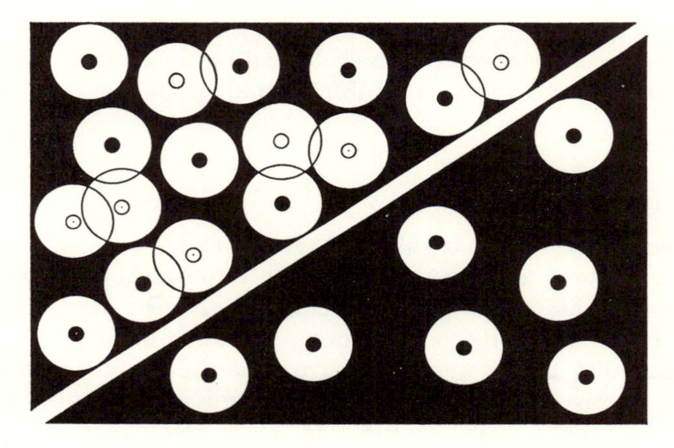

FIG. 257. Top: closely spaced individuals of a solitary species (black dots) each living in the centre of its critical space (white circular area). Where such critical spaces overlap one of the individuals dies (small circle). Bottom: widely spaced individuals of a solitary species. Critical spaces do not overlap, all individuals survive (from Schäfer, 1956d)

persists leads to the animal's death. In a critical situation, an essential difference appears between the behaviour of solitarily and socially living animals. Solitary individuals try to maintain themselves at the expense of their neighbours. There may be a fight or the general situation in the biotope causes the weakest or most exposed individuals to perish. The density of population in a given area thus is regulated by destroying individuals until the available space is sufficient for the survivors (Fig. 257). The size of the critical space determines the maximum density of population even under optimal living conditions. A striking example is the frequently regular spacing of benthonic organisms in the sediment.

Social animals respond differently to crowding. If the number of individuals has increased or the suitable biotope has been reduced in size the group compensates by organisational or technical changes. This can best be demonstrated by a few examples of marine social animals.

Botryllus schlosseri is a sessile tunicate that settles in colonies on stones, wood, groynes and ship hulls. While the colony is being formed the animals reproduce asexually. Individuals divide lengthwise so that they arrange themselves radially around a common egestive aperture. The resulting star-shaped groups grow larger as new individuals appear until they contain between eight and eighteen animals. In the meantime new stars are formed in the immediate vicinity of an earlier one by individuals that have separated themselves from them. Thus a colony grows until it covers an area the size of the palm of a hand. When all the available space is occupied and the continuously growing colony gets into a critical situation, no fight arises; rather, the colony responds as a whole, as a society. In such a case individual star-shaped groups detach themselves from the substrate and push upward; they support each other by opposite-facing stars joining their feet. This results in leaf-shaped and drop-like growths. In this way many more individuals can be accommodated in a settlement with the same base area. Nevertheless, the extension into the third dimension is not unlimited. Once such a colony spreads over a certain area, up to a diameter of approximately 30 cm and comprising 800 to 1000 star-shaped groups, the performance of the pumping mechanisms seems to become insufficient and incapable of transporting excreta through the common egestive opening and far enough away from the colony into the open water. Then excreta accumulate near the colony which perishes quite suddenly from self-poisoning. An increased flow of water can delay the death of the colony for a certain period, but not indefinitely. Less than a day after a colony that has taken four or five months to develop still seemed perfectly healthy, not a single animal survives, and only a few ragged pieces of tissue are left of the extensive colony. The animals do not contain any hard structural parts.

The *Botryllus* society is capable of overcoming a first overcrowding. It does so by a common reaction at the right time, by rearranging the individuals into a three-dimensional settlement. Such concerted action demonstrates that a *Botryllus* colony is a society. However, it is a primitive society which cannot overcome a second critical situation developing from the first. Lack of hygiene in the full-sized colony leads to catastrophic death of all members. Each *Botryllus* society has the same history; there is never an exception.

A colony of the sponge *Halichondria panicea*, common in the North Sea, is able to overcome more than one critical situation in the course of its life history. At first several individuals form a shallow crust that shows little differentiation. Through tiny, individual ingestive openings, sea water passes into the gastric cavities; from there waste water is expelled through larger, communal egestive openings (oscula). As the sponge settlement grows the quantity of polluted water increases proportionally. To prevent this water from poisoning the colony each osculum grows a collar-like extension made up of individual sponges. A developing critical situation is averted to the advantage of all. As the colony continues to grow the collars of the oscula lengthen into long tubes which remove the increasing quantities of polluted water further from the colony (Fig. 258, left and centre).

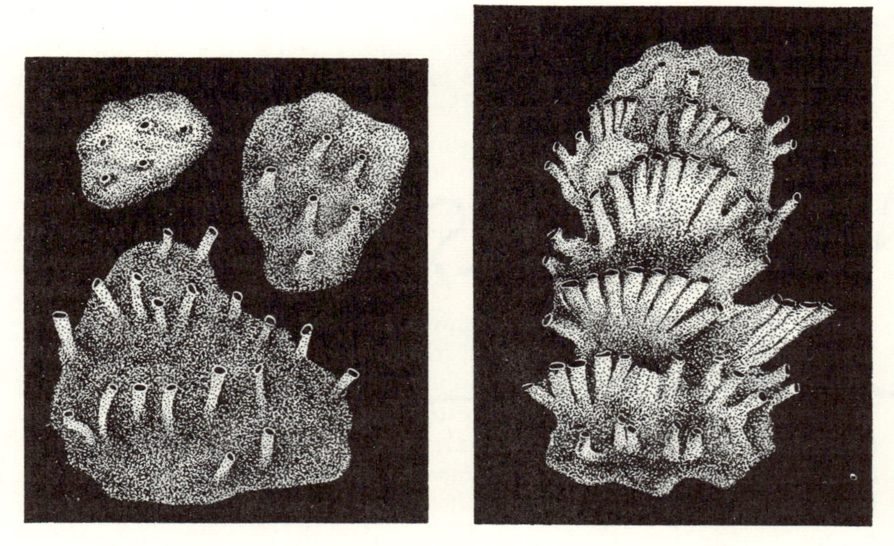

FIG. 258. Four stages in the development of a colony of *Halichondria panicea*. As area of the colony increases the egestive openings (oscula) become higher; right: oscula at the ends of batteries of tubes, arranged in several storeys (from Schäfer, 1956d). (Actual size × 0·3)

Eventually the chimney-like tubes reach a size which cannot be exceeded without danger of breaking or bending. Yet, the society can also overcome this new situation; the chimneys are now arranged in rows and interconnected lengthwise for greater strength. Such batteries of tubes can grow several times longer than single tubes without breaking under their own weight or in a current. The batteries of tubes are usually stacked above one another and point radially outward (Fig. 258, right). In this way a colony can grow 40 cm high and 15 cm wide and may contain many thousand individuals. At this stage, however, a critical situation arises which can no longer be mastered; the material of which the colony is built is not strong enough for the size of the colony, it breaks and both the broken-off part and the remnants which are still attached to the substrate but have lost their oscula die catastrophically. As the colony passes through successive stages of development, it attains several shapes each of which is typical for the species and functionally determined; the whole structure and details of its shape are indicative of certain definite functions of the society.

Both *Botryllus* and *Halichondria* are sessile social organisms, with little chance of being preserved as fossils. *Botryllus* has no hard parts, and *Halichondria* skeletons consist of separate, needle-shaped spicules which can hardly be preserved whole. But many other sessile social organisms have very strong hard parts and preserve well (e.g. coelenterates, polychaetes, bryozoans, balanids and lamellibranchs). They tend to form hard reefs. Once they have been studied and the critical situations in the growth histories of their colonies have been understood, the functional significance of specific forms of reefs could be interpreted in the same way as those of *Botryllus* and *Halichondria*.

As the example of *Botryllus* shows, the most important problem to arise sooner or later for sessile societies is the distribution of dwelling space with the increasing number of individuals. The means of surviving in crowded critical conditions is often to alter the overall shape of the colony while the structural plan is retained (e.g. *Balanus*, Schäfer 1952a, Gutmann 1960), or to shift the position of organs, or to use different building material. The phenomenon of change in forms under such conditions repeatedly puzzles palaeontologists. A single species occurs in different morphological (and taxonomic?) forms depending on the age of the society. The shapes of hard parts are subject to similar striking changes. The time in the growth history of a colony at which an individual and potential fossil becomes part of a society is therefore significant. During each new period of morphological change, individual members have new mutual relationships, and critical factors that affect the society as a whole also influence and mark each individual. Sessile aquatic species must feed at their dwelling place. Since the waste products are released into the same water that serves for feeding and breathing this usually becomes a critical problem when the society grows, and the hygienic requirements of the society cause many of the observed changes in the structural plan of the colony.

Regrettably little is known about the sequence and number of critical episodes that can be mastered by reef-forming and other sessile societies; the morphological structure of these societies is therefore not well understood, and the principles governing the growth of societies have been elucidated for few of them. Because social sessile organisms frequently preserve so well they deserve the special attention of palaeontologists.

Vagile as well as sessile organisms can form societies. They, too, pass through critical episodes. Any of these episodes may be resolved before it involves a part or all of the society in a catastrophe. However, the rate of continuing losses of a vagile society is usually so high, and the rate of increase so small that catastrophic destruction can occur only after a period of optimal living conditions. Thus field mice sometimes first proliferate and then perish catastrophically (Frank 1953a and b). A society of field mice can master four types of critical situation and stay intact. During the final collapse, however, all social values are lost and cannibalism takes over. Some societies achieve the necessary regulation

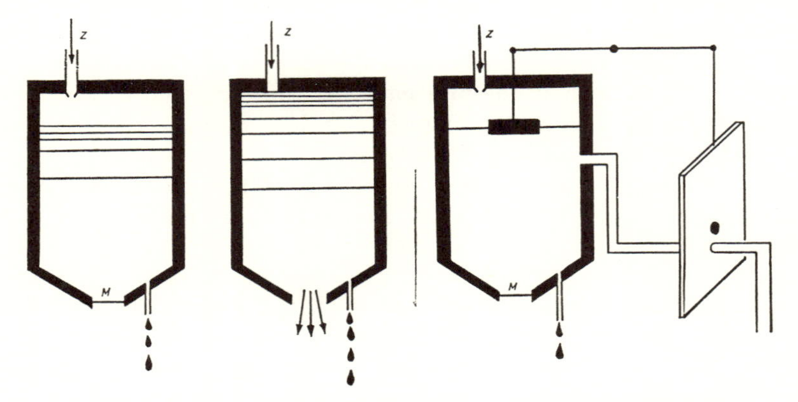

FIG. 259. Schematic models of societies of finite and potentially infinite age. Left: a continuous flow of liquid through inlet (Z) will burst the membrane (M) after the vessel is filled if the inflow exceeds the outflow (dripping tube); left centre: after the membrane bursts the vessel empties; the fluid represents the population, inflow the birth rate, the drip tube the normal death rate, and outflow through the burst membrane catastrophic death of the whole society. Right: a float opens a linked overflow valve when the fluid level rises to a given level; the membrane will not burst. The model represents a society whose potential infinite existence is safeguarded by emigration of parts of the population

by mass migration at the right time, and the society survives (as in lemmings and several social insects). Thus we can distinguish between societies of finite, and others of potentially infinite duration. Fig. 259 shows diagrammatically the difference between catastrophic termination of a society and its regulation by timely mass migration. More exact and more investigations of phenomena in animal societies from the point of view of systems control could clarify much that is not yet understood (see also Mittelstaedt 1956). Some aspects of insect pest control demonstrate the role of migrations. Mass accumulations of animal pests are never completely eliminated by chemical pesticides, and partial destruction has the same effect on the remaining population as active migration. Therefore such artificial measures generally do nothing but delay the natural end of a society by decreasing the population pressure in a restricted space before a natural catastrophic culmination is reached.

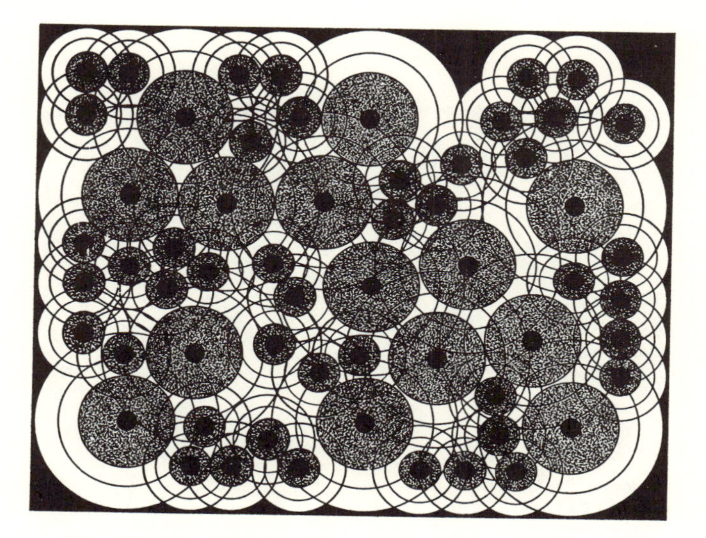

FIG. 260. Individuals of a society which can overcome two critical situations of crowding and is threatened with death only upon reaching a third critical situation. The successive critical spaces are symbolised by concentric circles, the final critical space being dotted. Position of individuals is in black (from Schäfer, 1956d)

Fig. 260 schematically represents the successive resolution of critical episodes by a society, analogous to the scheme used for solitary organisms (Fig. 257). Three concentric circles, from the outside inward, indicate subsequent critical situations of crowding. All but the situation shown by the innermost circle (filled with dots) can be overcome; thus the number of circles indicates at what stage an inescapable critical situation is in store for a particular society. If the individuals of the model are forced closer together more concentric circles intersect; but only when the (dotted) areas of the innermost circles begin to overlap with extreme crowding does the society perish. Fig. 261 shows the situation for a society with differentiated individuals in which specialists perform different tasks and have different space requirements. They are schematically rendered by concentric circles of different sizes. (Examples for differentiated societies are worker and soldier ants or specialised hydroid polyps.) Critical situations can be overcome more easily by differentiated societies, because specialist individuals usually differ considerably in their morphology, and this can be taxonomically significant for fossils.

Societies of vagile organisms are less interesting for palaeontologists because such a society is rarely preserved as a whole. However, one phenomenon in the behaviour of societies of higher vertebrates has palaeontological significance because it is frequently the cause of mass

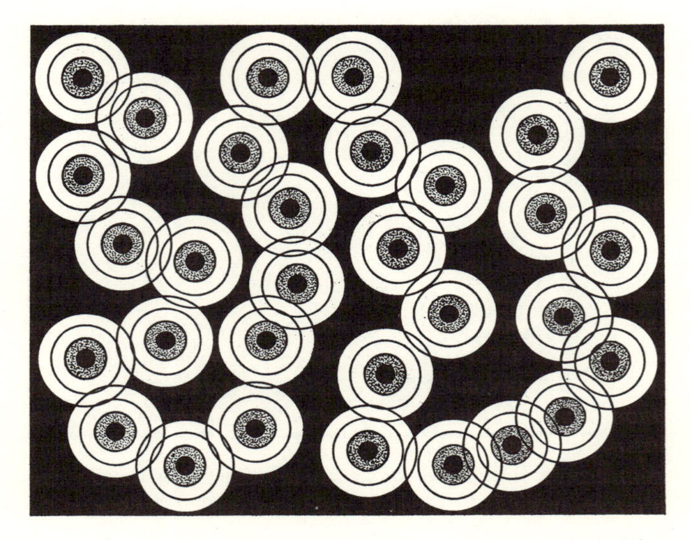

FIG. 261. Specialised individuals of a society with division of labour and with unequal space requirements for individuals of different classes. For explanation of symbols, see Fig. 260 (from Schäfer, 1956d)

death. Individual social birds and, more frequently, mammals try to compensate for lack of space by doing one after the other what they would all have done simultaneously were it not for overcrowding. An unhealthy proportion between the number of individuals and available space can lead to queuing (e.g. at watering holes or in narrow passes during migrations). Queuing becomes critical if it takes so much time that part of the society can no longer satisfy its needs. In vagile societies this can lead to panics or stampedes in which all social, regulating drives suddenly collapse and in which each individual acts as if it were a member of a solitary species. It 'forgets' all principles of social order and every obstacle to its blind will is suppressed. Panicking animals lose their sense of orientation, their locomotion is reduced to an inferior order of co-ordination, and they are driven by disparate nervous impulses. Worse, panic is contagious, and the sudden change of mood spreads to unaffected parts of the society by acoustical and optical signals from the panicking animals until all are in full-fledged panic. It has been demonstrated that whales do panic and commit mass suicide on beaches. Certain accumulations of skeletons of higher vertebrates are probably caused in this way.

Vagile, social organisms that build dwelling burrows, either of their own secretions or with foreign material (as some polychaetes), become secondarily sessile societies. Their multiple structures can be identified by palaeontologists, and they can study the social principles that govern them. Whether a society consists of primarily sessile individuals or of exclusively vagile animals, or whether vagile builders of dwelling burrows revert to a secondary, permanent or temporary sessile life, the same principle applies. All colonies are subject to successive critical episodes and each requires a new solution entailing abandonment of an earlier way of life. Increased crowding subjects individuals and community to one critical situation after another. Each new adjustment requires the co-operation of all individuals and often a complete reconstruction of the settlement. Yet, after new space for new individuals is created, the population grows and is pushed toward the next crisis until there is one it cannot master. The number of critical situations that can be overcome by any one vagile species is finite and constant. In the course of the growth of each animal society, the same forms and the same functions arise according to a predetermined, regular pattern which always ends with the same, inexorable catastrophe. Like individual organisms, societies are governed by instinct and depend on their surroundings.

The laws governing the space requirement of biocoenoses can be summarised briefly. Each organism requires for its existence a specific minimal area, the 'critical space'. A 'critical situation' arises when by crowding, the critical space of an individual begins to be overlapped by the critical space of a neighbour. The possible density of population of solitary organisms is regulated by elimination of individuals for whom the necessary living space (critical space) is no longer available. Social communities master critical situations by timely rearrangement of the individuals or by technical improvements that serve the whole. Only a finite and specifically constant number of critical situations can be overcome by the societies of each species until they end catastrophically. An exception are societies in which either migrations or death of part of the society regularly reduce the population pressure. Such societies have potentially infinite duration in contrast to most others which have finite duration.

II. Biofacies

Biocoenoses and the phenomenon of animal societies are but aspects of the larger-scale phenomenon which may be called Recent biofacies, by analogy to the accepted use of the term for the geological past.

(*a*) BIOFACIES AND BIOCOENOSIS

The study of biocoenoses of the sea floor, the subject of the previous section, demonstrates that (1) a biocoenosis can be considered as a process in time beginning with the foundation stage and ending in collapse; (2) while a biocoenosis lasts, materials and forms are created that can be preserved beyond the time of collapse of the biocoenosis; these include certain secondary sediment textures such as traces and burrows which can permanently record living activities; (3) remnants of a biocoenosis on the sea floor can only be preserved by sedimentation; of all the marine biocoenoses those of the sea floor are the most likely to be preserved. Terrestrial biocoenoses have the least chance.

What is the relation between a 'fossilised' biocoenosis and a biofacies? Ideally, a biofacies, that is the total imprint of organogenic material and of traces of life activities in a depositional unit formed under uniform conditions, owes all its features exclusively to a single former biocoenosis. Assuming such a direct and simple relationship one may consider biocoenosis and biofacies as synonyms as Hesse (1924) and Hagmeier (1925) have done. Normally, however, this assumption is incorrect, and the organogenic substances, structures and textures of a biofacies are derived from more than one single biocoenosis which existed at the sea floor at the time of deposition of a sedimentary unit. What then are common additional features of biofacies?

(1) A biofacies usually contains remnants of nekton or plankton. The organic products of the open water mix in the sediment with those of the bottom biocoenosis. Actually, all the information one has about ancient nekton and plankton is gained from fossils that have been transposed from their former biotope into the foreign one of the sea floor.

(2) A biocoenosis commonly contains remnants of organisms from neighbouring biocoenoses. The locality of permanent deposition is not necessarily identical with the former habitat. Lateral transportation may have been the effect of bottom currents, generally due to wave action or tides, or it may be due to slumping.

(3) Many of the sedimentary textures which characterise a biofacies are produced by physical forces that are effective at the sea floor, in the water body above the sea floor, or in the atmosphere. Tidal currents,

wave action, vertical thermal currents and storms affect the sediment directly or indirectly. The sediments themselves are usually not of local origin but have been transported. Rarely are they precipitates *in situ*. Sedimentary structures are part of the sea floor biotope. Although they are not effective elements of biocoenoses they are typical elements of biofacies. And although the sedimentary features themselves cannot be counted as elements of biocoenoses, the forces which cause them certainly are important elements of a biocoenosis.

Thus a biocoenosis is a complete system in which each element is interrelated with all the others, but a biofacies is not. Its elements are merely an odd collection of bodily remnants loosely connected by a certain coincidence of events in space and time. An individual biofacies is a unit, compared with other biofacies units, only because the effects of a certain set of forces and conditions for a certain set of biocoenoses happened to coincide at a given place and time, producing a fossil result with uniform aspect or 'face'. Facies is the general aspect of a local deposit, a concept which becomes meaningless if separated from that of the deposit. And just as the totality of the inorganic features of a single deposit constitute its lithofacies so does the totality of its organic components, no matter whether they originated locally or elsewhere, constitute its biofacies.

(*b*) TYPES OF BIOFACIES

One of the earliest users of the term 'facies' (Gressly 1838) already defines it clearly as what one would now call 'biofacies' by stating that a facies represents a unit in which the organisms and the physical environment have developed under uniform conditions. H. Schmidt (1958, see also 1935 and 1949) confirmed that real differences in facies are based on differences in biotopes and that a comprehensive classification of biotopes is necessary before a classification of biofacies can be attempted. He considered the term 'lithotope' as unsuitable because it places an abstract entity simultaneously into the mutually exclusive lithosphere and biosphere. He also emphasised what facies is not. The term facies cannot be used to designate properties such as jointing, degree of metamorphism or slaty cleavage. It cannot be applied to units that are defined in any but a stratigraphical sense; thus terms such as 'Bavarian facies' or 'orogenic facies' do not make sense. Neither can it be used to designate distance from the land (near or far from the coast), or depth of the sea (littoral, neritic or bathyal). According to Schmidt depth is neither a factor for the sediment nor for living organisms.

Schmidt proposed his own classification of biofacies. To do so he sought the most decisive factors that delimit a biotope. He accepted neither lighting nor temperature or salinity, important as they may seem to the observer, as sufficiently decisive for the sediment itself, and as only marginally significant for living organisms. Instead, he regarded the degree of water agitation as the paramount influence on the structure and texture of sediments because it determines the distribution of the clastic grains and drifting calcareous precipitates. Water agitation is equally critical for the spatial distribution of organisms because the oxygen supply is closely dependent on it. He maintained that for bottom-dwelling animals oxygen supply is even more important than food. Thus he made water agitation and oxygen supply the main criteria of his classification of marine facies, both for present biotopes and for the facies of the geological past. He thus applied an actualistic method although admitting that certain types of biotopes may have been more or less widespread than they are today.

Schmidt's terms are formed by adding a prefix which indicates the degree of aeration to 'pneuston' ('breathable'). His sixfold subdivision according to ranges of oxygen supply can be refined by the optional introduction of a twofold classification into sedimentary domains which are either predominantly clastic or predominantly calcareous. In order of decreasing oxygen supply his biofacies are either clastic or calcareous, hyperpneuston, plethopneuston, pliopneuston, miopneuston, oligopneuston, and apneuston. This scheme of Schmidt's is an ecological one; it is applicable to biotopes of the present as well as of the geological past; and it considers equally the sediment and the organic content of a facies. Thus theoretically it should be quite serviceable. However, it is questionable to what degree the partial pressure of oxygen leaves a permanent and measurable record in rocks. Possibly, a hyperpneuston can be distinguished unambiguously from an apneuston. But it is doubtful whether a distinction between adjacent types, such as plethopneuston and pliopneuston, will ever be possible in rock.

The author therefore proposes to modify Schmidt's classification so that it becomes simply descriptive of those aspects of marine biofacies which a geologist can examine and recognise in the field. Any biofacies is characterised by a number of material, individual bodies, such as sand grains or organogenic hard parts or their fragments. These individual bodies can be considered as the physical record of the biofacies and analysed as to their visible morphological properties, their internal structure, external shape and mutual position. Occasionally, these three types of morphological clues can be supplemented by chemical clues for a more complete reconstruction of the processes on the ancient sea floor and in the sediment. However, it seems to the author that the original chemistry can be reconstructed only in a few special cases, and when the original boundaries of the biotope can be accurately delimited. Chemical criteria thus play a role only for regionally limited low-order subfacies. Structure, shape and mutual position of objects in a sedimentary unit reveal several aspects of biofacies:

(1) The microscopic and submicroscopic structure of each object reveals whether its origin is organic or inorganic. If inorganic, the structure can occasionally furnish clues as to derivation; if organic, even fragments can frequently be specifically identified.

(2) The shape of an object also frequently identifies it as organic or inorganic before it was deposited.

(3) The mutual spatial relations of constituent objects of a sedimentary rock determine its texture. The objects themselves can be organic or inorganic, and so can the texture. The spatial relationship also helps to outline the bed as a unit; it helps to determine whether erosion has occurred or whether the beds are complete. If the deposited layers are complete this indicates absence of bottom currents and consequently poor oxygen supply or complete absence of free oxygen. Incomplete beds, on the other hand, indicate a good oxygen supply and thus favourable conditions for life. However, the presence of a single disconformity, and thus the proof of a single event of erosion, does not imply that the whole sequence of beds in which it is found belongs to the class of biofacies with strong bottom currents. Only regularly repeated loss of beds is significant. More generally, a biofacies can never be classified from the evidence in a single hand specimen; for this, confirming evidence is needed from a relatively large section. Only a sufficiently large domain can display all the characteristic shapes, structures and spatial relationships which are the full attributes of a biotope. The minimum size for a domain to be representative varies from case to case.

Combining the criteria of structure, shape and spatial relationships, five combinations result which describe the five major types of marine biofacies (biofacies domains of the first rank). Each of the five types can occur in either a predominantly clastic or predominantly calcareous variety, and each can be further subdivided into biofacies of the second and third ranks. The five biofacies of the first rank are:

(1) Vital-astrate biofacies, marked by the organogenic material of a permanent biocoenosis, not bedded.

(2) Vital-lipostrate biofacies, marked by multiple bottom-biocoenoses that are destroyed at an early stage, by taphocoenoses and by disconformities.

(3) Lethal-lipostrate biofacies, marked by rich taphocoenoses and disconformities and by the complete absence of biocoenoses of the sea floor.

(4) Vital-pantostrate biofacies, marked by stable biocoenoses of the sea floor, by taphocoenoses of nektonic and planktonic animals, and by

complete beds which, however, are occasionally destroyed by benthonic animals.

(5) Lethal-pantostrate biofacies, marked by taphocoenoses of nektonic and planktonic animals, complete beds, and the complete absence of biocoenoses of the sea floor.

Biocoenoses of the sea floor can be recognised by the presence of perturbed or shaped burrowing textures produced by benthonic and nektonic animals. Other significant features are benthonic animals fossilised in living position and locomotory and resting traces on bedding planes. The original bedding can be entirely replaced by burrowing textures or otherwise obliterated, indicating very slow sedimentation. Wherever taphocoenoses ('communities of graves') contain, or consist of, complete or partial carcasses of benthonic animals this is always an indication of disconformities; for only during erosion can carcasses of endobiontic animals be dislodged from the sediment and accumulated in taphocoenoses. Complete or fragmentary carcasses of plankton and nekton can occur in all marine sediments, whether they be hostile or favourable to life, complete, or full of disconformities.

Each of the five marine biofacies will now be explained in detail. An idealised, composite sketch for each of them is meant to combine symbolically the significant features of the biofacies, not to describe any real, observed sequence of beds on the Recent sea floor. Animals drawn in the open water above the sediment show constitutional types and thus are meant to indicate the conditions under which the sediment was formed. Each of the sketches also contains representative members of the Recent plankton, nekton and benthos that could eventually become fossils, and the types of bedding that can occur in its domain; most actual examples of a biofacies will therefore contain a smaller number of types of fossils and of different types of bedding than drawn in the composite sketches.

(1) Vital-astrate biofacies (Fig. 262). These biofacies are generally called 'reef-facies'. All the organogenic material with the possible exception of a few, rare nektonic and planktonic remnants, is produced locally; no organogenic material is ever added which is produced in neighbouring biocoenoses of the sea floor. Although currents capable of transportation exist and are often very strong, they either find nothing to transport, or the rough surface of the biocoenosis prevents accumulation. Loose organogenic material of local origin may be retained in place. Such conditions arise in the habitat of sessile organisms forming permanent structures such as corals, calcareous algae, sponges, bryozoans and reef-forming polychaetes. The oxygen concentration is very high as a consequence of water agitation. After death the organic material is rapidly decomposed; only the hard structural parts of the carcasses and the

FIG. 262. Vital-astrate biofacies. Characterised by hard material derived from the local permanent biocoenosis of sessile skeletal animals (biogenic reef). Note agitated, well-aerated but clear water with few suspended particles, populated with swarms of fish. Sediment only in reef crevices and consisting of hard skeletal fragments from the local, sessile, benthonic fauna. (1) living octo- and hexacorals; (2) lawn of living calcareous algae; (3) breccia of coral fragments; (4) reef limestone, containing organic material in form of thin, black lamellae; calcite fills cavities in reef limestones and interstices in rubble and breccias

structures and encrustations secreted by the organisms remain. Sedimentation is extremely slow or non-existent because this biocoenosis can only develop on sediment-free substrates. Each generation of sessile plants or animals uses the skeletons of their ancestors as substrate. The biofacies generally contains a large number of individual sessile societies. The texture of the biofacies is almost invariably massive and not bedded. This biofacies is usually particularly thick and massive where it occurs in littoral areas, but presumably it could exceptionally develop in the bathyal zone. A biofacies of this type is always composed exclusively of the products of a single biocoenosis and produces a complete material record of its existence, each period being represented.

(2) Vital-lipostrate biofacies (Fig. 263). Many different kinds of sediment occur in this biofacies, and the types of bedding are equally numerous, e.g. more or less perfect lenticular bedding, rippled bedding, cross-stratification and disconformities; these features indicate strong water agitation by currents, waves or tides. The oxygen concentration in the water is high. The bottom-biocoenoses consist of forms that are particularly well adapted to cope with abundant and sudden sedimentation of variable grain size: most of the organisms belonging to this biocoenosis can move through sediment, and they thus produce perturbed or shaped burrowing textures. Sessile species, for example balanids, polychaetes that build calcareous tubes, encrusting bryozoans, sponges or hydroid polyps, can only live on the shells of vagile species. During strong storms, parts of the biocoenosis can be covered by so much sediment that the organisms are killed in spite of their ability to escape through sediment; such sedimentation may be due to bottom-touching waves or to suddenly occurring or suddenly intensified currents. Other parts of the endobiontic population are frequently washed out and so destroyed. Losses caused in either of these ways are rapidly replaced by new settlements. The hard structural parts of the killed animals accumulate together with organogenic remnants imported from neighbouring biocoenoses and form extensive taphocoenoses. Similarly, many substances produced in the biocoenosis are carried outside its domain. Skeletons that consist of many individual parts are never embedded as a whole; the individual parts are scattered, often rolled and rounded, and many of them are completely destroyed. The material record furnished by the biofacies documents only a fraction of the events that have happened during the period of its existence because beds are lost and organogenic remnants are carried away.

Presumably this facies is not limited to the littoral zone; under special circumstances such biofacies also occupy parts of the bathyal, perhaps even of the abyssal zones (see Laughton 1959, who demonstrated the existence of ripple marks at a depth of 3000 m). If a biofacies with these

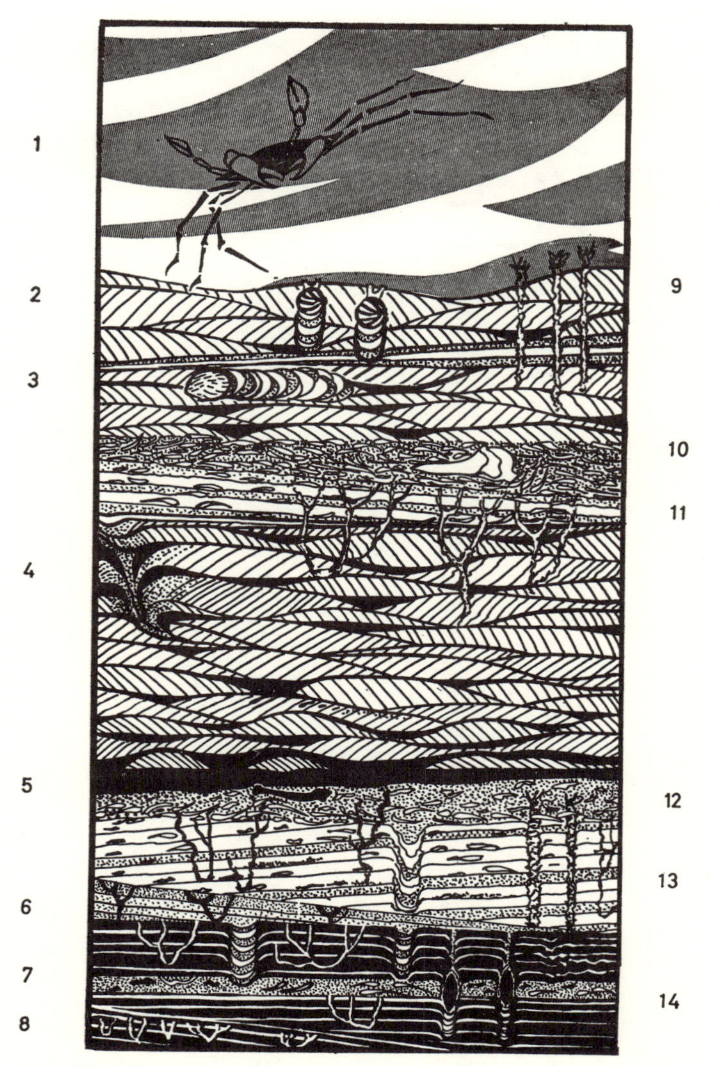

FIG. 263. Vital-lipostrate biofacies. Characterised by numerous, short-lived benthonic biocoenoses, by taphocoenoses and by disconformities. Note agitated, well-aerated water producing ripple marks in sandy sediment. Water and sea floor well populated. (1) swimming crustacean; (2) bivalves ascending to keep up with sedimentation, (3) *Echinocardium*, moving left and leaving cut-and-fill trail; layers of detritus and mud in ripple mark troughs; (4) escape trail of *Aphrodite* in ripple textured sand sediment; (5) bone of bird limb in layer of coarse, breccious sand covered by (black) mud layer; (6) escape trails of bivalves and burrows of worms; (7) otoliths; (8) worm burrows; (9) agglutinated polychaete tubes; (10) skull of marine mammal in layer of coarse breccia; (11) worm burrows, damaged in layer of coarse breccia; (12) agglutinated tubes, damaged in layer of coarse sand and breccia; (13) escape trail of benthonic gastropod; (14) bivalves in living position atop escape trails and surmounted by siphonal shafts

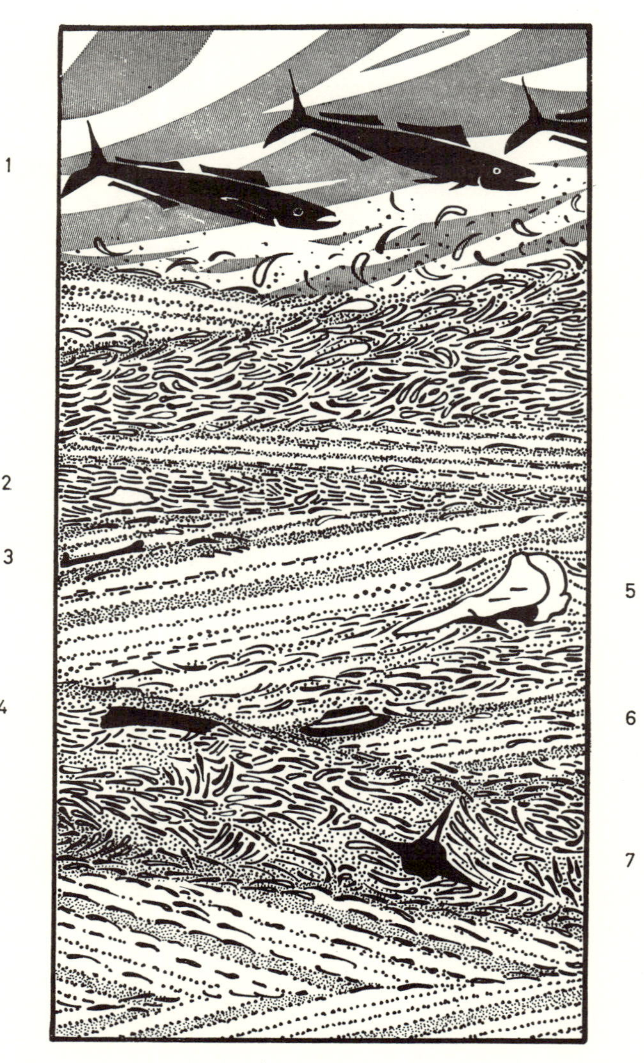

FIG. 264. Lethal-lipostrate biofacies. Characterised by abundant taphocoenoses and by disconformities due to considerable erosion; water is well aerated but continuous shifting of coarse material prevents establishment of a benthonic fauna and biocoenosis. What biogenic remains (heavy valves) can be found come from neighbouring benthonic biocoenoses. Note strongly agitated water transporting shell breccia; open water well populated. (1) swarm of fish (fast swimmers adapted to moving in currents); (2) clawed limb of crustacean; (3) limb bone of bird; (4) rib fragment of a Greenland whale; (5) dolphin skull; (6) peat pebble; (7) vertebra of marine mammal

characteristics contains complete, undamaged, composite skeletons alongside skeleton fragments the deposit must probably be a beach formation; criteria for this would be mummified carcasses, moulds of planktonic animals, desiccation cracks and flow marks.

(3) Lethal-lipostrate biofacies (Fig. 264). The disconformities typical of this biofacies already indicate temporary increases of water agitation. However, the continuous agitation is here such that only the coarsest grains are deposited together with organogenic hard parts. These parts come from surrounding bottom-biocoenoses of the open water, and they accumulate as taphocoenoses. No biocoenosis can form at the bottom because the constituent parts of the taphocoenosis are constantly moved and redeposited on a large scale. Living animals are rapidly killed in constantly shifting sediment, and their hard parts are incorporated into the taphocoenosis. Sediments being redeposited can grow to a thickness of several metres within hours. Cross- and ripple-bedding textures are pronounced; in pure taphocoenoses shells are occasionally stacked inside each other. These biofacies are not too extensive because zones of strong water agitation are usually laterally displaced, or periods of agitation do not last long or are followed by quieter periods in which a living community can develop. The material record of the biofacies reflects only a fraction of the events happening during its formation.

This biofacies occurs chiefly in littoral zones with strong tidal currents and frequent strong waves. Certain areas of the bathyal zone may possibly furnish similar conditions.

(4) Vital-pantostrate biofacies (Fig. 265). Complete beds of sediment occur in areas with little or no water agitation. Once deposited, the sediment remains in place, forming what Richter (1936) called 'calendar bedding'. Deposited as sediment are (1) products of the local bottom-biocoenosis such as faeces or skeletons, (2) organogenic remnants produced in biocoenoses of the open sea, again, possibly, faeces and skeletons, (3) inorganic products sinking from the top water layers and from the atmosphere. Sedimentation is light and the grain size small. The interstitial water of the sediment is not saturated with oxygen (gyttja). The local bottom-biocoenosis can exist through the entire period of formation of the biofacies. If a change in the fauna occurs this is due to biotic conditions. The number of the embedded species is large because it is supplemented by abundant addition of carcasses from the upper water layers. Composite skeletons remain more or less complete; there is no abrasion but solution marks may form. Fossil preservation is usually good. The material record of the biofacies is complete for the whole period of formation.

The degree of perturbation of the bedding by benthonic animals depends on the rate of sedimentation. The slower the sedimentation,

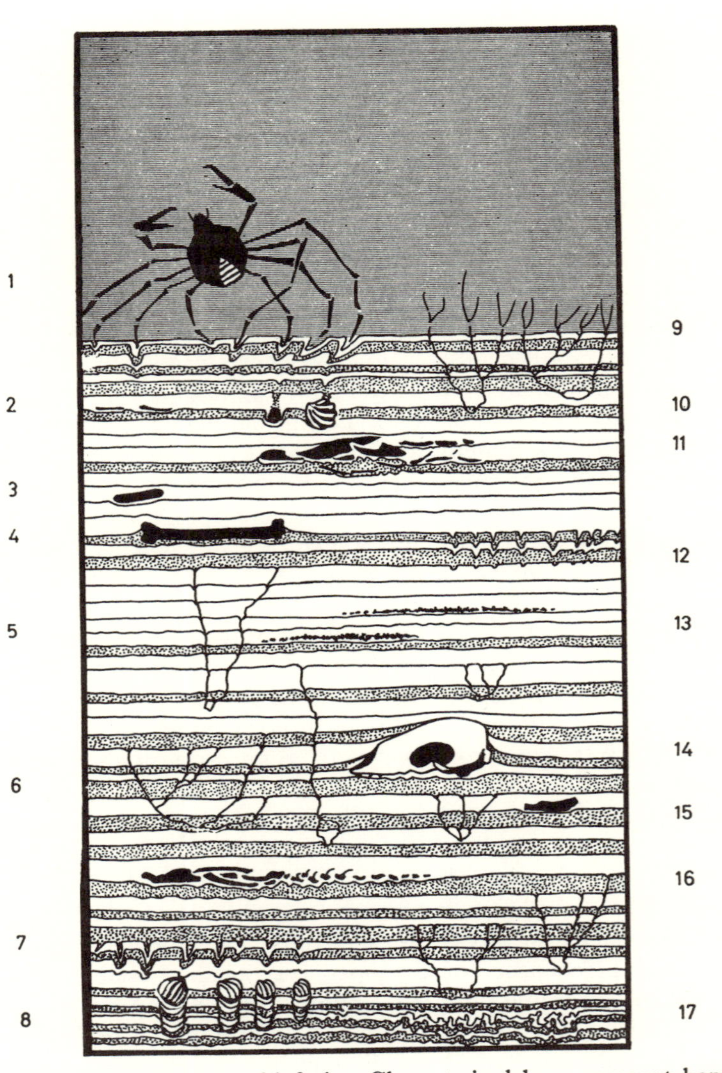

FIG. 265. Vital-pantostrate biofacies. Characterised by permanent benthonic biocoenoses, by taphocoenoses of nektonic and planktonic animals, and by complete, conformable bedding. Note quiet but adequately aerated water allowing benthonic life (walking brachyuran crustacean and polychaete worms in sediment). (1) sea spider, long-legged in adaptation for life in quiet water; (2) fragments of *Lepas* valves; (3) *Nautilus* shell; (4) limb bone of marine mammal, commonly first object to drop off a carcass drifting on the ocean surface; (5) and (6) dwelling structures of polychaete worms, prolonged upward with deposition of each new layer of sediment; (7) crustacean pacing trail, formed as in (1); (8) escape trails of bivalves, all simultaneously killed by temporarily anoxic water; (9) inhabited worm tubes; (10) bivalves in living position; (11) brachyuran crustacean; (12) crustacean pacing trail, formed as in (1); (13) brittle star skeleton, below skeleton of starfish; (14) skull of marine mammal; (15) coprolite; (16) skeleton of teleostean fish; (17) perturbed resting trace

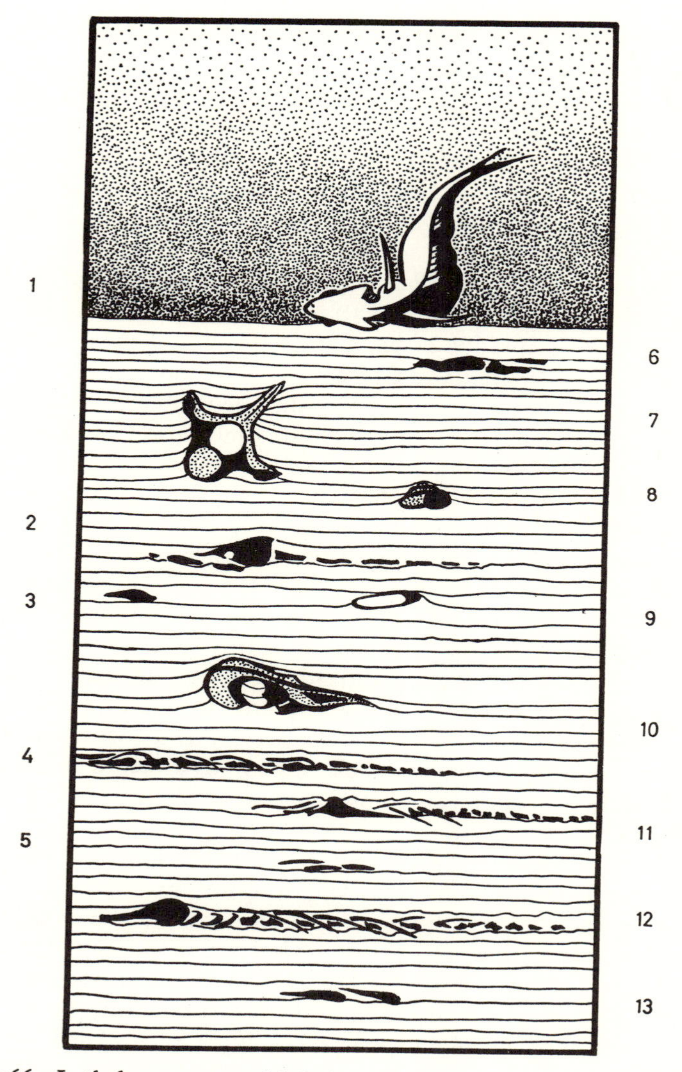

FIG. 266. Lethal-pantostrate biofacies. Characterised by taphocoenoses of nektonic and planktonic animals, and by complete, conformable bedding. Note quiet, anoxic water hostile to life. Multi-component skeletons of crustaceans, fishes and marine mammals lie undisturbed in their original position. (1) fish carcass having recently sunk to the sea floor, the abdominal cavity is still somewhat bloated by gas and thus not lying flat; (2) nektonic decapodan crustacean, spread out on bedding plane, abdominal portion lying on its side; (3) tooth of marine mammal; (4) skeleton of selachian fish; (5) scales of teleostean fish; (6) fragmented coprolite; (7) vertebra from proximal portion of marine mammal tail, commonly first part of drifting carcass to decay; (8) shell of *Janthina*, a prosobranch gastropod living as plankton by attaching itself to foam raft; (9) *Nautilus* shell (usually drift at the surface for 6 weeks after death, then sink); (10) bird skull; (11) teleostean fish skeleton; (12) skeleton of marine mammal; (13) fragments of *Lepas* valve

the more time the benthonic animals have to rework the deposits. As a result this biofacies appears in two forms; either the beds are clearly visible, with intercalated, equally clearly delimited, separate horizons of perturbed or shaped burrowing textures, or else, in spite of occasional fine laminations, which indicate beyond doubt the minimal rate of material addition, the bedding is obliterated by continuous and prolonged activity of benthonic animals.

The low rate of sedimentation and the absence of bottom currents which could rework the sediments indicate where such biofacies must develop; they form in wide areas of the bathyal and presumably also of the abyssal zone; exceptionally they can also develop in littoral areas. The presence of an isolated disconformity in a sequence of otherwise concordant beds is not to be taken as a contra-indication of this, or analogously of the fifth, biofacies.

(5) Lethal-pantostrate biofacies (Fig. 266). Under certain conditions of quiet stagnant water, beds can form in the absence of oxygen (sapropel) and therefore of life. No biocoenosis exists at the bottom. The sediments come from the upper water and from the atmosphere; sedimentation is light, the grain size fine. The sediments consist predominantly of organogenic substances. The taphocoenoses forming in this environment consist of complete or fragmented dead plants and animals belonging to constitutional types of the nekton and plankton. Hard structural parts, faeces and preservable organic substance are deposited with the sediment. The preservation of composite skeletons is good; there is no abrasion but solution marks occur. The biofacies as a whole produces a complete material record.

This biofacies usually develops in more or less enclosed basins without bottom circulation or in areas where upwelling causes an abundant supply of organic material. Such a situation can arise at any depth.

Transitions exist between these biofacies types except for the first, the 'vital-astrate biofacies', for which the conditions are extreme and whose constituent organisms are exclusive. Periodical or aperiodical alternations in time of any two or more types of biofacies occur frequently.

(*c*) BIOFACIES DOMAINS

The five marine biofacies types exist at the present day and are known from the geological past, sometimes as extremely large bodies of sediment. They transgress climatic boundaries and are independent of light, salinity or depth of the sea floor; the same holds for the constitutional types of preservable animals in them. To subdivide them, one must resort to the well-proven method of marine ecology, namely to classify physically well-delimited biotopes according to species and combinations of species found in them. To delimit a biotope, any available criteria can be used, grain size and shape, mineral composition and texture, chemistry and temperature, and the frequencies of autochthonous and allochthonous organisms. A biofacies of the second rank so characterised may have an areal extent of several square kilometres. As in ecological classification, subdivision should not be carried too far; trying to establish ever smaller sub-domains one ends up describing the accidental features of a single case rather than typical properties of a recognisable class. Thus, in contrast to the biofacies of the first rank which can occur throughout the world, their subdivisions, the biofacies of the second rank, are geographically restricted; they must be considered individually and not generally.

It is possible to classify biofacies by analogy to biocoenoses. Biocoenoses in separate locations are called isocoenoses if (1) they contain animals of the same constitutional types and if (2) the 'rosters of organisation' are also similar (i.e. similar niches are used by similar constitutional types). Isocoenoses are characterised by equivalent rosters of organisation but different species; in the ocean such differences usually depend on biotope differences in temperature, salinity or light. Where these physical features cannot be measured directly, the specific differences within a single constitutional type may be the only clue to such differences. This is almost invariably so in the case of fossil biocoenoses. One may therefore transfer the concept to biofacies and speak of 'isobiofacies'.

It is not difficult to subdivide a first-rank biofacies into second-rank domains and to establish boundaries between them, as long as a stratigraphic sequence contains reasonably preserved remains of several biocoenoses, each preserved *in situ*. Then a biofacies simply represents a single former bottom-biocoenosis, and the geographic boundaries of the biocoenosis become biofacies boundaries. More caution is needed in the normal case in which transportation has obliterated former ecological boundaries and brought in components from other biocoenoses. Remnants of nekton and plankton have no close ecological relation to the place where they are eventually embedded. They may have lived hundreds of nautical miles away. Desirable as such remnants may be as time indicators, they are useless as facies fossils. For this purpose all the attention must be concentrated on the remnants of benthonic organisms. To what degree then are their preserved hard parts suitable criteria for a subdivision into biofacies of the second rank? Fossils preserved in living position are the most useful, so that there is no doubt about possible transportation. However, hard parts are frequently transported before they are finally buried.

Several relevant observations have been made on transportation in the North Sea; lamellibranch valves, gastropod shells, *Dentalium* tubes, tests and parts of tests of echinoderms, skeletons of higher crustaceans, calcareous and agglutinated tubes of polychaete worms are actually transported over quite limited distances before they are destroyed; this is particularly so if numerous hard parts are transported together by bottom-touching waves and tidal currents. The remnants are then rapidly broken, abraded and finally comminuted into small splinters that become unrecognisable components of the sediment. The net drift is especially small in tidal currents which frequently carry the remnants back and forth. With increasing depth, tidal currents and waves usually become less effective transportation agents. Strong, unidirectional currents on the deep sea floor, however, may cause long-range transportation.

No direct measurements of the possible drifting distances of calcareous hard parts have been made in the North Sea or anywhere else; however, the distribution of hard parts of benthonic species with restricted habitats gives some indication about the range of transportation.

Venus gallina lives near the coast of Spiekeroog at a depth of at least 5 m with a population density of approximately 80 per square metre. The valves of this lamellibranch are rarely found on the beach of Spiekeroog, 3 km from their habitat; in a shell breccia 8 km east of their settlement off Spiekeroog, on the Minsener Oldeoog which lies right down the offshore current, hardly a shell of this species can be found. *Cyprina islandica* lives in Heligoland Bay to approximately 15 km south of Heligoland island. Only about 5 km south of this boundary, for example in the rich shell deposits at the mouth of the river Jade, only one or two fragments of *Cyprina* are found for each ton of *Cardium* shells. Plates of the tests of *Echinus esculentus*, common on the submarine cliffs of Heligoland, cannot be found anywhere on the sea floor outside the habitat of this sea urchin. The hard, calcareous tubes of *Dentalium* which is quite common in the waters off the Dutch coast hardly ever drift to the coast of Borkum island 25 km to the east. *Echinocardium cordatum* can barely withstand a transportation of 2 km along the sea floor, and the spines of *Psammechinus miliaris* drift no further than 100 m in Jade Bay.

From these examples one can conclude that hard parts do not travel far in the currents of the floor of the shallow sea but are rapidly destroyed and do not get far beyond the boundaries of their habitats. Caspers (1938 and 1950a) reached the same conclusions about the Heligoland channel and the oyster-bank off Heligoland. He thought that the composition of shell deposits faithfully reflects that of living communities nearby, and that a study of shell fragments alone would have had the same results as his study of the living animals and would have led to the establishment of the same faunal domains. Only under special conditions, therefore, should it be necessary for the demarcation of a second-rank biofacies domain to distinguish whether large fossils (generally bivalve shells) are in a living position or not.

Unfortunately, this rule holds only for hard parts that are large and heavy enough to be pushed and rolled over the ground during transportation. Very small and light hard parts, on the other hand, are destroyed much more slowly because currents and waves easily lift them from the ground and carry them in suspension. They therefore drift far beyond the boundaries of the habitat to which they belong and thus overlap facies boundaries. Several examples of this are known from the North Sea. The sea urchin *Echinocardium cordatum* occurs in Heligoland Bay near Wangerooge-West at depths of about 6 m and below. Yet its fine spines are found, frequently in thick, closely packed layers, in sediments at the head of Jade Bay at a distance of about 50 km. Foraminifera tests stay intact after similar or even longer drifts; they are frequently found in the sediments of tidal flats, far from the open sea habitats of the living Foraminifera. Schmidt (1953) discussed the whole problem of what proportion of a shell accumulation may represent nearby living communities and how much of it may be drifted material and Dr Gotthard Richter, Wilhelmshaven, is involved in similar investigations for the North Sea (see also G. Richter 1964, 1965). Lamellibranch fry with valves less than 3 mm long also drift far. So do the bristles of large polychaetes (for example *Aphrodite*, *Nereis virens*, or *Arenicola*).

The next question is whether biofacies of the second rank which may have been established by means of large hard parts of benthonic organisms, can meaningfully be subdivided into domains of the third rank. From Recent sea floors we know examples of closed biocoenoses developed within very small areas. An individual lamellibranch valve lying on the sea floor can become the substrate for a community of sessile plants and animals in which every niche is occupied by a certain constitutional type. The same is true of stones, peat pebbles or twigs of dead wood on the sea floor, each with its specific fauna. In fact, any carcass on the sea floor necessarily attracts a specific succession of communities during the short period of its existence. Similarly, any mud-filled basin in a sandy area carries a closed and specific fauna.

Any such biocoenosis may be extremely well preserved if it is composed of sessile species provided with hard parts; in fossil form it thus becomes a biofacies of the third rank, representing both a small area and a short time. However, the situation is quite different for the small objects that are studied by micropalaeontologists: radiolarians, diatoms, foraminifers, sponge spicules, ostracods, conodonts, mouth and epidermal teeth, scales, small vertebrae and many others. None of these are indicative of the specific conditions in the particular sediment in which they are now found. The reasons are several. Many of the organisms were part of the plankton and nekton, others vagile benthos, yet others were sessile but only produced small, disconnected hard parts. Furthermore, no matter what the living habits of the producers, all these parts are small and are easily scattered by very gentle water agitation. For all these reasons microfossils are widely distributed and there seems to be little point in using these fossils for the classification of 'micro-biofacies'. Even the sessile communities of small organisms which are indicative of local conditions are not overly helpful for such a classification because they become insignificant in a wider environment.

Microstratigraphical investigation of these same microscopically small objects, on the other hand, is all the more rewarding. Its purpose is not to subdivide space into possibly quite meaningless subunits but to subdivide time into precisely defined short periods with clearly distinct biotic and abiotic conditions.

Because first-rank biofacies can only be subdivided by variations in the distribution of species belonging to bottom-biocoenoses, the fifth or lethal-pantostrate biofacies which contains no fossils from this biocoenosis cannot be further subdivided. The third or lethal-lipostrate biofacies is indivisible for another reason; the fossils, although mostly derived from bottom-biocoenoses, are so intensely reworked and mixed that a subdivision according to species distribution is meaningless.

III. Biofacies domains of Heligoland Bay

The preceding definitions of biocoenosis and biofacies of the first and second ranks will now be applied to delimit biofacies domains in the best-known part of the southern North Sea, Heligoland Bay, including its southern extension, Jade Bay, an area 110 km from north to south and 65 km wide (Fig. 272). This is the only area for which a map of biofacies domains can be drawn with some confidence and for which detailed sedimentological and ecological data are available. The biofacies map is not a simple counterpart of the existing maps of sediment distribution (fishery maps, detailed charts and the *Neue Bodenkarte der südlichen Nordsee* by Jarke 1956), nor ought it to be, because sediment boundaries cannot be quite identical with the boundaries of biofacies domains. Biofacies depend not only on the material of the bottom sediment and the ratios of different grain sizes in it but also on sediment texture, organogenic remnants in the sediment and traces of life activities. Basic work in this field has been done for the last several years by

Reineck (1963), Reineck, Gutmann & Hertweck (1967), Reineck, Dörjes, Gadow & Hertweck (1968), Müller, Reineck & Staesche (1968), Dörjes, Gadow, Reineck & Singh (1969), and Hertweck & Reineck (1969).

Since all of the southern North Sea belongs to the littoral zone, the formation of deposits from shell and skeletal fragments plays a particularly important role. There is hardly a sediment body that is free from such components. A separate section dealing with the formation of shell deposits in the southern North Sea in general will therefore precede the description of the biofacies of Heligoland Bay.

(*a*) SHELL DEPOSITS AND THEIR FORMATION IN THE NORTH SEA

Shell deposits (or coquinas or lumachelles) are concentrations or exclusive accumulations of complete skeletons or of their parts brought together by some agent of transportation. Shell deposits may be due to water movement or to other causes. The components are mostly valves of lamellibranchs but may also be calcareous, chitinous or siliceous skeletons of other organisms such as balanid, hydrobian, echinoderm, brachiopod, or, less likely, bone or teeth coquinas. According to Richter (personal communication, 1956) the term originally implied lamellibranch shells but he thinks that to exclude other skeletons only deprives one of a necessary, more general term. The importance of shell deposits to humans is reflected in the name of a headland at the mouth of the river Jade, 'Schillig', which is descriptive of the large, sandspit-like accumulations of valves of which this headland consists and to which new shells are constantly added. They come from the surrounding tidal flats, especially from the northwest; during the past centuries the local people who live nearby, behind the dikes, have used these deposits to supply lime kilns and as road metal (see also Wasmund 1929).

The components of a shell deposit frequently lie so closely together that they touch, and a texture results. A single skeletal part such as a lamellibranch valve does not constitute a shell deposit, neither does a dense array of lamellibranchs in living position, dead or alive. Rather, the term implies by definition that the skeletal material must have been transported, short though the distance may have been. Thus shell deposits are composed of skeletal parts whose present location and composition are governed by the laws of sedimentation and not those of life. Krause (1950) cited Johansen (1901) who divided the shell deposits into 'shell mounds' and 'shell beds'. He considers the shell mounds as resulting from transportation, the shell beds as local accumulations of shells through time. According to our definition Johansen's shell mounds are shell deposits; his shell beds are not, because they are due to biological, not physical factors.

The relative proportions of constituent components of a shell deposit depend on their strength; and the strength of skeletons is strongly influenced by their functions. Thus, if the composition of a shell deposit is examined for the proportions of hard parts from planktonic, nektonic and benthonic animals, it is found that the benthos are dominant. Their skeletons are generally strong. The proportion of hard parts from nektonic organisms (teeth and vertebrae) is much smaller, and that from plankton is almost negligible unless one includes the calcareous and chitinous skeletons of very small organisms.

Vagile skeleton-bearing epibionts (molluscs, crustaceans, or echinoderms) are eaten by predators, e.g. *Rhombus*, *Hippoglossus*, *Molva*, *Octopus*, *Palinurus* and *Asterias* crush the skeletons of their prey but reject them before swallowing. Shells that pass through the intestines of other predators are, with few exceptions, mechanically and chemically comminuted and discarded as skeletal fragments. Predatory gastropods (e.g. *Lunatia*, *Murex* and *Nucella*) kill gastropods and lamellibranchs by boring through their shells. The most frequent cause of death of epibionts in the shallow sea, however, is a sudden covering by sediment (e.g. echinoderms, byssus-bearing and sessile lamellibranchs, see also Schäfer 1956a).

Vagile, skeleton-bearing endobionts (molluscs, crustaceans or echinoderms) are equally exposed to sedimentation. Thick covers of sediment caused by exceptional meteorological events frequently cause their death. However, they are equally sensitive to being washed out by the agitation of bottom-touching waves. Currents set up by such waves carry the skeletons both of dead animals and whole living animals which are then often eaten by enemies before they are able to re-enter the sediment. In the intertidal sea, washed-out organisms may also die of exposure to the sun and air. Intermittent but recurring tidal currents are less fatal for benthonic animals because the velocity is quite constant from one tide to the next. Animals living under their influence are adapted to them and do not suffer the catastrophical effects produced by sudden wave action. The water movement caused by wind and waves in the shallow sea is responsible not only for the death of many benthonic animals but also for the accumulation of the hard parts that remain after their death. In the shallowest zones of the shallow sea, however, transport is augmented by tidal currents so that they affect and modify the process of accumulation. Measurements of currents in Heligoland Bay during storm-raised tides have provided a survey of the velocity of the currents at various levels above the sea bottom (Gienapp & Tomczak 1968). The velocity of currents just above the sea floor is less than 40 cm/sec. Since both waves and tidal currents affect only the shallowest part even of a shallow sea, shell deposits can be considered as typical features of very shallow water.

R. Richter (1932, p. 370) and R. & E. Richter (1936, p. 235) defined a 'critical plane' as the 'effective lower boundary of agitated water'; the critical plane intercepts the sea floor along the 'critical line'. The author would consider the 100 m depth contour as the approximate critical line in the open shallow sea (see also Schäfer 1941a).

The paths of transportation and the places of deposition are best known in the intertidal sea. Walking over the open tidal flats one hardly sees any lamellibranch valves although they live there in abundance (densities of population are 150–3000 per square metre for *Cardium edule*, 110–200 per square metre for *Scrobicularia plana*, 15 per square metre for *Mya arenaria*; see also Kristensen 1957 for *Cardium*). Only where tidal flats are being eroded does one find individual lamellibranch valves, side by side with complete animals still sitting in the sediment. These are the places from where shells are transported. With each tide, bottom-touching waves reach the surface of the tidal flat, always affecting it more strongly than permanently immersed sea floors; and in the Jade area the wind force is 5 or more on 180 days of the year. Thus every loose valve is immediately transported in the direction of the momentary water flow. Valves shift continually until they either reach a coast where they accumulate with other hard parts along swash marks; or they drop into a tidal channel which intercepts their coastward journey. Most valves are moved from west to east under the influence of prevalent westerly winds. Shell accumulations form preferentially on swash marks on east coasts open to the west, and frequently produce eastward-pointing shell spits. The shell material that gets into the tidal channels either remains there or is transported down-channel into wider and deeper zones and basins and is deposited there.

During transportation many valves are mechanically broken, ground and polished or destroyed or weakened by calcium-carbonate-boring algae and by chemical reactions (for details about grinding and stages of the postmortal destruction of shells of *Cardium edule* see Hollmann 1968). Valves have a better chance of being preserved if they are quickly covered by sediment. Those ending up in the swash marks of the coasts are preserved only if there is sufficient sediment to cover the shell fragments soon after their arrival. The formation of beds formed during storm tides has frequently been described; such beds are usually deposited in the marshy zone above the mean high-water line along coasts that are particularly exposed to wind (Fig. 267; Häntzschel 1936b, see also Schütte 1929; Schwarz 1933; Reineck 1956a, p. 279). Generally the shells are rapidly destroyed on open beaches. Always the same quantity of newly deposited material is found on such a beach, the older shells being completely destroyed within a few weeks or months; thus the shells never accumulate into great piles even where they keep being deposited all year round.

FIG. 267. Origin of storm tide beds. A wave breaks against the low cliff of a salt marsh and throws sediment and shells on the marsh, raising its elevation. The stronger the surf, the more shells in a storm tide bed. (*NN*) standard sea level of topographic maps; (*HTHW*) maximum high tide; (*MTHW*) mean high tide; (*MTNW*) mean low tide; (*MSPTNW*) mean spring low tide, 2·18 m below standard sea level, datum level for sea charts.

In tidal channels the shells are also eventually destroyed unless they are preserved where the channel bed shifts sideways and the shells are covered and preserved under the obliquely stratified, newly deposited beds of the inside bank of a meandering channel (Fig. 268, van Straaten 1950 and 1954, Reineck 1958c).

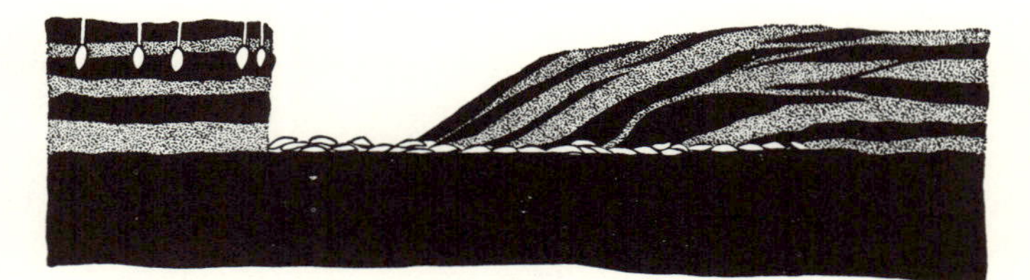

FIG. 268. Shifting tidal channel. Erosion on steep concave bank of meander, deposition on sloping convex bank. Shells deposited on channel floor (after transportation along it) incorporated as continuous bed in new sediments

The shells that are being transported across open tidal flats can be moved downward into the sediment by organisms and thus be preserved. The chief agent of this type is *Arenicola marina*; together with the sand running down into the feeding funnels of these worms, hard parts of gastropods, lamellibranchs and crustaceans also subside. However, being too large, these objects are not ingested by the worm but simply accumulate. The activity of thousands of densely settled worms thus can create a continuous, stratiform shell deposit, sometimes spreading over many square kilometres; it lies at a depth of approximately 30 cm, at the level of the bends of all the U-shaped burrows (Fig. 269, van Straaten 1952 and 1956). Densely packed shells of *Hydrobia ulvae* are especially frequent constituents of these worm-made shell beds.

Very large concentrations of shells occur at the mouths of the principal tidal channels flowing into the open sea, especially where large rivers flow into wide bays (Reineck 1958b). Here the calcareous hard parts collect at depths of 10 to 25 m; some have been transported down

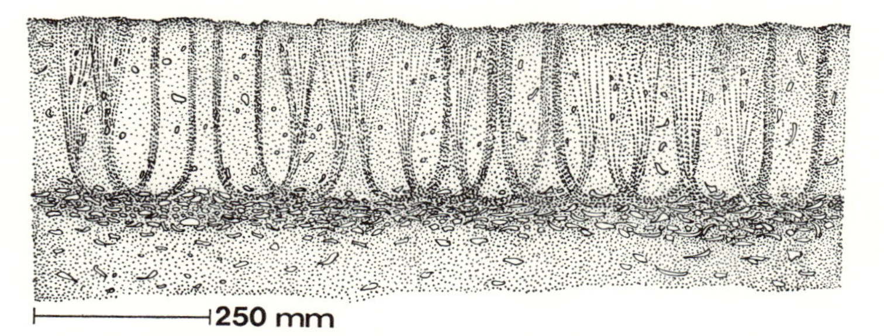

FIG. 269. Accumulation of continuous bed of shells at U-bend level of dense settlement of *Arenicola*

the tidal channels. Others have dropped from the open tidal flats into the tidal channels, again others have been trapped in the deep, washed-out pools and some have collected from the open sea, carried by waves and wind-driven currents. The channel floors are separated by sand flats and moving sandbanks and usually lie deeper than their surroundings, often even deeper than the floor of the open sea beyond. For this reason alone, these places must tend to trap the shells; in fact they are the most extensive and the purest accumulations of shells in the southern North Sea area.

For decades, these shell deposits have been dredged intensively (Jüngst 1942, Wasmund 1929). One dredging enterprise well known to the author hauls 70 000 cubic metres a year consisting mostly of *Cardium edule* shells from the mouths of tidal channels between Borkum island and the mouth of the river Jade (Fig. 270). When large-scale shifts of islands or sandbanks occur such shell deposits become immobilised and preserved. According to the dredging master the deposits are several metres thick (e.g. 4 m thick at one locality at a depth of 15 to 20 m). Schütte (1935) who worked on 630 samples obtained with a pumping dredge from the area around the mouth of the river Jade arrived at the same results (see also Lüders 1929). Measurements made by Lüders & Trusheim (1929) in the dredged shipping channel through Jade Bay show that coarse shell deposits begin to move at a current of 60 to 65 cm per second. The individual valves are usually deposited at random. Newly deposited valves fit themselves into the nooks and crannies of the rough surface of closely packed valves wherever they can. The force of the current can jam the valves against each other. It is not known whether shells of any shell deposits of Jade Bay have a preferred orientation. Krause (1950) has made an exact quantitative and qualitative investigation

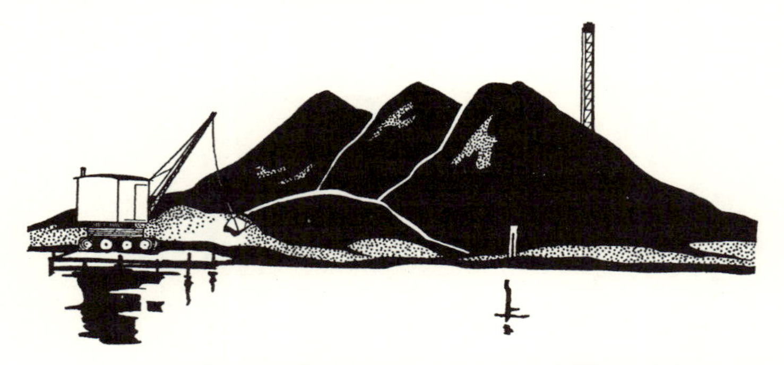

FIG. 270. Piles of shells at Vareler Hafen, dredged from outer Jade Bay. Almost exclusively valves of *Cardium edule* (after photograph by G. Richter)

of the shell deposits in the narrows between the islands Juist and Norderney. He stated that the proportion by weight of finest grain shell breccias to total shell deposit clearly indicates how much of the shell material has come from the sea; for the shells from the open sea are much more comminuted into breccia than those from the intertidal sea. At the time of his investigation the shell deposits included valves of 42 lamellibranch species, shells of 14 gastropod species, agglutinated tubes of 3 species of polychaetes, remnants of 3 echinoderm species and various components of other organisms with hard parts. Fig. 271 shows the transportation paths of the shells in the narrows between Juist and Norderney.

Endobionts with hard parts can be washed out not only in intertidal waters but also in the open shallow sea. Their shells are then transported with those of epibionts. Only bottom-touching waves create temporary currents in these waters, with speeds of 65 cm per second or more, capable of moving these shells. The tidal currents of the southern North Sea at depths below 30 or 40 m are hardly strong enough to transport shells. Since the bottom of the open shallow sea of that area is more or less planar with hardly any depressions (with the single exception of the

Heligoland channel) no shells can be trapped and accumulated. Hard parts are thus abundant and are embedded more or less closely together into the sandy or muddy sediments but are generally not accumulated (Wasmund 1926). Some are either undamaged or little rounded and have travelled only a short distance before being buried; others are reduced to fragments or splinters. It is possible that shells are locally trapped between glacial boulders which make up certain areas of the

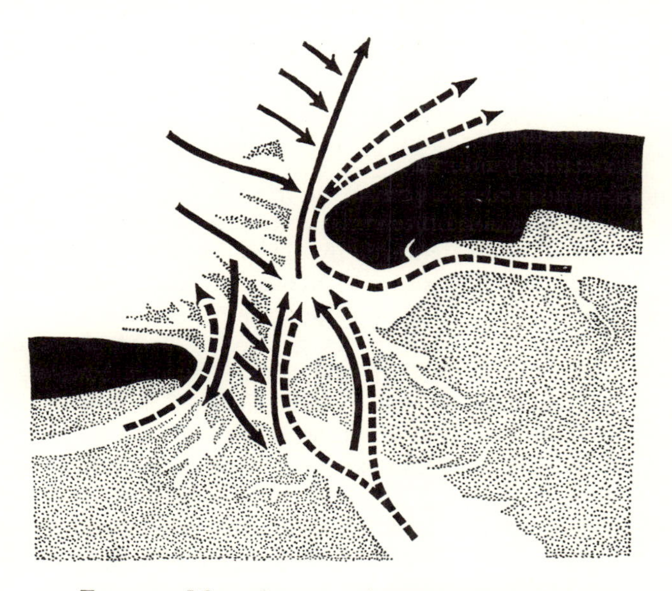

FIG. 271. Map of narrows between Juist (left) and Norderney islands. (Dotted area) tidal flats; (broken arrows) seaward transportation of shells from tidal flats; (black arrows) landward transportation of shells from sea floor (after Krause, 1950)

North Sea floor and thus form small, pure shell deposits. Since bottom-touching waves are not restricted to a certain direction the shells transported by them also may move in different directions. Once a valve has been lifted by a deep wave and transported over some distance it will then presumably be moved many times in different directions and describe a random zigzag course. Heincke (1894) thought that lines between the points where valves are found and those where the species live could indicate mean travelling paths; but this was disproved by Caspers (1938). The separately deposited valves of the open sea floor of the shallow sea seem, as a rule, to come to rest convex side up; they are rarely stacked but they do form pavements which may cover a surface almost completely. Distinct preferred orientation is observed of gastropod shells, lamellibranch valves and other hard parts of suitable shape (*Ensis*, *Solen*, *Phaxas*, *Petricola*, *Turritella*, *Scala*, stems of hydroid polyps, and agglutinated worm tubes; see also Seilacher 1959a). Because thick shell deposits free of matrix rarely occur in the open sea, random orientation of the type occurring in shell deposits at the mouth of tidal channels, the imbricated orientation typical of the bottoms of tidal channels, and the vertical orientation of the surf-produced swash marks are all equally unlikely on the open sea floor (see also Wasmund 1926).

(*b*) THE BOUNDARIES BETWEEN THE BIOFACIES DOMAINS OF HELIGOLAND BAY

The principal source for the map of the biofacies of Heligoland Bay (Fig. 272) are the investigations of the bottom sediments by the Deutsches Hydrographisches Institut; studies by the Biologische Anstalt Heligoland have established the composition of benthonic communities not only in Heligoland Bay but also in other large areas of the southern North Sea. Their studies have also shown the decisive role of the sediment in faunal composition and thus for the characterisation of communities. Remane (1940) established the general distinction between coenoses living on sand, soft and hard sea floor, and plants. Jarke (1956) refined the classification of the bottom sediments of the southern North Sea (see also Pratje 1931).

Maps showing merely the distribution of grain sizes are insufficient to form a picture of the biofacies of recent marine deposits. To obtain additional information about the texture it is necessary to sample marine sediments in undisturbed sequences. Reineck at the Research Institute Senckenberg, Wilhelmshaven (1957, 1958b, c, d, 1963), developed a box sampler and a hardening method. Although sediment texture, mutual position of hard parts, and living and dwelling traces of endobiontic animals can now be completely studied not enough of this kind of work has yet been done. The biofacies map as represented in Fig. 272 will certainly be improved in the near future.

Of the five first-rank biofacies only the second or vital-lipostrate biofacies, the third or lethal-lipostrate biofacies and the first or vital-astrate biofacies are represented in the southern North Sea; the last two are considerably less widespread than the first. Both types of biofacies with complete beds, the vital-pantostrate and lethal-pantostrate, are absent because in a shallow, open and stormy shelf sea sedimentation free of disconformities cannot exist. The first-rank biofacies that occur in Heligoland Bay will first be discussed and a tentative subdivision into second-rank biofacies, where this is possible, will follow.

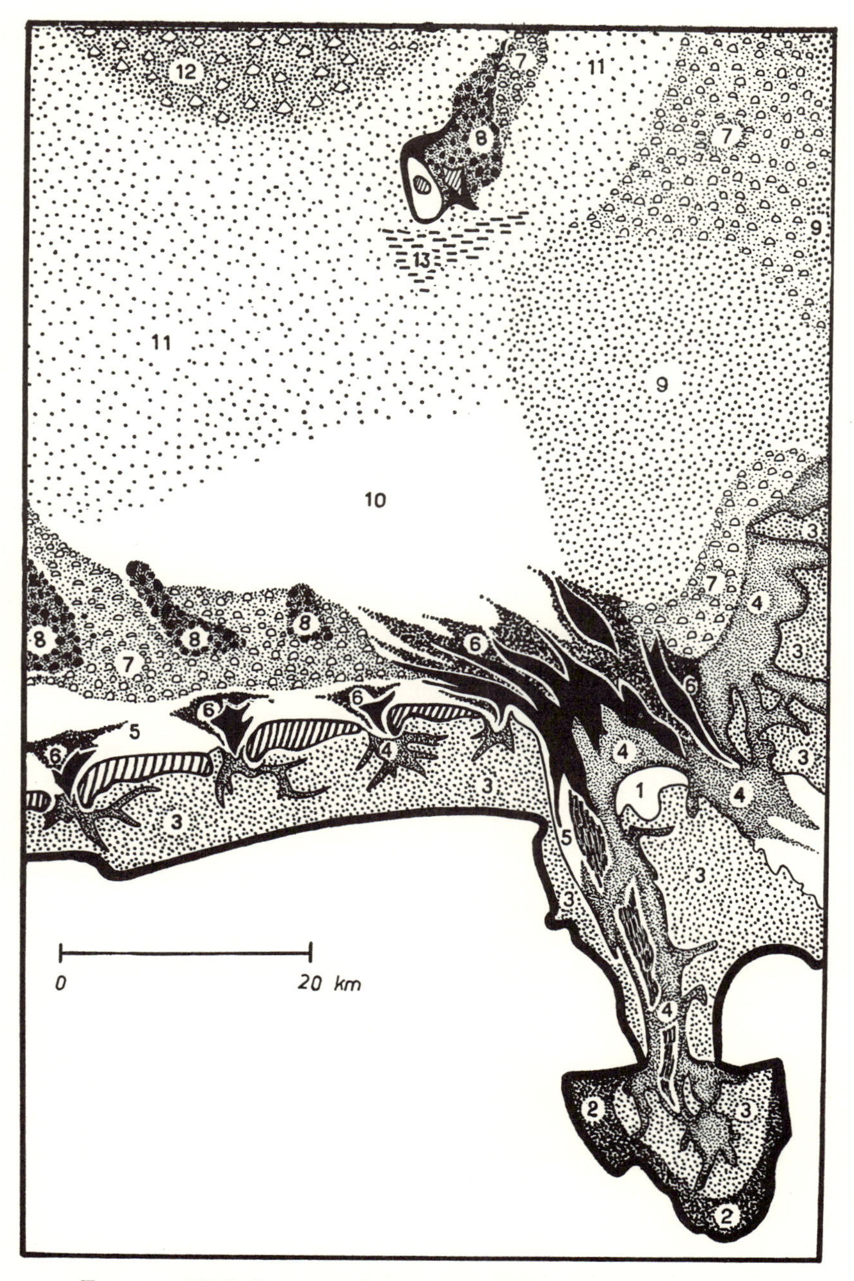

FIG. 272. Biofacies map of part of Heligoland Bay, including inner and outer Jade Bay, East Frisian Islands, Heligoland. (Black, outer Jade Bay) lethal-lipostrate biofacies; (vertical hatching, inner Jade Bay) vital-astrate biofacies; (diagonal hatching, islands); numbers (1) to (13) biofacies of second rank, numbered as in text

(1) Vital-lipostrate biofacies. This first-rank biofacies comprises almost the entire map area. It includes the eulittoral intertidal zone, the sublittoral shallow sea below the level of the tidal flats and extends beyond the map area to depths from 50 to 400 m. All these littoral depth zones are under the more or less continuous influence of tidal currents and bottom-touching waves, especially the map area which does not extend below depths of 60 m and includes only the uppermost portion of the eulittoral zone. Thus the sequences of beds show more or less pronounced loss of beds by disconformities; they are lipostrate, and are well aerated and therefore favourable for life, or vital. Even in the supralittoral zone where the ground is flooded only during unusually high tides, conditions are similar, the substrate is favourable for life, and disconformities do occur.

It is relatively easy to subdivide these littoral zones, all belonging to the same first-rank biofacies, into several biofacies of lower rank. Many sub-areas are clearly separable biotopes in which not only the grain sizes of the sediments but also textures and animal populations are reliably known. Each of these biotopes also has distinctive features that influence the biocoenosis, such as salinity, light and annual water temperatures. In Heligoland Bay the boundaries of depth zones do not always coincide with those of sediment types.

(2) Lethal-lipostrate biofacies. This biofacies occurs in the map area wherever the tidal currents are so confined in channels and narrows between islands that the current velocities become high, or where less confined tidal currents are modified by strong surf and wave action. Several combined factors prevent the settlement of an internal fauna: the sediments, coarse or medium sands and accumulated shells and shell fragments are frequently shifted and redeposited; the rate of sedimentation is frequently high and may be up to 2 m during a single tide; there may be extensive and deep erosion, and frequent displacement of shells or shell fragments. Vagile benthonic animals living at the surface of the sediment also occur only as rare stray individuals. The areas are hostile to life, or lethal, but certainly not because of lack of oxygen. Sediment textures show many disconformities, they are lipostrate, and giant ripples and sand waves occur. In the map area the lethal-lipostrate biofacies contains the richest and purest shell deposits, or taphocoenoses. The shells come from the lamellibranch settlements of the intertidal sea and from nearby portions of the vital-lipostrate biofacies seaward of the chain of the East Frisian Islands. The lethal-lipostrate biofacies is developed in the surf zones near the mouths of the rivers Ems, Jade and Weser, and, locally, in the rock debris below the cliffs of Heligoland off the 'Dune of Heligoland', where the mechanical forces are too destructive to allow organisms to settle permanently.

FIG. 273. 'Sand corals' of *Sabellaria spinulosa*, growing on shell bed, and later generations on abandoned *Sabellaria* structures. This vital-astrate biofacies covered by coarse sands and another shell deposit. *Syngnathus* fish adapted for fast currents. Composite drawing to summarise results of grab sampling over several years

These two varieties of the lethal-lipostrate biofacies, that near the East Frisian Islands and the other near Heligoland, constitute two separate second-rank biofacies.

(3) Vital-astrate biofacies. This first-rank biofacies would not exist in the North Sea, were it not for the polychaete worm *Sabellaria spinulosa*. It lives with an accompanying fauna on the bottom of the widest tidal channels with the most powerful currents. The reef facies produced by the settlements of these animals can be over a metre thick and consists of cemented sand tubes, hundreds of which are mutually joined into bundles (Fig. 273). They can extend over areas of several square kilometres. Being restricted to areas with powerful currents these reefs are almost free of sediment and of organogenic remnants derived from other biocoenoses.

The tree-like *Lophohelia* corals growing on stony ground and submarine cliffs in the northern and western North Sea and at the slope of the Continental shelf toward the Atlantic belong to the same vital-astrate biofacies (Gislen 1930). These coral banks extend beyond the North Sea along the Atlantic coasts of continental Europe, Scotland, and of the Shetland and Orkney Islands (Fig. 274). Their settlements roughly follow the 200 m contour, but at some points they descend to depths of 600 m and at others they ascend above the 100 m contour (Pratje 1924).

FIG. 274. Distribution of *Lophohelia* reefs (dotted areas) along Atlantic slope of the Continental shelf (after Pratje, 1924)

These reefs are built mainly by the madreporarians *Lophohelia prolifera* and *Amphihelia oculata* (Fig. 275); they produce a dense entanglement of hard and quite tough branches. The sessile accompanying fauna consists of the hydroid polyps *Stylaster gemmascens*, and *S. grandiflora*, *Primnoa resedaeformis* and *Paramuricea placomus*, all of which participate in the development of the reefs (Broch 1924, observations in the Trondhjemsfjord). The reefs are also settling substrates for other sessile organisms, mainly actinians, hydroid polyps and sponges, and hiding-places for many vagile organisms; as far as all these have hard parts they

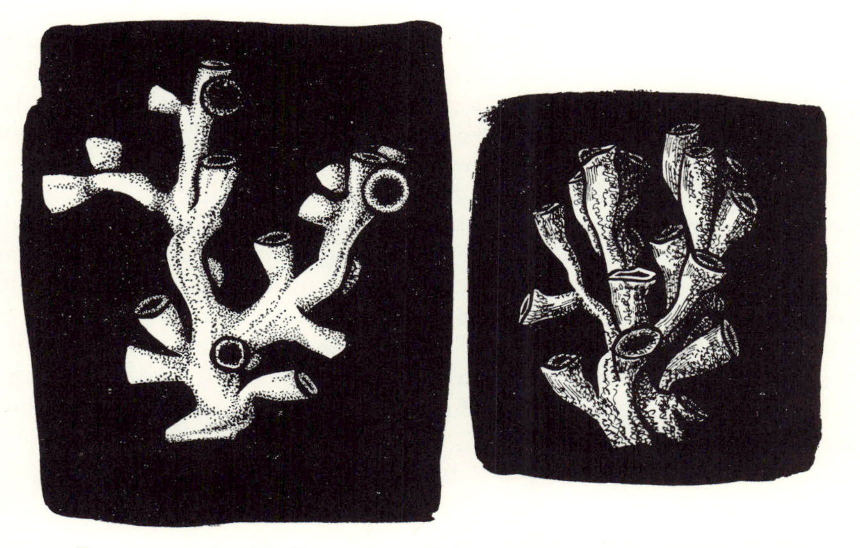

FIG. 275. *Amphihelia oculata* (left), *Lophohelia prolifera* (right). (Actual size)

contribute to the mass of the reef, generally in dispersed deposits that do not alter the non-bedded character of the reef.

In comparison to the other two first-rank biofacies, the vital-lipostrate biofacies occupies by far the largest areas on the map (Fig. 272); within the map area it can be divided into thirteen second-rank biofacies, each with a well-defined biotope and characteristic fossils. Unfortunately the sedimentary textures of some of these second-rank domains are not completely known; therefore their boundaries are uncertain and the descriptions of their textures are tentative. Listed as characteristic features are only members of the macrofauna that either have preservable hard parts or that produce traces which can mark the texture of the sediment. Wherever feasible, a list of characteristic features will be given at the end of the discussion of an individual biofacies. Such a feature could either be an organism, a trace, or a sedimentary texture; the most explicit of the possible connections is that between organism and sediment texture. Not every biofacies, however, can be characterised unambiguously by a single or a few features, and in these cases the list of features is long. If this is the case the biofacies should not be named after any arbitary organogenic or inorganic characteristics but simply after its geographical location.

Biofacies 1

This biofacies lies above the mean high-water line and extends up to 60 cm above it; only during spring tides is it flooded by sea water. Heavy rain lowers the salt concentration considerably. Sand drifts during periods of dry weather. *Salicornia herbacea* growth partly covers the area. Silt and fine shell beds are only occasionally interrupted by thin mud layers (diatom covers); there is dune bedding.

Inhabitants are the beetles *Bledius arenarius* living in 6 cm deep dwelling burrows; they are hunted by species of the genus *Dychirius*; occasionally there are larvae of flies, *Scatella subguttata*, *Hydrophorus praecox*. Bivalves are washed in from the surrounding tidal flats and from the region just below the beach, mainly *Mya arenaria*, fewer *Cardium edule*, *Macoma baltica*, *Mactra corallina cinerea*; occasionally they form shell pavements. Mummified carcasses of vertebrates and individual skeletal parts can be found, bird feathers and accumulations of vertebrae of fish.

Characteristic features: dune bedding and shell beds.

Biofacies 2

This biofacies extends downward from the mean high-water line and covers roughly the upper half of the intertidal area where it is sheltered from strong surf and where sedimentation is quiet. There are mud and thin silt layers, the bedding is typical of intertidal zones and also shows cross-bedding.

Burrowing textures can be found and individual burrows of worms, *Nereis diversicolor*, *N. pelagica*, *Nephthys hombergi*, *Scolecolepis squamata*, *Heteromastus filiformis*, *Polydora ciliata* and *Phylodoce maculata*, U-bend. shaped burrows with permanent and with continuously shifted U-bend. There are locomotory trails of lamellibranchs, walking traces of *Carcinides maenas* and *Crangon vulgaris*, burrowing textures of *Hydrobia ulvae*, *Gobius microps*, *Pleuronectes platessa*; *Scrobicularia plana*, *Macoma baltica*, *Cardium edule* and an occasional *Mya arenaria*, *Littorina littorea*, or *Corophium volutator*. The bottoms of tidal channels are covered with shell beds.

Characteristic features: finely layered muds, *Scrobicularia* and U-shaped burrows of *Corophium*.

Biofacies 3

This includes the lower intertidal area down to the mean low-water line; it is exposed to surf, and the tidal currents are strong. There is mostly silt but also mud layers consisting of faecal material, mud pebbles, rolled peat fragments, layers of peat debris and of *Echinocardium* spines. Lenticular bedding and slump structures occur, bedding planes show small or large-scale ripple marks.

Burrowing textures are produced by *Scoloplos armiger*, *Nereis pelagica*, *Scolecolepis squamata*, *Harmothoe sarsi* and *Eteone longa*, agglutinated tubes by *Lanice conchilega*, *Pectinaria coreni*, *Magelona papillicornis*, *Pygospio elegans*, *Ampharete grubei*, vertical escape trails by lamellibranchs, U-shaped burrows by *Arenicola marina*; *Cardium edule*, *Hydrobia ulvae*, *Littorina littorea*, *Maçoma baltica*, *Mya arenaria*, shell beds and shell breccia (mainly on bottoms and edges of tidal channels, below settlements of *Arenicola*) can be found.

Characteristic features: silt in lenticular bedding, bedding planes with ripple marks, *Mya arenaria* and *Arenicola marina*.

Biofacies 4

This biofacies covers a zone below the tidal flats and mainly contains tidal channels, deep channels and narrows to a depth of 20 m. The sediment is mostly sandy with occasional mud layers, and mud pebbles. Bedding planes show small and large-scale ripple marks, lenticular and cross-bedding occur and also shell pavement.

Bioturbate textures include burrowing textures that are resting traces or internal traces of nektonic or benthonic animals, vertical escape trails of lamellibranchs, starfish, gastropods, or polychaete worms, *Nereis virens*, *N. pelagica*, *N. succinea*, *Stylarioides plumosus*; also agglutinated tubes of *Lanice conchilega*, *Pectinaria coreni*, *Pygospio elegans*, *Sabellaria spinulosa*, *Magelona papillicornis*, *Ampharete grubei*, *Amphitrite cirrosa*, *A. johnstoni*; *Cardium edule*, *Donax vittatus*, *Mactra corallina cinerea*, *Mytilus edulis*, *Venerupis pullastra*, *Buccinum undatum* (with weakly sculptured shell), *Littorina littorea*; *Hyas araneus*, *Macropodia rostrata*, *Carcinides maenas*, *Portunus holsatus*, *Eupagurus bernhardus*, *Pycnogonum litorale*; *Asterias rubens*, *Ophiura albida* and *Psammechinus miliaris*.

Characteristic features: muds and silts in lenticular bedding, bedding planes with ripple marks, *Psammechinus miliaris* and *Venerupis pullastra*.

Biofacies 5

This biofacies occurs at depths of 2 to 12 m where the sea floor is affected by surf and tides so that silt is rapidly transported. Bedding planes show small and large-scale ripple marks; shell fragments and rounded pieces of wood are present (with boreholes by the lamellibranch *Zirfaea crispata*).

There are burrowing trails of lamellibranchs and starfish; *Angulus fabula*, *A. tenuis*, *Cardium edule*, *Donax vittatus*, *Ensis ensis*, *Macoma baltica*, *Mactra solida*, *M. subtruncata*, *Mytilus edulis*, *Bela turricula*, *Bittium reticulatum*, *Lunatia nitida*; *Echinocardium cordatum* occurs occasionally, *Asterias rubens* more frequently; *Cancer pagurus*; agglutinated tubes of *Lanice conchilega* occasionally form lawns; vertebrae of dolphins and fish are fairly common.

Characteristic features: silt beds with small and large-scale ripple marks and shell beds (shells of the lamellibranch species above).

Biofacies 6

There are wandering sand reefs at depths of 4 to 8 m in this biofacies, and the effect of surf and tides is strong. Medium grain sand contains shell fragments. The bedding shows small and large-scale ripple marks and reef structures.

There are few living organisms, correspondingly few burrowing textures, hardly any dwelling burrows and agglutinated tubes, few living lamellibranchs (the same species as in biofacies 5).

Characteristic features: medium and fine grain sands in reef bedding, rich in shells.

Biofacies 7

This biofacies covers areas at depths between 10 and 20 m, where the sea floor consists of medium grain sands with occasional coarser grains and small stones, also coarse sand with much shell breccia; rolled pieces of wood with boreholes from boring lamellibranchs occur, bedding is rippled.

Burrowing textures by *Echinocardium* are extensive. Other inhabitants are *Ensis ensis*, *Macoma baltica*, *Mactra corallina cinerea*, *Mytilus edulis*, *Angulus fabula*, *A. tenuis*, *Cardium edule*, *Donax vittatus*, *Spisula subtruncata*, *S. solida*, *Venus gallina*, *Ostrea edulis*, *Nucula nucleus*; *Bela turricula*, *Bittium reticulatum*, *Buccinum undatum*, *Hydrobia ulvae*; *Echinocardium cordatum* (in great numbers), *Ophiura texturata*, *O. albida*, *Asterias rubens*, *Lanice conchilega*, *Pectinaria coreni*.

Characteristic features: coarse and medium grain sands in ripple bedding, *Echinocardium cordatum.*

Biofacies 8

This biofacies occurs at depths of 15 to 25 m and corresponds to the 'reef bottom' of charts. There are medium grain and coarse sands, pebble beds, individual clasts of different sizes, many pieces of rounded wood, peat, locally richer shell beds and shell breccia. The bedding is generally coarse, with cross-bedding and occasional ripple marks.

There are a few burrowing textures by *Onuphis britannica* and species of the genera *Glycera* and *Lumbriconereis*. Other inhabitants are *Dosinia exoleta*, *Ensis ensis*, *Montaguta ferruginosa*, *Venus fasciata*, *V. ovata*, *Echinocyamus pusillus*, *Spatangus purpureus*, *Echinocardium flavescens*; *Cancer pagurus*, *Galathea dispersa*, *G. intermedia*, *Porcellana longicornis*; *Lanice conchilega*; *Branchiostoma lanceolatum.* (On larger rocks members of hard-bottom communities.)

Characteristic features: coarse sand, pebble beds, *Spatangus purpureus.*

Biofacies 9

This includes areas at depths of 10 to 28 m with silt and—mostly—muds, generally lenticular bedding and fine lamination, few shell beds, a few valves of the local lamellibranch inhabitants, peat and layers of spines of *Echinocardium.* (The salinity of the sea water can decrease to 18‰ due to fresh water from the rivers Weser, Elbe, Eider and Hever; temperature fluctuations are wide in some parts.)

There are many burrows of worms, *Scoloplos armiger*, species of the genus *Nephthys*, *Nereis virens*; internal traces of flatfish, burrowing textures of *Aphrodite aculeata*, escape trails of lamellibranchs; *Aloidis gibba*, *Angulus fabula*, *Mactra corallina*, *Nucula nitida*, *Phaxas pellucidus*, *Syndosmya alba*, *Ophiura texturata*, *O. albida.*

Characteristic features: finely laminated muds, lenticular bedding, *Syndosyma alba.*

Biofacies 10

This biofacies covers zones at depths between 20 and 30 m with detritus-rich silt in rippled bedding, probably rarely in lenticular bedding, and sparse fine shell breccia. Much slag occurs (because it is in the main shipping route), and members of hard-bottom communities live on slag.

Intensive burrowing textures are caused by lamellibranchs and gastropods, *Nucula*, *Phaxas*, *Lunatia*, by decapodan crustaceans, *Hyas araneus*, *H. coarctatus*, *Cancer pagurus*, *Corystes cassivellaunus* and *Eupagurus bernhardus*, internal traces by *Astropecten irregularis* and *Aphrodite aculeata*, fishes and macruran crustaceans, gastropods and lamellibranchs; resting traces of fishes and nektonic crustaceans, burrows of worms. Other inhabitants are *Aloidis gibba*, *Angulus fabula*, *A. tenuis*, *Macoma baltica*, *Mactra corallina cinerea*, *Mytilus edulis*, *Nucula nitida*, *Phaxas pellucidus*, *Spisula subtruncata*, *S. solida*; *Bela turricula*, *Bittium reticulatum*, *Buccinum undatum*, *Lunatia catena*, *L. nitida*; *Asterias rubens*, *Astropecten irregularis*, a few *Echinocardium cordatum*, *Ophiothrix fragilis*, *Ophiura texturata*, *Solaster papposus*; *Lanice conchilega* (in some places very frequent); *Pectinaria coreni* and *P. auricoma*.

Characteristic features: silt in rippled bedding, *Astropecten irregularis*, intensive burrowing textures and perturbation of the bedding.

Biofacies 11

This biofacies comprises zones at depths between 30 and 45 m with sandy muds in lenticular or laminar bedding, and few shell beds.

There are numerous and intensive bioturbate textures caused by lamellibranchs, gastropods, decapodan crustaceans, such as *Cancer pagurus*, *Eupagurus bernhardus*, *Hyas*, *Upogebia deltaura*, *Callianassa subterranea*, *C. helgolandica*, by *Echiurus echiurus* (mucous burrow, U-shaped burrow), and *Aphrodite aculeata*. Other inhabitants are *Aloidis gibba*, *Angulus fabula*, *A. tenuis*, *Arca lactea* (rare), *Cardium edule*, *C. echinatum*, *Cyprina islandica*, *Donax vittatus*, *Dosinia lincta*, *Ensis ensis*, *Mactra corallina cinerea*, *M. subtruncata*, *Modiolus modiolus*, *Montaguta ferruginosa*, *Mya truncata*, *Nucula nitida*, *Ostrea edulis*, *Phaxas pellucidus*, *Pecten varius*, *Psammobia ferroensis*, *Venerupis pullastra* (rare), *Venus gallina*, *V. ovata*; *Buccinum undatum*, *Bela turricula*, *Lunatia nitida*, *Neptunea antiqua*, *Scala clathrus*, *Turritella communis*; *Amphiura filiformis*, *Astropecten irregularis*, *Echinocyamus pusillus*, *Ophiura texturata*, *Ophiothrix fragilis* (and many individual skeletal parts of brittle stars). Vertebrae and skulls of teleosteans are frequent.

Characteristic features: sandy muds in lenticular and laminar bedding, *Cyprina islandica*, *Aloidis gibba*.

Biofacies 12

This biofacies is found at depths of 23 to 29 m, with sand of all grain sizes and pebble beds, but also layers rich in detritus, probably in lenticular and rippled bedding. Fauna and texture of this biofacies have not yet been investigated in detail.

There are few burrowing textures, mostly caused by *Spatangus purpureus*, and agglutinated tubes.

Characteristic features (?): sand of all grain sizes, *Spatangus*.

Biofacies 13

This includes deeper areas, from 45 to 60 m ('Deep Channel'), with muds, silt, medium and coarse sand, pebble beds and many shell beds. Bedding is lenticular and laminar. (Complete list of fauna published by Caspers 1938.)

Burrowing textures are caused by *Echiurus*, sipunculids and *Priapulus*, *Diastylis rathkei*, by decapodan crustaceans, *Homarus gammarus*, *Nephrops norvegicus*, *Galathea*, *Porcellana*, *Ebalia*; internal traces by echinoderms and crustaceans; there are also dwelling burrows of worms, decapodan crustaceans, *Upogebia*, *Callianassa*; agglutinated tubes of about eight species of polychaetes and of the coelenterates *Peachia* and *Cerianthus*. Twenty-one lamellibranch species, seven prosobranch species and fifteen echinoderm species also inhabit the zone.

Characteristic features: mud, silt, medium and coarse sand in laminar and lenticular bedding, *Cardium fasciatum*, *Venus ovata*.

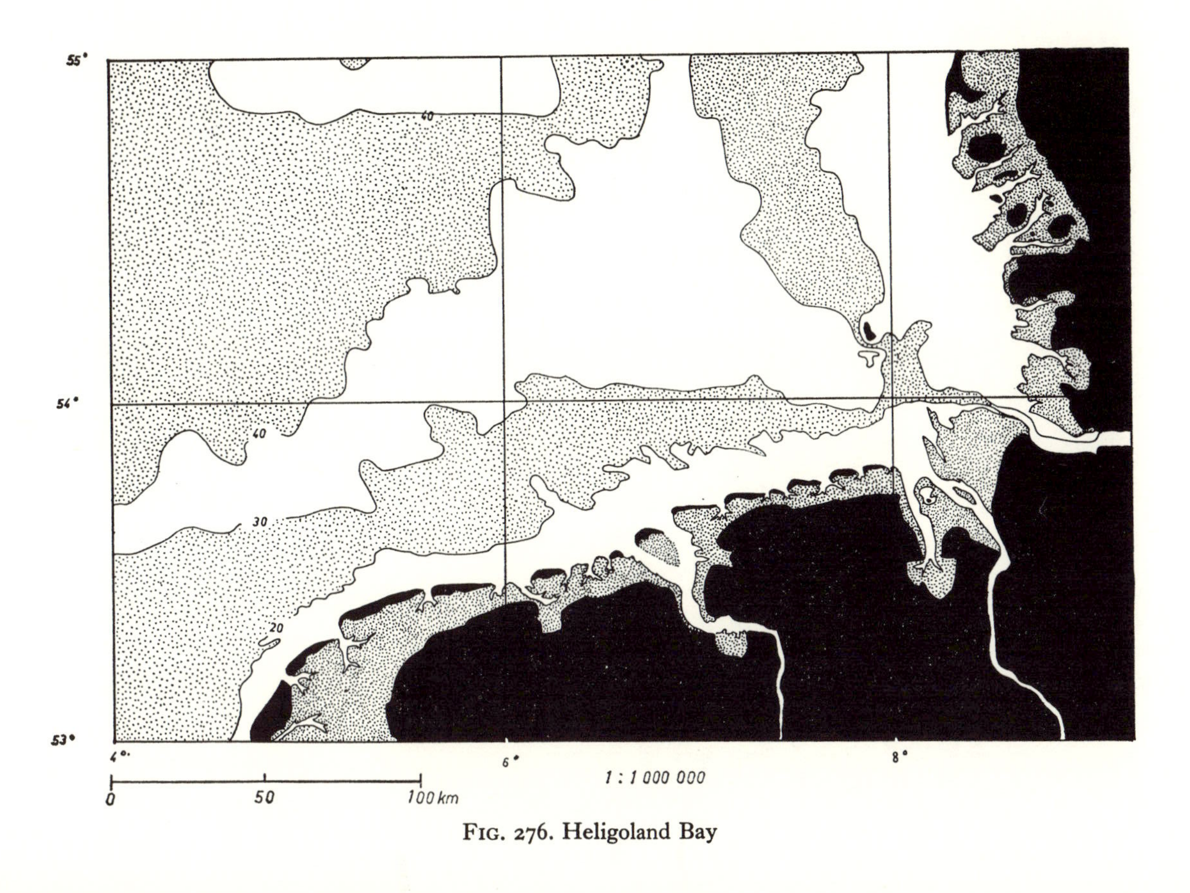

FIG. 276. Heligoland Bay

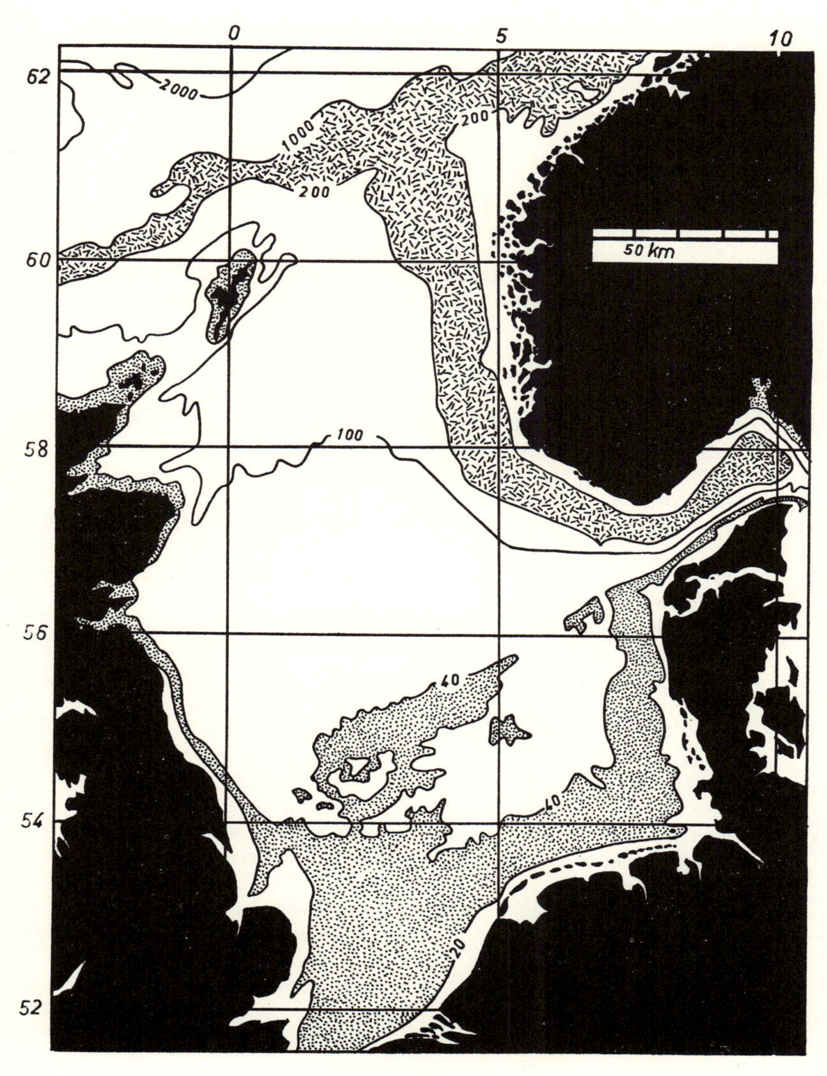

FIG. 277. North Sea

REFERENCES

AAGAND, B. 1933 *Den gamle Hvalfangst*. Oslo (Gyldendal Norsk Forlag)

ABEL, O. 1912 *Grundzüge der Palaeobiologie der Wirbeltiere*. Stuttgart (Schweizerbart)

— 1935 *Vorzeitliche Lebensspuren*. Jena (Fischer)

ALEXANDER, W. B. 1959 *Die Vögel der Meere* (Trs. Niethammer). Hamburg and Berlin (Parey)

ALLEN, E. J. 1905 The anatomy of *Poecilochaetus* (Claparède). *Quart. Journ. microsc. Sc.* N.S. **48**, 76, London

ANDREAE, J. 1882 Beiträge zur Anatomie und Histologie des *Sipunculus nudus*. *Z. wiss. Zool.* **36**, 201–258, Leipzig

ANDRÉE, K. 1915 Wesen, Ursachen und Arten der Schichtung. *Geol. Rundschau* **6**, 351–355, Leipzig

ANKEL, W. E. 1929 Hydrobienschill und Hydrobienkalk. *Natur u. Museum* **59**, 33–49, Frankfurt a. M.

— 1935 Das Gelege von *Lamellaria perspicua* L. *Z. Morph. u. Ökolog. d. Tiere* **30**, 635–647, Berlin

— 1936a Prosobranchia. In: Grimpe & Wagler: *Tierwelt Nord- und Ostsee*, Leipzig

— 1936b Die Netz-Reusenschnecke, ein Aasfresser im Watt. *Natur und Volk*, **66**, 341–345, Frankfurt a. M.

— 1936c Die Frasspuren von *Helcion* und *Littorina* und die Funktion der Radula. *Verhandl. Deutsch. Zool. Ges.* 174–182, Leipzig

— 1937a Der feinere Bau des Kokons der Purpurschnecke *Nucella lapillus* (L.) und seine Bedeutung für das Laichleben. *Verhandl. Deutsch. Zool. Ges.* 77–86, Leipzig

— 1937b Wie bohrt *Natica*? *Biol. Zentralbl.* **57**, 75–82, Leipzig

— 1937c Wie frisst *Littorina*? I. Radula-Bewegung und Fressspur. *Senckenbergiana* **19**, 317–333, Frankfurt a. M.

— 1938a Beobachtungen an Prosobranchiern der Schwedischen Westküste. *Arkiv f. Zool.* **330**, 1–27, Stockholm

— 1938b Erwerb und Aufnahme der Nahrung bei den Gastropoden. *Verh. deutsch. zool. Ges. Giessen*, 223–295, Leipzig

— 1950 Ein rankenfüssiger Krebs mit Schaumfloss (*Lepas fascicularis*). *Natur und Volk* **80**, 309–319, Frankfurt a. M.

— 1962 Das Märchen von den Entenmuscheln. *Natur u. Museum* **92**, 207–218, Frankfurt a. M.

APEL, W. 1885 Beiträge zur Anatomie und Histologie des *Priapulus caudatus* (Lam.) und des *Halicryptus spinosus* (v. Sieb.). *Z. wiss. Zool.* **42**, 459–529, Leipzig

ARNDT, W. 1943 Vier für Helgoland und vier für die deutschen Küstengewässer neue Schwammarten. *Zool. Anz.* **142**, 84–91, Leipzig

— 1952a Die Schwammfauna Helgolands. *Zool. Anz.* **148**, 1–12, Leipzig

— 1952b Ökologisches und Tiergeographisches über Spongien-Vorkommen bei Helgoland. *Zool. Anz.* **148**, 286–299, Leipzig

BACHMAYER, FR. & PAPP, A. 1951 Lebensspuren aus dem französischen Jura und dem Schlier Österreichs. *Sitzb. Österr. Akad. Wiss. math.-nat. Kl.* **160**, 199–206, Wien

BACKHAUS, H. 1937 Die natürliche Entwicklung der ostfriesischen Inseln. *Abh. Naturwiss. Ver. Bremen* **30**, 285–299, Bremen

BACQ, Z. M. & CHIRETTI, F. 1951 La sécrétion externe et interne des glandes salivaires postérieures des Céphalopodes octopodes. *Bull. Sci. Acad. roy. Belg.* **37**, 79–102, Brussels

— 1953 Physiologie des glandes salivaires postérieures des Céphalopodes octopodes isolées et perfusées in vitro. *Publ. della Staz. Zool. Napoli* **24**, 267–277, Naples

BALOGH, I. 1958 *Lebensgemeinschaften der Landtiere. Ihre Erforschung unter besonderer Berücksichtigung der zönologischen Arbeitsmethoden.* Berlin (Akad. Verl.) 2nd edition

BALSS, H. 1926 Decapoda. In: Grimpe & Wagler: *Tierwelt der Nord- und Ostsee*, Leipzig

BALTZER, F. 1931 Priapulida. *Kükenthal-Krumbach's Handbuch Zool.* **2** (2), 1–14, Berlin

BARNES, D. 1956 Underwater television and research in marine biology, bottom topography and geology. Part 2. Experience with the equipment. *Deutsch. Hydrograph. Z.* **8**, 213–236, Hamburg

BAUER, V. 1912 Zur Kenntnis und Lebensweise von *Pecten jakobaeus. Zool. Jahrb.* (*Abt. Allgem. Zool. u. Physiol.*) **33**, 127–150, Jena

BAUMBERGER, J. P. & DILL, D. B. 1931 A study of the glycogen and sugar content and the osmotic pressure of crabs during the molt cycle. *Physiol. Zool.* **4**, 122–187, Chicago

BECHER, S. 1913 Stachelhäuter. *Handwörterbuch Naturwiss.* **9**, 379–457, Jena (Fischer)

BECKER, G. & KAMPF, W. D. 1955 Die Holzbohrassel der Gattung *Limnoria* (Isopod) und ihre Lebensweise, Entwicklung und Umweltabhängigkeit. *Z. angew. Zool.* **4**, 477–517, Berlin

BENADE, W. 1938 Moore, Schlamme, Erden (Pleoide). *Der Rheumatismus, Sammlung v. Einzeldarstellungen a. d. Gesamtgebiet d. Rheumaerkrankungen*, **10**, 1–148, Dresden and Leipzig

BENNINGHOFF, A. 1935–36 Form und Funktion. *Z. ges. Naturwiss.* **1**, 149–160, Braunschweig

— 1938 Über Einheiten und Systembildungen im Organismus. *Deutsch. med. Wschr.* **64**, 1377–1382, Leipzig

— 1949 Über funktionelle Systeme. *Studium generale* **1**, 9–13, Berlin, Göttingen, Heidelberg

BENTHEM-JUTTING, VAN T. 1931 Scaphopoda. In: Grimpe & Wagler: *Tierwelt Nord- und Ostsee*, Leipzig

BERG, L. S. 1958 *System der rezenten und fossilen Fischartigen und Fische.* Berlin (VEB Deutsch. Verl. Wiss.)

BERNHÄUSER, A. 1956 Kann intravitaler Befall durch Bohrorganismen an fossilen Fischzähnen nachgewiesen werden? *Sitz.ber. Österr. Akad. Wiss. math.-nat. Kl. Abt. I* **165**, 383–396, Vienna

BETHE, A. 1930 Studien über die Plastizität des Nervensystems. I. Mitt. Arachnoiden und Crustaceen. *Pflüger's Archiv* **224**, 793–835, Bonn

— 1952 *Allgemeine Physiologie*. Berlin, Göttingen, Heidelberg (Springer)

BEURLEN, K. 1929 Parallelentwicklung und Iterationen bei Decapoden. *Palaeont. Z.* **11**, 50–52, Berlin

— 1932 Weshalb leben die Einsiedlerkrebse in Schneckenschalen? *Natur u. Museum* **62**, 74–78, Frankfurt a. M.

— 1933 Zur Entfaltung der Brachyuren und zu der Frage der explosiven Formbildung überhaupt. *Cbl. Min.* (*B*), 473–479, Berlin

— 1936 Das Gestaltproblem in der organischen Natur. *Z. f. d. ges. Naturwiss.* **1**, 445–457, Braunschweig

BIDDER, A. M. 1950 The digestive mechanism of the European squids *Loligo vulgaris, Loligo forbesi, Alloteuthis media* and *Alloteuthis subulata*. *Quart J. microsc. Sci.* **91**, 1–43, Oxford and London

BIEDERMANN, W. 1905 Studien zur vergleichenden Physiologie der peristaltischen Bewegungen. II. Die lokomotorischen Wellen der Schneckensohle. *Pflüger's Arch.* **107**, 1–56, Bonn

BISHOP, M. W. H. 1947 Establishment of an immigrant barnacle in British coastal waters. *Nature* **159**, 51, London

BOETTGER, C. 1907 *Petricola pholadiformis* Lam. *N. deutsch. Mal. Ges.* **39**, 206–217, Frankfurt a. M.

— 1907 *Petricola pholadiformis* Lam. im deutschen Wattenmeer. *Zool. Anz.* **31**, 268–270, Leipzig

BOHL, E. 1960 Otolithen-Charaktere des Sommer-Herbst laichenden Herings der Nordsee. *Arch. Fischereiwiss.* **10**, 169–178, Braunschweig

BOHN, G. 1897 Sur le renversement du courant respiratoire chez les Décapodes. *C. R. Acad. Paris* **125**, 539–542, Paris

— 1902 Des méchanismes respiratoires chez les crustacées décapodes. *Bull. sci. France Belg.* **36**, 178–551, Paris

— 1904 Sur les mouvements respiratoires des Annélides marines. *C. soc. Biol.* **56**, 23–33, Paris

BÖHNECKE, G. 1922 Salzgehalt und Strömungen der Nordsee. *Veröff. Inst. Meereskde., Berlin, N.F.* **10**, 5–34, Berlin

BÖKER, H. 1935 *Einführung in die vergleichende biologische Anatomie der Wirbeltiere*, 1. Jena (Fischer)

BORG, F. 1917 Über die *Spirorbis*-Arten nebst einem Versuch zu einer neuen Einteilung der Gattung *Spirorbis*. *Zool. Bidrag Uppsala* **5**, 15–38

BOURNE, D. W. & HEEGE, B. C. 1965 A wandering enteropneust from the abyssal Pacific, and the distribution of 'spiral' tracks on the sea floor. *Science* **150**, 60–63, Washington

BRATTSTRÖM, H. 1941 Studien über die Echinodermen des Gebietes zwischen Skagerrak und Ostsee, besonders des Öresundes, mit einer Übersicht über die physische Geographie. *Undersökningar över Öresund* **27**, 1–329, Lund

BRAUN, M. 1875 Über die histologischen Vorgänge bei der Häutung von *Astacus fluviatilis*. *Arb. zool. nat. Inst. Würzburg* **2**, 121–166, Würzburg

BREDER, C. M. 1942 The shedding of teeth by *Carcharias litoralis* (Mitchill). *Copeia Ann. Arbor*. 42–44

BROCH, H. 1922 Riffkorallen im Nordmeer einst und jetzt. *Naturwiss.* **10**, 804–806, Berlin

— 1928 Hydrozoa. In: Grimpe & Wagerl: *Tierwelt der Nord- und Ostsee*, Leipzig

BROCK, F. 1933 Analyse des Beute- und Verdauungsfeldes der Wellhornschnecke *Buccinum undatum* L. *Verh. deutsch. Zool. Ges. Bonn*, 243–250, Leipzig

— 1936 Suche, Aufnahme und enzymatische Spaltung der Nahrung durch die Wellhornschnecke *Buccinum undatum* L. *Bibliotheca Zoologica* **92**, 1–136, Leipzig

BROCKMANN, CHR. 1937 Küstennahe und küstenferne Sedimente in der Nordsee. *Abh. Nat. Ver. Bremen* **30**, 78–89, Bremen

— 1940 Das Plankton der Helgoländer Bucht im Sommer 1935. *Abh. Nat. Verh. Bremen* **31**, 712–749, Bremen

BRONGERSMA-SANDERS, M. 1944 Een H_2S-bevattend sediment met een hoog organisch gehalte uit open Zee. *Geol. en Mijnb.* **6**, 57–63, Groningen

— 1948 The importance of upwelling water to vertebrate palaeontology and oil geology. *Verh. Koninkl. Nederlandsche Akad. Wetensch. Afd. Natuurkunde* **45**, 1–112, Amsterdam

— 1949 On the occurrence of fish remains in fossil and recent marine deposits. *Bijdr. Dierk.* **28**, 65–76, Leiden

BUB, H. 1956 Eine Seevogelbestandsaufnahme an der ostfriesischen und oldenburgischen Küste. *Ornith. Mitt.* **8**, 49–50, Stuttgart

BUCHER, W. 1938 Key to papers by an institute for the study of modern sediments in shallow seas. *J. Geol.* **46**, 726–750, Chicago

BUDDENBROCK, W. V. 1912 Über die Funktion der Statozysten im Sande grabender Meerestiere. *Biol. Zentralbl.* **32**, 564–585, Leipzig

— 1913 Über die Funktion der Statozysten im Sand grabender Meerestiere. *Zool. Jahrb.* (*Abt. Allgem. Zool. u. Physiol.*) **33**, 441–482, Jena

— 1921 Die Schreitbeinbewegungen von *Dixippus morosus*. *Biol. Ctrbl.* **41**, 41–48, Leipzig

— 1928 *Grundriss der vergleichenden Physiologie*. Berlin (Borntraeger)

— 1930 Fortbewegung auf dem Boden bei Wirbellosen. *Hdb. norm. u. patholog. Physiol.* **15**, 1. Hälfte, 271–291, Berlin

— 1939 *Grundriss der vergleichenden Physiologie II*. Berlin (Borntraeger)

— 1953 *Vergleichende Physiologie II. Nervenphysiologie*. Basle (Birkhäuser)

— 1954 VII. Decapoda. *Bronn's Kl. u. O. d. Tierreichs* **5**, 1151–1283, Leipzig

BUDDENBROCK, W. V. & MOLLER-RACKE, J. 1953 Über den Lichtsinn von Pecten. *Pubbl. Staz. Zool. Napoli* **24**, 217–245, Naples

CARTHY, J. D. 1958 *An introduction to the behaviour of Invertebrates*. London

CASPERS, H. 1938 Die Bodenfauna der Helgoländer Tiefen Rinne. *Helgol. Wiss. Meeresunters.* **2**, 1–112, Heligoland

— 1940 Über Nahrungserwerb und Darmverlauf bei *Nucula*. *Zool. Anz.* **129**, 48–55, Leipzig

— 1949a Die tierische Lebensgemeinschaft in einem Röhricht der Unterelbe. *Verhl. Ver. naturw. Heimatforsch. Hamburg* **30**, 41–49, Hamburg

— 1949b Die Bewuchsgemeinschaft an der Landungsbrücke der Nordseeinsel Spiekeroog und das Formproblem von *Balanus*. *Zool. Jb.* **78**, 217–322, Jena

— 1950a Die Lebensgemeinschaft der Helgoländer Austernbank. *Helgol. Wiss. Meeresunters.* **3**, 119–169, List

— 1950b Der Biozönose- und Biotopbegriff vom Blickpunkt der marinen und limnischen Synökologie. *Biol. Zbl.* **69**, 43–63, Leipzig

— 1951 Bodengreiferuntersuchungen über die Tierwelt in der Fahrrinne der Unterelbe und im Vormündungsgebiet der Nordsee. *Verh. Deutsch. Zool. Ges. in Wilhelmshaven*, 404–418, Leipzig

CASPERS, H. 1952a Der tierische Bewuchs an Helgoländer Seetonnen. *Helgol. Wiss. Meeresunters.* **4**, 138–160, List

— 1952b Bodengreiferuntersuchungen über die Tierwelt in der Fahrrinne der Unterelbe und im Vormündungsgebiet der Nordsee. *Verh. Deutsch. Zool. Ges. in Wilhelmshaven*, 404–418, Leipzig

— 1954 Biologische Untersuchungen über die Lebensräume der Unterelbe und des Vormündungsgebietes der Nordsee. *Mitt. Geol. Staatsinst. Hamburg* **23**, 76–85, Hamburg

CHAINE, J. & DUVENGIER, I. 1934 Recherches sur les Otolithes des Poissons. *Act. Soc. Linn. Bordeaux* **86**, 1–254, Bordeaux

CHAPMANN, G. 1950 Of the movement of worms. *J. exp. Biol.* **27**, 29

COPELAND, M. 1919 Locomotion in two species of the gastropod genus *Alectrion* with observations on the behaviour of pedal cilia. *Biol. Bull.* **37**, 34, Lancaster

CORI, C. J. 1937 3. Ordnung der Tentaculata: Bryozoa. *Kükenthal's Handb. Zool.* **3**, 263–502, Berlin and Leipzig

COTTEN, A. C. 1912 Clare island survey, marine algae, Pt. 15. *Proc. Roy. Irish Acad.* **31**, 343–368, Dublin and London

COUCH, J. 1838 Bemerkungen über den Häutungsprozess der Krebse und Krabben. *Arch. Naturgesch.* **4**, 337–342, Berlin

COUGHLAN, L. 1969 The estimation of filtering rate from clearance of suspensions. *Marine Biol.* **2**, 356–358, Berlin, Heidelberg, New York

DAHL, FR. 1908 Grundsätze und Grundbegriffe der biocönotischen Forschung. *Zool. Anz.* **33**, 349–353, Leipzig

— 1928 Porifera-Coelenterata-Echinodermata. In: *Tierwelt Deutschlands und der angrenzenden Meeresteile* **4**, 1–332, Jena

DARBY, H. H. 1932, 1938 Moulting in the crustacean *Crangon annullatus*. *Anat. Rec.* **72** (Suppl.), 232–241, and **78**, 88–98, Philadelphia

DAVID, L. R. 1957 Fishes (other than Agnatha). *Treatise mar. Ecology and Paleoecology* **2** (Paleoecology), 999–1010, Baltimore, Maryland

DEEGENER, P. 1918 *Die Formen der Vergesellschaftung im Tierreich*. Leipzig (Veit)

DEECKE, W. 1915 Paläontologische Betrachtungen. VII. Über Crustaceen. *N. Jb. Min.* etc. **1**, 112–126, Stuttgart

DEHORNE, A. 1925 Observations sur la biologie de *Nereis diversicolor*. *C. R. Acad. Sc. Paris* **180**, 1441–1443, Paris

DEINSE, A. B. VAN 1931 De fossile en rezente Cetacea van Nederland. *Diss. Utrecht.*

DIENER, C. 1925 *Grundzüge der Biostratigraphie*. Leipzig and Vienna (Deuticke)

DIETRICH, G. D. 1957 *Allgemeine Meereskunde*. Berlin (Borntraeger)

DOFLEIN, F. & REICHENOW, E. 1953 *Lehrbuch der Protistenkunde*. 6th edition, Jena (Fischer)

DOHRN, A. 1875 *Der Ursprung der Wirbeltiere und das Prinzip des Funktionswechsels*. Leipzig (Engelmann)

DORE, W. H. & MILLER, R. C. 1923 The digestion of wood by *Teredo navalis*. *Univ. Calif. Publ. Zool.* **22**, 383–400, Berkeley

DÖRJES, J., GADOW, S., REINECK, H. E. & SINGH, J. B. 1969 Die Rinnen der Jade (Südliche Nordsee). Sedimente und Makrobenthos. *Senckenbergiana maritima*, [1] **50**, 5–62, Frankfurt a. M.

DRACH, P. 1935 Phénomènes de résorption dans l'endosquelette des Décapodes Brachyoures, au cours de la période qui précède la mue. *C. R. Acad. Sci. Paris* **201**, 1424–1426, Paris

— 1936 L'eau absorbée au cours de l'exuviation, donnée fondamentale pour l'étude physiologique de la mue. Définitions et déterminations quantitatives. *C. R. Acad. Sci. Paris* **202**, 1817–1819, Paris

DRAL, A. D. G. 1967 The movements of the latero-frontal cilia and the mechanism of particle retention in the mussel (*Mytilus edulis* L.). *Nth. J. Sea Res.* **3**, 391–422, Den Helder

DREVERMANN, F. 1931 Massensterben von Walen. *Natur u. Museum* **61**, 377–382, Frankfurt a. M.

DROST, R. 1925 Eine gewaltige Zugnacht auf Helgoland als Folge ungünstiger Wetterverhältnisse im Frühjahr 1924. *Ornith. Monatsber.* **33**, 11–13, Berlin

— 1928a Unermessliche Vogelscharen über Helgoland. *Ornith. Monatsber.* **36**, 3–6, Berlin

— 1928b Gewaltiger Vogelzug auf Helgoland. *Wild u. Hund* **34**, 216–217, Berlin

DROST, R. & SCHILLING, L. 1940 Über den Lebensraum deutscher Silbermöven, *Larus a. argentatus* Pontopp. auf Grund von Beringungsergebnissen. *Vogelzug* **11**, 1–22, Berlin

DUERDEN, J. E. 1902 Boring algae as agents in the disintegration of corals. *Bull. Amer. Mus. Nat. Hist.* **16**, 323–332, New York

DUNKER, G. (LADIGES, W.) 1960 *Die Fische der Nordmark.* Hamburg (Cram, De Gruyter & Co.)

EALES, N. B. 1950 *The littoral fauna of Great Britain.* Cambridge

EASTMANN, C. R. 1903 Sharks' teeth on cetacean bones from the red clay of the tropical Pacific. *Mem. Mus. comp. Zool. Harvard* **26**, 177–191, Cambridge, Mass.

— 1906 Sharks' teeth and cetacean bones. Rep. scient. Results Exped. eastern trop. Pacific. 'Albatross'. *Bull. Mus. comp. Zool. Harvard* **50**, 75–98, Cambridge, Mass.

EBERLE, G. 1929 Ein Massensterben von Heringen. *Natur u. Volk* **59**, 64–70, Frankfurt a. M.

EDGE, E. R. 1934 Faecal pellets of some marine Invertebrates. *Amer. Midland Naturalist* **15**, 78–84, Notre Dame, Ind.

EHLERS, E. 1868 Über eine fossile Eunice aus Solnhofen (*Eunicites avitus*) nebst Bemerkungen über fossile Würmer. *Z. wiss. Zool.* **18**, 421–443, Leipzig

— 1875 Beiträge zur Kenntnis der Verticalverbreitung der Borstenwürmer im Meere. *Z. wiss. Zool.* **25**, 1–102, Leipzig

EHLERT, W. 1957 Zur Ernährung der Silbermöwe (*Larus argentatus* Pont.) in der Vorbrutzeit. *Ornith. Mitt.* **9**, 201–203, Schweinfurt/M.

EHRENBAUM, E. 1907 Künstliche Zucht und Wachstum des Hummers. *Mitt. deutsch. Seefisch.-Ver.* **23**, 178–198, Berlin

EIBL-EIBESFELDT, I. 1955 Der Kommentkampf der Meerechse (*Amblyrhynchus cristatus* Bell.) nebst einigen Notizen zur Biologie dieser Art. *Z. Tierpsychol.* **12**, 49–62, Berlin, Hamburg

EIGENBRODT, H. 1941 Untersuchungen über die Funktion der Radula einiger Schnecken. *Z. Morph. u. Ökol. Tiere* **37**, 735–791, Berlin

EKMAN, S. 1935 *Tiergeographie des Meeres.* Leipzig (Akad. Verl. Ges.)

— 1953 Zoogeography of the sea. London (Sidgwick and Jackson)

ELSLER, E. 1907 Deckel und Brutpflege bei *Spirorbis*. Z. *wiss. Zool.* **87**, 603–643, Leipzig

ELTON, C. 1949 Population interspersion: an essay on animal community pattern. *J. Ecol.* **37**, 1–23, Cambridge

EMERY, K. O. 1952 Submarine photography with the Benthograph. *Sci. Monthly* **75**, 3–11, Iowa City

EMERY, K. O. 1953 Some surface features of marine sediments made by animals. *J. Sediment. Petrol.* **23**, 202–204, Menasha, Wis.

ENDERS, H. E. 1909 A study of the life history and basis of *Chaetopterus variopedatus*. *J. Morph.* **20**, 479–531, Boston

ERENKAMP, H. 1931 Morphologische Histologie und Biologie der Sabellidenspecies *Laonome kröyeri* Malmgr. und *Euchone papillosa* M. Sars. *Zool. Jb. Anat.* **53**, 405–534, Jena

FAIRBRIDGE, R. W. 1954 Stratigraphic correlation by microfacies. *Amer. J. Sci.* **252**, 683–694, New Haven, Conn.

FALKE, H. 1950 Das Fischsterben in der Bucht von Conception (Mittelchile). *Senckenbergiana* **31**, 57–77, Frankfurt a. M.

FAUVEL, P. 1903 Le tube des Pectinaires. *Mem. Pontif. Accad. Nouvi Lincei* **21**, 322–342, Rome

— 1927 *Faune de France* **16**, Polychètes sédentaires. Paris

FEDOTOV, D. M. 1924 Morphologie des axialen Organkomplexes der Echinodermen. *Z. wiss. Zool.* **123**, 209–304, Leipzig

FISCHEL, W. 1938 *Psyche und Leistung der Tiere*. Berlin (De Gruyter)

— 1947 *Die kämpferische Auseinandersetzung in der Tierwelt*. Leipzig (Barth)

FISCHER, W. 1928 Echiuridae, Sipunculidae, Priapulidae. In: Grimpe & Wagler: *Tierwelt Nord- und Ostsee*, Leipzig

FOERSTER, A. 1933 Die Luftröhrenäste in gefaulten Lungen von Neugeborenen. *Deutsch. Z. ges. gericht. Med.* **20**, 420–460, Berlin

FONDS, M. & EISMA, D. 1967 Upwelling water as a possible cause of red plankton bloom along the Dutch coast. *Nth. J. Sea Res.* **3**, 458–463, Den Helder

FRANK, F. 1953a Zur Entstehung übernormaler Populationsdichten im Massenwechsel der Feldmaus (*Microtus arvalis* Pallas). *Zool. Jb.* (*Abt. Syst. Ökol. Geogr. d. Tiere*) **81**, 610–624, Jena

— 1953b Untersuchungen über den Zusammenbruch von Feldmausplagen (*Microtus arvalis* Pallas). *Zool. Jb.* (*Abt. Syst. Ökol. Geogr. d. Tiere*) **82**, 95–136, Jena

FRANZ, E. 1931 Insektenbegräbnis am Meer. Bestimmung der Arten in der Insektendrift bei Wilhelmshaven vom 24. Mai 1931. *Senckenbergiana* **13**, 228–230, Frankfurt a. M.

FRANZ, V. 1924 *Geschichte der Organismen*. Jena (Fischer)

FREEMANN, B. 1954 Schicksalsstunde der Brandgänse auf dem Knechtsand? *Wild u. Hund* **57**, 168–169, Hamburg

— 1960 Der Knechtsand muss gehalten werden. *Deutsch. Jäger-Z.* **1**, 8–13, Melsungen

FRIEDRICH, H. 1938 Polychaeta. In: Grimpe & Wagler: *Tierwelt Nord- und Ostsee*, Leipzig

FRIEDRICH, H. & LANGELOH, H. P. 1936 Untersuchungen zur Physiologie der Bewegung und des Hautmuskelschlauches bei *Halicryptus spinulosus* und *Priapulus caudatus*. *Biol. Zbl.* **56**, 249–268, Leipzig

FRITSCH, R. H. 1938 Das 'Bauen' des *Octopus* und andere Beobachtungen an Cephalopoden. *Verh. Deutsch. Zool. Ges. 1937*, 119–126, Leipzig

— 1954 Zu Struktur und Funktion der Tapete in Gängen von *Arenicola marina* L. *Z. Morph. u. Ökol. Tiere* **43**, 94–98, Berlin

FUCHS, TH. 1895 Studien über Fucoiden und Hieroglyphen. *Denkschr. Akad. Wiss., math.-nat. Kl.* **22**, 369–448, Vienna

GÄTKE, H. 1900 *Die Vogelwarte Helgoland*. 2nd edition. Braunschweig

GALEINE, C. & HOULBERT, C. 1922 Les récifs d'Hermelles et l'assèchement de la baie du Mont-Saint-Michel. *Bull. Soc. Géol. et Min. de Bretagne* **2**, 319–324, Rennes

GARSTANG, W. 1896a On some modifications of structure, subservient to respiration in decapod crustacea, which burrow in sand. *Q. J. Micr. Sci.* **40**, 34–78, London

— 1896b The habits and respiratory mechanism of *Corystes cassivelaunus*. *J. mar. biol. Ass.* **4**, 223–232, Plymouth

GEISSLER, R. 1938 Zur Stratigraphie des Hauptmuschelkalkes in der Umgebung von Würzburg mit bes. Ber. der Ceratiten. *Diss.* Würzburg

GERLACH, S. A. 1968 Ein Schema der Produktionsverhältnisse im Meer. *Veröff. Inst. f. Meeresforsch. Bremerhaven* **11**, 61–64, Bremen

GERSCH, M. 1934 Zur experimentellen Veränderung der Richtung der Wellenbewegung auf der Kriechsohle von Schnecken und zur Rückwärtsbewegung von Schnecken. *Biol. Zbl.* **54**, 511–518, Leipzig

GEYR VON SCHWEPPENBURG, H. Freiherr 1930 Vogelverluste im Winter 1928–29. *Ornith. Monatsber.* **38**, 139–141, Berlin

GIENAPP, H. & TOMCZAK, G. 1968 Strömungsmessungen in der Deutschen Bucht bei Sturmfluten. *Helgoländer wiss. Meeresunters.* **17**, 94–107, Hamburg

GISLEN, T. 1930 Epibioses of the Gullmar Fjord, I. II. *Kristinebergs Zool. Stat. 1877–1927*, Uppsala

GLAESSNER, M. 1929 Zur Kenntnis der Häutung bei fossilen Krebsen. *Palaeobiologica* **2**, 49–56, Vienna

GOCHT, H. & GOERLICH, FR. 1957 Reste des Chitin-Skelettes in fossilen Ostracoden-Gehäusen. *Geol. Jb.* **73**, 205–211, Stuttgart

GOETHE, FR. 1939 Die Vogelinsel Mellum. Beiträge zur Monographie eines deutschen Seevogelschutzgebietes. *Abh. a. d. Geb. Vogelkde.* **4**, 1–110, Berlin

— 1956a Die Silbermöwe. *Die neue Brehm-Bücherei*, Wittenberg (Ziemsen)

— 1956b Fuchs, *Vulpes vulpes* (Linné, 1758), reisst Schlafgesellschaft von etwa sechzig jugendlichen Silbermöwen (*Larus argentatus* Pontopp.). *Säugetierkdl. Mitt.* **4**, 58–60, Stuttgart

— 1957 Über den Mauserzug der Brandenten (*Tadorna tadorna* L.) zum grossen Knechtsand. *Fünfzig Jahre Seevogelschutz*, 96–106, Verein Jordsand, Hamburg

— 1958 Anhäufungen unversehrter Muscheln durch Silbermöven. *Natur u. Volk* **88**, 181–187, Frankfurt a. M.

GÖTZE, L. 1938 Bau und Leben von *Caecum glabrum*. *Zool. Jb. Abt. Syst.* **71**, 55–122, Jena

GOULD, A. 1870 *Report on the Invertebrata of Massachusetts*. Boston (W. B. Binney)

GRÄF, I. 1956 Die Fährten von *Littorina littorea* Linné (Gastr.) in verschiedenen Sedimenten. *Senckenbergiana leth.* **37**, 305–318, Frankfurt a. M.

GRAHAM, A. 1931 On the morphology, feeding mechanisms, and digestion of *Ensis siliqua* (Schuhmacher). *Trans. Roy. Soc. Edinb.* **56**, 725–751, Edinburgh

GRASSÉ, P. P. 1948 *Traité de Zoologie. Anatomie, Systématique, Biologie* **11**, 932–1027, Paris

GRAY, J. 1933a The relations between waves of muscular contractions and propulsive mechanism of the eel. *J. exper. biol.* **10**, 386–390, London

— 1933b The movement of the fish with special reference to the eel. *J. exper. biol.* **10**, 88–104, London

GRELL, K. G. 1956 *Protozoologie*. Berlin, Göttingen, Heidelberg (Springer)

GRESSLY, A. 1838 *Le Jura Soleurois*. Neufchatel

GREVE, L. 1931 Eine Lederschildkröte an der Nordsee-Küste gestrandet. *Natur u. Museum* **61**, 30, Frankfurt a. M.

GRIMPE, G. 1920 Teuthologische Mitteilungen VII. Systematische Übersicht der Nordsee-Cephalopoden. *Zool. Anz.* **52**, 296–304, Leipzig

— 1921 Teuthologische Mitteilungen VIII. Die Sepiolinen der Nordsee. *Zool. Anz.* **53**, 1–16, Leipzig

— 1925 Zur Kenntnis der Cephalopodenfauna der Nordsee. *Wiss. Meeresuntersuch. Helgoland* **16**, 1–124, Kiel and Leipzig

GRIPP, K. 1944 Entstehung und künftige Entwicklung der Deutschen Bucht. *Arch. Deutsch. Seewarte u. Marineobservator.* **63**, 5–44, Hamburg

— 1927 Über einen 'geführten Mäander' erzeugenden Bewohner des Ostsee-Litorals. *Senckenbergiana* **9**, 93–99, Frankfurt a. M.

— 1958 Rezente und fossile Flachmeer-Absätze petrologisch betrachtet und gedeutet. *Geol. Rdsch.* **47**, 83–99, Stuttgart

GRIPP, K., STADERMANN, R., SCHMIDT, R. & JACOB FRIESEN, K. H. 1937 *Werdendes Land am Meer. Landerhaltung und Landgewinnung an der Nordseeküste.* Berlin (Mittler & Sohn)

GROHNE, U. 1956 Die Geschichte des Jadebusens und seines Untergrundes. *Natur u. Volk* **86**, 225–233, Frankfurt a. M.

GRUBER, A. 1933 Bohrorganismen im oberen Muschelkalk. *Geol. Rdsch.* **23a** (Salomon-Festschr.), 263, Berlin

GRUVEL, A. 1905 *Monographie des Cirripèdes, ou Thécostracés.* Paris

GRY, H. 1950 Das Wattenmeer bei Skallingen. Physiographisch-biologische Untersuchung eines dänischen Tidegebietes, I. Quantitative Untersuchungen über den Sinkstofftransport durch Gezeitenströmungen. *Medd. Skalling-Lab.* **10**, 1–137, Copenhagen

GUNTER, G. 1941 Death of fishes due to cold on the Texas coast, January 1940. *Ecology* **22**, 203–208, Brooklyn

— 1947 Paleoecological import of certain relationships of marine animals to salinity. *J. Paleont.* **21**, 77–79, Chicago

GUTMANN, W. F. 1960 Funktionelle Morphologie von *Balanus balanoides*. *Abh. senckenb. naturf. Ges.* **500**, 1–43, Frankfurt a. M.

— 1961 Die Siedlungsweise der Seepocke *Balanus balanus*. *Natur u. Volk* **91**, 171–178, Frankfurt a. M.

— 1962 Beobachtungen zum Formproblem der Seepockenschale. *Natur u. Museum* **92**, 193–200, Frankfurt a. M.

HAARNAGEL, W. 1950 Das Alluvium an der deutschen Nordseeküste. *Schr. Niedersächs. Landesst. Marsch.-Wurt. Forsch.* **4**, 1–148, Hildesheim

HAAS, F. 1926 Lamellibranchia. In: Grimpe & Wagler: *Tierwelt Nord- und Ostsee*, Leipzig

HAGMEIER, A. 1925 Vorläufiger Bericht über die vorbereitenden Untersuchungen der Bodenfauna der Deutschen Bucht mit dem Petersen-Bodengreifer. *Ber. Deutsch. wiss. Komm. Meeresforsch. N. F.* **1**, 247–272, Berlin

— 1930a Die Besiedlung des Felsstrandes und der Klippen von Helgoland. *Wiss. Meeresuntersuch. Helgoland* **15**, 1–35, Kiel

— 1930b Eine Fluktuation von *Mactra* (*Spisula*) *subtruncata* da Costa an der ostfriesischen Küste. *Ber. deutsch. Wiss. Komm. Meeresforsch.* (*N. F.*) **5**, 126–155, Berlin

HAGMEIER, A. & HINRICHS, J. 1931 Bemerkungen über die Ökologie von *Branchiostoma lanceolata* und das Sediment seines Wohnortes. *Senckenbergiana* **13**, 255–267, Frankfurt a. M.

HAGMEIER, A. & KÄNDLER, R. 1927 Neue Untersuchungen im nordfriesischen Wattenmeer und auf den fiskalischen Austernbänken. *Wiss. Meeresuntersuch. Abt. Helgoland (N. F.)* **16**, 1–90, Oldenburg

HAGMEIER, A. & KÜNNE, CL. 1950 Die Nahrung der Meerestiere. I. Einleitung. III. Die wirbellosen Boden- und Nektontiere. IV. Beziehungen der Ernährung zur Verbreitung der Arten und der Gemeinschaften der Bodentiere. *Handb. Seefisch. Nordeuropas* **1**, 1–9, 87–242, Hamburg

HALLSTRÖM, G. 1938 Monumental art of northern Europe from the stone age. I. The Norwegian localities. Stockholm (Thule)

HALTON, H. C. S. 1930 An account of *Priapulus caudatus* Lam. A rare Essex worm. *Essex Naturalist* **22**, 181–183, Stratford (Essex)

HÄNTZSCHEL, W. 1934 Sternspuren, erzeugt von einer Muschel: *Scrobicularia plana* (da Costa). *Senckenbergiana* **16**, 325–330, Frankfurt a. M.

— 1936a Seeigel-Spülsäume. *Natur u. Volk* **66**, 293–298, Frankfurt a. M.

— 1936b Die Schichtungs-Formen rezenter Flachmeer-Ablagerungen im Jade-Gebiet. *Senckenbergiana* **18**, 316–356, Frankfurt a. M.

— 1937 Erhaltungsfähige Abdrücke von Hydro-Medusen. *Natur u. Volk* **67**, 141–144, Frankfurt a. M.

— 1938a Quer-Gliederung bei rezenten und fossilen Wurmröhren. *Senckenbergiana* **20**, 145–154, Frankfurt a. M.

— 1938b Bau und Bildung von Gross-Rippeln im Wattenmeer. *Senckenbergiana* **20**, 1–42, Frankfurt a. M.

— 1938c Quer-Gliederung bei *Littorina*-Fährten, ein Beitrag zur Deutung von '*Keckia annulata* Glocker'. *Senckenbergiana* **20**, 297–304, Frankfurt a. M.

— 1939a Die Lebensspuren von *Corophium volutator* (Pallas) und ihre paläontologische Bedeutung. *Senckenbergiana* **21**, 215–227, Frankfurt a. M.

— 1939b Schlick-Gerölle und Muschel-Klappen als Strömungs-Marken im Wattenmeer. *Natur u. Volk* **69**, 412–417, Frankfurt a. M.

— 1955a Lebensspuren als Kennzeichen des Sedimentationsraumes. *Geol. Rdsch.* **43**, 551–562, Stuttgart

— 1955b Rezente und fossile Lebensspuren, ihre Deutung und geologische Auswertung. *Experientia.* **11**, 373–382, Basel

— 1956 Rückschau auf die paläontologischen und neontologischen Ergebnisse der Forschungsanstalt 'Senckenberg am Meer'. *Senckenbergiana leth.* **37**, 319–330, Frankfurt a. M.

HÄNTZSCHEL, W., FAROUK LE-BAZ & AMSTUTZ, G. C. 1968 Coprolites. An annotated bibliography. *Geol. Soc. America, Inc. Mem.* **108**

HARLEY, M. B. 1950 Occurrence of a filter-feeding mechanism in the polychaete *Nereis diversicolor. Nature* **165**, 734–735, London

HARTMANN, G. 1955 Zur Morphologie der Polycopiden. *Z. wiss. Zool.* **158**, 193–248, Leipzig

HARMS, G. 1950 Mellum in der Kriegs- und Nachkriegszeit. *Mellum ein Vogelparadies in der Nordsee,* Oldenburg

HARTUNG, W. 1950 Mellum in geographischer und geologischer Betrachtung. *Mellum ein Vogelparadies in der Nordsee,* Oldenburg

— 1951 Die Insel Wangeroog, ihre Entstehung und ihre Veränderung. *Wangerooge, wie es wurde, war und ist,* Oldenburg

HASE, A. 1911 Studien über das Integument von *Cyclopterus lumpus. Jena. J. Naturwiss.* **47**, N. F., 217–342, Jena

HAVINGA, B. 1933 Der Seehund (*Phoca vitulina* L.) in den holländischen Gewässern. *Tijdschr. Nederl. Dierkund. Vereen.* **3**, 79–111, Leiden

HECHT, F. 1933 Der Verbleib der organischen Substanz der Tiere bei meerischer Einbettung. *Senckenbergiana* **15**, 165–249, Frankfurt a. M.

— 1934 Die chemische Zersetzung der tierischen Substanz während der Einbettung in marine Sedimente. *Kali, verwandte Salze und Erdöl* **17**, 209–216

HEDIGER, H. 1942 *Wildtiere in Gefangenschaft. Ein Grundriss der Tiergartenbiologie.* Basle (Schwabe & Co.)

— 1946 Bemerkungen zum Raum-Zeit-System der Tiere. *Schweiz. Z. Psychol.* **5**, 241–269, Berne

— 1956 *Instinkt und Territorium.* Paris

HEIDER, J. 1925 Über *Eunice*, Systematisches, Kiefersack, Nervensystem. *Z. wiss. Zool.* **125**, 55–90, Leipzig

HEIDER, K. 1923 Über das Nervensystem der Eunicidae. *Sitz.-ber. Preuss. Akad. Wiss., phys.-math. Kl.* **27**, 298, Berlin

HEINCKE, F. 1894 Bericht über eine von der Sektion für Küsten- und Hochseefischerei im August und September 1890 veranstaltete Untersuchungsfahrt zur Aufsuchung laichreifer Herbstheringe in der Deutschen Bucht der Nordsee. *Mitt. Sekt. Küsten- u. Hochseefischerei*, 1–12, Berlin

HELDT, R. 1960 Tote Vögel im Spülsaum der Nordseeküste von Schleswig, insbesondere von Eiderstedt. *Mitt. faunist. Arb. Gem. Schleswig-Holstein, Hamburg u. Lübeck* (*N. F.*) **13**, 37–43, Hamburg

HELFER, H. 1933 Pantopoda. Nachtrag. *Kükenthal's Handb. Zool.* **6**, 67–72, Berlin and Leipzig

HEMPEL, C. 1957a Über den Röhrenbau und die Nahrungsaufnahme einiger Spioniden (Polychaeta sedentaria) der deutschen Küsten. *Helgol. Wiss. Meeresuntersuch.* **6**, 100–135, List

— 1957b Über die Ökologie einiger Spioniden (Polychaeta sedentaria) der deutschen Küsten. *Kieler Meeresforsch.* **13**, 275–288, Kiel

— 1960 Über das Festsetzen der Larven un die Bohrtätigkeit der Jugendstadien von *Polydora ciliata* (Polychaeta sedentaria). Helgol. *Wiss. Meeresuntersuch.* **7**, 80–92, Hamburg

HEMPELMANN, F. 1911 Zur Naturgeschichte von *Nereis dumerilii* And. et Edw. *Zoologica* **25**, 1–13, Stuttgart

— 1934 Archiannelida. In: *W. Kükenthal's Handb. Zool.* **2**, 1–212, Berlin and Leipzig

HERRICK, F. H. 1895 The American lobster: a study of its habits and development. *Bull. W. S. Fish.* **15**, 280–310, Washington

HERTER, K. 1926 Versuche über die Phototaxis von *Nereis diversicolor* O. F. Müller. *Z. vergl. Physiol.* **4**, 103–141, Berlin

— 1931 Untersuchungen über den Muskeltonus des Schneckenfusses. *Z. vergl. Physiol.* **13**, 709–739, Berlin

HERTLING, H. 1953 *Einführung in die Meeresbiologie.* Berlin (Dunker and Humblot)

HERTWECK, G. 1966 Möglichkeiten des Fossilwerdens von Quallen—im Experiment. *Natur u. Museum* **96**, 456–462, Frankfurt a. M.

HERTWECK, G. & REINECK, H. E. 1966 Untersuchungsmethoden von Gangbauten und anderen Wühlgefügen mariner Bodentiere. *Natur u. Museum* **96**, 429–438, Frankfurt a. M.

HERTWECK, G. & REINECK, H. E. 1969 Sedimentologie der Meeresbodensenke NW von Helgoland (Nordsee). *Senckenbergiana maritima* [1], **50**, 153–164, Frankfurt a. M.

HERTWIG, O. 1876, 1879, 1882 Über das Hautskelett der Fische, I-III. *Morph. Jb.* **2**, 328–395, **5**, 1–21, **7**, 1–42, Leipzig

HESSE, R. 1924 *Tiergeographie auf ökologischer Grundlage*. Jena (Fischer)

HESSE, R. & DOFLEIN, F. 1935 *Tierbau und Tierleben in ihrem Zusammenhang betrachtet*. 1. *Der Tierkörper als selbständiger Organismus*. Jena (Fischer)

HESSLAND, J. 1945 On the quaternary *Mya* period in Europe. *Ark. Zool.* **37**, 1–51, Stockholm

HESSLE, CH. 1917 Zur Kenntnis der terebellomorphen Polychaeten. *Zool. Bidr. Uppsala* **5**, 39–255, Uppsala and Stockholm

HILL, L. 1936 False killer whales in South Africa. *Nature* **138**, 541, London

HINCKS, TH. 1880 *A history of the British marine Polyzoa*. London

HINTON, M. A. C. 1928 Stranded whales at Dornoch Firth. *Nat. Hist. Mag. British Mus.* (Nat. Hist.) **1**, 131–138, London

HOFFMANN, BR. 1914 Über die allmähliche Entwicklung der verschieden differenzierten Stachelgruppen und der Fasciolen bei den fossilen Spatangiden. *Palaeont. Z.* **1**, 216–272, Berlin

HOFFMANN, H. 1930 Über den Fluchtreflex bei *Nassa*. *Z. vgl. Physiol.* **11**, 662–688, Berlin

HOFKER, J. 1930 Faunistische Beobachtungen in der Zuidersee während der Trockenlegung. *Z. Morph. Ökol. Tiere* **18**, 189–216, Berlin

HOFMANN, J. 1958 Einbettung und Zerfall der Ichthyosaurier im Lias von Holzmaden. *Meyniana* **6**, 10–55, Kiel

HOFSOMMER, A. 1913 Die Sabelliden-Ausbeute der 'Poseidon'-Fahrten und die Sabelliden der Kieler Bucht. *Wiss. Meeresuntersuch. N. F.* **15**, 306–364, Kiel

HOLLMANN, R. 1969 Die Entstehung fossilisationsfähiger Schalen-Frassreste, dargestellt am Nahrungserwerb von Homarus gammarus (Crustacea, Decapoda). *Helgoländer wiss. Meeresuntersuch.* **19**, 401–416, Hamburg

— 1968 Über Schalenabschliffe bei Cardium edule aus der Königsbucht bei List auf Sylt. *Helgoländer wiss. Meeresuntersuch.* **18**, 109–193, Hamburg

HOLST, E. V. 1957 Der Saurierflug. *Paläont. Z.* **31**, 15–22, Berlin

HOMMA, H. *Das Formproblem in der Biologie*. Vienna (Springer)

HOWELL, A. BR. 1930 *Aquatic mammals*. Springfield and Baltimore

HÜLSEMANN, J. 1966 Spiralfährten und 'geführte Mäander' auf dem Meeresboden. *Natur u. Museum* **96**, 449–455, Frankfurt a. M.

HUNT, O. D. 1925 The food of the bottom fauna of the Plymouth fishing grounds. *J. Mar. Biol. Assoc. N. S.* **13**, 560–598, Plymouth

IFFT, J. D. & ZINN, D. J. 1946 Tooth replacement in the dogfish, *Mustelus canis*. *Anat. Rec. Philadelphia* **96**, 512 (Abstract), Philadelphia

— 1948 Tooth succession in the smooth dogfish, *Mustelus canis*. *Biol. Bull. Woods Hole* **93**, 100–106, Woods Hole, Mass.

JAECKEL, S. G. A. 1948 Zur Cephalopodenfauna der Nordsee und westlichen Ostsee. *Verh. deutsch. Zool. Ges. in Kiel 1947*, 204–212, Leipzig

— 1951a Lamellibranchier der freien Nordsee (nach den 'Poseidon'-Fängen der Fahrten 1902–1912). *Verh. deutsch. Zool. Ges. Wilhelmshaven*, 221–241, Leipzig

— 1951b Prosobranchiaten der freien Nordsee (nach den 'Poseidon'-Fängen 1902–1912). *Verh. deutsch. Zool. Ges. Wilhelmshaven*, 207–220, Leipzig

— 1953 Über Scaphopoden der Nordsee (hauptsächlich nach den 'Poseidon'-Fängen 1902–1912). *Kieler Meeresforsch.* **9**, 293–299, Kiel

— 1958 Cephalopoda. In: Grimpe & Wagler: *Tierwelt der Nord- und Ostsee*, Leipzig

JÄGER, G. 1876 Über die Funktion der Kiemenspalten. *Jb. Ver. vaterl. Naturk. Württemberg* **32**, 23–98, Stuttgart

JAKOBI, K. 1939 *Kiefergebiss und Zähne der Rochen und Haie.* Marburg (Elwert)

JACOBSHAGEN, E. 1923 Placoidorgane und Selachierzähne. *Anat. Anz.* **57** (Erg.-H.), 174–179, Jena

JACOBSHAGEN, E. & WINKLER, A. 1951 Zur vergleichenden Morphologie der Ganoidzähne. *Z. mikr. anat. Forsch.* **56**, 559–601, Leipzig

JAMES, W. W. 1923 The succession of teeth in Elasmobranches. *Proc. Zool. Soc. Lond.* **123**, 419–474, London

JANSSEN, A. 1937 *Tausend Jahre deutscher Walfang.* Leipzig (Brockhaus)

JARKE, J. 1956 Der Boden der südlichen Nordsee. I. Beitrag. Eine neue Bodenkarte der südlichen Nordsee. *Deutsch. Hydrogr. Z.* **9**, 1–9, Hamburg

JENSEN, AD. S. 1951a Do the Naticidae drill by mechanical or by chemical means? *Nature* **167**, 901, London

— 1951b Do the Naticidae (Gastropoda Prosobranchia) drill by chemical or by mechanical means? *Videsk. Medd. Dansk. naturh. Foren.* **113**, 251–261, Copenhagen

JESSEN, W. 1932a Die postdiluviale Entwicklung Amrums und seine subfossilen und rezenten Muschelpflaster. (Unter Berücksichtigung der gleichen Vorgänge auf den Inseln Sylt und Föhr.) *Jb. Preuss. Geol. Landesanst.* **53**, 1–69, Berlin

— 1932b Über rezente und fossile Organismenpflaster. *Paläont. Z.* **14**, 67–77, Berlin

JOHANSEN, A. C. 1901 Om Aflegringen af Molluskernes Skaller i Indsøer og i Havet. *Vid. Medd. naturh. Foren Kobenhavn*

JOHANSSON, K. E. 1927 Beiträge zur Kenntnis der Polychaeten-Familien Hermellidae, Sabellidae und Serpulidae. *Zool. Bidr. Uppsala* **11**, 1–184, Uppsala and Stockholm

JONES, R. 1844 Über das Häuten des Krebses. *Iris*, 912, Frankfurt a. M.

JÖNS, D. 1965 Zur Biologie und Ökologie von *Ligia oceanica* (L.) in der westlichen Ostsee. *Kieler Meeresforsch.* **21**, 203–207, Kiel

JORDAN, H. 1904 Die physiologische Morphologie der Verdauungsorgane bei *Aphrodite aculeata. Z. wiss. Zool.* **78**, 165–189, Leipzig

— 1910a Die Leistungen des Gehirns bei den krebsartigen Tieren, bes. bei *Cancer pagurus. Biol. Ctrbl.* **30**, 310–316, Leipzig

— 1910b Die Leistungen des Gehirnganglions bei den krebsartigen Tieren. *Pflüger's Arch.* **131**, 317–386, Bonn

— 1918 Über die Physiologie der Muskulatur und des zentralen Nervensystems bei hohlorganartigen Wirbellosen, insb. bei Schnecken. *Erg. Physiol.* **16**, 87–227, Wiesbaden

JØRGENSEN, B. C. 1943 On the water transport through the gills of bivalves. *Acta physiolog. scand.* **5**, 297–304, Stockholm

JUNGFER, W. 1954 Die Entstehung eines Warftnestes der Silbermöve. *Natur u. Volk* **84**, 432–435, Frankfurt a. M.

JÜNGST, H. 1937 Fischsterben im Kurischen Haff. *Geol. Meere u. Binnengew.* **1**, 352–354, Berlin

— 1942 Schillkalk ('Schelpkalk') als nationale Industrie. *Geol. Meere u. Binnengew.* **5**, 220–231, Berlin

JUNKER, H. 1940a Die Aufzucht der Seehunde in den Tiergrotten der Stadt Wesermünde. *Zool. Garten N. F.* **12**, 306–315

— 1940b Zwillingsgeburten bei Seehunden. *Wild u. Hund* **46**, 234–235

JUST, B. 1924 Über die Muskel- und Nervenphysiologie von *Arenicola marina. Z. vergl. Physiol.* **2**, 155–183, Berlin

KAHL, A. 1933 Ciliata libera et ectocommensalia. In: Grimpe & Wagler: *Tierwelt der Nord- und Ostsee*, Leipzig

KAISER, E. 1930 Das Fischsterben in der Walfischbucht. *Palaeobiologica* 3, 14–20, Vienna and Leipzig

KEIL, W. 1952 Über die Natur der schmelzartigen Aussenschichten auf den 'Nägeln' von *Raja clavata* L. und den 'Dornen' von *Acanthias vulgaris* Risso. *Z. Zellforsch., mikrosk. Anat.* **37**, 350–376, Berlin

KELLOGG, J. L. 1915 Ciliary mechanisms of lamellibranchs with descriptions of anatomy. *J. Morphol.* **26**, 625–701, Philadelphia

KELLOGG, R. & WHITMORE, JR. F. C. 1957 Marine mammals. In: *Treatise mar. ecology and paleoecology* **1**, Ecology

KESSEL, E. 1936 Über Verfärbung mariner Molluskenschalen durch Einlagerung von Eisen. *Zool. Anz.* **115**, 129–139, Leipzig

— 1938a Beobachtungen an *Eupagurus*-Wohngehäusen mit *Hydractinia*-Bewuchs. *Verh. Deutsch. Zool. Ges. Bremen* 154–163, Leipzig

— 1938b Der Gelbrand als Schneckenfresser. *Natur u. Volk* **68**, 572–574, Frankfurt a. M.

KINNE, O. 1954a Zur Biologie und Physiologie von *Gammarus duebeni* Lillj. VIII: Die Bedeutung der Kopulation für Eiablage und Häutungsfrequenz. *Biol. Zbl.* **73**, 190–202, Leipzig

— 1954b Über das Schwärmen und die Larvalentwicklung von *Nereis succinea* Leuckart (Polychaeta). *Zool. Anz.* **153**, 114–126, Leipzig

— 1956 Über Temperatur und Salzgehalt und ihre physiologisch-biologische Bedeutung. *Biol. Zbl.* **75**, 314–327, Leipzig

KITZLER, G. 1941 Die Paarungsbildung einiger Eidechsen. *Z. Tierpsychol.* **4**, 353–401, Berlin

KLAATSCH, H. 1890 Zur Morphologie der Fischschuppen. *Morphol. Jb.* **16**, 97–258, Leipzig

KLÄHN, H. 1932 Der quantitative Verlauf der Aufarbeitung von Sanden, Geröllen und Schalen in wässrigem Medium. *N. Jb. Min. Bul.* **67** Abt. B., 313–412, Stuttgart

— 1936 Die Anlösungsgeschwindigkeit kalkiger anorganischer und organischer Körper innerhalb eines wässrigen Mediums. *Zbl. Min. Geol. Palaeontol.* Abt. A. 328–348, Stuttgart

KLIE, W. 1929 Ostracoda. In: Grimpe & Wagler: *Tierwelt Nord- und Ostsee*, Leipzig

— 1938 Zwei neue Ostracoden aus der Ostsee. *Kieler Meeresforsch.* **2**, 345–351, Kiel

KLINGHARDT, F. 1930 Über fossile und lebende Schlangensterne und Bemerkungen über eine Schlangenstern- und Seelilien-Brekzie. *Zt. Deutsch. Geol. Ges.* **82**, 711–718, Berlin

KLUGE, F. & GÖTZE, A. 1951 *Etymologisches Wörterbuch der deutschen Sprache.* 15th edition. Berlin (De Gruyter & Co.)

KNER, R. Über den Flossenbau der Fische. *Sitz. Ber. Akad. Wiss. Wien, Math.-nat. Kl.* **41**, 807–824, 1860; **42**, 759–786, 1861; **43**, 123–152, 1861; **44**, 49–80, 1862, Vienna 1860–3

KOLLMER, W. E. 1961 Über die Ursachen des Massensterbens von Meeresfischen. *Natur u. Volk* **91**, 371–374, Frankfurt a. M.

KÖNIG, D. 1948 Über die Wohnweise einiger im Boden lebender Tiere des Wattenmeeres. *Verh. Deutsch. Zool. Ges. Kiel*, 402–410, Leipzig

KOENIGSWALD, R. VON 1930 Die Arten der Einregelung ins Sediment bei den Seesternen und Seelilien des unterdevonischen Bundenbacher Schiefers. *Senckenbergiana* **12**, 338–360, Frankfurt a. M.

KÖLLIKER, A. 1860 Über das ausgebreitete Vorkommen von pflanzlichen Parasiten in den Hartgebilden niederer Tiere. *Z. wiss. Zool.* **10**, 215–232, Leipzig

KOLOSVARY, G. V. 1942 Zur Frage des Feindbewuchses und zur Ernährungsbiologie von *Balanus perforatus angustus* Gmelin aus der Adria. *Zool. Anz.* **139**, 149–159, Leipzig

— 1944 Über die Ernährungsbiologie der Cirripedia Thoracica. *Annales. Hist.-Nat. Musei National. Hungar.* **37**, 132–162, Budapest

KOPSCH, F. 1897 Ei-Ablage von *Scyllium canicula* in dem Aquarium der zoologischen Station zu Rovigno. *Biol. Zbl.* **17**, 885–893, Leipzig

KOWALSKI, R. 1955 Untersuchungen zur Biologie des Seesternes *Asterias rubens* L. in Brackwasser. *Kieler Meeresforsch.* **11**, 201–213, Kiel

KRAMER, E. 1960 Zur Form und Funktion des Lokomotionsapparates der Fische. *Z. wiss. Zool.* **163**, 1–36, Leipzig

KRAUSE, H. R. 1950 Quantitative Schilluntersuchungen im See- und Wattengebiet von Norderney und Juist und ihre Verwendung zur Klärung hydrographischer Fragen. *Arch. Molluskenkde.* **79**, 91–116, Frankfurt a. M.

KREGER, D. 1940 On the ecology of *Cardium edule* L. *Arch. Néerland. Zool.* **4**, 157–200, Leiden

KREJCI-GRAF, K. 1932 Senkrechte Regelung von Schneckengehäusen. *Senckenbergiana* **14**, 295–299, Frankfurt a. M.

KREY, JOH. 1956 Die Trophie küstennaher Meeresgebiete. *Kieler Meeresforsch.* **12**, 46–64, Kiel

KRISTENSEN, I. 1956 Een massale stranding van de makreelgeep. *De levende Natuur* **59**, 59–64, Amsterdam

— 1957 Differences in density and growth in a cockle population in the Dutch Wadden Sea. *Diss.* Leiden

KROGH, A. & SPÄRCK, R. 1936 On a new bottom-sampler for investigation of the micro-fauna of the sea bottom. *Kgl. Danske Vidensk. Selsk. Biol. Medd.* **13**, 1–27, Copenhagen

KRÜGER, F. 1958 Zur Atmungsphysiologie von *Arenicola marina* L. *Helgol. Wiss. Meeresunters.* **6**, 193–201, Kiel

— 1959 Zur Ernährungsphysiologie von *Arenicola marina* L. *Verh. Deutsch. Zool. Ges. in Frankfurt a. M.* 115–120, Leipzig

KRUMBACH, TH. 1918 Napfschnecken in der Gezeitenwelle und der Brandungszone der Karstküste. *Zool. Anz.* **49**, 113–123, Leipzig

KUHK, R. 1956 Hagelunwetter als Verlustursache bei Störchen und anderen Vögeln. *Vogelwarte* **18**, 180–182, Stuttgart

KÜHL, H. 1932 Beobachtungen zur Plastizität des Nervensystems bei langschwänzigen Krebsen. *Pflüger's Arch.* **229**, 636–641, Bonn

— 1950a Studien über die Sandklaffmuschel *Mya arenaria. Arch. Fischereiwiss.* **2**. 25–39, Braunschweig

— 1950b Vergleichende biologische Untersuchungen über den Hafenbewuchs. *Verh. Deutsche Zool. Ges. Marburg*, 233–244, Leipzig

— 1951 Über die Siedlungsweise von *Mya arenaria. Verh. Deutsch. Zool. Ges. Wilhelmshaven*, 385–391, Leipzig

— 1952a Über die Hydrographie von Wattenpfützen. *Helgol. Wiss. Meeresunters.* **4**, 101–106, List

— 1952b Über das Aufsuchen des Siedelplatzes durch die Cyprislarven von *Balanus improvisus* Darw. *Verh. Deutsch. Zool. Ges. Freiburg*, 189–200, Leipzig

— 1952c Unsere gegenwärtigen Kenntnisse über die Biologie der Terediniden. *Hansa* **89**, 111–128, Hamburg

— 1952–1953 Studien über die Klaffmuschel *Mya arenaria*. Die Myasiedlungen am Hakensand (Elbmündung). *Arch. Fischereiwiss.* **4**, 126–132, Braunschweig

— 1954 Über das Auftreten von *Elminius modestus* Darwin in der Elbmündung. *Helgol. Wiss. Meeresunters.* **5**, 53–56, List

— 1955 Studien über die Klaffmuschel *Mya arenaria*. 3. Das Junggut. *Arch. Fischereiwiss.* **6**, 33–44, Braunschweig

— 1957a Der Befall durch Bohrmuscheln und Bohrkrebse in Norderney, Wilhelmshaven, List a. Sylt und Kiel in den Jahren 1953–1955. *Z. angew. Zool.* **44**, 257–279, Berlin

— 1957b Das Auftreten mariner Holzschädlinge (Bohrmuscheln und Bohrkrebse) in Flussmündungen und Häfen in Abhängigkeit von den Wasserverhältnissen. *Fischereiblatt* **12**, 1–16

— 1958 Das Auftreten mariner Holzschädlinge (Bohrmuscheln und Bohrkrebse) in Flussmündungen und Häfen in Abhängigkeit von den Wasserverhältnissen. *Arb. Deutsch. Fischerei-Verb.* **10**, 19–34, Hamburg

— 1964–1965 Die Scyphomedusen der Elbmündung. *Veröff. Inst. f. Meeresforsch. Bremerhaven* **9**, 84–94, Bremen

KUHLEMANN, P. 1942 Springflut auf der Vogelinsel. *Natur u. Volk* **72**, 208–212, Frankfurt a. M.

KÜHN, H. 1952 *Die Felsbilder Europas*. Stuttgart (W. Kohlhammer)

KÜHNELT, W. 1930 Bohrmuschelstudien I. *Paläobiologica* **3**, 53–91, Vienna

— 1933 Bohrmuschelstudien II. *Paläobiologica* **5**, 371–408, Vienna

— 1954 Betrachtungen zum gegenwärtigen Stand in der Biozönotik. *Verh. Zool.-Bot. Ges. Wien* **94**, 29–39, Vienna

KÜKENTHAL, W. 1890 Über die Anpassung von Säugetieren an das Leben im Wasser. *Zool. Jb. Abt. Syst.* **5**, 32–97, Jena

— 1908 Über die Ursache der Asymmetrie des Walschädels. *Anat. Anz.* **33**, 609–618, Leipzig

— 1909 Untersuchungen an Walen. *Jenaische Z. Naturw.* **45**, 563, Jena

KÜKENTHAL, W. & KRUMBACH, T. 1927 Crustacea 3. *Handb. Zool. 1. Hälfte*, 840–1038, Berlin and Leipzig

KÜNNEMANN, CHR. 1936 *Meer und Mensch am Jadebusen*. Oldenburg

KYLIN, H. 1935 Über einige kalkbohrende Chlorophyceen. *Kgl. fysiogr. Sällsk. Lund. Förhandl.* **5**, 186–236, Lund

LADD, H. S. 1959 Ecology, Paleontology and Stratigraphy. Understanding of the habits of living organisms aids interpretations of fossiliferous sediments. *Science* **129**, 69–78, New York.

LAMBRECHT, K. 1933 *Handbuch der Palaeornithologie*. Berlin (Borntraeger)

LAMMENS, J. J. 1967 Growth and reproduction in a tidal flat population of *Macoma balthica* (L.). *Nth. J. Sea Res.* **3**, 315–382, Den Helder

LANDOLT, H. H. 1947 Über den Zahnwechsel bei Selachiern. *Rev. suisse Zool. Genève* **54**, 305–367, Geneva

LANE, F. W. 1957 *Kingdom of the* Octopus. *The life-history of the Cephalopoda*. London (Jarrolds)

LANG, A. 1888–1894 *Lehrbuch der vergleichenden Anatomie der wirbellosen Tiere*. Jena (Fischer)

LANG, K. 1949 Contribution to the ecology of *Priapulus caudatus* Lam. *Ark. Zool.* *41A*, 1–12, Stockholm

LARSEN, E. B. 1936 Biologische Studien über die Tunnel-grabenden Käfer auf Skallingen. *Medd. Skalling-Laboratorium* **3**, 1–231, Copenhagen

LASKER, R. & LANE, CH. E. 1953 The origin and distribution of nitrogen in *Teredo bartschi. Biol. Bull.* **105**, 316–319, Lancaster

LAUGHTON, A. S. 1959 Die Photographie des Meeresbodens. *Endeavour* **18**, 178–185, London

LEEGE, O. 1917 Die Nahrung der Silbermöven an der ostfriesischen Küste. *Ornith. M. schr. deutsch. Ver. Schutze d. Vogelwelt* **42**, 110–116, 123–134, Halle

— 1928 20 Jahre Vogelinsel Memmert. *Ornith. M. schr. deutsch. Ver. Schutze d. Vogelwelt* **53**, 2–24, Halle

— 1935 *Werdendes Land in der Nordsee.* Oehringen (Ferd. Rau)

LENGERKEN, H. V. 1929 *Die Salzkäfer der Nord- und Ostseeküste mit Berücksichtigung der angrenzenden Meere sowie des Mittelmeeres, des Schwarzen und des Kaspischen Meeres.* Leipzig (Akad. Verl. Ges.)

LEONARDO DA VINCI Tagebücher und Aufzeichnungen. *Nach den italienischen Handschriften übersetzt und herausgegeben von Theodor Lücke*, Leipzig, 1940

LIEBERKIND, J. 1928 Echinoderma, Stachelhäuter oder Echinodermen. *Dahl's Tierwelt Deutschlands und der angrenzenden Meeresteile etc.* **4**, 263–329, Jena

LINKE, O. 1939 Die Biota des Jadebusenwattes. *Helgol. Wiss. Meeresunters.* **1**, 201–348, Heligoland

— 1951 Neue Beobachtungen über Sandkorallen-Riffe in der Nordsee. *Natur u. Volk* **81**, 77–84, Frankfurt a. M.

— 1956 Quallen-Spülsäume. Ökologische Voraussetzung und aktuogeologische Ausdeutung. *Natur u. Volk* **86**, 119–127, Frankfurt a. M.

LOEB, J. 1891 *Untersuchungen zur physiologischen Morphologie der Tiere.* 1. *Über Heteromorphose.* Würzburg

LOMAN, J. C. C. 1907 Biologische Beobachtungen an einem Pantopoden. *Tijdschr. nederlandsche dierkundige Vereen. Leiden* **2**, 255–284, Leiden

LÖRCHER, E. 1931 Eine neue fossile Qualle aus den Opalinusschichten und ihre paläogeographische Bedeutung. *Jbr. u. Mitt. Oberrh. geol. Ver.* 44–46, Stuttgart

LORENZ, K. 1950 Ganzheit und Teil in der tierischen und menschlichen Gemeinschaft. *Studium generale* **3**, 455–499, Berlin, Göttingen, Heidelberg

LÜDERS, K. 1930 Entstehung der Gezeitenschichtung auf den Watten im Jadebusen. *Senckenbergiana* **12**, 229–254, Frankfurt a. M.

— 1939 Sediments of the North Sea. *Recent marine sediments, a symposium.* Amer. Assoc. Petrol. Geol. Tulsa, Oklahoma, U.S.A., London

— 1953 Die Entstehung der ostfriesischen Inseln und der Einfluss der Dünenbildung auf den geologischen Aufbau der ostfriesischen Küste. *Probl. Küstenforsch. südl. Nordseegeb.* **5**, 5–14, Hildesheim

— 1963 Katfische im Wilhelmshavener Seewasseraquarium. *Natur u. Museum* **94**, 165–168, Frankfurt a. M.

LÜDERS, K. & TRUSHEIM, F. 1929 Beiträge zur Ablagerung mariner Mollusken in der Flachsee. 1. K. Lüders: Entstehung und Aufbau von Grossrücken mit Schillbedekhung in Flut- bzw. Ebbetrichtern der Aussenjade. *Senckenbergiana* **11**, 123–142, Frankfurt a. M.

LUDWIG, H. & HAMANN, O. 1889–1907 Echinodermen. *Bronn's Kl. u. Ord. Tierr. II.* **3**, Leipzig

LÜHMANN, M. 1951a Gebiss und Zahnwechsel der Kattfische. *Verh. Anat. Ges.* **48**, Vers. Kiel, 22–25, Jena

— 1951b Gebiss und Zahnwechsel der Katfische. *Fischereiwelt H.* **5**, Kiel

— 1954 Die histologischen Grundlagen des periodischen Zahnwechsels der Katfische und Wasserkatzen (Fam. Anarhichidae, Teleostei). *Z. Zellforsch. u. mikr. Anat.* **40**, 470–509, Berlin

LUNDBECK, J. 1954 Gedanken zur Frage der Bildung und Veränderung natürlicher und genutzter Tierbestände, insbesondere vom Standpunkt der praktischen Fischerei. *Arch. Hydrobiol.* **49**, 225–257, Stuttgart

LUTHER, W. 1930 Versuche über die Chemorezeption der Brachyuren. *Z. vergl. Physiol.* **12**, 177–205, Berlin

— 1931 Zur Frage der Chemorezeption der Brachyuren und Anomuren. *Zool. Anz.* **94**, 147–153, Leipzig

LUTZE, I. 1938 Über Systematik, Entwicklung und Ökologie von *Callianassa*. *Helgol. Wiss. Meeresuntersuch.* **1**, 161–169, Heligoland

MAAS, O. 1911 Abgüsse rezenter Tiefseemedusen zum Vergleich mit Fossilien aus der Kreide. *Verh. Deutsch. Zool. Ges. Basel*, 186–192, Leipzig

MACGINITIE, G. E. 1934 The natural history of *Callianassa californiensis* Dana. *Amer. Midland Naturalist* **15**, 166–177, Notre Dame, Ind.

— 1941 On the method of feeding of four pelecypods. *Biol. Bull.* **80**, 18–25, Lancaster

MACGINITIE, G. E. & MACGINITIE, N. 1949 *Natural history of marine animals*. New York

MACINTOSH, W. C. 1886 On the boring of certain Annelids. *Ann. Mag. Nat. Hist., London* (*4*) **2**, 276–295, London

MÄGDEFRAU, K. 1937 Lebensspuren fossiler 'Bohr'-Organismen. *Beitr. naturkl. Forsch. SW-Deutschl.* **2** (1), 54–67, Karlsruhe

MANGOLD, E. 1908 Studien zur Physiologie des Nervensystems der Echinodermen. *Pflüger's Arch.* **123**, 1–40, Bonn

MANN, H. 1951 Vergleichende Untersuchungen an einigen Muscheln des Watts. *Verh. deutsch. Zool. Ges. Wilhelmshaven*, 374–378, Leipzig

MANN, H. & PIEPLOW, W. 1939 Der Kalkhaushalt bei der Häutung der Krebse. *S.-B. Ges. naturf. Fr. Berlin 1938*, 1–17, Berlin

MANNING, R. B. & KUMPF, H. E. 1959 Preliminary investigation of the faecal pellets of certain Invertebrates of the South Florida Area. *Bull. Marine Sc. of the Gulf of Caribbean* **9**, 291–309, Miami

MARCUS, E. 1921 Über die Verbreitung der Meeresbryozoen. *Zool. Anz.* **53**, 205–221, Leipzig

— 1926 Bryozoa. In: Grimpe & Wagler: *Tierwelt Nord- und Ostsee*, Leipzig

— 1934 Über den Einfluss des Kriechens auf Wirbelzahl und Organgestalt bei Apoden. *Biol. Zbl.* **54**, 518–523, Leipzig

MARKERT, F. 1896 Die Flossenstacheln von *Acanthias*. *Zool. Jb. Abt. Anat.* **9**, 665–722, Jena

MARQUARD, E. 1946 Beiträge zur Kenntnis des Selachiergebisses. *Rev. suisse Zool. Genève* **53**, 73–132, Geneva

MATHES, H. W. 1956 *Einführung in die Mikropaläontologie*. Leipzig (Hirzel)

MEIJERING, M. P. D. 1954 Zur Frage der Variationen in der Ernährung der Silbermöve, *Larus argentatus* Pontopp. *Ardea* **42**, 163–175, Leiden

MERTENS, R. 1926 Reptilia. In: Grimpe & Wagler: *Tierwelt Nord- und Ostsee*, Leipzig

MERTIN, H. 1941 Decapode Krebse aus dem subherzynen und Braunschweiger Emscher und Untersenon, sowie Bemerkungen über einige verwandte Formen in der Oberkreide. *Nova Acta Leopoldina, N. F.* **10**, 149–264, Halle

METALNIKOFF, S. 1900 *Sipunculus nudus*. *Z. wiss. Zool.* **68**, 261–322, Leipzig

MEYER, P. F. 1931 Ein Kabeljau und seine Malzeit. *Natur u. Mus.* **61**, 416–420, Frankfurt a. M.

MEYN, L. 1859 Wurmsandstein (*Mitteilungen des Vereins nördlich der Elbe für Verbreitung naturwissenschaftlicher Kenntnisse*, 102–104). Angeheftet an die *Jahrbücher für die Landeskunde der Herzogtümer Schleswig, Holstein und Lauenburg*, herausgegeb. v. d. S.H.L. Gesellschaft für vaterländische Geschichte. **2**, Kiel

MII, H. 1957 Peculiar accumulation of drifted shells. *Saito Ho-on kai Museum. Res. Bull.* **26**, 17–24

MILLER, R. C. 1924a The boring mechanism of *Teredo*. *Univ. Calif. Publ. Zool.* **26**, 41–80, Berkeley, Calif.

— 1924b The boring habits of the shipworm. *The Sci. Monthly* **19**, 434–440, New York

— 1925 A study of the nutrition of woodboring molluscs. *Anat. Rec.* **31**, 325, Philadelphia

MILLER, R. C. & BOYNTON, L. C. 1926 Digestion of wood by the shipworm. *Sci.* **63**, 524, New York

MINER, R. W. 1950 *Field book of seashore life*. New York (van Rees Press)

MITTELSTAEDT, H. 1956 Regelungsvorgänge in der Biologie. *Beihefte zur Regelungstechnik*, Munich (Oldenbourg)

MÖBIUS, K. 1871 Das Thierleben am Boden der deutschen Ost- und Nordsee. *Sammlg. gemeinverst. wiss. Vorträge* **6**, 3–32, Hamburg

— 1877 *Die Auster und die Austernwirtschaft*. Berlin

MOHR, E. 1937 Organisation der Meldungen von Walstrandungen in den Nordseeländern. *Verh. deutsch. Zool. Ges. Bremen 1937*, 118–121, Leipzig

— 1952 Der Stör. *Neue Brehmbücherei* (Akad. Verlagsges. G. & P.). Leipzig

— 1955 Der Seehund. *Neue Brehmbücherei*. Wittenberg (Ziemsen)

MOORE, H. B. 1931a The specific identification of faecal pellets. *J. mar. biol. Assoc.* **17**, 359–365, Plymouth

— 1931b The muds of the Clyde Sea area, III. Chemical and physical conditions; rate and nature of sedimentation and fauna. *J. Mar. Biol. Assoc.* **17**, 2, Plymouth

— 1931c The systematic value of a study of molluscan faeces. *Proc. malacol. Soc.* **19**, 6, London

— 1932a The faecal pellets of the Trochidae. *J. Mar. Biol. Assoc.* **18**, 235–241, Plymouth

— 1932b The faecal pellets of the Anomura. *Proc. Roy. Soc. Edinbg.* **52**, 296–308, Edinburgh

— 1933a The faecal pellets of *Hippa asiatica*. *Proc. Roy. Soc. Edinbg.* **53**, 243–258, Edinburgh

— 1933b Faecal pellets from marine deposits. *Discovery Rep.* **7**, 17–26

— 1939 Faecal pellets in relation to marine deposits. *Amer. Assoc. Petrol. Geol.* 'Recent marine sediments', 516–524

MOORE, H. B. & KRUSE, P. 1956 *A review of present knowledge of faceal pellets*. Marine Laboratory, Univ. of Miami, July 1–25

MORTENSEN, TH. 1938 Über die stratigraphische Verwendbarkeit der mikroskopischen Echinodermen-Reste. *Senckenbergiana* **20**, 342–345, Frankfurt a. M.

MORTENSEN, TH. & LIEBERKIND, J. 1928 Echinoderma. In: Grimpe & Wagler: *Tierwelt der Nord- und Ostsee*, Leipzig

MÜLLER, A. H. 1951 Grundlagen der Biostratonomie. *Abh. Deutsch. Akad. Wiss.* **3**, 3–147, Berlin

MÜLLER, C. D. 1966 Seltene Bryozoen-Kugelform in einem Spülsaum. *Natur u. Museum* **96**, 176–179, Frankfurt a. M.

MÜLLER, CL. D. 1956 Die Epifauna auf den Hölzern der *Teredo*-Untersuchungsstation in Norderney. *Jber. 1955* **7**, 106–130, Norderney

MÜLLER, G., REINECK, H. E. & STAESCHE, W. 1968 Mineralogisch-sediment-petrographische Untersuchungen an Sedimenten der Deutschen Bucht (südöstliche Nordsee). *Senckenbergiana lethaea* **49**, 347–365, Frankfurt a. M.

MÜLLER, G. W. 1894 Ostracoda. *Fauna u. Flora des Golfes von Neapel*, Berlin

— 1912 Ostracoda. *Das Tierreich*, Berlin

NAIR, N. B. 1958 The marine timber-boring molluscs and crustaceans of western Norway. *Publ. Biol. Stat., Espegrend* **25**, 1–23, Bergen

NATHORST, A. G. 1881 Om aftryck af medusor i Sveriges kambriska lager. *Kgl. Svenska Vsk.-Akad. Handlingar N. F.* **19**, 1–34, Stockholm

— 1886 *Mémoire sur quelques traces d'animaux sans vertébres etc. et de leur portée palèontologique*. Stockholm

NEBESKI, O. 1880 Beiträge zur Kenntnis der Amphipoden der Adria. *Arb. Uni. Wien* **3**, 27–78, Vienna

NEWELL, G. E. 1948 A contribution to our knowledge of the life history of *Arenicola marina* L. *J. Mar. Biol. Ass.* **27**, 554–580, Plymouth

— 1949 The later larval life of *Arenicola marina* L. *J. Mar. Biol. Ass.* **28**, 635–639, Plymouth

NICOL, E. A. T. 1931 The feeding mechanism, formation of the tube, and physiology of digestion in *Sabella pavonina*. *Trans. Roy. Soc. Edinb.* **56**, 537–598, Edinburgh

NIERSTRASZ, H. F. & SCHUURMANS-STEKHOVEN, J. H. 1930 Anisopoda. In: Grimpe & Wagler: *Tierwelt Nord- und Ostsee*, Leipzig

NIETHAMMER, G. 1938, 1942 *Handbuch der deutschen Vogelkunde*. **2** and **3**, Leipzig (Akad. Verl. Ges.)

NILSSON 1912 Beiträge zur Kenntnis des Nervensystems der Polychaeten. *Zool. Bidr. Uppsala* **1**, 85–160, Uppsala and Stockholm

NILSSON, E. & CANTELL, C. A. 1921–1922 Cirripedien-Studien zur Kenntnis der Biologie, Anatomie u. Systematik dieser Gruppe. *Zool. Bidrag* **7**, 75–130, Uppsala

NITSCHE, H. 1871 Beiträge zur Kenntnis der Bryozoen: Über die Morphologie der Bryozoen. *Z. wiss. Zool.* **21**, 415–498, Leipzig

NÜMANN, W. 1957a Natürliche und künstliche 'redwater' mit anschliessendem Fischsterben im Meer. *Arch. Fischereiwiss.* **8**, 204–209, Braunschweig

— 1957b Kälteschocks als natürliche Ursachen periodisch auftretenden Fischsterbens im Bosporus. *Arch. f. Fischereiwiss.* **8**, 210–212, Braunschweig

OPITZ, R. 1930 Über das Präparieren von Versteinerungen im Hunsrück-Dachschiefer. *Natur u. Museum* **60**, 135–140, Frankfurt a. M.

— 1931 Seltene Seesternfunde von Bundenbach. *Natur u. Museum* **61**, 352–354, Frankfurt a. M.

ORTMANN, W. 1901 Crustacea. *Bronn's Kl. u. Ord.* **5**, Leipzig and Heidelberg

ORTON, J. H. 1930 On the Oyster-drills in the Essex estuaries. *Essex Naturalist* **22**, 298–306, Stratford (Essex)

OSBURN, R. C. 1914 Movements of sea anemones. *Bull. Zool. Soc. New York* **17**, 1163–1166

OSTENFELD, C. H. 1931 Concluding remarks on the plankton on the quarterly cruises in the years 1902–1908. *Cons. perm. internat. Explor. de la mer. Bull. trimestr.* IV

OWEN, D. M. 1958 Photography underwater. *Oceanus* **6**, 22–39
PAPP, A. 1939 Über das Vorkommen von Austern und Balanen in der Gaadener Bucht. *Paläobiologica* **7**, 212–216, Vienna
— 1944 Die senkrechte Einregelung von Gastropodengehäusen in Tertiärschichten des Wiener Beckens. *Paläobiologica* **8**, 144–153, Vienna
PAPP, A., ZAPFE, H., BACHMAYER, F. & TAUBER, A. F. 1947 Lebensspuren mariner Krebse. *Sitz.-Ber. Akad. Wiss. Math.-nat. Kl. Abt. I.* **115**, 281–317, Vienna
PARKE, MW. & MOORE, H. B. 1935 The biology of *Balanus balanoides*. II. Algae infection of the shell. *J. mar. biol. Assoc. unit. Kingd.* **20**, 49–56, London
PARKER, G. H. 1915 The locomotion of actinians. *Science* **41**, 471, New York
— 1917 Pedal locomotion in actinians. *Science* **22**, 111–124, New York
PAUL, J. H. & SHARPE, J. S. 1916 Studies in calcium metabolism. I. The deposition of lime salts in the integument of decapod Crustacea. *J. Physiol.* **50**, 183–192, London
PAX, F. 1925 Hexacorallia. *Kükenthal-Krumbach's Handb. Zool.* **1**, 889–901, Berlin (De Gruyter)
— 1928 Anthozoa. *Dahl's Tierwelt Deutschlands u. angrenzenden Meeresteile* **4**, 189–240, Jena (Fischer)
— 1934 Anthozoa. In: Grimpe & Wagler: *Tierwelt Nord- und Ostsee*, Leipzig
PEACOCK, A. D., COMRIE, L. & GREENSHIELDS, F. 1936 The false killer whales stranded in the Tay Estuary. *Scottish Naturalist* **220**, 93–104, Edinburgh
PELSENEER, P. 1925 Comment mangent divers gastropodes aquatiques. *Ann. Soc. Royal Zool. de Belg.* **55**, 31–45, Brussels
PETERS, H. M. 1948 *Grundfragen der Tierpsychologie. Ordnungs- und Gestaltprobleme.* Stuttgart
— 1957 Über die Beziehungen der Tiere zu ihrem Lebensraum. *Studium generale* **10**, 523–531, Berlin, Göttingen, Heidelberg
PETERS, N. & PANNING, A. 1933 Die chinesische Wollhandkrabbe (*Eriocheir sinensis* H. M. Edw.) in Deutschland. *Zool. Anz. Ers.-Bd.* **104**, 1–156, Leipzig
PETERSEN, C. G. JOH. 1915 On the animal communities of the sea bottom in the Skagerrak, the Christiana Fjord and the Danish waters. *Rep. Dan. Biol. Stat.* **23**, 3–28, Copenhagen
— 1918 The sea-bottom and its production of fishfood. *Rep. Dan. Biol. Stat.* **25**, 1–62, Copenhagen
— 1925 The motion of whales during swimming. *Nature* **116**, 327–329, London
PETERSEN, C. G. JOH. & BOYSEN-JENSEN, P. 1911 Valuation of the sea I. Animal life of the sea-bottom, its food and quantity. *Rep. Dan. Biol. Stat.* **20**, 1–80, Copenhagen
PEUS, F. 1954 Auflösung der Begriffe 'Biotop' und 'Biozönose'. *Deutsch. Entomol. Z. N. F.* **1**, 271–308, Berlin
PEYER, B. 1957 Über bisher als Fährten gedeutete problematische Bildungen aus den oligozänen Fischschiefern des Sernftales. *Schweiz. Paläont. Abh. Serie Zoologie* **73** (164), 1–33, Basle
PHILLIPS, J. 1853 Notes on a living specimen of *Priapulus caudatus*, dredged off the coast of Scarborough. *Rep. 23 Meeting Brit. Ass.* 70–71, London
PIA, J. 1933 Die rezenten Kalksteine. *Mineralogische u. petrographische Mitteilungen*, Erg. Bd., Leipzig (Akad. Verlagsges.)
— 1937 Die kalklösenden Thallophyten. *Arch. Hydrobiol. u. Planktonkde.* **31**, 264–328 and 341–398, Stuttgart
PLATE, L. 1896 Über den Habitus und die Kriechweise von *Caecum auriculatum* de Fol. *Ges. Naturfreunde Berlin* **7**, 130–133, Berlin

PLATH, M. 1965 Ein im Gezeitenbereich des Wattenmeeres selbsttätig arbeitendes Sinkstoff-Schöpfgerät und die Bedeutung der Wattfauna für die Bildung von Sinkstoffen. *Die Küste. Arch. Forschung u. Technik Nord- u. Ostsee* **13**, 117–132, Heide i. H.

POKORNY, V. 1958 *Grundzüge der zoologischen Mikropaläontologie.* **2**, Berlin (VEB Deutsch. Ver. Wiss.)

POMPECKJ, J. F. 1922 Das Ohrskelett von *Zeuglodon. Senckenbergiana* **4**, 43–100, Frankfurt a. M.

POOLE, D. F. G. 1956 The fine structure of the scales and teeth of *Raja clavata. Quart. J. micr. Sci.* **97**, 99–107, London

POPTA, C. M. L. 1901 Les appendices des ancs branchiaux des poissons. *Ann. sc. nat.* (7) *Zool.* **12**, 139–215, Paris

PORTMANN, A. 1926 Die Kriechbewegung von *Aiptasia carnea.* Ein Beitrag zur Kenntnis der neuromuskulären Organisation der Aktinien. *Z. vergl. Physiol.* **4**, 659–667, Berlin

— 1948 *Einführung in die vergleichende Morphologie der Wirbeltiere.* Basle (Schwabe & Co.)

— 1952 *Die Tiergestalt. Studien über die Bedeutung der tierischen Erscheinung.* Basle (Reinhardt)

POUVOT, G. 1885 Recherches anatomiques et morphologiques sur le système nerveux des Annélides polychètes. *Arch. Zool. exper. et gén.* **3**, 210–336, Paris

PRATJE, O. 1924 Korallenbänke in tiefem und kühlem Wasser. *Cbl. Min. etc. 1924*, 410–415, Berlin

— 1929 Fazettieren von Molluskenschalen. *Palaeont. Z.* **11**, 151–169, Berlin

— 1931 Die Sedimente der Deutschen Bucht. Eine regional-statistische Untersuchung. *Wiss. Meeresuntersuch. Helgol.* **18**, 1–126, Oldenburg

— 1934 Die Schlickgebiete der Deutschen Bucht und die Beziehungen zwischen Strömung und Sediment. *Geol. Rundschau* **25**, 145–160, Berlin

— 1937 Das Werden der Nordsee. *Bremer Beitr.* **4**, 64–94, Bremen

— 1949a Der Meeresboden als Lebensgrundlage. *Fischereiwelt, Beih.* **1**, 31–33, Bremerhaven and Hamburg-Altona

— 1949b Bodenbedeckung. *Nordsee-Handbuch*, 1st edition, 1–15, Hamburg

— 1950 Die Bodenbedeckung des Englischen Kanals und die maximalen Gezeitenstromgeschwindigkeiten. *Dt. Hydrogr. Z.* **3**, 201–205, Hamburg

— 1951 Die Deutung der Steingründe in der Nordsee als Endmoränen. *Deutsch. Hydrogr. Z.* **4**, 106–114, Hamburg

— 1951b Die Fortsetzung der Endmoränen am Boden der Nordsee. *Z. Deutsch. Geol. Ges.* **103**, 75–77, Hanover

PRELL, H. 1911 Beiträge zur Kenntnis der Lebensweise einiger Pantopoden. *Bergens Mus. Aarbok, naturvidensk. Raebbe Nr. 10*, Bergen

PREYER, O. 1886 Über die Bewegungen der Seesterne. *Mitt. Zool. Stat. Neapel* **7**, 27–127, Naples

QUENSTEDT, W. 1927 Beiträge zum Kapitel Fossil und Sediment vor und bei der Einbettung. *N. Jb. Min. etc.* (*Pompeckj-Festb.*) *Abt. B.* **58**, 353–432, Stuttgart

RAUSCHENPLAT, E. 1901 Über Nahrung von Tieren aus der Kieler Bucht. *Wiss. Meeresuntersuch. Kiel, N. F.* **5**, 85–151, Kiel and Leipzig

RAUTHER, M. 1925 Die Syngnathiden des Golfes von Neapel. *Fauna e Flora del Golfo die Napoli*, 36a Monogr.

— 1940 Echte Fische. Teil 1. Anatomie, Physiologie und Entwicklungsgeschichte.

1. Hälfte. *Bronn's Kl. u. Ord. Tierreichs* **6** (Wirbeltiere) I. Abt. 2. Buch, Leipzig (Akad. Ver. Ges.)

REIBISCH, J. 1927 Amphipoda. *Kükenthal's Handb. Zool.* **3**, 767–808, Berlin

REINECK, H. E. 1956a Der Wattenboden und das Leben im Wattenboden. Ein geologischer. Streifzug. *Natur u. Volk* **86**, 268–284, Frankfurt a. M.

— 1956b Wattenmeer im Winter. *Senckenbergiana leth.* **37**, 129–146, Frankfurt a. M.

— 1957 Stechkasten und Deckweiss, Hilfsmittel des Meeresgeologen. *Natur u. Volk* **87**, 132–134, Frankfurt a. M.

— 1958a Wühlbau-Gefüge in Abhängigkeit von Sediment-Umlagerungen. *Senckenbergiana leth.* **39**, 1–23, Frankfurt a. M.

— 1958b Über Gefüge von orientierten Grundproben aus der Nordsee. *Senckenbergiana leth.* **39**, 25–36, Frankfurt a. M.

— 1958c Über das Härten und Schleifen von Lockersedimenten. *Senckenbergiana leth.* **39**, 49–54, Frankfurt a. M.

— 1958d Kastengreifer und Lotröhre 'Schnepfe'. Geräte zur Entnahme ungestörter, orientierter Meeresgrundproben. *Senckenbergiana leth.* **39**, 45–48, Frankfurt a. M.

— 1958e Longitudinale Schrägschichtung im Watt. *Geol. Rdsch.* **47**, 73–82, Stuttgart

— 1959 Wenn eine Seehunds-Spur versteinerte. *Natur u. Volk* **89**, 47–53, Frankfurt a. M.

— 1960 Einige Beispiele von heutzeitlichen Tieren, die ihre Gänge teilweise oder ganz verfüllen. *Natur u. Volk* **90**, 282–288, Frankfurt a. M.

— 1960 Über eingeregelte und verschachtelte Röhren des Goldköcher-Wurmes (*Pectinaria koreni*). *Natur u. Volk* **90**, 334–337, Frankfurt a. M.

— 1963 Der Kastengreifer. *Natur u. Museum* **93**, 102–108, Frankfurt a. M.

— 1963 Sedimentgefüge im Bereich der südlichen Nordsee. *Abh. senckenberg. naturforsch. Ges.* **505**, 1–138, Frankfurt a. M.

— 1968 Lebensspuren von Herzigeln. *Senckenbergiana lethaea* **49**, 311–319, Frankfurt a. M.

REINECK, H. E., GUTMANN, W. F. & HERTWECK, G. 1967 Das Schlickgebiet südlich Helgoland als Beispiel rezenter Schelfablagerungen. *Senckenbergiana lethaea* **88**, 219–275, Frankfurt a. M.

REINECK, H. E., DÖRJES, J., GADOW, S. & HERTWECK, G. 1968 Sedimentologie, Faunenzonierung und Faziesabfolge vor der Ostküste der inneren Deutschen Bucht. *Senckenbergiana lethaea* **49**, 261–309, Frankfurt a. M.

REINECK, M. 1959 Neue Erfahrungen bei der Aufzucht von Seehunds-Säuglingen. *Natur u. Volk* **89**, 43–46, Frankfurt a. M.

REINECK, M. & REINECK, H. E. 1956 Ein Heuler wird aufgezogen. *Natur u. Volk* **86**, 397–407, Frankfurt a. M.

— 1958 Junge Seehunde, ein Aufzuchtproblem. *Natur u. Volk* **88**, 37–44, Frankfurt a. M.

REMANE, A. 1928 Kinorhyncha. In: Grimpe & Wagler: *Tierwelt Nord- und Ostsee*, Leipzig

— 1934 Die Brackwasserfauna. *Verh. Deutsch. Zool. Ges.* **7**, 34–74, Leipzig

— 1936 *Monobryozoon ambulans* n. g. n. sp., ein eigenartiges Bryozoon des Meeressandes. *Zool. Anz.* **113**, 161–167, Leipzig

— 1938 Ergänzende Mitteilungen über *Monobryozoon ambulans* A. Remane. *Kieler Meeresforsch.* **2**, 356–358, Kiel

— 1940 Einführung in die zoologische Ökologie der Nord- und Ostsee. In: Grimpe & Wagler: *Tierwelt Nord- und Ostsee*, Leipzig

— 1943 Die Bedeutung der Lebensformtypen für die Ökologie. *Biologia General.* **17**, 164–182, Vienna

— 1950 Ordnungsformen der lebenden Natur. *Studium generale* **3**, 404–410, Berlin, Göttingen, Heidelberg

— 1951 Marine Schillablagerungen im Süsswasser und aeolischer Hydrobienschill. *Kieler Meeresforsch.* **8**, 98–101, Kiel and Leipzig

— 1952 Die Besiedlung des Sandbodens im Meere und die Bedeutung der Lebensform-Typen für die Ökologie. *Verh. deutsch. zool. Ges. Wilhelmshaven 1951*, 327–359, Leipzig

— 1956 *Die Grundlagen des natürlichen Systems, der vergleichenden Anatomie und Phylogenetik.* Leipzig (Akad. Verl. Ges.)

REMANE, A. & SCHLIEPER, C. 1958 Die Biologie des Brackwassers. From: *Die Binnengewässer* (12.), *Einzeldarstellungen aus der Limnologie und ihren Nachbargebieten.* Dr A. Thienemann, Plön, Stuttgart

REMMERT, H. 1957 Aves. In: Grimpe & Wagler: *Tierwelt der Nord- und Ostsee*, Leipzig

REQUATE, H. 1951 Über Brutausfälle durch den Befall mit der Schmeissfliege *Lucilia sericata* Meig. bei einigen Seevogelarten. *Vogelwelt* **72**, 33–34, Berlin and Munich

— 1954 Die Entenvogelzählung in Deutschland (1948 bis April 1953). *Biolog. Abh.* **10**, 1–40, Würzburg

RHUMBLER, L. 1928 Amoebozoa und Reticulosa. In: Grimpe & Wagler: *Tierwelt der Nord- und Ostsee*, Leipzig

RICHTER, G. 1962 Beobachtungen zum Beutefang der marinen Bohrschnecke *Lunatia nitida. Natur u. Museum* **92**, 186–192, Frankfurt a. M.

— 1964 Zur Ökologie der Foraminiferen. I. Die Foraminiferen-Gesellschaften des Jadegebietes. II. Lebensraum und Lebensweise von *Nonion depressulum*, *Elphidium excavatum* und *Elphidium selseyense. Natur u. Museum* **94**, 343–353 and 421–430, Frankfurt a. M.

— 1965 Zur Ökologie der Foraminiferen. III. Vertriftung und Transport in der Gezeitenzone. *Natur u. Museum* **95**, 51–62, Frankfurt a. M.

RICHTER, RUD. 1916 Vom Bau und Leben der Trilobiten. I. Das Schwimmen. *Senckenbergiana* **1**, 213–238, Frankfurt a. M.

— 1920a Flachseebeobachtungen zur Paläontologie und Geologie. I. Ein devonischer 'Pfeifenquarzit' verglichen mit der heutigen 'Sandkoralle' (*Sabellaria*, Annelidae). *Senckenbergiana* **2**, 215–235, Frankfurt a. M.

— 1920b Vom Bau und Leben der Trilobiten II. Der Aufenthalt auf dem Boden. Der Schutz. Die Ernährung. *Senckenbergiana* **2**, 23–43, Frankfurt a. M.

— 1921 *Scolithus*, *Sabellarifex* und Geflechtquarzite. *Senckenbergiana* **3**, 49–52, Frankfurt a. M.

— 1922a Flachseebeobachtungen zur Paläontologie und Geologie. III. Die Lage schüsselförmiger Körper bei der Einbettung. *Senckenbergiana* **4**, 103–126, Frankfurt a. M.

— 1922b Flachseebeobachtungen zur Paläontologie und Geologie. IV. Gesonderte Verbreitung der rechten und linken Klappe einer Muschelart. *Senckenbergiana* **4**, 127–132, Frankfurt a. M.

— 1924a Flachseebeobachtungen zur Paläontologie und Geologie. VII. *Arenicola* von heute und '*Arenicoloides*' eine Rhizocorallide des Buntsandsteins, als Vertreter verschiedener Lebensweisen. *Senckenbergiana* **6**, 119–140, Frankfurt a. M.

RICHTER, RUD. 1924b Flachseebeobachtungen zur Paläontologie und Geologie. VIII. Geflechtquarzite aus einzelnen Vertikalröhren nachträglich zusammengeballt. *Senckenbergiana* **6**, 140–141, Frankfurt a. M.
— 1924c Flachseebeobachtungen zur Paläontologie und Geologie. IX. Zur Deutung rezenter und fossiler Mäanderfiguren. *Senckenbergiana* **6**, 141–157, Frankfurt a. M.
— 1924d Flachseebeobachtungen zur Paläontologie und Geologie. X. Weiteres zur Verschieden-Häufigkeit der beiden Klappen einer Spezies bei Muscheln und Brachiopoden. *Senckenbergiana* **6**, 157–163, Frankfurt a. M.
— 1926 Flachseebeobachtungen zur Paläontologie und Geologie XII-XIV. XIV. Abdrücke lebendiger Tiere (Fische und Würmer). *Senckenbergiana* **8**, 221–224, Frankfurt a. M.
— 1927a 'Sandkorallen'-Riffe in der Nordsee. *Natur u. Museum* **57**, 49–62, Frankfurt a. M.
— 1927b Die fossilen Fährten und Bauten der Würmer, ein Überblick über ihre biologischen Grundformen und deren geologische Bedeutung. *Paläont. Z.* **9**, 193–240, Berlin
— 1929a Gründung und Aufgaben der Forschungsstelle für Meeresgeologie 'Senckenberg' in Wilhelmshaven. *Natur u. Museum* **59**, 1–30, Frankfurt a. M.
— 1929b Grundsätzliches zur Erweiterung der Forschungsanstalt für Meeresgeologie und Meerespaläontologie 'Senckenberg' in Wilhelmshaven. *Natur u. Museum* **59**, 250–253, Frankfurt a. M.
— 1931 Tierwelt und Umwelt im Hunsrückschiefer; zur Entstehung eines schwarzen Schlammsteines. *Senckenbergiana* **13**, 299–342, Frankfurt a. M.
— 1936 Marken und Spuren im Hunsrück-Schiefer. II. Schichtung und Grund-Leben. *Senckenbergiana* **18**, 215–244, Frankfurt a. M.
— 1937a Schneckenlaich als mögliche Versteinerung. *Natur u. Volk* **67**, 236–239, Frankfurt a. M.
— 1937b Marken und Spuren aus allen Zeiten I. and II. I. Wühlgefüge durch kotgefüllte Tunnel (*Planolites montanus* n. sp.) aus dem Ober-Karbon der Ruhr. *Senckenbergiana* **19**, 151–159, Frankfurt a. M.
— 1937c Vom Bau und Leben der Trilobiten. 8. Die 'Salter'sche Einbettung' als Folge und Kennzeichen des Häutungsvorgangs. *Senckenbergiana* **19**, 413–431, Frankfurt a. M.
— 1942 Die Einkippungsregel. *Senckenbergiana* **25**, 181–206, Frankfurt a. M.
— 1950 Massensterben im Meer, Auftriebwasser und Erdöl-Bildung. *Natur u. Volk* **80**, 21–28, Frankfurt a. M.
— 1952 Fluidal-Textur in Sediment-Gesteinen und über Sedifluktion überhaupt. *Notizbl. Hess. L.-Amt Bodenforsch.* **3**, 67–81, Wiesbaden
— 1956 Die Jahresversammlung der 'Paläontologischen Gesellschaft' in Wilhelmshaven im August 1956. *Senckenbergiana leth.* **37**, 147–148, Frankfurt a. M.
RICHTER, RUD. & RICHTER, E. 1932 Unterlagen zu Fossilium Catalogus Trilobitae. VI. *Senckenbergiana* **14**, 359–371, Frankfurt a. M.
— 1939 Marken und Spuren aus allen Zeiten IV. Die Kotschnur *Tomaculum* Groom (*Syncoprulus* Rud. & E. Richter), ähnliche Scheitelplatten und beider stratigraphische Bedeutung. *Senckenbergiana* **21**, 278–291, Frankfurt a. M.
— 1941 Das stratigraphische Verhalten von *Tomaculum* als Beispiel für die Bedeutung von Lebensspuren. *Senckenbergiana* **23**, 127–132, Frankfurt a. M.

RICKLEFS, O. 1908 *Petricola pholadiformis* in the North Sea. *Nachr. Bl. deutsch. Malac. Ges.* **40**, 41, Frankfurt a. M.

RITTER, P. 1900 Beiträge zur Kenntnis der Stacheln von *Trygon* und *Acanthias*. Inaug. Diss. Rostock

ROCH, F. 1926 Die Holzschädlinge der Meeresküsten und ihre Bekämpfung. *Z. Ver. deutsch. Ingenieure* **70**, 89–93, Berlin

— 1927 Die Holz- und Steinschädlinge der Meeresküsten und ihre Bekämpfung. *Veröff. Medicinalverw.* **24**, 1–78, Berlin

ROUX, W. 1895 Struktur eines hochdifferenzierten bindegewebigen Organes (der Schwanzflosse des Delphin). *Ges. Abh. üb. Entwicklungsmech. Organism.* **1**, 458–574, Leipzig

RÜGER, L. & HAAS, P. 1925 *Palaeosemacostoma geryonides* v. Huene, eine sessile Meduse aus dem Dogger von Wehingen in Württemberg und *Medusina lianica* nov. sp., eine Coronaten-ähnliche Meduse aus dem mittleren Lias von Hechingen in Württemberg. *Sitz.-Ber. Heidelberg, Akad. Wiss. math.-nat. Kl.* **15**, 3–22, Berlin and Leipzig

RÜHMER, K. 1954 *Fische und Nutztiere des Meeres, deren Fang und Verwertung*. Ebenhausen

SAINT-JOSEPH, DE 1894 Les annélides polychètes des côtes de Dinard. *Ann. Sci. nat. Zool.* **17**, 1–395, Paris

SANDER, K. 1950 Beobachtungen zur Fortpflanzung von *Assiminea grayana* Leach. *Arch. Molluskenkde.* **79**, 147–149, Frankfurt a. M.

— 1952 Beobachtungen zur Fortpflanzung von *Assiminea grayana* Leach. *Arch. Molluskenkde.* **81**, 133–134, Frankfurt a. M.

SCHÄFER, W. 1937 Bau, Entwicklung und Farbenentstehung bei den Flitterzellen von *Sepia officinalis*. *Z. Zellforsch. u. mikroskop. Ant.* **27**, 221–245, Berlin

— 1938a Über die Zeichnung in der Haut einer *Sepia officinalis* von Helgoland. *Z. Morph. Ökol. Tiere* **34**, 129–134, Berlin

— 1938b Palökologische Beobachtungen an sessilen Tieren der Nordsee. *Senckenbergiana* **20**, 323–331, Frankfurt a. M.

— 1939a Fossile und rezente Bohrmuschel-Besiedlung des Jadegebietes. *Senckenbergiana* **21**, 227–254, Frankfurt a. M.

— 1939b Polypen-Kolonien im Watt. *Natur u. Volk* **69**, 408–411, Frankfurt a. M.

— 1941a Zur Fazieskunde des deutschen Wattenmeeres. 1. Dangast und die Ufersäume des Jadebusens. *Abh. senckenb. naturforsch. Ges.* **457**, 1–33, Frankfurt a. M.

— 1941b Zur Fazieskunde des deutschen Wattenmeeres. 2. Mellum, eine Düneninsel der deutschen Nordsee-Küste. *Abh. senckenberg. naturforsch. Ges.* **457**, 34–54, Frankfurt a. M.

— 1941c Fossilations-Bedingungen von Quallen und Laichen. *Senckenbergiana* **23**, 189–216, Frankfurt a. M.

— 1941d *Assiminea* und *Bembideon*, Fazies-Leitformen für MHW-Ablagerungen der Nordseemarsch. *Senckenbergiana* **23**, 136–145, Frankfurt a. M.

— 1943 Weichkörperbewegungen von *Buccinum undatum*. *Senckenbergiana* **26**, 459–466, Frankfurt a. M.

— 1948 Wuchsformen von Seepocken. *Natur u. Volk* **78**, 74–78, Frankfurt a. M.

— 1949 Sandkorallen. *Natur u. Volk* **79**, 244–245, Frankfurt a. M.

— 1950a Über Nahrung und Wanderung im Biotop bei der Strandschnecke *Littorina littorea*. *Arch. Molluskenkde.* **79**, 1–8, Frankfurt a. M.

— 1950b Nahrungsaufnahme und ernährungsphysiologische Umstimmung bei *Aeolis papillosa*. *Arch. Molluskenkde.* **79**, 9–14, Frankfurt a. M.

SCHÄFER, W. 1950c Klaffmuschel-Spülsäume am Wattenstrand. *Natur u. Volk* **80**, 173–176, Frankfurt a. M.
— 1950d Der 'Sipho' der Klaffmuschel (*Mya arenaria*). *Natur u. Volk* **80**, 142–146, Frankfurt a. M.
— 1951a Der 'kritische Raum', Masseinheit und Mass für die mögliche Bevölkerungsdichte innerhalb einer Art. *Verh. Deutsch. Zool. Ges. in Wilhelmshaven*, 391–395, Leipzig
— 1951b Fossilisations-Bedingungen brachyurer Krebse. *Abh. senckenberg. naturforsch. Ges.* **485**, 221–238, Frankfurt a. M.
— 1952a Biologische Bedeutung der Ortswahl bei Balaniden-Larven. *Senckenbergiana* **33**, 235–246, Frankfurt a. M.
— 1952b Biogene Sedimentation im Gefolge von Bioturbation. *Senckenbergiana* **33**, 1–12, Frankfurt a. M.
— 1953a Zur Fortpflanzung der Rochen. *Natur u. Volk* **83**, 245–292, Frankfurt a. M.
— 1953b Zur Unterscheidung gleichförmiger Kot-Pillen meerischer Evertebraten. *Senckenbergiana* **34**, 81–93, Frankfurt a. M.
— 1954a Form und Funktion der Brachyuren-Schere. *Abh. senckenberg. naturforsch. Ges.* **489**, 1–65, Frankfurt a. M.
— 1954b Mellum: Inselentwicklung und Biotopwandel. *Abh. naturw. Ver. Bremen* **33**, 391–406, Bremen
— 1954c Modell-Versuch zur Formänderung der Mellum-Plate. *Natur u. Volk* **84**, 426–432, Frankfurt a. M.
— 1954d Über das Verhalten von Jungheringsschwärmen im Aquarium. *Arch. Fischereiwiss.* **64**, 276–287, Hamburg
— 1954e Dehnungsrisse unter Wasser im meerischen Sediment. *Senckenbergiana leth.* **35**, 87–99, Frankfurt a. M.
— 1955a Über die Bildung der Laichballen der Wellhorn-Schnecken. *Natur u. Volk* **85**, 92–97, Frankfurt a. M.
— 1955b Wale auf norwegischen Felsbildern, vom Meeresbiologen betrachtet. *Germania* **33**, 333–339, Frankfurt a. M.
— 1955c Fossilisations-Bedingungen der Meeressäuger und Vögel. *Senckenbergiana leth.* **36**, 1–25, Frankfurt a. M.
— 1956a Wirkungen der Benthos-Organismen auf den jungen Schichtverband. *Senckenbergiana leth.* **37**, 183–263, Frankfurt a. M.
— 1956b Gesteinsbildung im Flachseebecken am Beispiel der Jade. *Geol. Rdsch.* **45**, 71–84, Stuttgart
— 1956c Wale auf norwegischen Felsbildern im Lichte meerespaläontologischer Beobachtungen. *Natur u. Volk* **86**, 233–240, Frankfurt a. M.
— 1956d Der kritische Raum und die kritische Situation in der tierischen Sozietät. *Aufsätze u. Reden senckenberg. naturforsch. Ges.* Frankfurt a. M.
— 1957 Aufgaben und Ziele der Meerespaläontologie. *Naturwissenschaften* **44**, 294–299, Berlin, Göttingen, Heidelberg
— 1959 Gibt es eine Überspezialisierung im Laufe der stammesgeschichtlichen Entwicklung? *Natur u. Volk* **89**, 65–73, Frankfurt a. M.
— 1965 Aktuopaläontologische Beobachtungen. 4. Spiralfährten und 'geführte Mäander'. *Natur u. Museum* **95**, 83–90, Frankfurt a. M.
— 1966 Aktuopaläontologische Beobachtungen. 6. Otolithen-Anreicherungen. *Natur u. Museum* **96**, 439–444, Frankfurt a. M.
SCHARFF, R. F. 1900 A list of the Irish Cetacea (Whales, Porpoises, Dolphins). *Irish Naturalist* **9**, 83–91, Dublin

SCHELLENBERG, A. 1928 Krebstiere oder Crustacea II. Decapoda. *Dahl's Tierwelt Deutschlands u. angrenz. Meeresteile* **10**, 78–79, Jena (Fischer)

SCHERF, H. 1957 Der Goldköcherwurm *Pectinaria koreni. Natur u. Volk* **87**, 108–111, Frankfurt a. M.

SCHIEMENZ, P. 1884 Über die Wasseraufnahme bei Lamellibranchiern und Gastropoden. *Mitt. Zool. Stat. Neapel* **5**, 509–543, Naples

SCHINDEWOLF, O. 1950 *Grundfragen der Paläontologie.* Stuttgart (Schweizerbart)

SCHLIEPER, K. 1955 *Praktikum der Zoophysiologie.* 2nd edition, Stuttgart (Fischer)

SCHLOEMER, A. 1949 *Sagartia* und *Metridium,* zwei 'Seerosen' der Gezeitenzone. *Natur u. Volk* **79**, 237–243, Frankfurt a. M.

SCHMIDT, BR. 1913 Das Gebiss von *Cyclopterus lumpus. Jena Z. Naturwiss.* **49**, 313–372, Jena

SCHMIDT, H. 1935 Die bionomische Einteilung der fossilen Meeresböden. *Fortschritte geol. Pal.* **38**, 1–54, Berlin

— 1944 Ökologie und Erdgeschichte. *Z. deutsch. Geol. Ges.* **96**, 113–128, Berlin

— 1949 Die fazielle Einstufung fossil-führender Sedimente. *Erdöl u. Technik in NW-Deutschland. Amt f. Bodenforschung Hannover-Celle. Sammelband 1948,* 96–97, Hanover

— 1952 Erkennbarkeit fossiler Brackwasserabsätze. *Z. deutsch. Geol. Ges.* **103**, 9–16, Berlin

— 1953 Ökologische Beobachtungen an den Foraminiferen des Golfes von Neapel. *Paläont. Z.* **27**, 123–128, 3/4, Stuttgart

— 1958 Zur Rangordnung der Faziesbegriffe. *Mitt. Geol. Ges. Wien* **49** (1956), 333–345, Vienna

SCHMIDT, W. J. 1924 *Die Bausteine des Tierkörpers im polarisierten Licht.* Bonn (Cohen)

— 1929 Die Kalkschale der Sauropsideneier als geformtes Sekret. *Z. Morph. Ökol. d. Tiere* **14**, 400–420, Berlin

— 1942 Polarisationsoptische Beobachtungen an *Amoeba proteus. Protoplasma* **36**, 371–380, Berlin

— 1951 Die Unterscheidung der Röhren von Scaphopoda, Vermetidae und Serpulidae mittels mikroskopischer Methoden. *Mikroskopie* **6**, 373–381, Vienna

— 1954 Über Bau und Entwicklung der Zähne des Knochenfisches *Anarhichas lupus* L. und ihren Befall mit '*Mycelites ossifragus*'. *Z. Zellforsch. u mikr. Anat.* **40**, 25–48, Berlin

— 1955 Die tertiären Würmer Österreichs. *Österr. Akad. Wiss.* **109**, 1–121, Vienna

SCHMIDT, W. J. & KEIL, A. 1958 *Die gesunden und die erkrankten Zahngewebe des Menschen und der Wirbeltiere im Polarisationsmikroskop. Theorie, Methodik, Ergebnisse der optischen Strukturanalyse der Zahnsubstanzen samt ihrer Umgebung.* Munich (Hansen)

SCHNAKENBECK, W. 1947 Tiere und Pflanzen des Salz- und Süsswassers. *Fischwirtschaftskunde* **1**, Lief. 1. Hamburg (Keune)

— 1955 Der Kiemenreusenapparat vom Riesenhai (*Cetorhinus maximus*). *Zool. Anz.* **154**, 99–108, Leipzig

SCHREMMER, FR. 1954 Bohrschwammspuren in Actaeonellen aus der nordalpinen Gosau. *Sitz-Ber. österr. Akad. Wiss. math.-nat. Kl. Abt. I.* **163**, 297–300, Vienna

SCHRÖDER, CHR. 1925 *Handbuch der Entomologie.* 3. (Geschichte, Literatur, Technik, Paläontologie, Phylogenie, Systematik von A. Handlirsch). Jena (Fischer)

SCHUBERT, K. 1938 Häutung, Wachstum und Alter der Wollhandkrabbe. From:

Untersuchungen über die chinesische Wollhandkrabbe in Europa. *Mitt. hamb. zool. Mus. Inst.* **47**, 83–104, Hamburg

SCHUBERT, K. 1954 Walfang und Walbestand. *Fette, Seifen, Anstrichmittel* **56**, 568–573, Hamburg

— 1955 *Der Walfang der Gegenwart*. Stuttgart (Schweizerbart)

SCHULZ, B. 1932 Einführung in die Hydrographie der Nord- und Ostsee. In: Grimpe & Wagler: *Tierwelt Nord- und Ostsee*, Leipzig

SCHULZ, E. 1931 Kurze Notiz zur Biologie von *Priapulus caudatus* Lam. *Zool. Anz.* **96**, 61–63, Leipzig

SCHUSTER, O. 1952 Die Vareler Rinne im Jadebusen. Die Bestandteile und das Gefüge einer Rinne im Watt. *Abh. senckenberg. naturforsch. Ges.* **486**, 1–38, Frankfurt a. M.

SCHÜTTE, H. 1905 Ein neu entstandenes Düneneiland zwischen Aussenjade und Aussenweser. *Jahrb. Ver. Naturkd. Unterweser 1903–4*, Bremerhaven

— 1929 Über Sedimentbildung an der Küste des norddeutschen Wattenmeeres. *Senckenbergiana* **11**, 345–352, Frankfurt a. M.

— 1935 *Das Alluvium des Jade-Weser-Gebietes*. Oldenburg (Stalling)

— 1939 Sinkendes Land an der Nordsee? Zur Küstengeschichte Nordwestdeutschlands. *Schriften deutsch. Naturkundever. N. F.* **9**, 3–144, Öhringen

SCHÜTZ, L. 1961 Verbreitung und Verbreitungsmöglichkeiten der Bohrmuschel *Teredo navalis* L. und ihr Vordringen in den NO-Kanal bei Kiel. *Kieler Meeresforsch.* **17**, 228–236, Kiel

SCHWARZ, A. 1929a Ein Verfahren zum Härten nichtverfestigter Sedimente. *Natur u. Museum* **59**, 204–208, Frankfurt a. M.

— 1929b Schlickfall und Gezeitenschichtung. *Senckenbergiana* **11**, 152–155, Frankfurt a. M.

— 1930 Ein Seeigelstachel-Gestein. *Natur u. Museum* **60**, 502–506, Frankfurt a. M.

— 1931 Insektenbegräbnis im Meer. *Natur u. Museum* **61**, 453–465, Frankfurt a. M.

— 1932a Der Lichteinfluss auf die Fortbewegung, die Einregelung und das Wachstum bei einigen niederen Tieren. *Littorina, Cardium, Mytilus, Balanus, Teredo, Sabellaria. Senckenbergiana* **14**, 429–454, Frankfurt a. M.

— 1932b Der tierische Einfluss auf die Meeressedimente (besonders auf die Beziehungen zwischen Frachtung, Ablagerung und Zusammensetzung von Wattensedimenten). *Senckenbergiana* **14**, 118–172, Frankfurt a. M.

— 1932c Mövengewölle. *Natur u. Museum* **62**, 305–310, Frankfurt a. M.

— 1933 Meerische Gesteinsbildung I. *Senckenbergiana* **15**, 69–160, Frankfurt a. M.

SCHWENKE, W. 1953 Biocönotik und angewandte Entomologie. *Beitr. Ent. Sonderheft* **3**, 86–162, Berlin

SCOTT, T. 1902 Observations on the food of fishes. *Ann. Rep. Fish. Board Scotland* 1901, Part III, 486–538, Edinburgh

SEIBOLD, E. 1955 Beobachtungen zur Tätigkeit von Bohrmuscheln. *N. Jb. Geol. u. Paläont.* **6**, 248–251, Stuttgart

SEILACHER, A. 1951 Der Röhrenbau von *Lanice conchilega* (Polychaeta). Ein Beitrag zur Deutung fossiler Lebensspuren. *Senckenbergiana* **32**, 267–280, Frankfurt a. M.

— 1953a Studien zur Palichnologie I. Über die Methoden der Palichnologie. *N. Jb. Geol. u. Paläont. Abh.* **96**, 421–452, Stuttgart

— 1953b Studien zur Palichnologie II. Die fossilen Ruhespuren (Cubichnia). *N. Jb. Geol. u. Paläont. Abh.* **98**, 87–124, Stuttgart

— 1954a Die geologische Bedeutung fossiler Lebensspuren. *Z. deutsch. geol. Ges.* **105**, 214–227, Hanover
— 1954b Ökologie der triassischen Muschel *Lima lineata* (Schloth.) und ihrer Epöken. *N. Jb. Geol. Paläont. Abh.* **4**, 163–183, Stuttgart
— 1956a Der Beginn des Kambriums als biologische Wende. *N. Jb. Geol. u. Paläont. Abh.* **103**, 155–180, Stuttgart
— 1956b *Ichnocumulus* n. g., eine weitere Ruhespur des schwäbischen Jura. *N. Jb. Geol. u. Paläont. Abh.* **103**, 153–159, Stuttgart
— 1957 An-aktualistisches Wattenmeer? *Paläont. Z.* **31**, 198–206, Stuttgart
— 1959a Fossilien als Strömungsanzeiger. *Aus. der Heimat* **67**, 170–177, Öhringen
— 1959b Vom Leben der Trilobiten. *Naturwiss.* **46**, 389–393, Berlin, Göttingen, Heidelberg
— 1960 Lebensspuren als Leitfossilien. *Geol. Rundsch.* **49**, 41–48, Stuttgart
SICK, H. 1937 Morphologisch-funktionelle Untersuchungen über die Feinstruktur der Vogelfeder. *J. Ornithol.* **85**, 206–372, Berlin
SIEDENTOP, W. 1927 Die Kriechbewegung der Aktinien und Lucernariden. *Zool. Jb.* (*Allg. Zool.*) **44**, 149–210, Jena
SIMROTH, H. 1896–1907 Gastropoda prosobranchia. *Bronn's Kl. u. Ord. Tierr.* **3**, Leipzig
SKUTCH, A. F. 1926 On the habits and ecology of the tubebuilding Amphipod *Amphithoe rubricata* Montagu. *Ecology* **7**, 481–502, Brooklyn
SLIJPER, E. J. 1936 Die Cetaceen vergleichend-anatomisch und systematisch. *Capita Zoologica* **7**, 1–590, The Hague
— 1958 *Walvissen*. Amsterdam
— 1958 Das Verhalten der Wale (Cetacea). From: W. Krüger: Bewegungstypen. *Handb. Zool.* **8**, 1–30, Berlin
SMIDT, E. L. B. 1951 Animal Production in the Danish Waddensea. *Medd. Komm. Danmarks Fiskeri- og Havundersogelser* (Serie: Fisk.) **11**, 3–151, Copenhagen
SMITH, J. E. 1932 The shell gravel deposits and the infauna of the Eddystone grounds. *J. mar. biol. Assoc. Plymouth* **18**, 243–278, Plymouth
SÖDERSTRÖM, A. 1920 Studien über die Polychätenfamilie Spionidae. Inaug. Diss. Uppsala
— 1923 Über das Bohren der *Polydora ciliata*. *Zool. Bidrag Uppsala* **8**, 319–340, Uppsala and Stockholm
SPEYER, W. 1937 *Entomologie mit besonderer Berücksichtigung der Biologie, Ökologie und Gradationslehre der Insekten.* Dresden and Leipzig (Steinkopf)
STADTMÜLLER, F. 1927 Über das Kiemenfilter der Dipnoer. *Morph. Jb.* **57**, 489–529, Leipzig
STEINER, G. 1941 Untersuchungen über die Kältewiderstandsfähigkeit der Eier und Larven von Fleischfliegen. *Anz. Schädlingskde.* **17**, 133–139, Berlin
STEPHENSEN, K. 1929 Amphipoda. In: Grimpe & Wagler: *Tierwelt der Nord- und Ostsee*, Leipzig
STOCKS, TH. 1956 Der Boden der südlichen Nordsee, 2. Beitrag. Eine neue Tiefenkarte der südlichen Nordsee. *Deutsch. Hydrograph. Z.* **9**, 265–280, Hamburg
STRESEMANN, E. 1930 Die mörderische Wirkung des harten Winters 1928–9 auf die Vogelwelt. *Ornithol. Monatsber.* **38**, 37–43, Berlin
— 1934 Sauropsida Aves. *Kükenthal und Krumbach's Handb. Zool.* **7**, 2, Berlin
STÜBEL, H. 1909 Studium zur vergleichenden Physiologie der peristaltischen Bewegungen. I, II, III, IV. *Pflüger's Arch.* **129**, 1–34, Bonn
SUNKEL, W. 1925 Mellum 1924. *J. Ornith.* **73**, 110–127, Berlin

TANSLEY, A. G. 1935 The use and abuse of vegetational concepts and terms. *Ecology* **16**, 284–307, Brooklyn

TEICHERT, C. 1958 Concepts of facies. *Bull. Amer. Ass. Petrol. Geologists* **42**, 2718–2744, Tulsa, Oklahoma

TEICHERT, C. & SERVENTY, D. L. 1947 Deposits of shells transported by birds. *Amer. J. Sci.* **245**, 322–328, New Haven, Conn.

THAMDRUP, H. M. 1935a Beiträge zur Ökologie der Wattenfauna auf experimenteller Grundlage. *Medd. Skalling-Laboratorium* **2**, 1–125, Copenhagen

— 1935b Beiträge zur Ökologie der Wattenfauna auf experimenteller Grundlage. *Medd. Komm. Danmarks Fisk. og Havundersögelser.* (Ser. Fisk.) **10**, 181–246

TEN CATE, J. 1930 Beiträge zur Physiologie des Zentralnervensystems der Einsiedlerkrebse. *Arch. néerl. Physiol.* **15**, 242–252, Haarlem

THEEDE, H. 1963 Experimentelle Untersuchungen über die Filtrationsleistung der Miesmuschel *Mytilus edulis* (L.). *Kieler Meeresforsch.* **19**, 20–41, Kiel

THEEDE, H., PONAT, A., HIROBI, K. & SCHLIEPER, C. 1969 Studies on the resistance of marine bottom invertebrates to oxygen-deficiency and hydrogen sulphide. *Marine Biol.* **2**, 324–337, Berlin, Heidelberg, New York

THENIUS, E. 1954a Die Säugetierfauna aus dem Torton von Nendorf an der Marsch (CSR). *N. Jb. Geol. u. Paläont. Abh.* **99**, 189–230, Stuttgart

— 1954b Die Caniden (Mammalia) aus dem Altquartär von Hundsheim (Niederösterreich) nebst Bemerkungen zur Stammesgeschichte der Gattung Cyon. *N. Jb. Geol. u. Paläont. Abh.* **99**, 230–286, Stuttgart

THIELE, J. 1926 Scaphopoda. *Kükenthal und Krumbach's Handb. Zool.* **5**, 156–160, Berlin and Leipzig (De Gruyter)

THIENEMANN, A. 1918 Lebensgemeinschaft und Lebensraum. *Naturwiss. Wschr. N. F.* **17**, 281–290, Jena

— 1939 Grundzüge einer allgemeinen Ökologie. *Arch. Hydrobiol.* **35**, 267–285, Stuttgart

— 1941 Leben und Umwelt. *Bios. Abh. theoret. Biol. u. Geschichte* **12**, 1–122, Leipzig

— 1942 Vom Wesen der Ökologie. *Biol. General.* **15**, 312–331, Vienna

— 1954 'Lebenseinheiten'. Ein Vortrag. *Abh. naturw. Verein Bremen* **33**, 303–326, Bremen

THIENEMANN, A. & KIEFER, J. J. 1916 Schwedische Chironomiden. *Arch. Hydrobiol. u. Planktonkde. Suppl. Bd.* **2**, 483–554, Stuttgart

THIENEMANN, J. 1925 Wieder ein Massensterben von Vögeln in der Ostsee. *Ornith. Monatsber.* **33**, 73–76, Berlin

THORSON, G. 1951 Zur jetzigen Lage der marinen Bodentier-Ökologie. *Verh. deutsch. zool. Ges. Wilhelmshaven*, 276–327, Leipzig

— 1966 Some factors influencing the recruitment and establishment of marine benthic communities. *Nth. J. Sea Res.* **3**, 267–293, Den Helder

TINBERGEN, L. & VERWEY, J. 1945 Zur Biologie von *Loligo vulgaris* Lam. *Arch. Néerland. Zool.* **7**, 213–286, Haarlem

TINBERGEN, N. 1936 Zur Soziologie der Silbermöve (*Larus argentatus* Pontopp). *Beitr. Fortpflanzungsbiol. Vögel* **12**, 89–96, Berlin

— 1952 *Instinktlehre. Vergleichende Erforschung angeborenen Verhaltens.* Berlin and Hamburg (Parey)

TISCHLER, W. 1951 Zur Synthese biozönotischer Forschung. *Acta Biotheor.* **9**, 135–162, Leiden

— 1955 *Synökologie der Landtiere.* Stuttgart (G. Fischer)

TORNQUIST, A. 1903 Die Beschaffenheit des Apicalfeldes von *Schizaster* und seine geologische Bedeutung. *Z. deutsch. geol. Ges.* **55**, 375–392, Berlin

— 1911 Die biologische Bedeutung der Umgestaltung der Echiniden im Paläozoikum. *Z. induktive Abstammungs- u. Vererbungslehre* **6**, 29–60, Berlin

TREUENFELS, P. 1896 Die Zähne von *Myliobatis aquila*. Inaug. Diss. Breslau

TRIEBEL, E. 1941 Zur Morphologie und Ökologie der fossilen Ostracoden. *Senckenbergiana* **23**, 294–400, Frankfurt a. M.

TROJAHN, E. 1913 Über Hautdrüsen des *Chaetopterus variopedatus* Ren. *Sitz.ber. Akad. Wiss. Wien, math.-nat. Kl. Abt.* 1/**122**, 565–596, Vienna

TROSCHEL, B. 1913 Ein neuer Feind unserer Wasserbauhölzer. *Ztrbl. Bauverwaltung*, 273–274, Berlin

TRUSHEIM, F. 1929a Zur Bildungsgeschwindigkeit geschichteter Sedimente im Wattenmeer, besonders solcher mit schräger Parallelschichtung. *Senckenbergiana* **11**, 47–55, Frankfurt a. M.

— 1929b Massentod von Insekten. *Natur u. Museum* **59**, 54–61, Frankfurt a. M.

— 1932 Paläontologisch Bemerkenswertes aus der Ökologie rezenter Nordsee-Balaniden. *Senckenbergiana* **14**, 70–87, Frankfurt a. M.

TULLBERG, T. 1822 Studien über den Bau und das Wachstum des Hummerpanzers und der Molluskenschale. *Kgl. Svenska Vet.-Akad. Handl.* **19**, No. 3, Stockholm

ULRICH, W. 1927 Bemerkungen zu einer ökologischen Erklärung zweier verschiedener Wuchsformen bei Balaniden. *Sitz.-Ber. Ges. naturforsch. Freunde* **60**, Berlin

UEXKÜLL, J. V. 1905 Die Bewegungen der Schlangensterne. *Z. Biol.* **46**, 1–37, Munich and Berlin

— 1907 Studien über den Tonus (IV. Die Herzigel). *Z. Biol. N. F.* **31**, 307–332, Munich and Berlin

— 1909 *Umwelt und Innenwelt der Tiere*. Berlin (Springer)

— 1913 Studien über den Tonus der Pilgermuschel. *Z. Biol.* **58**, 305–332, Munich and Berlin

— 1921 *Umwelt und Innenwelt der Tiere*. 2nd edition, Berlin (Springer)

— 1933 Studien über den Tonus. I. Der biologische Bauplan von *Sipunculus nudus*. *Z. Biol.* **44**, 269–344, Munich and Berlin

UEXKÜLL, TH. V. 1949 Der Begriff der Funktion und seine Bedeutung für unsere Vorstellung von der Wirklichkeit des Lebensvorganges. *Studium generale* **2**, 13–21, Berlin, Göttingen, Heidelberg

VAN DAM, L. 1935 On the utilisation of oxygen by *Mya arenaria*. *J. exper. Biol. Cambridge* **12**, 86–94, Cambridge

— 1937 Über die Atembewegungen und das Atemvolumen von *Phryganea*, *Arenicola* und *Nereis virens*. *Zool. Anz.* **118**, 122–128, Berlin

VAN DEINSE, A. B. 1931 *De fossiele en recente Cetacea van Nederland*. Amsterdam

VAN STRAELEN, V. 1928 Les œufs de Reptiles fossiles. *Palaeobiologica* **1**, 295–312, Vienna and Leipzig

VAN STRAATEN, L. M. J. U. 1950 Environment of formation and facies of the Wadden Sea sediments. *Tijdschr. Koninkl. Nederl. Aardvijkskd. Genootschap* **67**, 94–108, Leiden

— 1952 Biogene textures and the formation of shell beds in the Dutch Wadden Sea I. *Koninkl. Nederl. Akad. Wettensch. Ser.* **55**, 500–516, Amsterdam

— 1954 Composition and structure of recent marine sediments in the Netherlands. *Leidse Geol. Med.* **19**, 1–110

VAN STRAATEN, L. M. J. U. 1956 Composition of shell beds formed in tidal flat environment in the Netherlands and in the Bay of Arcachon (France). *Geologie en Mijnbouw* **18**, 209–226, Groningen

VAN WEEL, P. B. 1937 Die Ernährungsbiologie von *Amphioxus lanceolatus*. *Publ. Staz. zool. Napoli* **16**, 221–272, Rome, Berlin

VERWEY, J. 1952 On the ecology of distribution of Cockle and Mussel in the Dutch Wadden Sea, their role in sedimentation and the source of their food supply. *Arch. Néerland. Zool.* **10**, 171–239, Leiden

VITZOU, A. N. 1882 Recherches sur la structure et la formation des téguments chez les Crustacées décapodes. *Arch. Zool. exp. gén.* **10**, 451–576, Paris

VLÈS, F. 1908 Remarques diverses sur la reptation des Mollusques. *Bull. Soc. Zool. France* **33**, 170, Paris

VOIGT, E. 1933 Bryozoa (Paläontologie). *Handw. Buch Naturwiss.* **2**, 280–286. 2nd edition, Jena

— 1959 Die ökologische Bedeutung der 'Hartgrounds' in der oberen Kreide Europas. *Kurzbericht in Paläont. Z.* **33**, 5, Stuttgart

VOLZ, P. 1939 Die Bohrschwämme (Clioniden) der Adria. *Thalassia* **3**, 3–64, Venice, Jena

VOLLBRECHT, K. 1954 Zur Frage der Sedimentbewegung im litoralen Gürtel. *Acta Hydrophysica* **2**, 43–80, Berlin

WALCHER, K. 1933 Eindringen von Maden in die Spongiosa der grossen Röhrenknochen. *Deutsch. Z. ges. gerichtl. Med.* **20**, 469–471, Berlin

WALTHER, J. 1893–1894 *Einleitung in die Geologie als historische Wissenschaft*. Jena (Fischer)

— 1904 Die Fauna der Solnhofer Plattenkalke, bionomisch betrachtet. *Festschr. z. 70. Geb. v. E. Haeckel, Deutsch. med.-naturwiss. Ges.* **11**, 135–214, Jena

— 1919 *Allgemeine Paläontologie*. Berlin

— 1930 Die Methoden der Geologie als historische und biologische Wissenschaft. In: Abderhalden: *Handb. biol. Arbeitsmeth., Abt. X. Methoden der Geologie, Mineralogie, Paläontologie und Geographie*, 529–658, Berlin and Vienna

WASMUND, E. 1926 Biocoenose und Thanatocoenose. Biosoziologische Studie über Lebensgemeinschaften und Totengesellschaften. *Arch. Hydrobiol. u. Planktonkde.* **17**, 1–116, Stuttgart

— 1929 Schalenfischerei an Meeresküsten. *Mitt. deutsch. Seefischerei-Ver.* **45**, 55–70

WATSON, A. T. 1890 The tube-building habits of *Terebella litoralis*. *J. Roy. Microscop. Soc.* **2**, 685–689, London and Edinburgh

— 1927 Observations on the habits and life-history of *Pectinaria* (*Lagis*) *koreni* Mgr. *Proc. and Transac. Liverpool Biol. Soc.* **42**, 25–60, Liverpool

WEBB, J. E. 1969 On the feeding and behaviour of the larva of *Branchiostoma lanceolatum*. *Marine Biol.* **3**, 58–72, Berlin, Heidelberg, New York

WEBER, H. 1925 Über die arhythmische Fortbewegung bei einigen Prosobranchiern. Ein Beitrag zur Bewegungsphysiologie der Gastropoden. *Z. Physiol.* **2**, 109–121, Berlin (Springer)

WECKMANN-WITTENBURG, P. N. 1931 *Norderoog, ein deutsches Vogelparadies*. Berlin-Lichterfelde

WEISMANN, A. 1892 *Vorlesungen über Deszendenztheorie*. Jena

WEIGELT, J. 1927 *Rezente Wirbeltierleichen und ihre paläobiologische Bedeutung*. Leipzig (Max Weg)

— 1928 Ganoidfischleichen im Kupferschiefer und in der Gegenwart. *Palaeobiologica* **1**, 323–356, Vienna and Leipzig

— 1930 Vom Sterben der Wirbeltiere. Ein Nachtrag zu meinem Buch 'Rezente Wirbeltierleichen und ihre paläobiologische Bedeutung'. *Nov. Act. Leopoldina* **6**, 281–340, Leipzig

— 1935a *Lophiodon* in der oberen Kohle des Geiseltales. *Nov. Act. Leopoldina N. F.* **3**, 369–402, Halle

— 1935b Was bezwecken die Hallenser Universitätsgrabungen in der Braunkohle des Geiseltales. *Natur u. Volk* **65**, 347–356, Frankfurt a. M.

WEIGOLD, H. 1930 Der Vogelzug auf Helgoland, graphisch dargestellt (Abh. a. d. Gebiet d. Vogelzugforschung 1.). *Vogelwarte d. Staatl. biol. Anst. Helgoland* Berlin (Friedländer)

WEISMANN, A. 1882 *Die Dauer des Lebens*. Jena

WELLS, G. P. 1951 On the behaviour of *Sabella*. *Proc. Roy. Soc. B.* **138**, 278–299, London

— 1952a The proboscis apparatus of *Arenicola*. *J. mar. biol. Ass. Unit. Kingd.* **31**, 1–28, Plymouth

— 1952b The respiratory significance of the crown in the polychaete worms *Sabella* and *Myxicola*. *Proc. Roy. Soc. B.* **140**, 70–82, London

— 1953 Defaecation in relation to the spontaneous activity cycles of *Arenicola marina* L. *J. mar. biol. Ass. Unit. Kingd.* **32**, 51–63, Plymouth

— 1954 The mechanism of proboscis movement in *Arenicola*. *J. Microsc. Sc.* **95**, 251–270, London

— 1957a Variation on *Arenicola marina* L. and the status of *Arenicola glacialis* Murdach (Polychaeta). *Proc. zool. Soc. London* **129**, 397–419, London

— 1957b The life of the lugworm. *New Biology* **22**, 1–19

— 1966 The lugworm (*Arenicola*)—study in adaption. *Nth. J. Sea Res.* 3, 294–313, Den Helder

WELLS, G. P. & ALBRECHT, E. B. 1951a The integration of activity cycles in the behaviour of *Arenicola marina* (L.). *J. exp. Biol.* **28**, 41–50, Cambridge

— 1951b The role of oesophageal rhythms in the behaviour of *Arenicola ecaudata* Johnston. *J. exp. Biol.* **28**, 51–56, Cambridge

WELLS, M. M. 1926 Collecting *Amphioxus*. *Science*, **64**, 187–188, New York

WERNER, B. 1951 Über die Bedeutung der Wasserstromerzeugung und Wasserstromfiltration für die Nahrungsaufnahme der ortsgebundenen Meeresschnecke *Crepidula fornicata* L. (Gastropoda Prosobranchia). *Zool. Anz.* **146**, 97–113, Leipzig

— 1954a Eine Beobachtung über die Wanderung von *Arenicola marina* (Polychaeta sedentaria). *Helgol. Wiss. Meeresuntersuch.* **5**, 93–102, List

— 1954b Über die Winterwanderung von *Arenicola marina* L. (Polychaeta sedentaria). *Helgol. Wiss. Meeresuntersuch.* **5**, 353–378, List

WESENBERG-LUND, C. (in E. Warning) 1904 Bidrag til Vadernes, Sandernes og Marskens Naturhistorie. *Kgl. Danske Vid. Selsk. Skrifter*, (7) *Nath.-Math. Afd.* **2**, 1–47, Copenhagen

— 1905 Umformungen des Erdbodens. *Prometheus* **16**, 577–582, Berlin

— 1929 Some remarks on the biology and anatomy of the genus *Priapulus* and its distribution in Danish and adjacent waters. *Videnskab, Meddel. Dansk. naturh. Foren* **88**, 165–202, Copenhagen

— 1951 Polychaeta. **2**, Part 19, *The zoology of Iceland*. Copenhagen and Reykjavik

WETZEL, O. 1956 Massenproduktion, Sedimentation und Fossilisation von Mikro-Organismen im Gebiet der Nord- und Ostsee, — ein Beitrag zur 'Paläo-Planktontologie' Europas. *Int. Kongr. Zool. Copenhagen* **14**, 102–103, Copenhagen

WETZEL, W. 1937a Die koprogenen Beimengungen mariner Sedimente und ihre diagnostische und lithogenetische Bedeutung. *N. Jahrb. Min. etc. Beil. Bd.* **78**, 109–122, Stuttgart

— 1937b Die Schalenzerstörung durch Mikroorganismen. Erscheinungsform, Verbreitung und geologische Bedeutung in Gegenwart und Vergangenheit. *Kieler Meeresforsch.* **2**, 254–266, Kiel

WILDVANG, D. 1938 *Die Geologie Ostfrieslands*. Berlin (Preuss. Geol. L.anst.)

WILKE, D. E. 1952 Beobachtungen über den Bau und die Funktion des Röhren- und Kammersystems der *Pectinaria koreni* Malmgren. *Helgol. Wiss. Meeresuntersuch.* **4**, 130–137, List

WILLEM, V. 1927 Observations sur la locomotion des Actinies. *Bull. sci. Acad. Roy. Belgique* (5) **13**, 630–650, Brussels

WILLEMSEN, J. 1952 Quantities of water pumped by mussels (*Mytilus edulis*) and cockles (*Cardium edule*). *Arch. Néerland. Zool.* **10**, 153–160, Leiden

WILLEMOES-SUHM, R. V. 1871 Biologische Beobachtungen über niedere Meeresthiere. *Z. wiss. Zool.* **21**, 380–396, Leipzig

WILSON, D. P. 1950 *Life of the shore and shallow sea*. London

WIMAN, K. 1914 Über die paläontologische Bedeutung des Massensterbens unter den Tieren. *Paläont. Z.* **1**, 145–154, Berlin

WOHLENBERG, E. 1937 Die Wattenmeer-Lebensgemeinschaften im Königshafen von Sylt. *Helgol. Wiss. Meeresunters.* **1**, 1–92, Kiel

— 1950 Entstehung und Untergang der Insel Trischen. *Mitt. Geograph. Ges. Hamburg* **49**, 158–187, Hamburg

WOHLFAHRT, TH. A. 1956 Analogie als Begriff und Methode der vergleichenden Anatomie. *Studium Generale* **3**, 136–142, Berlin, Göttingen, Heidelberg

WOLANSKY, D. 1956 Über echte Fossilien und Pseudofossilien. *Geol. Rdsch.* **45**, 327–332, Berlin

YONGE, C. M. 1925 The digestion of cellulose by invertebrates. *Science Progr.* **20**, 242–248, New Haven, Conn.

— 1936–1937 The biology of *Aporrhais pes pelicani* (J.) and *A. serresiana* (Mich.). *J. mar. biol. Ass. Plymouth* **21**, 687–703, Plymouth

— 1949 *The sea shore* (The New Naturalist series). London (Collins)

ZANDER, E. 1903 Das Kiemenfilter bei Süsswasserfischen. *Z. wiss. Zool.* **75**, 232–257, Leipzig

— 1906 Das Kiemenfilter der Teleosteer. Eine morpho-physiologische Studie. *Z. wiss. Zool.* **84**, 619–713, Leipzig

ZAVATTARI, E. 1920 Osservazioni etologiche sopra l'anfipodo tubicola *Ericthonius brasiliensis* (Dana). *R. Com. Talassogr. Ital.* **142**, 200–217

ZIEGELMEIER, E. 1952 Beobachtungen über den Röhrenbau von *Lanice conchilega* (Pallas) im Experiment und am natürlichen Standort. *Helgol. Wiss. Meeresuntersuch.* **4**, 107–129, List

— 1954 Beobachtungen über den Nahrungserwerb bei der Naticide *Lunatia nitida* Donovan (Gastropoda Prosobranchia). *Helgol. Wiss. Meeresuntersuch.* **5**, 1–33, List

— 1957 Die Muscheln (Bivalvia) der deutschen Meeresgebiete (Systematik und Bestimmung der heimischen Arten nach ihren Schalenmerkmalen). *Helgol. Wiss. Meeresuntersuch.* **6**, 1–51, Hamburg

— 1958 Zur Lokomotion bei Naticiden (Gastropoda Prosobranchia) (Kurze Mitteilung über Schwimmbewegungen bei *Polynices josephinus* Risso). *Helgol. Wiss. Meeresuntersuch.* **6**, 202–206, List

— 1961 Zur Fortpflanzungsbiologie der Naticiden (Gastropoda Prosobranchia). *Helgol. Wiss. Meeresuntersuch.* **8**, 94–118, Hamburg

— 1963 Das Makrobenthos im Ostteil der Deutschen Bucht nach qualitativen und quantitativen Bodengreiferuntersuchungen in der Zeit von 1949–1960. *Veröff. Inst. f. Meeresforsch. Bremerhaven. Sonderband*, 101–114, Bremen

ZIMMER, C. 1932 Beobachtungen an lebenden Mysidaceen und Cumaceen. *Sitz.-Ber. Ges. naturf. Freunde Berlin*, 326–347, Berlin

— 1933a Cumacea. In: Grimpe & Wagler: *Tierwelt Nord- und Ostsee*, Leipzig

— 1933b Mysidacea. In: Grimpe & Wagler: *Tierwelt Nord- und Ostsee*, Leipzig

INDEX